Forest Plans of North America

Forest Plans of North America

Edited by

Jacek P. Siry
Warnell School of Forestry and Natural Resources, University of Georgia, Athens, Georgia, USA

Pete Bettinger
Warnell School of Forestry and Natural Resources, University of Georgia, Athens, Georgia, USA

Krista Merry
Warnell School of Forestry and Natural Resources, University of Georgia, Athens, Georgia, USA

Donald L. Grebner
Department of Forestry, Mississippi State University, Mississippi, USA

Kevin Boston
Department of Forest Engineering, Resources and Management, Oregon State University, Corvallis, Oregon, USA

Christopher Cieszewski
Warnell School of Forestry and Natural Resources, University of Georgia, Athens, Georgia, USA

AMSTERDAM • BOSTON • HEIDELBERG • LONDON
NEW YORK • OXFORD • PARIS • SAN DIEGO
SAN FRANCISCO • SINGAPORE • SYDNEY • TOKYO
Academic Press is an imprint of Elsevier

Academic Press is an imprint of Elsevier
32 Jamestown Road, London NW1 7BY, UK
525 B Street, Suite 1800, San Diego, CA 92101-4495, USA
225 Wyman Street, Waltham, MA 02451, USA
The Boulevard, Langford Lane, Kidlington, Oxford OX5 1GB, UK

British Library Cataloguing in Publication Data
A catalogue record for this book is available from the British Library

Library of Congress Cataloging-in-Publication Data
A catalog record for this book is available from the Library of Congress

ISBN: 978-0-12-799936-4

For information on all Academic Press publications
visit our website at store.elsevier.com

Cover image: Photograph provided by M. Huntsman. View of Summit Lake
from Capitol State Forest, Washington, USA.

14 15 16 17   10 9 8 7 6 5 4 3 2 1

Contents

Contributors

Mark S. Andre City of Arcata, Arcata, California, USA

Steve Andringa Yakama Nation Forestry, Toppenish, Washington, USA

Mark S. Ashton Yale School of Forestry and Environmental Studies, New Haven, Connecticut, USA

Jeff Barkley Wisconsin County Forests Association, Rhinelander, Wisconsin, USA

Alex L. Barrett Yale School of Forestry and Environmental Studies, New Haven, Connecticut, USA

Greg Bentley Campbell Global, Forest and Natural Resources Investments, Mississippi, USA

Pete Bettinger Warnell School of Forestry and Natural Resources, University of Georgia, Athens, Georgia, USA

Bud Bigelow Weyerhaeuser Company, Timberlands Strategic Planning, Federal Way, Washington, USA

Gerardo Bocco Centro de Investigaciones en Geografía Ambiental, Universidad Nacional Autónoma de México, Morelia, Michoacán, Mexico

Jason Bodine Bayfield County Forestry and Parks Department, Washburn, Wisconsin, USA

Kevin Bollefer Revelstoke Community Forest Corporation, Revelstoke, British Columbia, Canada

Kevin Boston Department of Forest Engineering, Resources and Management, Oregon State University, Corvallis, Oregon, USA

Javier L. Bretado-Velázquez Universidad Juárez del Estado de Durango, Durango, Mexico

Angus Brodie Washington Department of Natural Resources, Olympia, Washington, USA

Cam Brown Forsite Consultants Ltd., Salmon Arm, British Columbia, Canada

Paul Casey Northern Forest Land Management Research and Demonstration Program, Umbagog National Wildlife Refuge, Errol, New Hampshire, USA

Juan Manuel Cassian-Santos Col. Periferia, Durango, Mexico

Cathy Chauvin Washington Department of Natural Resources, Olympia, Washington, USA

Chris Cieszewski Warnell School of Forestry and Natural Resources, University of Georgia, Athens, Georgia, USA

J. Javier Corral-Rivas Instituto de Silvicultura e Industria de la Madera, Universidad Juárez del Estado de Durango, Durango, Mexico

Alex Cousins Trinity County Resource Conservation District, Weaverville, California, USA

Kristofer Covey Yale School of Forestry and Environmental Studies, New Haven, Connecticut, USA

Kevin Crowe Lakehead University, Thunder Bay, Ontario, Canada

Tamara L. Cushing Department of Forest Engineering, Resources and Management, Oregon State University, Corvallis, Oregon, USA

Scott Danskin South Carolina Forestry Commission, Wedgefield, South Carolina, USA

Barry Davidson Westwind Forest Stewardship Inc., Parry Sound, Ontario, Canada

Adam Davis DPW Environmental Division, Fort Wainwright, Alaska, USA

Liane Davis The Nature Conservancy, Astoria, Oregon, USA

Brian Davis Sand Hills State Forest, South Carolina Forestry Commission, Patrick, South Carolina, USA

Héctor M. de los Santos-Posadas Colegio de Postgraduados, Postgrado en Ciencias Forestales, Montecillo, Estado de México, Mexico

Mehmet Demirci Forest Management and Planning Department, General Directorate of Forestry, Ministry of Forest and Water Affairs, Ankara, Turkey

Duane Dippon Bureau of Land Management, Department of the Interior, Washington, District of Columbia, USA

Ian Drew Northern Forest Land Management Research and Demonstration Program, Umbagog National Wildlife Refuge, Errol, New Hampshire, USA

Marlyse C. Duguid Yale School of Forestry and Environmental Studies, New Haven, Connecticut, USA

Helge Eng California Department of Forestry and Fire Protection, Sacramento, California, USA

Robert A. Ewing Weyerhaeuser Company, Timberlands Strategic Planning, Federal Way, Washington, USA

Peter Farrell New England Forestry Consultants, Inc., Alton, New Hampshire, USA

Sean Flint Northern Forest Land Management Research and Demonstration Program, Umbagog National Wildlife Refuge, Errol, New Hampshire, USA

David Foster Harvard Forest, Harvard University, Petersham, Massachusetts, USA

Patrick Frost Shasta College-Trinity Campus, Weaverville, California, USA

Felipe Aguilar Gómez Comunidad Indígena de Nuevo San Juan Parangaricutiro, Municipio de Nuevo San Juan, Nuevo San Juan Parangaricutiro, Michoacán, Mexico

Donald L. Grebner Department of Forestry, Mississippi State, Mississippi, USA

Oliver Grimm Pike Lumber Company, Akron, Indiana, USA

Anne Hairston-Strang Maryland Department of Natural Resources, Forest Service, Annapolis, Maryland, USA

Jim Hawkins Green Diamond Resource Company, Korbel, California, USA

Kirsten Held Wisconsin Department of Natural Resources, Madison, Wisconsin, USA

J. Ciro Hernández-Díaz Instituto de Silvicultura e Industria de la Madera, Universidad Juárez del Estado de Durango, Durango, Mexico

Chris Hoffman Ashland County Forestry Department, Butternut, Wisconsin, USA

William Hunter U.S. Marine Corps, Hampstead, North Carolina, USA

Bob Izlar Center for Forest Business, Warnell School of Forestry and Natural Resources, University of Georgia, Athens, Georgia, USA

Gary W. Johnson Blue Ridge Parkway, Asheville, North Carolina, USA

Kenneth Jolly Maryland Department of Natural Resources, Forest Service, Annapolis, Maryland, USA

Steven W. Kallesser Gracie & Harrigan Consulting Foresters, Inc., Far Hills, New Jersey, USA

Scott Kelly North Coast Forest Conservation Program, The Conservation Fund, Caspar, California, USA

Robert Keron Ontario Ministry of Natural Resources, Forests Branch, Sault Ste. Marie, Ontario, Canada

Anna M. Klepacka Department of Production Management and Engineering, Warsaw University of Life Sciences, Warsaw, Poland

Dave Knight Sakâw Askiy Management Inc., Prince Albert, Saskatchewan, Canada

Steven Koehn USDA Forest Service, Washington, District of Columbia, USA

Tom Kollasch The Nature Conservancy, Astoria, Oregon, USA

Jeremy Koslowski Polk County Forestry Department, Balsam Lake, Wisconsin, USA

Venkatesh Kumar Weyerhaeuser Company, Timberlands Strategic Planning, Federal Way, Washington, USA

Thomas LaPointe Northern Forest Land Management Research and Demonstration Program, Umbagog National Wildlife Refuge, Errol, New Hampshire, USA

Kerry Livengood Tennessee Division of Forestry, Nashville, Tennessee, USA

Héctor M. Loera-Gallegos Universidad Juárez del Estado de Durango, Durango, Mexico

Adolfo Chavez Lopez Comunidad Indígena de Nuevo San Juan Parangaricutiro, Municipio de Nuevo San Juan, Nuevo San Juan Parangaricutiro, Michoacán, Mexico

Pat Mackasey Government of Saskatchewan, Ministry of Environment, Forest Service, Prince Albert, Saskatchewan, Canada

Gretchen Marshall LEAF Program, University of Wisconsin-Stevens Point, Wisconsin Center for Environmental Education, Wisconsin, USA

Norris Mattox Sessoms Timber Trust, Homerville, Georgia, USA

Erika Mavity Chattahoochee-Oconee National Forest, Gainesville, Georgia, USA

Heather McPherson Washington Department of Natural Resources, Olympia, Washington, USA

John Paul McTague Rayonier, Inc., Yulee, Florida, USA

Krista Merry Warnell School of Forestry and Natural Resources, University of Georgia, Athens, Georgia, USA

Jose Carlos Monarrez-Gonzalez Instituto Politécnico Nacional, CIIDIR Durango, Durango, Mexico

Eusebio Montiel-Antuna Universidad Juárez del Estado de Durango, Durango, Mexico

Scott Mueller Wisconsin Department of Natural Resources, Madison, Wisconsin, USA

Ian Munn Department of Forestry, Mississippi State University, Mississippi, USA

Robert Nall Oregon Department of Forestry, Salem, Oregon, USA

Michael Newton College of Forestry, Oregon State University, Corvallis, Oregon, USA

Abu Nurullah Washington Department of Natural Resources, Olympia, Washington, USA

John O'Keefe Harvard Forest, Harvard University, Petersham, Massachusetts, USA

Colleen O'Sullivan Trinity County Resource Conservation District, Weaverville, California, USA

Michael J. Oppenheimer Rayonier, Inc., Fernandina Beach, Florida, USA

Aaron Palmer Haileybury, Ontario, Canada

Gustavo Perez-Verdin Instituto Politécnico Nacional, CIIDIR Durango, Durango, Mexico

Markian Petruncio Yakama Nation Forestry, Toppenish, Washington, USA

Audrey Barker Plotkin Harvard Forest, Harvard University, Petersham, Massachusetts, USA

Dotty S. Porter Sessoms Timber Trust, Homerville, Georgia, USA

Kip Powers Maryland Department of Natural Resources, Forest Service, Salisbury, Maryland, USA

Dan Rees DPW Environmental Division, Fort Wainwright, Alaska, USA

Don Reimer D.R. Systems Inc., Nanaimo, British Columbia, Canada

Russ Richardson Appalachian Investments, Arnoldsburg, West Virginia, USA

Dick Rightmyer Chattahoochee-Oconee National Forest, Gainesville, Georgia, USA

John Ross John Ross Tree Farm, Savannah, Tennessee, USA

Dan Rouillard Ontario Ministry of Natural Resources, Forests Branch, Sault Ste. Marie, Ontario, Canada

Gary Rynearson Green Diamond Resource Company, Korbel, California, USA

Carlos Antonio López Sánchez Instituto de Silvicultura e Industria de la Madera, Universidad Juárez del Estado de Durango, Durango, Mexico

James Savage SUNY-ESF Ranger School, Wanakena, New York, USA

Joseph Schwantes Wisconsin Department of Natural Resources, Madison, Wisconsin, USA

John Sessions College of Forestry, Oregon State University, Corvallis, Oregon, USA

Jane Severt Wisconsin County Forests Association, Rhinelander, Wisconsin, USA

Edward W. Shepard Bureau of Land Management, Department of the Interior, Newberg, Oregon, USA

Jacek P. Siry Warnell School of Forestry and Natural Resources, University of Georgia, Athens, Georgia, USA

Kyle M. Smith The Nature Conservancy, Astoria, Oregon, USA

Jeremy Solin University of Wisconsin-Stevens Point, Wisconsin Center for Environmental Education, Wisconsin, USA

José Encarnación Luján Soto Unidad de Prestación de Servicios Ejidales de El Salto Dgo. A. C., Durango, Mexico

Randy Spyksma Forsite Consultants Ltd., Salmon Arm, British Columbia, Canada

Larry Stevens Vilas County Forestry Department, Eagle River, Wisconsin, USA

Ron Stevens Chattahoochee-Oconee National Forest, Gainesville, Georgia, USA

Thomas J. Straka School of Agricultural, Forest, and Environmental Sciences, Clemson University, Clemson, South Carolina, USA

Sarah Sullivan Tembec, Timmins, Ontario, Canada

Jon Swae City & County of San Francisco Planning Department, San Francisco, California, USA

Alejandro Torres Comisión Nacional de Áreas Nacionales Protegidas, Andador Vicente Guerrero, Jalpan de Sierra, Querétaro de Arteaga, Mexico

Juan Manuel Torres-Rojo Centro de Investigación y Docencia Económicas, Col. Lomas de Santa Fe, México DF, Mexico

Yenie Tran Center for Forest Business, Warnell School of Forestry and Natural Resources, University of Georgia, Athens, Georgia, USA

J. René Valdez-Lazalde Colegio de Postgraduados, Postgrado en Ciencias Forestales, Montecillo, Estado de México, Mexico

Laird Van Damme KBM Resources Group, Thunder Bay, Ontario, Canada

Alejandro Velázquez Centro de Investigaciones en Geografía Ambiental, Universidad Nacional Autónoma de México, Morelia, Michoacán, Mexico

Klaus von Gadow Burckhardt Institut, Georg-August University, Göttingen, Germany

Robin G. Willhoite United Timber Management Company, Timpson, Texas, USA

Del Williams Revelstoke Community Forest Corporation, Revelstoke, British Columbia, Canada

William C. Wright Department of Forestry, Mississippi State University, Mississippi, USA

Preface

Forest Plans of North America presents case studies of contemporary forest management plans developed for federal, state, county, and city governments, nongovernmental organizations, private individuals, families, indigenous communities and tribes, the timber industry, investment organizations, and other landowners in Canada, Mexico, and the United States. The chapters, written mainly by people who have developed these forest plans, address the objectives, constraints, issues, and methods involved in the development of the plans, the outcomes of the plans, and relevance to sustainability and unique situations in the management of forestlands. *Forest Plans of North America* offers forestry students, practitioners, policy makers, and the general public an opportunity to greatly improve their appreciation of forest management, and, more importantly, it fosters an understanding of the history of certain forests and what forces and tools may shape their future development.

In soliciting contributions for *Forest Plans of North America*, we have contacted the members of the Society of American Foresters E2 Working Group (Land Use Planning, Organization, and Management). We also used our extensive professional contacts to reach out to a wide range of forest owners, forest managers, and forestry consultants with the goal to provide readers with a robust and diverse representation of forest owners and their management plans. Although we have tried to provide a broad representation of forest owners and their management approaches in North America, this book is by no means exhaustive. We would like to thank chapter authors for their efforts in developing their contributions to the book and many others who have made this book possible.

Given a very broad range of forest owners and their management objectives, ordering the book chapters proved challenging. In the end, we have chosen to organize the chapters by the five forest size classes presented in Table 1 and then to arrange them alphabetically within these classes. The size of a forest is, among other considerations, an important management characteristic as it determines, to a certain extent, the forest management opportunities and constraints faced by landowners. There are, for example, many different management aims between the smallest, 58 ac Pike Lumber Company's Sam Little Forest in Indiana (Chapter 4), and the largest, 17 million acre Tongass National Forest in

Alaska (Chapter 46), forests described in this book. The management of small forest properties may be more challenging because per unit area management costs are likely to be higher, it may be more difficult to achieve consistent cash flows to fund a full range of management activities (if timber production is one of the management objectives), and it also may be more difficult to achieve certain multiple-use and environmental goals. Conversely, very large forests and their management affect people and environmental conditions on vast scales, with their own set of management opportunities and constraints. The focus of management also is much different, from consideration of individual trees or stands to the consideration of expansive landscapes. The impact of forest size on management becomes evident even before one considers different ownership, management objectives, guiding laws, regulations, and policies, and even the type of forest resources that are being managed.

The result is 48 chapters representing case studies of contemporary forest plans developed for a very diverse group of forest owners, followed by the synopsis chapter developed by the editors. Each of the 48 chapters contains color photographs, maps, and figures representing a forest managed and the outcomes of planning processes. The location of all forests included in the book is illustrated in Figure 1. To help grasp forest plan settings and conditions that they represent, we created a common template for all of the authors to use. Chapter sections include: (1) Management Setting and Background (overview and history of an area, resource statistics, a photograph illustrating the landscape, and an area map), (2) Planning Environment and Methodology (objectives, constraints, methods for developing a plan, regulatory environment, certification requirements, and organizational policies), (3) Outcomes of the Plans (example outcomes and plan direction, perhaps a table with important outcomes), and (4) Discussion and Conclusions (learning and insights, sustainability issues, plan development and implementation challenges, and other unique issues). Due to space constraints, only the most relevant information about the forest plans is presented in the chapters. In several chapters, however, extensive information about the forest plans can be obtained from the Internet sites that are noted, and from other published sources that have been identified. When available, the relevant

TABLE 1 Forest property size classes used in *Forest Plans of North America*

Size Class	Property Size in ac
1. Very small	Less than 1,000
2. Small	From 1,000 to less than 10,000
3. Medium	From 10,000 to less than 100,000
4. Large	From 100,000 to less than 1,000,000
5. Very large	1,000,000 and more

reference information is provided at the end of the chapters in Additional Reading and Resources.

The book uses the U.S. customary units (also termed English units) for chapters originating from the United States and metric system units for chapters originating from Canada and Mexico. Conversions of specific values between these two unit systems are provided throughout the text. Book tables and figures typically use only one unit system. All units used in the book and their conversions to the other system were gathered and presented on the *Units Page* in the beginning of the book. If, for some reason, chapter authors have used

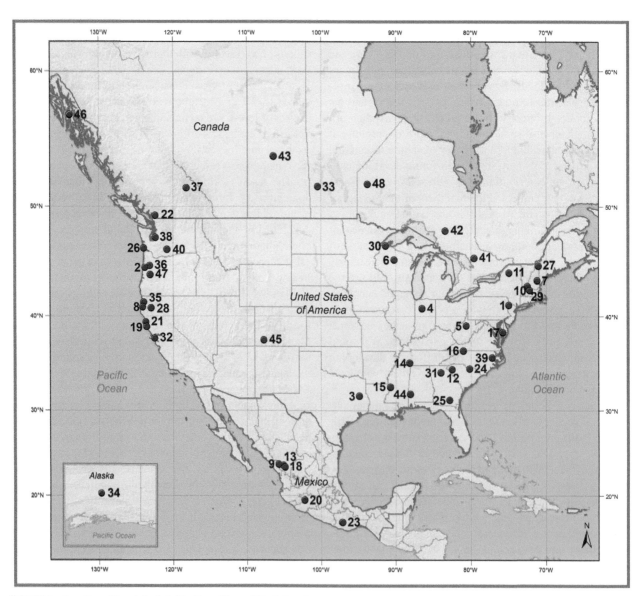

FIGURE 1   Location of forests included in *Forest Plans of North America*.

1   Camp No-Be-Bo-Sco, New Jersey, United States of America
2   Eddyville Tree Farm, Oregon, United States of America
3   Michael Family Forest, Texas, United States of America
4   Pike Lumber Company's Sam Little Forest, Indiana, United States of America
5   Poole and Pianta Woodlot, West Virginia, United States of America
6   Rib Lake School Forest, Wisconsin, United States of America
7   Shirley Town Forest, New Hampshire, United States of America
8   Arcata Community Forest, California, Unites States of America
9   Ejido Borbollones, Durango, Mexico
10   Harvard University Forest, Massachusetts, United States of America
11   Dubuar Memorial Forest, New York, United States of America
12   McPhail Tree Farm, South Carolina, United States of America
13   Molinillos Private Forest Estate, Durango, Mexico
14   Ross Forests, Tennessee, United States of America
15   Willow Break LLC, Mississippi, United States of America
16   Blue Ridge Parkway, Virginia and North Carolina, United States of America
17   Chesapeake Forest Lands, Maryland, United States of America
18   Durango State University Forest Las Bayas, Durango, Mexico
19   Garcia River Forest, California, United States of America
20   Indigenous Community of Nuevo San Juan Parangaricutiro, Michoacán, Mexico
21   Jackson Demonstration State Forest, California, United States of America
22   Mission Municipal Forest, British Columbia, Canada
23   San Pedro El Alto Community Forest, Oaxaca, Mexico
24   Sand Hills State Forest, South Carolina, United States of America
25   Sessoms Timber Trust, Georgia, United States of America
26   South Willapa Bay Conservation Area, Washington, United States of America
27   Unbagog National Wildlife Refuge, New Hampshire, United States of America
28   Weaverville Community Forest, California, United States of America
29   Yale School Forests, New England, United States of America
30   Bayfield County Forest, Wisconsin, United States of America
31   Chattahoochee-Oconee National Forest, Georgia, United States of America
32   City of San Francisco, California, United States of America
33   Forest Management Unit 13 — Forest Management License (FMU) 3, Manitoba, Canada
34   Fort Wainwright, Alaska, United States of America
35   Green Diamond Resource Company, California, United States of America
36   Northwest Oregon State Forests, United States of America
37   Revelstoke Community Forest—Tree Farm License (TFL) 56, British Columbia, Canada
38   South Puget Planning Unit, Washington, United States of America
39   Weyerhaueser, North Carolina, United States of America
40   Yakama Reservation, Washington, United States of America
41   French-Severn Forest, Ontario, Canada
42   Martel Forest, Ontario, Canada
43   Prince Albert Forest Management Agreement (FMA), Saskatchewan, Canada
44   Rayonier, Inc., Southern United States of America
45   San Juan National Forest, Colorado, United States of America
46   Tongass National Forest, Alaska, United States of America
47   Western Oregon Districts, Bureau of Land Management, United States of America
48   Whitefeather Forest, Ontario, Canada

FIGURE 1—CONT'D

different unit conversions from those provided on the *Units Page*, this fact has been noted in that particular chapter. Likewise, common abbreviations were gathered on the *Abbreviations Page* of the book. Abbreviations specific to particular chapters are also provided in the beginning of these chapters.

Forest Plans of North America arose from our passion for forestry and our desire to maintain lush, productive, diverse, healthy, verdant, and thriving forests for generations to come. Forests are very complex living things, meeting untold scores of our material, environmental, recreational, and spiritual needs, and so their management can become quite a complex affair. Developing decisions about the allocation of forests to consumptive, conservation, and preservation uses and implementing these decisions in the forest can be challenging and can require substantial knowledge and resources. Yet, forests are very responsive and dynamic living things and can provide tremendous opportunities for enjoyment and satisfaction that can arise from the fruits of one's labor and the jobs well-done within one's lifetime, regardless of whether one is a caretaker of a very small or a very large forest.

We hope that the readers of this book will find it both a useful learning tool as well as a valuable reference. Our goal is to provide you with the tools to become a confident and competent forest resource manager and an informed participant in private and public debates about forest planning and management.

J.P.S.
P.B.
K.M.
D.L.G.
K.B.
C.C.

Units

ac	acre (0.4047 hectares)
bf	board foot (0.002359737 cubic meters)
cm	centimeter (0.394 inches)
ft	foot (0.305 meters)
ft²	square foot (0.092903 square meters)
ft³	cubic foot (0.0283168 cubic meters)
ha	hectare (2.471 acres)
in	inch (2.54 centimeters, 254 millimeters)
km	kilometer (0.621 miles)
L	liter (0.264 gallons)
m	meter (3.281 feet)
m²	square meter (10.7369 square feet)
m³	cubic meter (35.315 cubic feet, 424 board feet)
m³rta	cubic meter of standing timber including stem and branches (all woody material above the ground)
mi	mile (1.609 kilometers)
mm	millimeter (0.0393701 inches)
MBF	thousand board feet of wood (2.36 cubic meters of lumber)
MCF	thousand cubic feet of wood (28.317 cubic meters of wood)
MMBF	million board feet of wood (2,360 cubic meters of lumber)
MMCF	million cubic feet of wood (28,317 cubic meters of wood)
yd³	cubic yard (27 cubic feet, 0.764554858 cubic meters)

Abbreviations

2,4-D	Two, Four Dichlorophenoxyacetic Acid		**FLPMA**	Federal Land Policy and Management Act
2,4,5-T	Two, Four, Five Trichlorophenoxyacetic Acid		**FML**	Forest Management License
AAC	Annual Allowable Cut		**FMP**	Forest Management Plan
AHA	Available Harvest Area		**FMP**	Forest Management Program
AMZ	Aesthetic Management Zone		**FMU**	Forest Management Unit
AR	Annual Report		**FPA**	Forest Practices Act
ArcGIS	ESRI GIS software		**FSC**	Forest Stewardship Council
ATFS	American Tree Farm System		**FWS**	United States Fish and Wildlife Service
BC	British Columbia		**GIS**	Geographic Information Systems
BCF	Bayfield County Forest		**GMP**	General Management Plan
BLM	Bureau of Land Management		**GPV**	Greatest Permanent Value
BMPs	Best Management Practices		**GUI**	Graphical User Interface
BOF	Oregon Board of Forestry		**GVW**	Gross Vehicle Weight
BRP	Blue Ridge Parkway		**HBU**	Higher and Better Use
BSA	Boy Scouts of America		**HCP**	Habitat Conservation Plan
CCF	Continuous Cover Forestry		**HCVF**	High Conservation Value Forest
CCP	Comprehensive Conservation Plan		**HEC**	Habitat Element Curves
CEQA	California Environmental Quality Act		**HMP**	Habitat Management Plan
CFR	Code of Federal Regulations		**HYF**	High Yield Forestry
CFSA	Crown Forest Sustainability Act		**ICNSJP**	Indigenous Community of Nuevo San Juan Parangaricutiro
CIAC	Communities of Interest Advisory Committee		**INRMP**	Integrated Natural Resource Management Plan
COFF	Community-owned Forest Firm			
CONAFOR	National Forestry Commission		**IP**	Implementation Plans
CUVA	Conservation Use Valuation Assessment		**IRMU**	Integrated Resource Management Unit
DBH	Diameter at Breast Height		**iSharp**	Integrated Sustainable Harvest and Resource Planner
DCHS	Dynamic Caribou Habitat Schedule			
DEP	Department of Environmental Protection		**K-12**	Kindergarten through Twelfth Grade
DFC	Desired Forest Condition		**LEAF**	Wisconsin K-12 Forestry Education Program (formerly Learning, Experiences & Activities in Forestry)
DFS	Delmarva Fox Squirrel			
DMAP	Deer Management Assistance Program			
DNR	Department of Natural Resources			
EA	Environmental Analysis		**LLC**	Limited Liability Corporation
EIS	Environmental Impact Statement		**LMVJV**	Lower Mississippi Valley Joint Venture
EQIP	Environmental Quality Incentives Program, New Jersey		**LP**	Linear Programming
			L-P	Louisiana-Pacific Canada, Ltd.
EPA	Environmental Protection Agency		**LSMA**	Late Successional Management Areas
ESA	Ecologically Significant Area		**LTMD**	Long Term Management Direction
ESA	Endangered Species Act		**MAI**	Mean Annual Increment (average wood volume per unit area divided by average age of trees in that area)
FCE	Forest Community Enterprise			
FEMAT	Forest Ecosystem Management Assessment Team			
			MAV	Mississippi Alluvial Valley
FIDS	Forest Interior Dwelling Species (bird species)		**MBF**	Thousand Board Feet
FLI	Forest Lands Inventory		**MD DNR**	Maryland Department of Natural Resources
FLPA	Forestland Protection Act		**MDS**	Silvicultural Development Method

MDS	Method for Silvicultural Development		SFL	Sustainable Forest Licence
MDS	Método de Desarrollo Silvícola		SFMM	Strategic Forest Management Model
MDWFP	Mississippi Department of Wildlife Fisheries and Parks		SHSF	Sand Hills State Forest
MMCF	Million Cubic Feet		SICODESI	System for Conservation and Forest Development
MMFR	Mexican Method of Forest Regulation		SiPlaFor	Sistema de Planeación Forestal para Bosque
MMOB	Mexican Method of Forest Regulation			Templado
MMOBI	Mexican Forest Management Method		SJNF	San Juan National Forest
MMBF	Million Board Feet		SMZ	Streamside Management Zone
MMIFR	Mexican Method of Irregular Forests Regulation		SPF	San Pedro Forest
MNRF	Ministry of Natural Resources and Forestry		SUNY-ESF	State University of New York College of Environmental Science and Forestry
MNRF	Ontario Ministry of Natural Resources and Forestry		TCF	The Conservation Fund
MNR	Ministry of Natural Resources		TCRCD	Trinity County Resource Conservation District
MPS	Mathematical Programming System		TDEC	Tennessee Department of Environment and Conservation
NCASI	National Council for Air and Stream Improvement		TDF	Tennessee Division of Forestry
NCDWQ	North Carolina Division of Water Quality		TFL	Tree Farm License
NEPA	National Environmental Policy Act		TFS	Texas A&M Forest Service
NJ	New Jersey		THLB	Timber Harvesting Land Base
NJSA	New Jersey Statutes Annotated		THP	Timber Harvesting Plan
NPS	National Park Service		TIMO	Timberland Investment Management Organization
NRCS	U.S. Department of Agriculture, Natural Resources Conservation Service		TMA	Timber Management Areas
NSO	Northern Spotted Owl		TRFO	Tres Rios Field Office
NWR	National Wildlife Refuge		TSI	Timber Stand Improvement
NW FMP	Northwest Oregon State Forest Management Plan		UCODEFO	Unidad de Conservación y Desarrollo Forestal
			UJED	Universidad Juárez del Estado de Durango
NWFP	Northwest Forest Plan		UNAM	National Autonomous University of Mexico
OAR	Oregon Administrative Rule		U.S.	United States of America
ODF	Oregon Department of Forestry		UMA	Wildlife Management Unit
OSB	Oriented Strand Board		USDA	United States Department of Agriculture
PDO	Pacific Decadal Oscillation		USFS	United States Forest Service
PEFC	Programme for the Endorsement of Forest Certification		USFWS	United States Fish and Wildlife Service
			USGS	United States Geological Survey
PRC	California Public Resource Code		WCF	Weaverville Community Forest
RCW	Red-cockaded Woodpecker		WCFA	Wisconsin County Forests Association
REIT	Real Estate Investment Trust		WDFW	Washington Department of Fish and Wildlife
RMA	Riparian Management Areas		WDNR	Wisconsin Department of Natural Resources
RMPs	Resource Management Plans		WDR	Waste Discharge Requirement
RMS	Resource Management System		WFCRMA	Whitefeather Forest Community Resource Management Authority
RSPS	Remsoft® Spatial Planning System			
SAC	Stakeholders' Advisory Committee		WisFIRS	Wisconsin Forest Inventory and Reporting System
SAS	Statistics Analysis Software			
SBM	Structure Based Management		WOPR	Western Oregon Plan Revision
SCFC	South Carolina Forestry Commission		WREDCO	Weyerhaeuser Real Estate Development Company
SEMARNAT	Ministry of Environment and Natural Resources (Secretaria del Medio Ambiente y Resursos Naturales)		WLFW	Working Lands For Wildlife
			WRP	Wetland Reserve Program
SFI	Sustainable Forestry Initiative		yr	Year

Chapter 1

Camp No-Be-Bo-Sco, New Jersey, United States of America

Steven W. Kallesser

Gracie & Harrigan Consulting Foresters, Inc., Far Hills, New Jersey, USA

ABBREVIATIONS

BSA   Boy Scouts of America
DEP   Department of Environmental Protection
EQIP   Environmental Quality Incentives Program
NJ   New Jersey
NJSA   New Jersey Statutes Annotated
WLFW   Working Lands for Wildlife

MANAGEMENT SETTING AND BACKGROUND

Camp No-Be-Bo-Sco is the oldest continuously operating Boy Scout summer camp in the state of New Jersey (Jenkins, 2012). It is currently owned and operated by the Northern New Jersey Council, Boy Scouts of America (hereafter "the Council"), having been purchased by a predecessor entity in several transactions in 1928 and 1930. The property is currently understood to be 380.8 acres (ac) (154.1 hectares (ha)), of which 342.9 ac (138.8 ha) are woodland, with 17.3 ac (7.0 ha) of pond, 12.7 ac (5.1 ha) under an electrical transmission right-of-way, and 7.9 ac (3.2 ha) being non-forested camp facility areas.

The camp is located within the Ridge and Valley geologic province of northwest New Jersey and at the foot of the Kittatinny Ridge. The camp has an extensive history of past management. The forest was likely within seasonal hunting routes of the Minisink subgroup of the Delaware tribe, given its close proximity to known villages along the Delaware River and proximity to an important gap in the Kittatinny Ridge. As such, the forest would have had a history of low- and moderate-intensity wildfires as part of its hunting regimen. Four or five early homesites were located on the property, with two of these being associated with very small sustenance farms. These farms were abandoned prior to 1878. Aside from a minor amount of land clearing associated with the activities and from wood extracted for fuel and farm buildings, very little history of management is apparent until about 1854. At that time, the railroad was extended to Newton, the nearby seat of Sussex County. At about that time, demand was ramping up for fuel for iron furnaces and other industrial developments. The extensive forests of the Ridge and Valley and Highlands provided this fuel and other valuable wood products. By 1874, three sawmills were located within one-half mile of the camp boundary. One of those sawmill owners had title to most of the property and owned a second sawmill on the Paulins Kill that made axe handles. One of the old farmsteads has evidence of being used as a logging camp, likely during the 1880s. In the 1910s or early 1920s, the property was likely heavily cut for charcoal production. The forest is currently composed of mature hardwood trees, the vast majority of which are about 90 to 95 years old. This relatively recent history is confirmed through the earliest photographs and postcards of the camp showing mostly saplings with occasional larger trees of poor growth form.

In the summer of 1927, the camp was leased from a contract purchaser. At that point, campsites were cleared, trails were built, and a dining hall was constructed mostly from American chestnut (*Castanea dentata*) salvage. The title to the land was obtained by the Boy Scouts from the contract purchaser in 1928 and, with some additional land purchases, totaled about 980 ac (396.6 ha) in 1930. The Council continued to construct log cabins from chestnut logs until 1931 when it was forced to lay off its camp ranger. By 1942, the Council had begun to barter timber with local sawmills in exchange for dimensional lumber and other wood products in order to build cabins, lean-tos, and other buildings throughout the camp. Such records of annual forestry reports exist through 1967. Following this period, there appears to have been several individual tree selection harvests, the last of which was primarily a salvage of trees in 1989 due to a gypsy moth (*Lymantria dispar*) outbreak. Generally speaking, the Council appears to have obtained informal advice from government agents during most of this time. The Council's promotional material routinely touted the involvement of persons from the NJ Department of Conservation, the USDA Soil Conservation Service, as well as a wide variety of educators, academics, and people from the conservation not-for-profit community.

The forest at the camp was first assessed by Gracie & Harrigan Consulting Foresters, Inc. in 1999 during a forest inventory in advance of a forest management plan that was written in 2000. The 2000 Plan was most likely the first written forestry plan for the camp. The reason for developing a plan for the property at that time involved an effort to reduce the property taxes for the Council. In accordance with the Farmland Assessment Act of 1964, as amended (NJSA 54:4-23.1 et seq.), a property may qualify for assessment based on an agricultural valuation if the property is able to meet certain agricultural activity and income tests annually. Wood products are considered agricultural in nature. However, for properties that are principally forested, a "woodland management plan" must be prepared and approved by a forester who therein is approved to practice forestry by the NJ Department of Environmental Protection Forest Service. A forester must also annually attest to the fact that the landowner is adhering to the recommendations in the plan.

At the time of the preparation of the 2000 Plan, mapping problems became apparent to the Council and its consulting foresters. Although the camp had been surveyed at the time of its purchase, and portions of the camp were resurveyed in 1965, no unified map existed. To complicate matters, in 1970 the federal government condemned nearly 600 ac (242.81 ha) of the camp property to be used for the Tocks Island dam project (now known as the Delaware Water Gap National Recreation Area). What had existed in terms of maps were never updated, and the Council's internal maps from that point on apparently relied on amateurish deed plots and municipal tax maps. Given the unclear boundary, forest management activities were focused initially on areas well within the camp boundaries.

Forestry activities during the term of the 2000 Plan focused on projects that could be implemented by volunteers, or by contractors who were compensated through federal cost-share programs such as the Forest Land Enhancement Program, the Wildlife Habitat Incentive Program, and the Environmental Quality Incentives Program. Examples of these activities included 15 ac (6.1 ha) of precommercial thinning, 26 ac (10.5 ha) of non-native invasive plant control, 2 ac (0.8 ha) where seedlings of various tree and shrubs were underplanted, and 1.25 ac (0.5 ha) in a group selection harvest system. In addition, 300 ac (121.4 ha) of the upland forest were sprayed to control gypsy moths during two successive growing seasons. Access throughout the property was improved, and intense research on resolving boundary issues was conducted, concluding with a definitive remarking of boundary lines in 2006. Unfortunately for wildlife species requiring elements of young forests, the limited age distribution of Camp No-Be-Bo-Sco is characteristic of upland forests within heavily forested areas of the Ridge and Valley and northern Highlands provinces within northern New Jersey.

Just as important to the Council were educational materials made available to scouts and scoutmasters. These were improved following the 2000 Plan. A permanent self-guided ecology and forestry interpretive trail was established in 2006. This trail identifies and interprets various forestry practices on the property and other important ecological features. Also, the Council and the consulting foresters maintain a set of "From the Forester" fact sheets that are designed to explain potentially difficult topics to merit-badge instructors and to guide them to useful areas within camp to teach on the topic.

The forest was re-inventoried in 2010 to prepare for the development of the management plan. The Council opted to have a Forest Stewardship Plan developed in order to pursue cost-sharing assistance for plan development through the U.S. Department of Agriculture Natural Resources Conservation Service. At the time of the 2010 inventory, 342.9 ac (138.8 ha) of forest were identified, of which 47.2 ac (19.1 ha) are wetlands or are immediately adjacent to wetlands. Most of these lowland areas are extremely mucky and are currently impacted by beavers. Of the remaining 295.7 ac (119.7 ha), 99.6% are mid-successional, mostly having years of origin of between 1900 and 1925. Most of the upland forest is oak-hickory (76%) (Figure 1.1), with some mesic areas dominated by yellow poplar (*Liriodendron tulipifera*) (24%). Areas unthinned during the 2000 Plan were at high levels of full stocking, with relative densities of 84% to 100%. Other areas that had been thinned, had experienced high mortality during the 1987–1989 gypsy moth infestations, or had experienced mortality from hemlock wooly adelgid (*Adelges tsugae*), were at lower levels of full stocking with relative densities ranging from 63% to 70%. The forests are generally considered two-aged. A typical age class and species distribution of the upland forests is illustrated in Figure 1.2.

FIGURE 1.1   Hardwood forests typical of the Camp No-Be-Bo-Sco property. Photograph courtesy of William Kallesser.

FIGURE 1.2   A typical two-aged distribution of the upland forests of the Camp No-Be-Bo-Sco property.

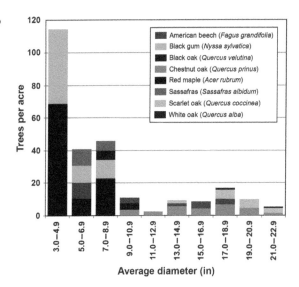

PLANNING ENVIRONMENT AND METHODOLOGY

Local councils of the Boy Scouts of America are organized to serve youth within a defined geographic area. The Council is governed by a board of directors. The directors, some of whom are officers, are elected annually by representatives of institutions that sponsor scouting units (e.g., Boy Scout troops, Cub Scout packs, etc.), directors, and members-at-large. The board establishes certain standing committees whose members include directors and other volunteers. These committees are often composed of various subcommittees with more specific purposes. Each Council's board of directors also hires a scout executive who functions as the chief executive officer. The scout executive manages a professional staff whose responsibilities are to serve as professional advisors to the various committees and subcommittees. Each local Council is overseen by the National Council of the Boy Scouts of America, which sets policies and programs at a national level.

Although each individual Council's committees may be organized in unique ways, the committees of the Northern New Jersey Council who have oversight over conservation matters are the camping committee and the properties committee. The camping committee's responsibility is to oversee camping programs and operations. It has a conservation subcommittee whose responsibility is to administer conservation programs such as awards programs for youth, produce and update camp conservation plans for summer camps, and other similar responsibilities as may be assigned. The conservation subcommittee also reports to the properties committee, whose responsibilities include administration of capital projects and certain issues related to real estate assets (including timber). The chairman of the camping committee and the chairman of the properties committee are each vice presidents of the Council. Each of the committees has the Council's chief operating officer for its professional advisor.

The camp conservation plan is a required document for any property serving the high-intensity active recreation use of a summer camp operation. This plan is optional for camps with short-term weekend camping operations. Its purpose is to identify conservation issues of concern for a particular camp and suggest projects to solve identified problems and minimize impacts from recreational use. Where 5-year camp conservation plans and 10-year forest stewardship plans are in place for the same property, it has been the policy of the conservation subcommittee to have a unified document developed as a 10-year plan, with a minor update at the middle of the term.

Like most, if not all, private forestlands in New Jersey or nearby southern New York, production of wood products is not a primary consideration of the landowner. As such, a clear indication of the Council's objectives for forest management was needed. The consulting foresters worked with the conservation subcommittee and the chief operating officer to develop a short list of objectives. These objectives were informally brought to the chairmen of the camping and properties committees for consensus to ensure there was no strong opposition to them. Had there been strong opposition, the conservation committee would have re-evaluated the objectives.

The 2010 Plan's objectives begin with the mission statement of the BSA, followed by the National Council's policy on conservation program. From there, four objectives are defined, namely (1) to operate a multipurpose recreational facility for camping and other scouting activities in order to foster good citizenship, foster physical fitness, and build character; (2) to develop a healthy, diverse, sustainable forest with a variety of wildlife to act as an outdoor laboratory to provide learning opportunities for scouts; (3) to be a good neighbor to adjacent property owners and society in general by providing clean surface water; and (4) to satisfy the planning requirements of this property to qualify for farmland assessment under the Farmland Assessment Act and regulations of 1964 as amended by Chapter 201, Laws of 1986 and regulations implemented in conjunction therewith, and associated case law.

The second objective was written with specific goals in mind. First, although more than 99% of the upland forest is mid-successional, the camp had provided habitat for early successional species within the utility rights-of-way and the swamp, some of which were emergent wetlands. Prior to the regional power blackout of 2003, the right-of-way was maintained at what appeared to be a 15-year interval, at which time woody vegetation was cut, treated with herbicide, or burned. Less management was conducted in proximity to campsites and wetlands and thus a significant variety of early successional habitat types existed within the 150 foot (ft) wide (47.7 meter (m)) rights-of-way. After the 2003 blackout, right-of-way management appeared to shift to a 3-year interval in accordance with stricter regulation by the Federal Energy Regulatory Commission, and the NJ Board of Public Utilities. Right-of-way management also appears to focus more on herbicide treatment of woody vegetation that could exceed a maximum height of three feet. This has significantly changed the habitat for certain animal species. In addition, as beaver populations continue to expand within the camp and elsewhere throughout the region, many of the wetland areas are significantly ponded and too wet for some wildlife species who lately have utilized these areas as scrub/shrub or grassland habitat. In order to accommodate these and other declining wildlife, and in order to provide observation areas for nature studies for scouts, the Council felt that it was important to create early-successional forest within the matrix of the older forest.

There is also a strong concern that much of the oak-dominated forest within northern and central New Jersey is converting to northern hardwood forest, given the last 100 years of fire exclusion and lack of active management (Brose et al., 2008). Oak-dominated forests are disturbance-dependent ecosystems that provide myriad ecosystem services and habitat to a tremendous variety of wildlife (Brose et al., 2008). The plan would address these concerns through forest stand improvement programs designed to create better growing conditions for existing oak overstory trees; reduce seed sources of black birch (*Betula lenta*), red maple (*Acer rubrum*), and black gum (*Nyssa sylvatica*); and control competing understory vegetation, especially exotic invasive plants, in order to improve understory light conditions for oak seedlings.

The 1999 forest inventory noted widespread and significant white-tailed deer (*Odocoileus virginianus*) browse damage to the forest understory. The 2010 forest inventory showed that significant deer browse damage was confined to certain areas of the camp property, with some forest stands on xeric soils having large amounts of oak seedlings three feet in height or less. The plan recommended managing the white-tail deer population. The 2010 forest inventory also noted a small area of hemlock-hardwood forest where the eastern hemlock (*Tsuga canadensis*) component had been killed by hemlock wooly adelgid (*Adelges tsugae*). The plan recommended restoration of this area through exotic invasive plant control, the planting of eastern white pine (*Pinus strobus*) seedlings, and the erection of deer exclusion fencing.

The ability to use forestry as a tool to satisfy these objectives is constrained in several ways. First, as is apparent through the wording of the objectives, the property is first and foremost a Boy Scout camp. The aesthetics of the core of the property containing the various cabins, campsites, and program area is of primary concern. In order to address aesthetic concerns, a 100 ft (30.5 m) buffer was placed around the campsites, summer camp program areas, the lake, and marked hiking trails. A 200 ft (61 m) buffer was placed on the camp entrance road and all areas used extensively during the winter season. Second, the Council is very sensitive to issues connected to water quality at the camp. Effectively, this is a minor constraint as the important stream crossings were bridged more than 80 years ago and have been adequately maintained since then, and pre-existing stream crossings exist in two other places where their use would be helpful. A 150 ft (45.8 m) buffer was thus placed on all streams, wetlands, and suspected wetlands. Third, in accordance with minimum guidelines set by the NJ Forest Stewardship Committee, proper considerations need to be given to threatened or endangered plant and animal species. The consulting foresters enlisted the assistance of the U.S. Fish & Wildlife Service with regard to federally listed species, the NJ DEP Division of Fish & Wildlife with regard to state-listed wildlife species, and the NJ DEP State Forestry Services Office of Natural Lands Management with regard to state-listed plant species and rare ecological communities. Using the GIS layers, 10 state-listed animal species were identified as having potential habitat on the property by NJ DEP Division of Fish & Wildlife. The U.S. Fish & Wildlife Service also stated that the camp is located within the range of the Indiana bat (*Myotis sodalis*), but this species of bat has not been recorded within either municipality that the camp is located in since at least 1970.

Although considerations were given to all of the species identified through consultation, four species were identified for further consideration due to their known or strongly suspected occurrence on or near the camp property. Most of the wildlife species were given consideration through timing restrictions on tree cutting not related to camper safety issues. The timing restrictions were given to avoid disruption to the courting, breeding, and feeding cycles of many of these species. Reptile and amphibian species strongly associated with open water or wetlands were given considerations by adhering to the current DEP forestry and wetlands Best Management Practices manual (Cradic, 1995) in order to prevent impairment of water quality and hydrological functions of the waterways. Habitat specialists were addressed by ensuring that an area of their habitat would either be retained or created during the term of the plan (e.g., areas of dense native brush, trees with cavities and snags, and dominant trees in proximity to the lake).

Finally, forest products markets are a challenge in the state, particularly in the northern half of the state. Aside from local firewood processors, there are no economical outlets for low-quality wood. This is extremely frustrating when executing regeneration harvests, as the higher quality wood can be sold, usually to out-of-state sawmills, but cutting of low-quality trees must be paid for by the landowner and is usually accomplished by loggers not using mechanized methods.

In order to provide the readers of the 2010 Plan with a more complete and transparent view of how the plan was developed, the consulting foresters first divided the property into six different land-use areas, based on current use. Three of the areas were intuitively obvious—the lake, the utility rights-of-way, and the swamp. No further actions were recommended within the swamp, given constraints related to mucky soils and the uncertainty related to local beaver populations and associated damage. The remaining land-use areas are Camp (85.8 ac (34.6 ha) Infrastructure and Buffers (95.5 ac (38.7 ha)), and Passive Recreation (132.0 ac (53.4 ha)) (Figure 1.3). The Camp land-use area represents areas in and around campsites,

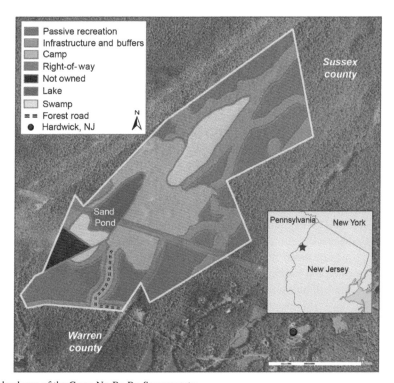

FIGURE 1.3   A map of the land area of the Camp No-Be-Bo-Sco property.

buildings, facilities, activity areas, access roads, and their associated buffers as previously described. Management within the Camp land-use area excludes regeneration harvests. The Infrastructure and Buffers land-use area includes marked hiking trails, waterways, the camp entrance road, and all associated buffers as previously described (including buffers around the swamp land-use area). Management within the Infrastructure and Buffers land-use area excludes even-aged regeneration harvest methods. The Passive Recreation land-use area includes all land not allocated to other land uses, and it is described as currently being used for off-trail hiking, orienteering, wilderness camping, wildlife and forestry studies, and other passive recreation uses. Even-aged regeneration harvest methods are deemed appropriate in the Passive Recreation land-use area.

Maximum harvest rates for the Passive Recreation land-use area were determined using simple area regulation, using 70 years as a future rotation age and a cutting interval of once every 10 years. Maximum harvest rates for the Infrastructure and Buffers land-use area were determined using simple area regulation, using 100 years as a future rotation age and a cutting interval of once every 10 years. Group selection harvests of 0.25 ac (0.1 ha) each were selected due to their ability to properly regenerate either oak or yellow poplar forest (Burns, 1983).

Having regeneration harvests or cuts on the camp property was very important to the Council as certain conservation/ecology merit badges require learning about forestry. It had previously struck some volunteers within the Council as odd that scouts were visiting the camp in order to learn about what forestry and wildlife habitat improvement projects look like from books instead of on-the-ground activities. In addition, the plan recommended improvement of educational materials for campers, improvement of ruffed grouse (*Bonasa umbellus*) habitat within the utility rights-of-way by dragging drumming logs and large rocks into the cleared area, tree planting, treatment of gypsy moths on an as-needed basis, boundary line marking, and access maintenance and improvement.

In terms of the regulatory environment, forestry operations within and near wetlands, open waters, and floodplains are regulated by the NJ Freshwater Wetlands Protection Act and the NJ Flood Hazard Area Control Act. The Freshwater Wetlands Protection Act was written to assume Section 404 responsibilities under the federal Clean Water Act. If forestry operations are being conducted in accordance with a forest management plan approved by the state forester and those operations are being conducted in accordance with the current forestry and wetlands Best Management Practices manual, then those activities are exempted under the Freshwater Wetlands Protection Act and qualify as a permit-by-rule under the Flood Hazard Area Control Act. Best Management Practices require considerations for threatened and endangered species within jurisdictional areas. Jurisdictional areas under these acts cover 150 ft (45.7 m) from stream banks and edges of wetlands. Since the Council chose to pursue a cost-sharing option for a Forest Stewardship Plan, it has voluntarily restricted itself to give considerations to threatened and endangered species in forested areas outside of these jurisdictional areas during the term of the plan as well.

Shortly before the 2010 Plan was written, the property was enrolled as a Certified Tree Farm within the American Tree Farm System based on its stewardship during the term of the 2000 Plan. Third-party verification of sustainable management was not a primary concern of the Council, but American Tree Farm System was highly recommended by the consulting foresters. Following the production of the first draft of the 2010 Plan, copies were provided to the camping committee and the properties committee. Minor changes were made based on their review. At that point, the scout executive approved the plan on behalf of the Council, and it was filed with the municipal tax assessors and with the NJ DEP Forest Service. The plan underwent minor edits based on Forest Service review. Following this, the plan was posted to the Council's website and made available to camp staff and other stakeholders.

OUTCOMES OF THE PLAN

Shortly after the 2010 Plan was finalized, the Council and its forester were approached by NJ Audubon and the NJ Chapter of the Nature Conservancy in regard to applying for cost-share funding with the Common Waters Fund being administered by the Pinchot Institute for Conservation. A meeting was facilitated at the camp, including representatives of these groups, the Council's chief operating officer, camp ranger, and a representative of the USDA Natural Resources Conservation Service. As a result of that meeting, the Council decided to expedite 71.5 ac (28.9 ha) of exotic invasive plant control, 5 ac (2.0 ha) of forest stand improvement thinning, and restoration of 1.7 ac (0.7 ha) of hemlock-hardwood forest (Table 1.1) that had been recommended in the plan. This would be paid for by leveraging cost-share funds in the Environmental Quality Incentives Program (EQIP) against grants from the Common Waters Fund. Work began shortly after approvals were gained from each program. Inspections were made of the management areas by the Council, its consulting foresters, the USDA NRCS, and representatives of the Common Waters Fund. Results were satisfactory for all involved, although the consulting foresters took note that in future forest stand improvement programs, designed to stimulate or promote oak regeneration, additional focus needs to be placed on cutting trees in the two-inch DBH class.

TABLE 1.1 Important Outcomes of the 2010 Camp No-Be-Bo-Sco Forest Plan and Subsequent Negotiations

Years	Activity	Acres Accomplished	Acres Remaining
2011–2012	Conclude forest stand improvement previously marked	5.0	–
	Control exotic invasive plants	71.5	–
2013–2015	Conduct regeneration harvest	17.9	1.0
	Conduct first harvest of shelterwood method	5.0	13.9
	Conduct group selection harvests	–	19.0
	Plant white pine seedlings in former hemlock-hardwood swamp	1.7	–
	Exclude deer in former hemlock-hardwood swamp	1.7	–
	Monitor exotic invasive plant control areas	Ongoing	–
2016–2018	Forest stand improvement activities	–	5.0
	Monitor exotic invasive plant control areas	–	71.5
2019–2020	Forest stand improvement activities	–	5.0
	Monitor exotic invasive plant control areas	–	71.5
2010–2020	Continue to maintain or improve property access trails	Ongoing	
	Continue to maintain property boundary lines	Ongoing	
	Monitor for gypsy moth infestations and spray when necessary	Ongoing	
	Improve educational materials for campers	Ongoing	

In 2012, the NJ DEP Division of Fish & Wildlife updated its listing of threatened and endangered species. Significant to the camp, the state listed golden-winged warbler (*Vermivora chrysoptera*) as endangered. This warbler's habitat is a very specialized subset of early-successional forest, and because early-successional forest is becoming uncommon in northern New Jersey, its numbers have declined to just 25 breeding pairs statewide. Specifically, this bird species requires forests aged generally 0 to 20 years that are located at relatively high elevation (Bakermans et al., 2011). Another reason for its decline is hybridization with the blue-winged warbler (*Vermivora cyanoptera*). To attempt to differentiate habitat enough to attract golden-winged warbler, but exclude blue-winged warbler, the habitat descriptions are as follows:

- at least 70% of the landscape within one-half mile of the project area should be forested, and
- the project area should contain residual trees of less than 20% of forest canopy, or
  - no more than 10–40 ft^2 per ac (2.3–9.2 m^2 per ha) of basal area, or
  - no more than 10–15 trees per ac > 9 inches (in) (22.9 centimeters (cm)) DBH

The project area would also include, or be immediately adjacent to, a wetland area or stream and contain a mixture of forbs and grasses at about 45% cover with saplings, seedlings and other small woody vegetation giving the area a very diverse and stratified vegetation pattern (Bakermans et al., 2011; Patton et al., 2010; Petzinger, personal communication). To have a high degree of success, a habitat area should contain a large amount of edge, be generally complex in shape, and be located within one mile from another suitable stand (Bakermans et al., 2011).

Later in 2012, USDA NRCS announced the Working Lands For Wildlife program, which would focus funding of habitat improvement work on private lands to benefit bog turtle (*Glyptemys muhlenbergii*) and golden-winged warbler. Many other species use a similar habitat and would benefit from this program. However, most of the heavily forested areas at high elevations within the Ridge and Valley are publicly owned with a recent history of little to no forest management. To rapidly and effectively promote habitat for the critically imperiled warbler, NJ DEP Division of Fish & Wildlife collaborated with USDA NRCS to share its models of areas favorable for habitat creation. This information was overlaid with local parcel data, and private forest landowners seen as holding high-priority forests were solicited for their willingness to participate in the program. Working Lands For Wildlife was also promoted through NJ DEP Forest Service and the New Jersey Division Society of American Foresters. Thus, the Council and its forester submitted interest in the program.

In 2012 and 2013, the consulting foresters toured the camp with representatives of USDA NRCS, NJ DEP Division of Fish & Wildlife, and National Wild Turkey Federation to select an area that was compatible with the 2010 Plan and had a high likelihood of success for the golden-winged warbler. Due to unrelated internal issues within the Council, a proposed harvest during the winter of 2012–2013 was not possible. As of March 31, 2014, 17.9 ac (7.2 ha) of habitat creation work was accomplished by a logger after a competitive bidding process. Timber quality within the habitat creation area was less than optimal, largely due to past gypsy moth damage and damage from Hurricane Sandy in 2012. The vast majority of the trees within the habitat creation area were low-quality, and no large market existed. Thus, the harvest portion of the habitat creation was being done at a net cost to the Council.

Cost-sharing funds have been allocated under WLFW to assist in the cutting of low-quality trees, herbicide treatment of existing understory to meet targets for grass and forb coverage, forest stand improvement along the perimeter of the area to soften the edge, and post-harvest treatment of any potential exotic invasive plant invasions. Wildlife biologists from NJ DEP Division of Fish & Wildlife have agreed to monitor the site for golden-winged warblers and other birds for at least 5 years post-harvest. The Council incorporated the habitat creation area into the instruction of certain merit badges during the 2014 summer camp season. The Council's public relations committee is also considering sharing the project with the public within their service area as an example of the benefits of scouting.

DISCUSSION AND CONCLUSIONS

Working with small, low-intensity private forest landowners in the northeastern United States is a challenge. These types of landowners are typical of northern New Jersey and southern New York. In northern New Jersey, the median size of Farmland Assessed forestland under a forest management plan is less than 45 ac. It is no great surprise given the parcelization and suburbanization of this area that timber production is not a primary concern of the average landowner. Because many in the general public and in the preservation community associate forestry only with the harvest of timber, foresters are constantly working with clients and others to illustrate how forestry activities—commercial and noncommercial—can act as proxies for natural processes and natural disturbances necessary for maintaining existing forest types. Without the ability to communicate the benefits of forestry and silvicultural activities in this way, we would not have had the opportunity for success on Camp No-Be-Bo-Sco.

In many situations, foresters working for private organizations or public agencies should take the time to incorporate the overarching policies promoting conservation, or the statutory authority for forest management into forest management plans. This was of significant benefit to the Council to be able to better explain itself to its members, many of whom are preservationists. It also allowed a teachable moment to show how forestry is not only compatible with the scouting program, but how it is taught to scouts within such programs as Forestry Merit Badge and Fish & Wildlife Management Merit Badge.

A discussion of any forest issue in the mid-Atlantic United States would not be complete without some discussion of white-tailed deer or exotic invasive plant impacts. The forest plan developed for Camp No-Be-Bo-Sco takes a reasonable approach to address both, although there is a vocal minority within the local preservation community who will insist that any Forest Stewardship Plan, regardless of length or detail, does not adequately address either. At this point in the recent management history of the camp, it is safe to say that the only serious exotic invasive plant issue remaining on the camp is Japanese stiltgrass (*Microstegium vimineum*). The consulting foresters and the Council are awaiting a reasonably effective biocontrol method for this plant. Regarding deer, the current hypothesis is that American black bear (*Ursus americanus*) and other forms of predation on fawns are reducing deer populations in heavily forested areas of the state. If this is true, it would suggest a cyclical, not a structural, reduction in the deer herd, placing more emphasis on effective deer management. Deer population was not measured directly, but quality and quantity of oak regeneration was used as a proxy to determine if the deer population was at levels that would retard forest regeneration (Brose et al., 2008; Gould, 2005; Steiner et al., 2008). However, these concerns need to be balanced with the need to accommodate the early-successional forests required of the golden-winged warbler. These measurements were not taken during the latest forest inventory, but were recommended during pre-harvest planning. During the 2020 forest inventory, we hope to make standardized measurements part of the inventory procedure.

Of significant concern to foresters is the expected impact of global warming to the forests of northern New Jersey. As previously mentioned, a lack of recent disturbance events is causing a gradual change from oak-dominated cover type to a northern hardwood forest cover type. Disturbances caused by recent events such as the 2011 ice storm and Hurricane Sandy in 2012 caused significant damage to overstory trees (mostly oaks), but did little to the midstory (mostly black birch, red maple, and black gum). The concern is that increasing temperatures and other climactic conditions will shift the range of northern hardwoods further north (Prasad et al., 2007-ongoing) after they have displaced oaks as the dominant cover type. Given the regional difficulties associated with properly regenerating oak forests, the potential does exist for significant stresses to ecosystems.

Given the benefits derived from the Farmland Assessment program, the decision to develop a new plan following the expiration of the 2000 Plan was an easy one. While reporting to two separate committees was mildly redundant, it did allow for increased buy-in by Council directors to the plan once it was finalized. Proper plan implementation is always a challenge regardless of ownership. The consulting foresters have had experience preparing forest management plans for charitable not-for-profit corporations that were minimally executed, if at all. They credit the relative lack of challenges in implementing the 2010 Plan to a combination of enhanced engagement with the Council and the offers of financial assistance from the non-governmental organization community and government agencies. While it is not expected that offers of financial assistance will continue, most of the critical management recommendations have been implemented or are imminent.

Since the initiation of forestry activities under formal forest management plans in 2000, the Council has developed an improved relationship with the NJ Chapter of the Nature Conservancy. The Nature Conservancy manages a preserve adjacent to the camp, but no plan exists for that property and management has focused purely on deer management and exotic invasive species control, with significant investments in deer exclusion fencing. At current, the two properties are an interesting juxtaposition of management styles. It is anticipated that if current trajectories hold true, during the next 5 years the difference between outcomes on the two properties will diverge dramatically as reductions in canopy closure and reduction in competing understory vegetation drive improvements in oak regeneration at the camp, and successful creation of early successional forest boost biodiversity.

REFERENCES

Bakermans, M.H., Larkin, J.L., Smith, B.W., Fearer, T.M., Jones, B.C., 2011. Golden-Winged Warbler Habitat Best Management Practices for Forestlands in Maryland and Pennsylvania. American Bird Conservancy, The Plains, VA. 26 p.

Brose, P.H., Gottschalk, K.W., Horsley, S.B., Knopp, P.D., Kochenderfor, J.N., McGuiness, B.J., Miller, G.W., Ristau, T.E., Stoleson, S.H., Stoudt, S.L., 2008. Prescribing regeneration treatments for mixed-oak forests in the Mid-Atlantic region. U.S. Department of Agriculture, Forest Service, Northern Research Station, Newtown Square, PA. General Technical Report NRS-33. 100 p.

Burns, R.M., 1983. Silvicultural Systems for the Major Forest Types of the United States. U.S. Department of Agriculture, Forest Service, Washington, DC, Agricultural Handbook 445. 191 p.

Cradic, A. (Ed.), 1995. New Jersey Forestry and Wetlands Best Management Practices Manual. New Jersey Department of Environmental Protection, Trenton, NJ. 31 p.

Gould, P.J., 2005. Regenerating Oak-Dominated Stands: Descriptions, Predictive Models, and Guidelines. Ph.D dissertation, Pennsylvania State University, College Park, PA. 156 p.

Jenkins, M., 2012. Camp No-Be-Bo-Sco: An Adventure in Scouting Excellence Since 1927. Lulu Press, Inc, Raleigh, NC. 266 p.

Patton, L.L., Maehr, D.S., Duchamp, J.E., Fei, S., Gassett, J.W., Larkin, J.L., 2010. Do the golden-winged warbler and blue-winged warbler exhibit species-specific differences in their breeding habitat use? Avian Conserv. Ecol. 5 (2), 2. http://www.ace-eco.org/vol5/iss2/art2/ (Accessed December 17, 2013).

Prasad, A.M., Iverson, L.R., Matthews, S., Peters, M., 2007. A climate change atlas for 134 forest tree species of the eastern United States [database]. ongoing, U.S. Department of Agriculture, Forest Service, Northern Research Station, Delaware, OH. http://www.nrs.fs.fed.us/atlas/tree (Accessed December 17, 2013).

Steiner, K.C., Finley, J.C., Gould, P.J., Fei, S., McDill, M., 2008. Oak regeneration guidelines for the Central Appalachians. Northern Journal of Applied Forestry 25, 5–16.

Chapter 2

Eddyville Tree Farm, Oregon, United States of America

Michael Newton

College of Forestry, Oregon State University, Corvallis, Oregon, USA

ABBREVIATIONS

2,4-D   Two, Four Dichlorophenoxyacetic Acid
2,4,5-T   Two, Four, Five Trichlorophenoxyacetic Acid
ODF   Oregon Department of Forestry

MANAGEMENT SETTING AND BACKGROUND

The Eddyville Tree Farm is a near-sea-level 200 acre (ac) (81 hectare (ha)) tract situated 30 miles (48 km) west of Corvallis, Oregon, on the north side of Highway 20 in the middle of Oregon's Coast Range. The current owner purchased the tract in 1961 from the last survivor of a family who had homesteaded the property in 1905. The first settlers found a landscape with deep, fertile soil that had been burned many times, leading to domination by bracken ferns (*Pteridium aquilinum*), herbs, and shrubs of many species. There were virtually no conifers or merchantable hardwoods on the land when it was homesteaded. The homesteaders valued the open space for grazing purposes and found it relatively easy clearing for pasture. In the five decades after settlement, the open space was maintained by winter burning of dormant bracken fern, leading to a couple of years of marginally sufficient grass for livestock. At the outbreak of World War II, most of the family left to work in factories or to pursue military service. Burning ceased and forest succession restarted. The process that began with a few residual Douglas-fir (*Pseudotsuga menziesii*) trees provided for invasion of about 40 ac (16 ha) of clearings by Douglas-fir seedlings. At the time the property was purchased, this young stand had been high-graded to an 8 inch (in) (20.3 centimeter (cm)) diameter at breast height (DBH) limit. Thus, there were no conifers with a DBH over 8 in anywhere on the property. These were specimens too poor to market and presumably not worth growing.

Here, the inherent high quality of the soil and climate took over. A few million years ago, this area was beneath the Pacific Ocean, collecting sediments sliding off the western North American Continental slope to form what became the Tyee Formation, a layer of sedimentary rocks now underlying several thousand square miles on the coastal side of the Oregon Coast Range. This formation was made up of silt and clay particles in its northerly parts, and in the process of weathering, this rock has been rapidly decomposing into deep silt loam soils of remarkable water-holding capacity and fertility. In the past few thousand years, the area has been dominated by species that typically follow large disturbances—namely, those that demand light to survive their regeneration phase and sprouting shrubs. Douglas-fir and red alder (*Alnus rubra*) were among the most successful pioneer species following periodic lightning- or native American-induced, large-scale, stand-replacing fires.

This property has been subjected to a number of fortuitous events leading to the formation of deep soil layers. Significant annual precipitation (80+ in (2,032 millimeters (mm))) and multiple generations of forests that either fixed nitrogen (red alder) or provided extreme amounts of biomass, have led to the formation of deep organic layers. The result is that a large area became one of the most productive forested areas in the world. Without occupation by conifers, however, it also grows and is occupied by noncommercial species of great resiliency, among which are many species capable of sprouting vigorously and quickly suppressing Douglas-fir regeneration. Those species had occupied the Eddyville Tree Farm for many decades where the settlers had not cleared patches and burned them repeatedly.

Forest Plans of North America. http://dx.doi.org/10.1016/B978-0-12-799936-4.00002-3

The owner (and author of this chapter) graduated from Oregon State University College of Forestry in 1958, where his thesis work embraced various ways of converting brush fields to conifer plantations. Eventually, herbicides and large planting stock emerged as dominant components in establishing conifers in diverse shrub and hardwood cover. At that time, two herbicides (2,4-D and 2,4,5-T), common crop and lawn weed-control products, were the only ones known to be reasonably effective in controlling deciduous shrubs and hardwoods. Subsequent research soon found these herbicides to be selective in favoring conifers, if applied in early spring while conifers are dormant. The addition of diesel fuel added to efficacy on dormant shrubs. Since 1960, many herbicide products have been developed that are safe and effective in providing a wide range of selectivity for managing forest vegetation for many purposes. They were especially valuable for favoring Douglas-fir, western hemlock (*Tsuga heterophylla*), and other highly valuable timber species. Selectivity among the nonconiferous species, ecosystem management at a new level, permitted multiple benefits from sophisticated vegetation management. The Eddyville Tree Farm became a very useful test case because so many noncommercial species were prevalent and resilient after treatment, leading to early interpretation and management of ecosystems following treatment.

The property's purchase price reflected no obvious value for forestry. Establishment of a coniferous forest meant careful replacement of very dense, diverse, and vigorous stands of large shrubs and scattered hardwoods. Shrubs were generally shade tolerant, and included vine maple (*Acer circinatum*), Oregon hazel (*Corylus cornuta californica*), salal (*Gaultheria shallon*), cascara buckthorn (*Rhamnus purshiana*), red elderberry (*Sambucus callicarpa*), salmonberry (*Rubus spectabilis*), and escaped Himalaya blackberry (*Rubus procerus*). Other shrubs, ferns, and large herbaceous species were abundant, and sometimes reached heights of 10–14 feet (ft) (3.0–4.3 meters (m)) in dense stands. Hardwood trees included bigleaf maple (*Acer macrophyllum*), a very vigorous species that sprouts profusely when cut reaching heights of more than 100 ft (30 m) in multiple stems. Each mature maple can occupy a 0.25 ac (0.1 ha) space, yet have little commercial value. Scattered red alder and bitter cherry (*Prunus emarginata*) were abundant on the moister sites, and also were capable of reaching more than 80 ft (24 m) high in 25 years without commensurate commercial value.

PLANNING ENVIRONMENT AND METHODOLOGY

The management plan for the Eddyville Tree Farm seeks to optimize opportunities for providing amenities to a major degree while also providing the yields that this environment makes possible. This plan has goals of very high yield of the most valuable conifers combined with a wide variety of age classes and structure classes that follow disturbance of such shrub communities over a period of at least 70 years following disturbances that had favored Douglas-fir for millennia. This plan was built on the principle that the Douglas-fir would prosper only in even-aged stands in this area, and that its regeneration requires one or two years of respite from early-seral shrubs and herbs, leading to a decade or so of dense early-seral habitat. This dense seral habitat would give way to five to six decades of dense conifers during which thinnings capable of supporting some understory browse would provide both revenue and habitat. The end result of the management plan is the tract reaching a stage with features similar to late-seral natural stands of far greater age in its oldest stands. This would require periodic entries to provide clearings every 5–10 years, commercial thinning at ages 20–45 years, and regeneration harvest at age 70, markets permitting, with vegetation control to favor restarting the cycle.

This plan required the use of herbicides, products requiring registration and application by certified applicators, controlled fire to reduce slash, and observation of a number of forest practice rules overseen by the Oregon Department of Forestry (ODF).

Guiding Laws, Regulations, and Policies

Oregon has strict land-use laws guided by Oregon Ballot Measure 37, in 2004, and Ballot Measure 49, in 2007, which define land-use goals and restrict the conversion of lands from forest to other uses. Most of the average or better quality forest soils are to be used exclusively for forest use, with building limited to structures used for equipment. The Eddyville property cannot be subdivided; one residence may be constructed on individual purchases of forest properties in certain zones. The Eddyville Tree Farm also includes 3 ac (1.2 ha) where a house once stood. Building another residence there would be complicated. Taxes on such land, however, are generally kept low to permit long-term affordable management of the land where revenue may be infrequent. If there is a structure on the land, the structure and the 1 ac (0.4 ha) on which it stands is taxed as residential property.

There are strict laws pertaining to water (Oregon Secretary of State, 2013a) that define regulations for managing streamside forest cover. Oregon is home to a valuable salmonid cold-water fishery, and forest practice rules require wide buffers to remain along any fish-bearing stream. The Eddyville Tree Farm has about 6 ac (2.4 ha) of land in riparian buffers, where no harvest is permitted. Permission to harvest will not be granted for substantial cutting in riparian buffers. The very small upper tributaries at the Eddyville Tree Farm do support spawning coho salmon (*Oncorhynchus kisutch*), and offer some opportunities for enjoyment when salmon runs are occurring (Figures 2.1 and 2.2).

Land is becoming valuable as a nonrenewable resource, and hence, it is increasingly scarcer than potential buyers. Over 53 years, the Eddyville Tree Farm has increased in value, by a factor of 500 times or more with timber, but 100 times as just bare land. Most of this reflects the tremendous growth of the Douglas-fir and excellent markets. Table 2.1 summarizes harvest activity since purchase, noting that initially little merchantable timber was present. Following planting and intensive brush control, it has become feasible to maintain harvests of about 300 thousand board feet (MBF) per year from the original 200 ac (81 ha) that had previously had almost no salable timber on it. Land in 1961 sold for $25–$30 U.S. dollars (USD) per ac ($61.78–$74.13 per ha) without timber. In 1961, logs from nearby land sold for $70 USD per MBF ($18 per m³), netting less than $25 USD per MBF ($6.0 per m³) after deducting the cost of logging and hauling three miles to a local mill. Today, this high site land is being sold for roughly $2,000 USD per ac ($4,942 per ha) plus timber value. In other words, it is being sold for about $24,000 USD per ac ($59,304 per ha) for a well-tended forest that is 60 years old. The temptation to sell land and timber will be high for short-term investors. But for long-term investors, people can still be paid $22,000 USD

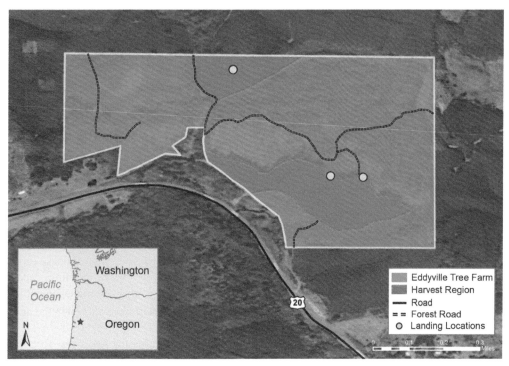

FIGURE 2.1   Map of the Eddyville Tree Farm in western Oregon.

FIGURE 2.2   Interface of patch occurrences of early- and late-seral environments that harbor different groups of wildlife while all participate in the continuous flow of maturing Douglas-fir. In the foreground is a 2-year-old clear-cut, planted, and sprayed twice. In the near background is a 9-year-old clear-cut, with planted Douglas-fir and scattered western hemlock. In the background is a 45-year-old planted Douglas-fir/western hemlock stand that is about 135–140 ft wall.

TABLE 2.1 Newton Forest Harvests for the Eddyville Tract, 1964–2014

Year	Wood Harvested (MBF)	Tree Species, Type of Harvest	Revenue (USD)	Type of Revenue
1964	45.0	Red alder	0	net stumpage
1966	60.0	Douglas-fir, salvage	1,114	gross
1968	5.0	Douglas-fir, salvage	67	gross
1977	103.0	Douglas-fir, scattered selection	10,893	net stumpage
1986	56.0	Douglas-fir, scattered selection	5,910	net stumpage
1987	108.0	Douglas-fir, thinning	14,807	net stumpage
1988	31.0	Douglas-fir, salvage	5,868	net stumpage
1989	16.0	Douglas-fir, salvage	3,298	gross
1990	22.0	Douglas-fir, salvage	5,563	gross
1994	34.5	Douglas-fir, salvage	18,595	gross
1995	19.2	Douglas-fir, salvage	8,728	gross
1996	40.8	Douglas-fir, thinning	14,053	gross
2001	152.0	Douglas-fir, sanitation	47,822	net stumpage
2003	862.0	Douglas-fir, clear-cut	303,800	net stumpage
2004	660.3	Douglas-fir, clear-cut	270,163	net stumpage
2006	3.4	Douglas-fir, salvage	1,236	gross
2007	12.7	Douglas-fir, salvage	3,568	net stumpage
2011	330.0	Douglas-fir, clear-cut	135,000	net stumpage
2014	1,470.0	Douglas-fir, clear-cut	610,000	net stumpage

per ac ($54,362 per ha) for the timber and have their potential heirs reap another very large harvest in the next rotation from the same sites without having to put out cash to buy land and start over.

Forest certification was not available at the time of purchase; however, soon after the American Tree Farm System did begin certification. The Eddyville Tree Farm has been certified for more than 50 years at the time of writing. It is also managed under rules defined in the various amendments of the Oregon Forest Practices Act of 1971. This certification involves periodic inspection to ensure certain criteria are met for stand stocking, reforestation, observance of riparian protection, condition and maintenance of roads, and proper storage of chemicals. The observance of rules germane to fire and fuel management is also part of the inspection process, along with a review of the management plan.

Silvicultural Issues

Oregon is a land of dry summers. Establishing a coniferous plantation, even in zones with 80 or more inches (2,032 mm) of precipitation, almost always requires use of artificial regeneration after a clear-cut harvest. Clear-cutting is the standard practice in final harvests owing to the shade intolerance of the Douglas-fir, and it is often coupled with weed control. Plantations will require herb and shrub or hardwood control for one or more years if planted seedlings are to survive and grow well. The right to spray with registered herbicides is protected. These practices must be conducted according to clearly laid out (and practical) regulatory rules and applied in time to control competition *before* stresses occur in planted seedlings. These practices represent a small fraction of the cost of establishing a new forest, but the dry summers and heavy regrowth of weeds and brush will severely threaten the growth of a planted area without control. After clear-cut harvesting, Oregon rules require almost immediate reforestation and that a required number of seedlings per unit area be *free to grow* (200 per ac (494 per ha) in western Oregon) within 5 years after harvest (Oregon Secretary of State, 2013b).

The dry summers offer risks of very large fires. Reduction of slash and other fuel following timber harvest is the owner's responsibility that can become an extreme expense if a fire is propagated by fuels left by management operations. Additionally, leaving dry slash (debris) on a harvest unit can expose property owners to damages should a fire start in the slash and escape to adjacent properties. In high risk areas, slash management is an important part of harvest planning for landowner protection. Fortunately, the Eddyville Tree Farm is located in a lower than average fire risk area of western Oregon. In this area, the common practice today is machine piling of slash and burning such piles after onset of fall rains. The ODF must be notified for each harvest, spray, or burn activity, and will issue a "Notification" for each of these practices. The Notification must describe proximity to streams (avoidance of riparian corridors), degree of slope, expected harvest volume, and many other elements of forest practices. The ODF determines when air conditions and movement allow burning without exposing population centers to smoke. All weed-control chemicals also must have been registered by the Oregon Department of Agriculture and the Environmental Protection Agency. The wide use of such chemicals in Oregon ensures that a suitable suite of vegetation-control chemicals suitable for forests is available.

At the time the Eddyville Tree Farm was purchased, there was virtually no regeneration of conifers present. The first challenge in managing the property was in learning how to establish valuable conifers (mostly Douglas-fir) under circumstances where controlling this diverse community was due to the dominance of nonconiferous cover and its diverse composition. Moreover, in 1961, forest weed control was in its infancy, and few chemicals were available that would be effective enough to avoid severe suppression of planted conifers. Considering the diversity and vigor of the nonconiferous cover (some hardwoods over 100 ft (30.5 m) tall), simply lowering the canopy that intercepted most of the sunlight was difficult, yet essential. By 1963, data were available indicating that many of the shrubs could be at least suppressed, even if not controlled, with 2,4,5-T at rates of application reasonably safe to use over dormant Douglas-fir. But this diverse plant cover also hosted a wide variety of herbivores that have a strong preference for Douglas-fir and other conifer seedlings, leading to re-suppression or mortality. The control of cover at levels useful for controlling wildlife damage to seedlings was a serious challenge. Of interest is that these same chemical treatments were then being evaluated for enhancement of wildlife habitat in Idaho and the midwestern United States.

Planting small nursery-grown, bare-root conifer seedlings did not facilitate a highly competitive stand. Their slow juvenile growth and susceptibility to browsing, especially under recovering sprayed shrubs, led to prolonged suppression without constant brush suppression. Moreover, it became clear immediately that wildlife, such as mountain beaver (*Aplodontia rufa*), snowshoe hares (*Lepus borealis*), and various other rodents, black-tailed deer (*Odocoileus hemionis hemionis*), and elk (*Cervus elaphus*) were attracted to sprayed brush that was resprouting and providing browse much closer to the ground than in an untreated brush field. These observations agreed with those observed in the Midwest by wildlife scientists studying ways of *enhancing wildlife forage of short stature* while clearing power line rights-of-way. Thus, the process of controlling the competition was also creating an environment that increased herbivory. This posed a challenge for locating a source of seedlings that could handle the pressure of herbivory and still gain in height growth. They had to be cheap and large enough in height and diameter to escape or tolerate browsing and overtopping by brush.

Reforestation trials, using large transplant stock, had been conducted in the Tillamook burns that deforested about 350,000 ac (141,643 ha) of land about 60 miles (mi) (97 kilometers (km)) north of Eddyville in four overlapping fires that ended in 1951. The ODF performed many tests of nursery technology, including growth of large transplants. It found these large, transplanted seedlings to be an important feature of successful reforestation in areas of early-seral stage vegetation. However, commercial sources of large transplants were, and remain, expensive. The owner did find an alternative source in natural regeneration occurring along forest service road rights-of-way where bare soil had been left following road construction projects conducted a few years previously. Between 1966 and 1972, the owner's family, with permits from the U.S. Forest Service requiring that seedlings be left on an 8 × 8 ft (2.4 × 2.4 m) spacing, were able to pull and collect about 2,000–4,000 seedlings a day—one or two pickup loads—from a 4- to 6-year-old roadside right-of-way in the Siuslaw National Forest. These roadsides furnished roughly 100,000 hand-pulled seedlings during this period.

The species composition of the hand-pulled seedlings was about two-thirds Douglas-fir and one-third western hemlock. The Douglas-fir seedlings ranged in height from 24 to 48 in (61–122 cm); the hemlock seedlings were shorter and more slender. Hemlock was not attractive to most of the herbivores, and it did not survive as well as Douglas-fir because it was sensitive to root damage while being pulled from road banks. The smaller hemlock seedlings were observed to be less subject to herbivory and were also tolerant of shade; therefore, many of these seedlings did survive. After several years, mixed-species plantations that contained approximately 200–300 conifers per ac (494–741 per ha) were established. Many remained suppressed beneath the sprouting brush despite one or more aerial applications of 2,4,5-T. It often required three aerial applications of phenoxy herbicides, and sometimes as many as five, to allow the conifers to grow to a dominant position in this brush. In the late 1970s, glyphosate was registered for commercial use, and in test plots was observed to selectively reduce the brush problem enough to allow Douglas-fir and hemlock to dominate with hemlock comprising about

10–15% of the stands. No known examples of vegetation species *eradication* were observed in the process of releasing the planted seedlings with this spraying program.

Between herbivory and rapid growth of damaged brush, the residual shrub-dominated cover was a major constraint on seedling survival. The planted conifers were not yet free to grow, even 10 years after planting. At this time, mountain beavers became very abundant. These rodents can climb 10–20 ft (3–6 m) up small conifers, and clip off the branches and tops, leaving bare poles with sprigs of greenery. Both Douglas-fir and hemlock were damaged as a result, and large seedlings did not prevent this type of damage as well as they had in minimizing deer and elk browse damage. It was essential to learn that the mountain beaver burrows in heavy cover, place traps in them, and monitor the traps to ensure that the colonies had been reduced or eliminated. This approach was successful in reducing mountain beaver damage without eradicating any wildlife species. Deer provided another problem when the hemlock (especially) approached 6–10 ft (1.8–3 m) in height. Antler rubbing by black-tailed deer became a serious problem as it caused girdling and top-killing in the conifers, less serious with Douglas-fir than with hemlock, but still a problem. There was not a good remedy to prevent this type of wildlife damage because the spraying program had provided such productive wildlife habitat. A planting effort that led to 200–300 trees per ac (494–741 per ha) of live seedlings that had escaped damage; however, provided enough trees on these productive sites to provide maximum timber yields in high-quality trees. Rubbing scars on the saplings remained a few years, and then healed over; Douglas-fir trees survived this damage, and the scars healed without significant damage. Hemlock tended to develop heart rot if the bark was damaged, however. Eventually, this damage decreased the net volume in some hemlocks and was evident even after 40 years.

Establishment of conifers at the spacing of 12 × 12 ft (3.7 × 3.7 m) (300 per ac; 741 per ha) of the large wild seedlings left enough space between trees to provide about 15 years of early-seral habitat and diverse wildlife communities when maintained by repeated applications of herbicides. The registration and use of glyphosate in late 1970s sharply reduced deciduous cover and browse availability for a few years. After 15 years of phenoxy herbicides had favored those species while releasing shade tolerant forbs and ferns of lesser palatability, the populations of wildlife decreased. The dual effect of reduced damage and release from competition accelerated conifer development markedly. By 1980, most of the tree farm was adequately stocked with conifers. Once the seedlings had escaped browsing and overtopping, height increment soon reached 2–3 ft (almost 1 m) per year, accelerating in a few years to 4–5 ft (122–152 cm) per year, a pace that has continued.

By 1990, most of the land was covered by young conifers growing at very high rates. Plantations established between 1965 and 1972 are, in 2014, averaging 124–140 ft (38–43 m) tall, and dominant trees are more than 20 in (51 cm) in diameter and are of excellent form. Most have been thinned, with yields as shown in Table 2.1. Commercial thinning activities have removed virtually all of the trees with stem defects in periodic small operations. All of the thinning practices have been implemented with small crawler tractors using single-tree selection that was based on leaving trees of large size and quality, and an eventual goal of roughly 100 trees per ac (247 per ha), after a final thinning about 25 years before an expected final clear-cut harvest. Based on harvest data and inventories at 6-year intervals, net growth after the last thinning is expected to approximate 45 MBF per ac (about 667 m^3 per ha)[1] in just their last 25 years. Total yield for a 70-year rotation will exceed 90 MBF per ac (about 1,334 m^3 per ha), and the average tree diameter at final harvest is projected to be 24–26 in (61–66 cm) with basal area of 300 ft^2 per ac (69 m^2 per ha), with dominant height of Douglas-fir trees of about 180 ft (55 m).

The option of changing the rotation length will depend on markets. Periodic annual increment in these sites and in these stands will peak just after 50 years and remain at about 2100 board feet per ac per year (roughly 31 m^3 per ha per year) in sawlog net yield if left standing until they are nearly 100 years old. The thinning program must be maintained so that there is 25–30 years between the last thinning and final harvest to obtain maximum yield and quality, with minimum mortality. Absolute maximum mean annual yield would require rotations of at least 100 years, based on biology and maintenance. But profitability will be dependent on markets for very large logs, some of which will have scaling diameters of 40 in (102 cm). However, the scarcity of such logs from most private lands is likely to sharply reduce the milling capacity for such timber, hence markets. The owner's experience is that rotations for the maximum value growth in Douglas-fir on such good sites will peak between 55 and 90 years, depending on markets. The owner's experiments with rotation length and stocking control show that maximum yield per unit area per year is reached at age 60, and is then maintained well beyond 100 years. Dr. Robert Curtis, of the U.S. Forest Service, Pacific Northwest Forest Research Station, has published numerous reports showing the prolonged high growth rates of Douglas-fir on good sites. Dr. Carl Berntsen, also of the U.S. Forest Service, reported high rates of gross growth past the age of 250 years in Douglas-fir (Berntsen, 1960). Mortality in dense stands tends to put a ceiling on accumulation of live, healthy biomass.

Selecting an optimum age at which to conduct a regeneration harvest does not need to be based exclusively on internal rate of return on investments if one wishes to make money growing trees on these very productive sites. Short rotations

---

1. *Scribner scale.* In contrast to the conversion applied in other chapters, this chapter assumes 5,000–6,000 m^3 per MMBF depending on the size of the logs.

never reach the period of maximum growth of high-value products and offer less diverse habitats. The owner believes that when combined with higher logging costs, short rotations are simply less efficient at growing net dollar yield per unit area per year. Short rotations also require a higher frequency of reforestation costs with higher cost per unit area associated with higher stand density needed for short rotations. Moreover, in ownerships with goals beyond internal rate of return on investment, short rotations do not provide opportunities for late-seral habitat or high-quality sawlogs.

Thinning in stages to about 25 years prior to rotation offers significant intermediate revenue while creating small patches of early-seral habitat where significant sunlight can reach the forest floor. It also provides crown structures approaching features of much older stands and their habitats. Older, taller trees, are also more cylindrical than young trees, hence less volume is outside the scaling cylinder of the Scribner scaling system used in western Oregon. Including the occasional groups of hemlocks during the reforestation effort also resulted in deep, multiple-layered crowns commonly found in late-seral habitats, without sacrificing the total yield of the stand. This kind of intensive, multiple-use management provides an abundance of features of late-seral habitats on 35–40% of moving patches in the forest as part of a management system that can yield 1,400 or more board feet per ac (roughly $21\,m^3$ per ha) per year over an entire life cycle when the yield computation includes thinnings and final harvest. Most of the revenue still accrues during the regeneration harvest, but much of the habitat preparation will occur with thinning while providing some revenue in the process. Most of the operations providing habitat will, at some point, provide some revenue.

The harvest of the original stands of scattered residual Douglas-fir in 2003–2004 led to about 40 ac (16 ha) of quite persistent early-seral habitat while yielding about 1.3 million board feet (MMBF) (about 7,800 cubic meters (m^3)) of harvests. It was planted with large transplant nursery stock instead of wild roadside stock, and it is now all on course to become a pure Douglas-fir stand suitable for thinning and other stand management. Since harvest and planting; however, the young stand has had several different weed-control strategies applied, depending on the nature of competing shrubs and forbs and the type of early-seral habitat desired. Thus far, it has some areas that have been superior deer and elk habitat for 10 years. Some patches still do not have enough shrub cover to be the most desirable for large game, but they are excellent for birds nesting in small sprout clumps and as foraging sites for avian predators while growing conifer seedlings with very little animal damage by pocket gophers (*Thomomys* spp.), wood rats (*Neotoma* spp.), mountain beavers, brush rabbits (*Sylvilagus bachmani*), or snowshoe hares. Vegetation management is often the most effective way to manage populations of such animals while accelerating seedling growth toward a wildlife-proof size. These stands, where the ground is not too steep, will support deer and elk for a few more years.

The eventual silvicultural plan is to have a full array of age classes—from zero to about 70 years—of even-aged stands in patches of 15–30 ac (6–12 ha), with a full array of good-sized habitat patches with productive early-seral and other patches with intermediate or features of late-seral cover. One cannot have early- and late-seral habitat on the same area all the time, but some late-seral features have been preserved by allowing one small Sitka spruce (*Picea sitchensis*) per clearcut. The three spruces (diameters 16–19 in, in 1961) remaining through all the afforestation efforts are now over 6 ft (1.8 m) in diameter, clearly outstanding old-forest features that do not take up much space. The phases of cover development that even-aged management offers are ideal for promoting age-dependent plant communities that fade in heavy shade or when full sunlight is present. In western Oregon, having a wide variety of stages present all the time is most easily provided by patches of even-aged stands each being harvested at a planned rotation age. A mosaic of even-aged stands ranging from 0 to 70 years on this property is now providing a small landscape with diverse habitats while having reached a plateau of incredible timber productivity. In another 15–20 years, the existing age-class distribution will have all age classes of habitat represented within 20% of the theoretical goal.

Forests ideally are allowed to persist well beyond the lifespan of their owners. There are various ways of defining a managed property as a separate survivable administrative unit that will last many generations, such as a trust or as a partnership organized as a Limited Liability Company (LLC), with identified managers and broad achievable goals agreed upon by the beneficiaries. LLCs with charters that clearly spell out long-term objectives can produce benefits as long-term capital gains, but which, by direction, may be isolated from disagreements among beneficiaries in the planning process. There are few families not affected over several generations by calamity or individual partners who do not value long-term assets. And, in this region of immensely productive and valuable forests, opportunities to accumulate large value offer many temptations to liquidate assets that are far from mature. At this point, the current management plan has been in place for 53 years, and has been successful in reaching a plateau of both revenue and habitat objectives. At this stage, heirs to the estate are committed to maintaining all goals. The tree farm was organized as Newton Family Forest, LLC in 2010 and Newton offspring are already majority owners, working with the original owner, who also serves as the manager while being involved in the process of temporary ownership and maintaining a long-lived valuable estate. The LLC structure provided a mechanism for passing along certain amounts of valuable shares tax-free when needed and for protecting against various issues of lack of interest, divorce, and other mishaps common to many families with multigenerational structure. Involvement of all shareholders in the business is an important activity in this LLC. An occasional very large paycheck in the form of a timber sale is a great incentive for maintaining enthusiasm for a forestry family. In this case, a very large paycheck every 4 or 5 years has been received, and all the shareholders know how to maintain the resource.

OUTCOMES OF THE PLAN

Over the five-plus decades that the property has been a Certified Tree Farm, it has used modern technology to provide a fine opportunity to establish forests with a *natural* look. With the addition of 95 ac (38 ha) of similarly productive land being managed as part of the even-aged management scheme, and the purchase of 140 ac (57 ha) by children of the original owners with forest age classes deficient in the original 200 ac (81 ha), the family ownership has virtually every phase of maturity in well-stocked Douglas-fir-dominant stands.

For obvious reasons, harvesting in the first 40 years was spotty and revenues during the first 25 years barely covered the costs of management (Table 2.1). As of 1990, most of the scattered conifers present outside the high-graded young growth had been harvested, after which harvested sites were sprayed and replanted. The plantings, with the road bank seedlings, are 42–48 years old, and have been commercially thinned. The high-graded 40 ac (16 ha) residual stand was clear-cut and replanted, and has provided 10 years of productive early-seral habitat, while producing a net yield of about 1.5 MMBF (9,000 m^3) of large, high-value logs, Scribner scale. Total volume salvaged from rehabilitation of hardwood patches with scattered Douglas-fir trees has added another 1 MMBF (5,000 m^3) in small increments of mostly small logs. The landowner invented an injector hatchet (Hypo-Hatchet) that he has used on the Eddyville Tree Farm to kill unmerchantable hardwoods everywhere except in riparian zones. In 2014, the final loads of a 1.5 MMBF (9,000 m^3) of high-value Douglas-fir sawlogs were harvested from 30 ac (12 ha) that had been suppressed by Scotch broom (*Cytisus scoparius*) shrubs in 1962, when the stand was sprayed to release knee-high volunteer conifer seedlings.

The periodic inventories have revealed a steady exponential increase in standing volume. The decision to maintain an inventory of 5–6 MMBF (25,000–30,000 m^3) in 40-year-old or more stands of trees has plateaued on the first 200 ac (81 ha), and with the addition of family lands now having reached 440 ac (178 ac), will soon rise to a steady state of about 8 MMBF (40,000 m^3). This will maintain the full range of age classes. It will also take an allowable annual harvest of about 600-700 MBF (3,000–3,500 m^3) annually to maintain this balance. The tremendous productivity of these sites provides a unique opportunity to keep an incredible asset in families and so provide stability for generations to come. Even modest management with 50-year rotations will provide net revenues of well above $300 USD per ac ($741 per ha) per year indefinitely with careful planning and reasonable culture. The Eddyville property and subsequent additions could, therefore, support two or three families very handily if managed somewhat intensively.

There are other economies of scale associated with matching the timber produced to market opportunities. A recent harvest yielded 1.25 MMBF (7,500 m^3) of very high-quality sawlogs from 21 ac (8.5 ha) of an intensively managed 55-year-old stand. A prior thinning treatment (1986–1987) had accelerated the diameter growth of this stand while yielding more than $20,000 USD net capital gain. The median stand diameter in 2014 was 24 in (61 cm) that resulted in 35% of the volume being in sawlogs over 18 in (46 cm) scaling diameter and of high quality. This brought a maximum price per 30-ton truckload, providing high ratios of volume to weight, hence, low price per board-foot-mile. The logs in all size ranges sold were consistently near the highest pay grade for Douglas-fir sawlogs, facilitated by tall, cylindrical boles and high volumes per unit area. Both of these factors will increase in importance as the percentage of yields from 70-year-old stands approach culmination with this even-aged management plan. This plan was made feasible by having more than eight sawmills located within 50 mi (80 km) that were capable of handling a wide variety of log qualities and sizes, and by chemical tools that allowed high level of species composition control.

DISCUSSION AND CONCLUSIONS

Western Oregon is an outstanding location for growing conifers and benefits from a thriving industry that will absorb a wide variety of products at maximum prices. With very productive climates and soil resources, a family tree farm offers superb opportunities for enjoyment and income while growing trees and ecosystems. The region is well-served by professional loggers, it has excellent trucking companies and highways for transport of logs, and its mills are well-served by railways that cover the nation. It is not unlikely that a similar approach on good sites in any region with decent markets would prove successful.

Learnings and Insights

Managing a productive tree farm provides many lessons. They are the sources of incredible satisfaction over long periods:

1. There may be no activity as effective in developing powers of observation as managing a forest through a lifetime with tools in hand.
2. The value of something depends on what one wants and whether one is willing to wait.
3. A quick return on investment is usually a small return when one's product depends on quality and efficiency of harvest.
4. Always having forests in a full array of age and condition classes is excellent maintenance of high wildlife populations and near-maximum yields.
5. Including offspring in growing and harvesting trees, and eventually sharing ownership with them, inspires the next generation to maintain the system. Especially when they see the size of their pay checks.

Some simple guidelines can save a lot of money and avoid many risks to people and habitats:

1. Weeds will conflict with management on productive sites. One should not procrastinate about them.
2. Herbicides are more effective and safer than cutting tools for managing plant cover/habitat for trees and wildlife. Learn about them from experts, including professional foresters and weed-control experts. Use these tools wisely.
3. Reliance on large seedlings reduces hazards from wildlife and brush.
4. Sustainability of this system is guaranteed by the relatively long rotation, in small patches, that maintain a maximum array of habitats while maintaining maximum yield.
5. Patches need to be large enough to offer economic harvesting, ideally <15 ac (6 ha).

Plan Development Challenges

It is easier for one person to maintain control of a management plan when the rest of the family agrees with it. This allows the person to follow the plan's goals while understanding the details of the work that is necessary. It is another matter when there is disagreement about the goals to be pursued when all goals are still valid. This is a common problem for family tree farms. Among the most important events are the preparations of wills and insistence that wills preserve the integrity of the management plan. The best way to handle this issue, which is often difficult to ensure, is to use constant communication within a family once the plan has been made, keeping each member informed about planned action, distribution of funds, and maintenance of cash reserves and noncash goals.

Plan Implementation Challenges

Societal influences often conflict with establishment of tree farms. Economic challenges are increasing with increasing prices on land. Forests have enduring value that can be managed investments for generations. Among the most serious social challenges are first keeping with one management system long enough to make it work, and, second, bringing up future generations to learn importance of work and patience while learning the value of natural landscapes. They will treasure what they have and bring up their own children with the same level of commitment if each generation grows with the forest.

Landowners should be prepared with cash reserves to maintain flexibility. The owners nearly missed an opportunity in winter 2013 when timber prices rose markedly, because the roads were not ready for transporting logs from the property. The family was fortunately able to borrow $79,000 USD to winterize roads, leading to about a $150,000 USD increase in the sale prices of logs ready for sale, an extra payment of about $71,000 USD after payback of the loan.

REFERENCES

Berntsen, C.M., 1960. Productivity of a Mature Douglas-fir Stand. U.S. Department of Agriculture, Forest Service, Pacific Northwest Forest and Range Experiment Station, Portland, OR. Research Note 188. 4p.

Oregon Secretary of State, 2013a. Department of Forestry, Division 635, Water protection rules: Purpose, goals, classification and riparian management areas. Oregon Secretary of State, Salem, OR. http://arcweb.sos.state.or.us/pages/rules/oars_600/oar_629/629_635.html (Accessed July 5, 2014).

Oregon Secretary of State, 2013b. Department of Forestry, Division 610, Forest practices reforestation rules. Oregon Secretary of State, Salem, OR. http://arcweb.sos.state.or.us/pages/rules/oars_600/oar_629/629_610.html (Accessed July 5, 2014).

Chapter 3

Michael Family Forest, Texas, United States of America

Robin G. Willhoite

United Timber Management Company, Timpson, Texas, USA

ABBREVIATIONS

AMZ   Aesthetic Management Zone
PDO   Pacific Decadal Oscillation
TFS   Texas A&M Forest Service

MANAGEMENT SETTING AND BACKGROUND

The Michael Family Forest is a 347-acre (ac) (140 hectare (ha)) forest located in western Nacogdoches County in central East Texas. Commonly referred to as Deep East Texas, this area of the state is in the Southern Tertiary Uplands of the South Central Plains, which is a part of the Gulf Coastal Plains. The South Central Plains ecoregion (Bureau of Economic Geology, 2010), also known as the *Piney Woods*, includes portions of eastern Texas, western Louisiana, southwest Arkansas, and the extreme southeast portion of Oklahoma. This territory represents the western extent of the southern pine region of the United States. The southern pine region of the United States is often referred to as the *wood basket* of the country because it generates more than 55% of the country's timber harvests by volume and 25% of world pulpwood production (World Resources Institute, 2014). The 43 county East Texas Piney Woods region contains about 12 million ac (about 4.86 million ha) of productive timberland (Cooper and Bentley, 2012). According to data from 2006 published in the 2008 National Woodland Owner Survey (Butler, 2008), 8 million ac (3.24 million ha) in East Texas are owned by about 210,000 nonindustrial private forest landowners. About 70% of the nonindustrial private forests in Texas are in holdings of 50 ac (20 ha) or more and are represented by some 30,000 landowners like the Michael family. Timber is consistently ranked as one of the top 10 agricultural products in Texas, and was ranked third when timber prices were higher than they are currently. The climate of Deep East Texas is warm, humid, and temperate. The average temperatures range from a high of the mid-90°s Fahrenheit (F) (35 °Celsius (C)) in July to a low in the mid 30°s F (2 °C) in January, with an average annual precipitation of 45–60 inches (in) (1,143–1,270 millimeters (mm)), and an average annual growing season of 245 days.

This area of East Texas has been inhabited by humans for at least several thousands of years. Several Native American archaeological sites have been documented on the property. The Hasinai tribe, members of the Caddo confederacy, occupied a region including this portion of western Nacogdoches County. The Caddo were agriculturalists and enjoyed a highly developed culture, living in grass huts and constructing flat-topped earthen mounds. The Hasinai tribe was also known as the Tejas, meaning *friends*, which is the origin of the name Texas (Long, 2011). The Michael Family Forest is believed to be located on a portion of the historic El Camino Real, also known as the Old San Antonio Road, The King's Highway, and the San Antonio–Nacogdoches Road. Based partly on a network of ancient Indian trails, the route was actually a group of intermingling roads and trails. Used by early Anglo explorers as early as the 1690s, this well-travelled thoroughfare eventually became an important trade route between Natchitoches, Louisiana and Guerrero, Coahuila, Mexico. The property is adjacent to the site of the Presidio Nuestra Senora de Los Dolores, built by the Spanish government in 1716 as a fort and headquarters for soldiers to guard the East Texas missions and the borders of the New Philippines. The site was abandoned

Forest Plans of North America. http://dx.doi.org/10.1016/B978-0-12-799936-4.00003-5

around 1730. The first sawmill in Texas associated with a particular person was in Nacogdoches County, operated in 1829 on Carrizo Creek by Peter Ellis Bean.

The Michael Family Forest (Figure 3.1) has excellent accessibility, with more than one mile (mi) (1.6 kilometers (km)) of paved highway frontage and excellent interior roads. The main woods roads into the interior of both the eastern and western blocks are gated and locked; all weather roads were built to access two gas-well facilities on the property. Elevations on the property range from about 220–400 feet (ft) (67–122 meters (m)) above sea level. The terrain ranges from gently sloping areas to steep areas. Forest cover types on the property include native mixed hardwood and pine forests as well as intensively managed pine plantations. The largest two parcels of land, comprising about 208 ac (84 ha), were owned and managed as industrial timberland since at least 1970, and were sold to nonindustrial private landowners in 2003. Most of the property on the north side has primarily been used for pasture and cattle production for a number of years, but also includes significant areas of timberland. The Michaels acquired the two larger tracts in 2005 and subsequently acquired the balance of the properties between 2008 and 2013. At the time the larger tracts were acquired, a selective timber sale conducted by United Timber Management, a local independent consulting forestry firm, was in progress on the pine plantations. This thinning harvest was suspended during the acquisition process and was subsequently completed by United Timber Management in 2010 and 2011.

The Michael Family Forest cover types (Figure 3.2) consist of about 153 ac (61.9 ha) of older pine plantations of sawtimber size (Figure 3.3), 7 ac (2.8 ha) of newly established pine plantations of seedling and sapling size, 42 ac (17.0 ha) of mixed pine-hardwood of pulpwood and sawtimber size, 7 ac (2.8 ha) of native loblolly (*Pinus taeda*) -shortleaf pine (*Pinus echinata*) of pulpwood and sawtimber size, 14 ac (5.7 ha) of various hardwood species of pulpwood and sawtimber size, and 35 ac (14.2 ha) of streamside management and riparian zones with mixed hardwood and pine of pulpwood and sawtimber size. In addition, there are about 87 ac (35.2 ha) currently classified as pasture and open land.

All of the soils on the Michael tracts are suitable for commercial pine timber production. Using a base age of 50, the soils on about 25% of the property have site indexes of 95–102 ("excellent"), and the soils on 73% of the property have site indexes of 80–89 ("good"). Site index is a general indication of potential soil productivity and represents the expected average height of dominant and codominant trees at the referenced base age.

The property is divided east to west by a creek (Legg Creek). The creek and its tributaries constitute an especially important and valuable water resource and add tremendously to the appeal of the Michael property. Legg Creek originates as an intermittent stream about 5 mi (8 km) north of this property, and is then joined by Mill Creek. Legg Creek passes through the property, and continues to flow southwesterly about 3 mi (4.8 km) to its mouth on the Angelina River. Even during periods of extreme drought, Legg Creek and its tributaries provide bountiful clean, cool water as they pass through the Michael property

FIGURE 3.1 A map of the Michael Family Forest.

FIGURE 3.2 Michael Family Forest land cover.

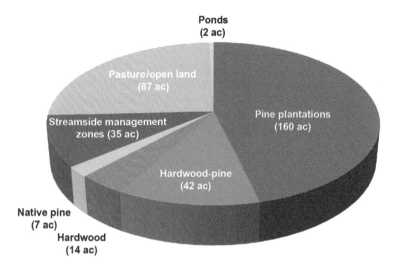

FIGURE 3.3 Loblolly pine plantation on the Michael Family Forest. Photograph courtesy of Robin G. Willhoite.

as a perennial stream. These waterways provide important habitat for waterfowl, neotropical migratory song birds, other birds, fish, invertebrates, and an abundance of other wildlife. They also provide excellent recreational and aesthetic values.

The Michael Family Forest is strategically located in a healthy and competitive timber marketplace. Multiple outlets exist for pulpwood and sawtimber products from both pine and hardwood species. Destinations within economically viable hauling distances for the harvested forest products from this property include paper mills, chip mills, OSB and plywood structural panel board mills, lumber mills, post and pole mills, and tie and pallet mills. A biomass-fueled electrical generating plant is located just a few miles north of the property, and existing and planned wood-fuel pellet plants are within reach.

In December 2011, a management plan and a timber inventory and appraisal were completed by United Timber Management Company, and the property was subsequently approved as a Certified Family Forest by the American Tree Farm System® (ATFS). Such accreditation requires an adaptive, sustainable forest management plan developed by a qualified tree farm inspector addressing air, water, soil quality, wildlife, special sites, invasive species, and integrated pest management as described in the American Forest Foundation's Standards for Sustainability for Forest Certification. These requirements also qualify the management plan to be suitable for use with the U.S. Department of Agriculture's (USDA) Forest Stewardship Program (U.S. Forest Service, 2013) and Natural Resource Conservation Service (NRCS) programs.

PLANNING ENVIRONMENT AND METHODOLOGY

The Michael family has maintained a secondary residence on some nearby timberland for many years. Lloyd and Barbara began acquiring the subject properties as a long-term investment and recreational opportunity in 2005, as they were approaching retirement as faculty members from the Baylor College of Medicine. In Lloyd's own words:

Our family forest in East Texas, affectionately referred to as El Refugio, meaning place of refuge in Spanish, is comprised of several parcels acquired over a number of years from landowners with different agendas. Thus, we decided to establish a

Certified Family Forest to achieve a number of objectives: preserve the watershed; provide an exceptional wildlife habitat; promote educational recreational activities; ensure financial self-sufficiency through timber production; preserve identified archeological and historical sites; and to be a haven of peace and solace distinct from urban landscapes, now and in the future.

In order to establish a baseline for action, we had a forest management plan developed with rigorous attention to standards and supporting documentation. The plan was a logical first step as I was raised on a farm in the Appalachian Mountains and helped my family to harvest a variety of trees that grew on the property to build barns and outbuildings. With this background, at an early age, I was indoctrinated with the importance of forest and land management to ensure that timber and other natural resources would be available for the next generation.

Our vision for our family forest is to preserve it for future generations with the engagement and support of family members. This will, of course, depend on careful adherence to an intergenerational management plan that will hopefully continue our family legacy. (Lloyd Michael, personal communication, May 29, 2014).

The ATFS template (as described in the McPhail Family Forest chapter (Straka and Cushing, 2014)) was used as a guide to develop the management plan for the Michael Family Forest. Although the template was not strictly followed, the content and character of the resulting plan are in compliance with requirements for the ATFS. The data and observations from the recent timber inventory and appraisal were utilized to better understand the current state of the forest and to assist in formulating future needs and directions.

The primary constraint on implementation of the forest management plans on the Michael Family Forest have so far been limited to attempts to convert an existing upland pasture area on the north side of the property to a longleaf pine (*Pinus palustris*) plantation. These operations were partially funded through the USDA NRCS Wildlife Habitat Improvement Program. Following recommendations from the Longleaf Alliance, in early 2011, the site was scalped, that is, shallow strips of sod about 3 ft (0.9 m) wide were peeled back, in this case on 12 ft (3.7 m) centers. Scalping removes the competing grasses and weeds as well as the pent up weed seed bed, resulting in excellent herbaceous weed control and the ability to plant directly in the mineral soil. Scalping is also thought to improve available moisture. The scalps were subsequently hand planted with containerized longleaf pine seedlings. The seedlings were off to a great start during spring 2011 before experiencing near 100% mortality during the ensuing near-record summer drought. In fall 2012, the scalps were band-sprayed with glyphosphate as a site preparation operation, because they had since recolonized with herbaceous and some woody competition. In late 2012, the scalps were again planted with containerized longleaf pine seedlings. The seedlings were doing very well until early summer 2013 when they again experienced near 100% mortality.

The longleaf planting site is essentially on top of a hill that is subjected to desiccating wind and sunshine. The soil on this site is mapped as a Briley fine sandy loam that is moderately well drained, has a high available water capacity, and is rated as moderate in terms of management concerns for seedling mortality (Dolezel, 1980). Based on the idea that the longleaf seedlings had so far been planted on a bare strip of mineral soil without the benefit of significant dead vegetation to act as a mulch to protect the seedlings from getting "baked" on top of the hill, in early 2014, a trial planting of about 2 ac (0.8 ha) of hand-planted containerized longleaf seedlings was established on an adjacent pasture using alternative treatments. This trial site received no site preparation. A post-planting broadcast herbaceous weed control was conducted, thus leaving a fresher mat of dead vegetation to act as a mulch to protect the seedlings. In addition, Dr. Michael enlisted the assistance of some family members to place handfuls of hay around each of the longleaf seedlings to act as additional mulch. At the time of this writing, precipitation in 2014 has been very favorable, and the results of this mulching experiment have been positive. This mulching trial can be considered a part of the adaptive management process of the forest certification.

The 103 ac (41.7 ha) western part of the tract is mostly occupied by a well-stocked loblolly pine plantation on a predominantly upland site. The soils, access, and topography make this management unit ideal for a pine plantation. This tract was formerly owned and managed as an industrial timberland property since at least 1970. A second thinning on most of the stand was conducted in 2005, and the balance of the stand was second thinned in early 2011. Mature hard mast hardwoods were retained throughout the unit, and a generous streamside management zone (SMZ) was designated and implemented along Legg Creek. The topography of this tract is mostly level, gently sloping from the west down toward Legg Creek, a change of about 110 ft (34 m) in elevation in a distance of about 3,000 ft (914 m). This tract has an excellent all-weather road running west–east down the middle of the property almost its entire length. The basal area of these plantation areas as a whole averages about 70–80 ft² per ac (16.1–18.4 m² per ha), which at this stage in its rotation is consistent with industrial pine plantation management practices in this region. This measure of stocking levels indicates that the site is fully stocked but is several years away from a possible third thinning. The trees are generally ideally spaced, and the stand could be left to grow for an additional 6–8 years before overcrowding and growth reduction occurs. Portions of the management unit that were thinned in 2005 could be thinned again now. A traditional pine plantation management prescription would be to clear-cut the stand within 10 years when markets are favorable. At that time the stand would be regenerated with improved loblolly pine or longleaf pine, implementing an appropriate site preparation prescription.

The 102 ac (41.2 ha) eastern part of the tract also was formerly owned and managed as an industrial timberland property since at least 1970. During that ownership, the tract was mostly planted with loblolly pine, in many cases right up to Legg Creek and its tributaries. This complex management unit ranges from bottomland to upland and contains a mosaic of timber types. Topography ranges from nearly level to steep. The bottomland in the western portion of the compartment rises about 20 ft (6.1 m) in elevation from Legg Creek, then rises steeply up to an additional 120 ft (36.6 m) in elevation in the eastern portion of the management compartment. Accessibility to this management unit is also excellent, with an improved woods road running north from the paved Farm-Market highway to an old gas-well location on the northern edge of the property. Other woods roads, logging trails, and newly constructed all-terrain-vehicles trails are maintained by Dr. Michael. The current diversity of this management unit is very attractive from a mixed forest resource standpoint, providing bountiful timber production, wildlife habitat ideal for a variety of species, numerous recreational opportunities, and pleasing aesthetic values. Dr. Michael has constructed foot trails and wooden benches near "babbling brooks" that make ideal spots for meditation, reflection, and relaxation. The pine plantation on this eastern management unit is interspersed with stands of mixed pine-hardwood and hardwood. This is likely a result of the relatively rugged terrain in those areas that may have hindered a clear-cut during the last final harvest, and a lack of hardwood control with the use of herbicides within the plantation. Portions of this unit were selectively harvested in 2005, and the entire unit was "manicured" in 2010–2011, leaving a well-stocked stand of the better quality pines and retaining plentiful hardwoods for wildlife and aesthetics. Mature hardwoods were retained throughout the unit, and pure hardwood stands were largely left without manipulation. Generous SMZs were designated and implemented along Legg Creek and its tributaries.

Guiding Laws, Regulations, and Policies

Generally, Texas is considered a de facto "right to practice forestry" state; although, a few municipalities have burdensome requirements for timber harvesting. Effective September 1, 1999, the Texas Reforestation and Conservation Act of 1999 (Senate Bill 977) provided important timberland property tax incentives. Timberland owners may apply for a special appraisal of their property for taxation purposes that is based on potential productivity of timber types and soil types rather than being taxed on full market value. The burden of proof required to be furnished by the landowner needs to reflect that the property is actually being used for commercial timber production, and this proof varies significantly between county appraisal districts. The Act also permits a reduced special appraisal (or restricted-use timberland appraisal) for qualified forest zones and regenerated timberland to encourage reforestation and preservation of nontimber forest values (such as water quality and critical wildlife habitat protection). These incentives are in the form of reductions in property taxes on those areas that have been reforested or designated as an aesthetic management zone (AMZ) on a public right-of-way, an AMZ on a special or unique area, a critical wildlife habitat zone, or a SMZ. The SMZs in all management units are scheduled to be maintained in perpetuity to ensure good watershed stewardship and conservation principals. The potential tax savings from these assessed valuations is significant and helps to make commercial timberland ownership affordable and profitable.

In 1989, the Texas A&M Forest Service (TFS) promulgated a set of recommended Best Management Practices (BMPs) for silviculture and wetlands (Texas Forest Service and Texas Forestry Association, 2010). These conservation practices are designed and implemented to help protect soil and water resources and to reduce on-site and downstream sedimentation and pollution of waterways. In Texas, compliance with these BMPs is voluntary, but a 2011 TFS study estimated a 94% BMP compliance rate within the forest community. Although the state of Texas treats these BMPs as voluntary, their adherence is considered mandatory for participation as a Certified Family Forest or Tree Farm with the ATFS. In addition, most loggers are required to adhere to BMPs as a part of their supply agreements with mills participating in the Sustainable Forestry Initiative (SFI), which was originally launched in 1994 by the American Forest and Paper Association. As a measure to prevent timber theft, Title 6 of the Texas Natural Resources Code provides that a seller of timber must provide, and a purchaser of timber must retain, a Bill of Sale for Purchase of Trees and Timber.

OUTCOMES OF THE PLAN

The Michael family intends to afforest additional areas within this property, which are currently classified as pasture and open land. After the two failed attempts at the longleaf pine plantation establishment, the family is understandably reluctant to invest additional resources without better assurances of mitigating the associated risks. Dr. Michael has a particular fondness for the concept of establishing at least some viable longleaf pine stands on the property.

Those portions of the older pine plantations that were thinned in 2005 are due for another selective harvest soon to optimize spacing and promote maximum growth on a fully stocked stand of the largest and best quality trees.

A small trial of about 2 ac (0.8 ha) of containerized varietal loblolly pine was established in early 2014 on a portion of the upland pasture area. An additional 14 acres of longleaf pine and eight acres of loblolly pine are scheduled to be planted December 2014 in the pasture areas.

As should be any forest management plan, this is a living, flexible, adaptive, and continuing process. As the composition of the timber and other attributes of the property continue to change over time, so shall this document. In particular, any proposed schedules in the plan will be adjusted to reflect changes in the timber marketplace.

DISCUSSIONS AND CONCLUSIONS

The Michael Certified Family Forest management plan is designed so that any other family member can read and refer to it to understand the current ownership goals and expectantly continue good stewardship as a future legacy. The Michael Certified Family Forest management plan was more of an affirmation of existing conservation and sustainable stewardship values and directions rather than the establishment of a new policy. The Michaels and their consulting forester had already deliberated and implemented a comprehensive forest management strategy intended to maximize a broad range of ownership goals, including profitable timber production, wildlife habitat preservation and enhancement, protection of special areas in the forest, protection of soil and water resources, recreational opportunities, asset protection and estate planning, aesthetic values, and other interests.

Sustainability Issues

Sustainability is a requisite of continued recognition as an ATFS Certified Family Forest. Sustainability is also a primary tenet of ownership by the Michael family. The pine plantations will continue to be managed as even-aged stands, and will be promptly regenerated at the end of their rotation age. Mixed pine-hardwood and hardwood areas will likely be managed as uneven-aged stands using the single tree and group selection silvicultural systems. Some areas of the native stands will likely be left in peace for the foreseeable future.

Plan Implementation Challenges

A challenge that is likely to affect the implementation of the Michael Family Forest management plan is fluctuations in the timber marketplace. Timber prices in this marketplace continue to be well below their pre-recession levels, particularly when adjusted for inflation. Demand for southern pine products is expected to rebound in the coming years due to forecasted improvements in the U.S. housing construction market, declining availability of competing Canadian timber, increases in export opportunities, and development of outlets that are relatively new to this region, such as pellet mills. The southeastern United States has an excellent opportunity to expand mill and export capacity to meet the projected increased demand. However, a major factor that will hinder associated increases in stumpage prices that private landowners realize for their timber will be the pent up local and regional supply of timber, whose growth rates have been exponentially exceeding removals for several years and continue to do so (Stewart, 2012). Should timber prices increase significantly, there will likely be no shortage of eager sellers on the market, which will impede any stumpage price increases. It would likely take many years of increased demand to deplete the currently expanding inventory sufficient to significantly improve stumpage prices.

Increases in transportation charges are another critical factor limiting the stumpage prices that landowners receive. For example, in the past, a 140 mi (225 km) one-way haul did not seem to affect stumpage prices in this region, even for low-margin products, such as pulpwood. Currently, any haul distance over 50–60 mi (80–97 km) starts affecting stumpage prices. A shortage of logging capacity is also challenging, as many loggers have dropped out of the business in Texas in recent years.

Another factor that may affect the implementation of the Michael management plan is variability in seasonal and annual precipitation patterns during periods of stand establishment, especially long-term cyclical changes. As experienced with the attempts to establish the longleaf pine plantation, severe seedling mortality is possible even after taking all usual precautions. During the recent drought years, other tracts in the near vicinity of the subject property experienced severe mortality of even mature pine and hardwood timber. Drought-stressed pines in East Texas are more susceptible to pathogens, such as the *Ips* bark beetle. Drought-stressed hardwoods in East Texas are more susceptible to pathogens, such as *Hypoxylon* canker. Texas has historically experienced a 30-year drought cycle correlated to periods when the Pacific Decadal Oscillation (PDO) is in its cool or negative phase and the Atlantic Multi-Decadal Oscillation is in its warm or positive phase (Nielsen-Gammon, 2011). Based on the historical evidence, the current drought cycle in Texas could last another 15 or more years. Recurring and prolonged drought could have significant impacts on the forest health in the region in terms of forest growth and survival, risk of wildfires, wildlife, and water resources.

REFERENCES

Bureau of Economic Geology, 2010. EcoRegions of Texas 2010. University of Texas, Jackson School of Geosciences, Bureau of Economic Geology, Austin, TX.

Butler, B.J., 2008. Family forest owners of the United States, 2006. U.S. Department of Agriculture, Forest Service, Northern Research Station, Newtown Square, PA, General Technical Report NRS-27. 72 p.

Cooper, J.A., Bentley, J.W., 2012. East Texas, 2011 forest inventory and analysis factsheet. U.S. Department of Agriculture, Forest Service, Southern Research Station, Asheville, NC e-Science, Update SRS-052.

Dolezel, R., 1980. Soil survey of Nacogdoches County, Texas. U.S. Department of Agriculture, Soil Conservation Service, Washington, DC, 146 p.

Long, C., 2011. Nacogdoches County. Handbook of Texas Online. Texas State Historical Association, Denton, TX. http://www.tshaonline.org/handbook/online/articles/hcn01 (Accessed May 29, 2014).

Nielsen-Gammon, J., 2011. Texas drought and global warming. Houston Chronicle, Houston Texas. http://blog.chron.com/climateabyss/2011/09/texas-drought-and-global-warming/ (Accessed May 29, 2014).

Stewart, P., 2012. When will the recovery in sawtimber prices start in the US South? Forest2Market, Charlotte, NC. http://www.forest2market.com/blog/When-Will-the-Recovery-in-Sawtimber-Prices-Start-in-the-US-South (Accessed May 29, 2014).

Straka, T.J., Cushing, T.L., 2014. McPhail tree farm, South Carolina, United States of America. In: Siry, J.P., Bettinger, P., Merry, K., Boston, K., Grebner, D.L., Cieszewski, C. (Eds.), Plans of North America. Academic Press, New York. P. 87.

Texas Forest Service and Texas Forestry Association, 2010. Texas forestry best management practices. Texas Forest Service and Texas Forestry Association, Lufkin, TX. http://texasforestservice.tamu.edu/uploadedFiles/Sustainable/bmp/Publications/BMP%20Manual_Aug2010%20-%20web.pdf (Accessed May 29, 2014).

U.S. Forest Service, 2013. Forest Stewardship Program. U.S. Department of Agriculture, Forest Service, Washington, DC. http://www.fs.fed.us/spf/coop/programs/loa/fsp.shtml (Accessed May 29, 2014).

World Resources Institute, 2014. Southern forests: the wood basket of the nation. World Resources Institute, Washington, DC. http://www.seesouthern-forests.org/case-studies/southern-forests-wood-basket-nation (Accessed May 30, 2015).

Chapter 4

Pike Lumber Company's Sam Little Forest, Indiana, United States of America

Oliver Grimm
Pike Lumber Company, Akron, Indiana, USA

ABBREVIATIONS

ATFS American Tree Farm System
BMP Best Management Practices
DBH Diameter at Breast Height
TSI Timber Stand Improvement

MANAGEMENT SETTING AND BACKGROUND

The Sam Little Forest is situated at the northern edge of the Tipton Till Plain Natural Region on the banks of the Wabash River in north-central Indiana. The underlying soil consists of very productive loam composed of fine sands and silt loams formed on loess, limestone, and glacial outwash. The topography is flat to sloping with interspersed steep limestone out-croppings, streams, abandoned quarries, and springs. The 58 acre (ac) (23.5 hectare (ha)) property is bounded on its north side by the Wabash River and is dissected by a county road (Figure 4.1). The property was acquired in 1959 at public auction by the owners of Pike Lumber Company and named after the previous owner, Sam Little. In 1972, ownership of the land was transferred to Pike Lumber Company.

The forest is fairly typical for north-central Indiana (Figure 4.2), and is described as a sugar maple (*Acer saccharum*) and American beech (*Fagus grandifolia*) type. Due to its terrain, it was never cleared and farmed, but the understory was grazed. A relatively level area in the southeast portion of the property was once pastured intensively and is today stocked with mostly post-size black walnut (*Juglans nigra*). Two small-scale limestone quarries were established and later abandoned on the property. During the last 55 years, the forest has seen seven harvests, which were single tree and group selection harvests or salvage harvests after storm events. A small walnut enrichment planting was established in an opening created during a harvest. The harvests were followed up by timber stand improvement (TSI) measures, consisting of grape vine (*Vitis* spp.) control and crop tree release. In 2008, an Asian bush honeysuckle (*Lonicera* spp.) infestation was addressed with an inten-sive control measure, which included the cutting of larger plants with chainsaws, stump treatment, as well as foliar herbicide application. Since that time, periodical follow-up treatments have become necessary to control this prolific, invasive plant.

In 1972, all trees 10 inches (in) (25.4 centimeters (cm)) in diameter at breast height (DBH) and larger were inventoried. The species consisted mainly of sugar maple, American beech, and northern red oak (*Quercus rubra*). In 1990 and in 2014, follow-up sample inventories were conducted. By 2014, the species composition had shifted considerably, although sugar maple continued to be the dominant single species. The forest was overstocked in 1959 and the initial harvests were light. Not until the 1990s were mature and over-mature trees harvested in significant numbers. This was reflected in depressed growth rates, which are rebounding since the 1990s due to the harvests and TSI activities, but have changed the stocking level dramatically resulting in a forest in a developmental stage. The volume of beech, a nonpreferred species, has been reduced over time. Other undesirable species, including sycamore (*Platanus occidentalis*) and hackberry (*Celtis occiden-talis*), have increased because of their ability to regenerate well on certain sites available to them. Red oak volume has decreased for the opposite reason, namely, due to its inability to regenerate well under a single-tree and group selection harvest regime and due to herbivory from white-tailed deer (*Odocoileus virginianus*). The increase in volume of the desir-able black walnut has been due to the enrichment planting and ingrowth of the post-size walnuts in the former pasture.

Forest Plans of North America. http://dx.doi.org/10.1016/B978-0-12-799936-4.00004-7

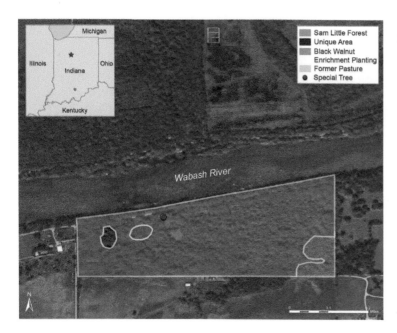

FIGURE 4.1 A map of the Pike Lumber Company's Sam Little Forest.

FIGURE 4.2 The entrance to the Sam Little Forest in winter.

The overall reduction in volume from trees in larger diameter classes from 1972 to 2013 is in favor of smaller but more vigorously growing trees.

PLANNING ENVIRONMENT AND METHODOLOGY

Pike Lumber Company's objectives for owning and managing forestland are (Pike Lumber Company, 1995a):

- To provide sound, diversified investment and a profit for the company and its stockholders
- To act as insurance or a protective measure against a timber (raw material) famine
- To grow vigorous, high quality, principally hardwood trees, on a perpetual basis
- To encourage forest land ownership and management by example
- To provide wildlife habitat, clean water, clean air, recreation, and biological diversity

To achieve these goals, a forestry department was established and charged with managing the forestland so that high-quality sawlogs and veneer are produced more rapidly than would be possible in an unmanaged forest.

Up until the 1990s, timber harvests were conservative and guided by the above described goals. The emphasis was that the company-owned timber was to provide an "insurance policy" against raw material shortages. This resulted in holding timber past its economic maturity and, in some cases, its biological maturity. Harvest methods employed included single-tree selection and salvage logging. After the company realized that maintaining high inventories of large timber resulted in financial losses due to deteriorating lumber quality and declining growth rates, a period of harvesting large, over-mature timber began. This allowed the company to utilize group selection harvests, and resulted in overall reductions of stocking levels. TSI activities have since been used intensively, as it became understood that these activities were a good investment for the purpose of growing high-value trees faster, or to grow larger volumes. TSI activities include the deadening of grape vines, crop tree release measures, precommercial thinning, and pruning. Using controlled fire is not a management option due to the small size of the parcel and the close proximity of the parcel to neighboring homes. Controlled fire could be a valuable tool in stands that are more oak-dominated than the Sam Little Forest.

Within the Sam Little Forest, black walnut trees were planted in an opening that was about 2 ac (0.8 ha). Larger hardwood tree planting establishment has taken place on other properties, typically when entire farms were acquired and the tillable ground was afforested. The high costs associated with farmland acquisition and plantation establishment, the lower lumber quality produced by plantation timber, the long pay-back periods, and excessive deer browse damage all suggest that hardwood species tree planting is not economically feasible as an investment in north-central Indiana.

Pike Lumber Company's focus is on producing high-grade hardwood lumber. Pallet-grade sawlogs are not processed through Pike Lumber Company's sawmills, but they are shipped to local pallet mills. Veneer-grade timber is typically sold in log form to veneer processing mills. The lack of an outlet for pulpwood in the operating area makes the harvest of small-diameter trees financially infeasible. For many years, there was very little interest in the sale of firewood. In recent years, more firewood demand has been observed, which is satisfied mostly through logging debris (tree tops and cut-offs) located at landing sites. The majority of undesirable, small-diameter trees are deadened (killed) during TSI measures.

When developing a new management plan for a property with a long tenure of ownership and a long history of management, the current plan is obviously a good starting point and resource. Previously planned and executed activities are updated and their efficacy evaluated. Care has to be taken to use source materials to verify the statements made in previous plans. The desired future condition of the forest along with the timber products that might become available during the plan period are two important markers in plan development. The base information includes an inventory indicating the current status of the resources; comparisons with previous inventories facilitate the analysis (Table 4.1). The inventory is further described in terms of volume by diameter classes (Figure 4.3) and species as well as by indications of maturity and over-maturity of

TABLE 4.1 Timber Inventories of Trees 12 inches (30.5 centimeters) in Diameter at Breast Height or Larger (Board feet, Doyle Rule) in Pike Lumber Company's Sam Little Forest

	Year		
Tree Species	1972	1990	2014
Ash (*Fraxinus* spp.)	8,549	3,325	8,493
American basswood (*Tilia americana*)	5,226	6,292	4,730
American beech (*Fagus grandifolia*)	69,244	23,582	4,929
Black cherry (*Prunus serotina*)	22	421	3,925
Cottonwood (*Populus deltoides*)	201	2,333	7,042
Elm (*Ulmus* spp.)	6,431	2,785	5,099
Hickory (*Carya* spp.)	17,683	6,432	4,573
Sugar maple (*Acer saccharum*)	209,614	186,511	36,938
Red maple (*Acer rubrum*)	24	3,332	3,129
Bur oak (*Quercus macrocarpa*)	4,230	4,263	1,653
Chinkapin oak (*Quercus muehlenbergii*)	6,103	10,525	5,848
Northern red oak (*Quercus rubra*)	58,342	28,139	3,026
Yellow-poplar (*Liriodendron tulipifera*)	455	1,213	6,521
Sycamore (*Plantanus occidentalis*)	14,568	23,367	10,103
Black walnut (*Juglans nigra*)	12,961	10,726	11,035
Other species[a]	26,470	8,337	11,270
Total	440,123	321,583	129,314

[a]*Mainly composed of hackberry (Celtis occidentalis), white oak (Quercus alba), Ohio buckeye (Aesculus glabra), ironwood (Ostrya virginiana), and Kentucky coffeetree (Gymnocladus dioicus).*

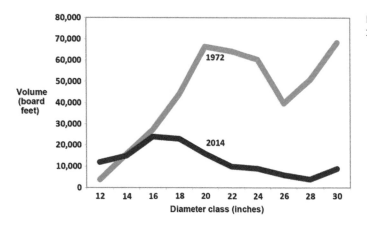

FIGURE 4.3 Volume distribution by diameter class, 1972 and 2014, in the Sam Little Forest.

trees and threats of loss due to damage from insects, such as the emerald ash borer (*Agrilus planipennis*) and gypsy moth (*Lymantria dispar dispar*). Short-term changes in timber markets typically do not influence the management plans.

Guiding Laws, Regulations, and Policies

The Sam Little Forest has been entered into two state-level forest stewardship programs. One is the Indiana's Classified Forest and Wildlands program (Indiana Department of Natural Resources, Division of Forestry, 2014a) and the other is Indiana's Classified Forest Certified Group (Indiana Department of Natural Resources, Division of Forestry, 2014b). Under both programs, forest landowners are subject to audits of their management practices, especially Best Management Practices (BMPs) (Ernst, 2004). Enrollment in the Classified Forest and Wildlands Program requires the landowner to follow a written stewardship plan and to commit to keeping the property from being converted to other land uses including livestock grazing. In return the landowner principally receives a lower property tax assessment. Enrollment in the Classified Forest Certified Group provides the landowner third-party certification through the Forest Stewardship Council, a benefit Pike Lumber Company subsequently has chosen to opt out of due to lack of financial return. The forest has been certified by the American Tree Farm System (ATFS) since 1989.

A management plan developed purely for internal use would be quite different from a plan that has to meet both the requirements of the Classified Forest and Wildlands Program and the American Tree Farm System. In an effort to synergize the plan development process, the topics, scope, and data required by Pike Lumber Company, Classified Forest and Wildlands Program, and ATFS are all combined into one plan. Adherence to the management plan's prescriptions and their timing are being monitored by the Indiana Department of Natural Resources (DNR) and ATFS inspectors. Some Indiana counties have enacted, or are in the process of enacting, logging ordinances as well. The professed goal of such ordinances is to protect local communities from the negative effects of logging operations, such as soil erosion, damage to roads, aesthetical impairment of neighboring properties, and detrimental effects to certain plant or wildlife species. Securing a permit under a logging ordinance requires the preparation of a specialized plan, several office visits, bonds, as well as other significant fees. It can be argued that the creation of a logging ordinance is redundant, as protections are already in place through federal and state laws, and that the result is to discourage timber utilization.

To protect the genetic material of trees that exhibit the size, straightness, taper, vigor, and quality deemed vital to the management of the forest, older trees have been designated as Special Trees, and seeds of these are made available to genetic repositories and nurseries. Other protection measures include buffer zones that limit road construction, timber harvests, and herbicide use. One large red oak has been designated as a Special Tree (Pike Lumber Company, 1995b). For the maintenance of areas significant for geological, archaeological, historical, biological, or aesthetic reasons—and areas that are habitat for flora or fauna with unique site requirements—a Unique Area Policy has been created. Black-seeded rice grass (*Oryzopsis racemosa*) has been identified in an area of the Sam Little Forest and enjoys the same protections as mentioned for the Special Trees (Pike Lumber Company, 1995c).

Pike Lumber Company's forestland is to be managed in a sustainable fashion, which specifically means that no more than the volume equivalent to the growth rate may be harvested as a sustainable yield. Of course, with holdings made up of relatively small parcels, this rule is not meant to be followed on a parcel-by-parcel basis but on a portfolio scale. The annual allowable cut is based on the following information:

- The baseline inventory of all forests
- The average annual growth rate determined by comparing baseline and subsequent inventories
- Tree ring analyses derived from core samples and harvested logs

The annual allowable cut, despite its name, does not require the harvest to be realized in any one specific year but over a longer period of time (yet not more than 10 years). This allows the company some flexibility in its ability to react to logging and timber market conditions.

OUTCOMES OF THE PLAN

The plan for the Sam Little Forest suggests that conditions will continue to be monitored and that activities may be implemented as markets and budgets change. The accumulation of over-mature timber having been reduced over the last decade has led to future management activities that will include a harvest to remove the remnants of the overstory and allow the younger timber to grow unimpeded. This harvest will be followed by precommercial crop tree release. The combination of these tools will contribute to further reduce the undesirable tree species (i.e., hackberry, beech, sycamore). The control of invasive species will require vigilance and frequent treatments, a process that has no determinable point of completion.

DISCUSSION AND CONCLUSIONS

The management of Pike Lumber Company's Sam Little Forest, a hardwood forest located in north-central Indiana, is guided by a number of goals that the company has developed to address its economic, environmental, and social objectives for owning the land. A forest management plan fulfilling the requirements of company policies, the Indiana DNR-Forestry Division, and the American Tree Farm System has been developed that provides direction for timber stand improvement activities as well as periodic harvests.

Sustainability Issues

The development of the timber resource of this forest between 1972 and today clearly shows a decline in standing timber volume. The small size of the parcel does not allow a management approach that has sustainable yield of timber products as a measurable goal. Sustainable yields across the portfolio of forest lands owned by Pike Lumber Company, however, have been achieved. More broadly, the protection and recreation functions of the forest have been sustained continuously over the tenure of ownership.

The regeneration of the timber resource on the Sam Little Forest is an area of concern, largely because young regenerating trees are being continually consumed by white-tailed deer, whose populations have increased due, in part, to restrictive hunting regulations and a general decline of interest in hunting. White-tailed deer predators also are largely absent or ineffective as a population control mechanism. Diseases such as epizootic hemorrhagic disease have occurred in recent years, but herbivory remains a significant problem.

Plan Development Challenges

The collection of inventory data, which is the basis for the management plan, constitutes a significant cost. It is comprised of the cost to train personnel in data collection, including handheld computer and GIS hardware and software as well as the man-hours spent in the forest collecting data. TCruise® software (World Wide Heuristic Solutions, 2014) was utilized for data analysis. The initial acquisition cost is significant; additionally, a high customization and training expenditure was necessary until serviceable reports could be produced. Over time, the use of GPS receivers in combination with the gridbuilder function of the ExpertGPS® software (Topografix, 2014) have proven to be significant time and cost savers. This allows finding sample plot centers much more quickly than past methods. Other data had to be acquired from established sources and integrated into the planning process. For example, information about the soils was gathered using the U.S. Department of Agriculture's Web Soil Survey (U.S. Department of Agriculture, Natural Resources Conservation Service, 2013).

Plan Implementation Challenges

Invasive plant species, Asian bush honeysuckle and tree-of-heaven (*Ailanthus altissima*), require control measures. Total eradication does not seem possible. Surrounding forests have developed a dense understory of seed-producing bushes, and these seeds will spread continuously throughout the Sam Little Forest by different bird species. The frequency of control activities necessary to allow the natural regeneration of desirable hardwood trees, and the costs necessary to ensure this, will have to be studied. Therefore, it is difficult to prescribe in detail in a plan with a 10-year life span.

Additionally, the forest and the managers face challenges from invasive insect and fungal pests whose scope and severity is difficult to predict. The emerald ash borer is currently decimating the ash trees in the Sam Little Forest. The gypsy moth, Asian longhorned beetle (*Anoplophora glabripennis*), and thousand canker disease (U.S. Department of Agriculture, Forest Service, 2013) are in threatening proximity. As a result, management activities have recently included the timely salvage of ash (*Fraxinus* spp.) trees and the monitoring of other pests.

Finally, BMPs make it necessary to maintain a riparian buffer. In this forest, this means that approximately 3 ac (1.2 ha) of land situated between the road and the Wabash River produce no income because no harvests can take place there.

REFERENCES

Ernst, D., 2004. Indiana Forestry Best Management Practices. Indiana Department of Natural Resources, Indianapolis, IN.

Indiana Department of Natural Resources, Division of Forestry, 2014a. The Classified Forest and Wildlands Program. Indiana Department of Natural Resources, Indianapolis, IN. http://www.in.gov/dnr/forestry/files/fo-ClassifiedForestBrochure.pdf (Accessed May 11, 2014).

Indiana Department of Natural Resources, Division of Forestry, 2014b. Indiana Classified Forest Certified Group. Umbrella Management Plan. Indiana Department of Natural Resources, Indianapolis, IN. http://www.in.gov/dnr/forestry/files/fo-ICFCG_Umbrella_plan_11_10.pdf (Accessed May 11, 2014).

Pike Lumber Company, 1995a. Forestland Policy. Pike Lumber Company, Akron, IN.

Pike Lumber Company, 1995b. Special Tree Classification Policy. Pike Lumber Company, Akron, IN.

Pike Lumber Company, 1995c. Unique Area Policy. Pike Lumber Company, Akron, IN.

Topografix, 2014. ExpertGPS. Topografix, Stow, MA. http://www.expertgps.com/ (Accessed June 11, 2014).

U.S. Department of Agriculture, Forest Service, 2013. Pest alert: thousand cankers disease. U.S. Department of Agriculture, Forest Service, Northeastern State and Private Forestry, Newtown Square, PA, NA-PR-02-10.

U.S. Department of Agriculture, Natural Resources Conservation Service, 2013. Welcome to Web Soil Survey (WSS). U.S. Department of Agriculture, Natural Resources Conservation Service, Washington, D.C. http://websoilsurvey.sc.egov.usda.gov/App/HomePage.htm (Accessed May 13, 2014).

World Wide Heuristic Solutions, 2014. Timber Cruise Suite. World Wide Heuristic Solutions, Starkville, MS. http://www.timbercruise.com/index.php/software-overview/desktop-solutions/16-hmstc (Accessed June 11, 2014).

Chapter 5

Poole and Pianta Woodlot, West Virginia, United States of America

Russ Richardson

Appalachian Investments, Arnoldsburg, West Virginia, USA

MANAGEMENT SETTING AND BACKGROUND

The property owned by Denise Poole and Lisa Pianta (Figure 5.1), consisting of 194 ac (78.5 ha), is located in Courthouse Magisterial District of Lewis County, West Virginia. The property lies on Cove Lick, a moderately large stream that flows into Sand Fork, a minor tributary of the Little Kanawha River. The woodlot has been owned by Poole and Pianta for more than 20 years and has been in their family for several generations. This property was the site of a typical Appalachian subsistence farm that was abandoned from agriculture in the later part of the twentieth century. Modern access has not been developed to the interior or upland locations, and most farm structures are either in disrepair or have begun rotting into the ground. There is no mention of any historic activities having taken place on the farm, and no cemeteries or burial grounds are known to be present on the property. Scattered evidence of small-scale coal extraction activities by earlier residents can be identified at higher elevations near the northeastern-most property corner. All public roads accessing the farm appear to be on long established rights-of-way and are not known to possess historically important significance.

The woodlot has been under active forest management for nearly 20 years, and recent activities on the property include completion of a boundary survey, 100 ac (40.5 ha) of grapevine (*Vitus rotundifolia*) treatment, 25 ac (10.1 ha) of forest improvement cuts, and a 100 ac (40.5 ha) timber sale. Nearly all portions of the property contain trees of commercial size, and the harvest of timber is a moderately high priority for the woodland owners at this time. Elevation on the woodlot ranges from approximately 980 ft (299 m) near Cove Lick, to over 1,500 ft (457 m) in several locations along the ridge, which comprises the northern boundary of the property. Soil types range from deep fertile soils in the valley bottoms to erosive and moderately productive soils on sloping lands.

The principal timber species include yellow-poplar (*Liriodendron tulipifera*), chestnut oak (*Quercus prinus*), white oak (*Quercus alba*), black oak (*Quercus velutina*), northern red oak (*Quercus rubra*), scarlet oak (*Quercus coccinea*), red maple (*Acer rubrum*), sugar maple (*Acer saccharum*), hickory (*Carya* spp.), and white ash (*Fraxinus americana*). American beech (*Fagus grandifolia*), black walnut (*Juglans nigra*), sourwood (*Oxydendrum arboreum*), sassafras (*Sassafras albidum*), redbud (*Cercis canadensis*), elm (*Ulmus* spp.), black gum (*Nyssa sylvatica*), black locust (*Robinia pseudoacacia*), and hophornbeam (*Ostrya virginiana*) occur as associated species throughout the stands. The majority of the timber on the property is in the small to medium sawtimber size classes. Currently, the standing inventory of the woodlot is estimated to be about 1.6 million board feet (MMBF) (4,248 m³). The sites are of good to very good quality for the production of forest products. The woodlot has historical evidence of fire, but no fires have occurred within the past 60 years. Yellow-poplar trees were partially defoliated during an outbreak of a native forest pest (yellow-poplar weevil, *Odontopus calceatus*) in 2012, but no serious insect or disease problems were detected during this evaluation. During the past 15 years, invasive species such as multifloral rose (*Rosa multiflora*) and autumn olive (*Elaeagnus umbellata*) have become established in the understory of the most productive part of the woodlot, and Japanese stiltgrass (*Microstegium vimineum*) is a developing problem along all roads and right-of-ways on the property.

The woodlot consists of a variety of tree size classes and tree species that provide the diversity needed by game and nongame wildlife species. Brushy open areas surrounding an abandoned beaver (*Castor canadensis*) swamp, the Cove Lick wetlands, rocky areas, cliffs, and scattered patches of mature beech and hickory are all very important in maintaining diverse wildlife populations. Recent development of gas wells (Figure 5.2) has removed 5 ac (about 2 ha) of forest cover from the property, and this has provided the opportunity for additional wildlife habitat enhancement. Road work and gas-well

Forest Plans of North America. http://dx.doi.org/10.1016/B978-0-12-799936-4.00005-9

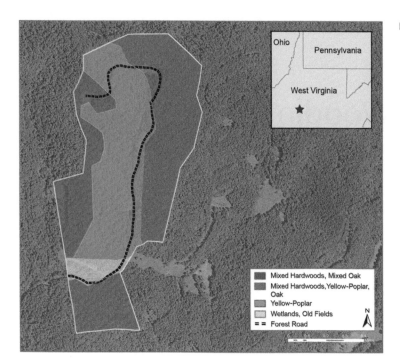

FIGURE 5.1 A map of the Poole/Pianta woodlot.

FIGURE 5.2 A recently developed well location within the mixed oak and mixed hardwood stand.

development are needed to accommodate the wells. The recommended cutting practices, plantings, and other cultural treatments will help regulate, on a continuing basis, the kind, amount, and arrangement of food and cover needed by wildlife during all seasons of the year.

PLANNING ENVIRONMENT AND METHODOLOGY

The most recent Forest Stewardship Plan was developed in 2013 under the auspices of the West Virginia Division of Forestry (Richardson, 2013). Through the planning process, it was learned that the main objectives of Poole and Pianta are to continue a forest management program that will improve the long-term value of the timber growing on the property while providing for an increased diversity of wildlife species along with enhanced opportunities for woodland recreation. For the owners, the highest priority is to manage for multiple-use forestry with primary consideration given to the scenic and aesthetic qualities of the forest. Water quality is also a high priority, and timber harvesting is a moderately high priority. Any harvesting to be planned or undertaken should have silvicultural improvement as the objective. The property contains five management areas or stands, and these are classified as mixed oaks/hardwoods (2), wetlands, yellow-poplar, and other mixed hardwoods. Management activities have recently been concentrated on the better quality sites where maximum returns from wood products and associated forest benefits are being realized. Employment of cultural treatments on the better sites has shortened the time necessary to produce high-quality sawlog or veneer products.

Given that there are five general management areas (stands) within this woodlot, stand-level silvicultural decisions were developed through an analysis of the potential of each to meet the objectives of the landowners. For example, the mixed hardwood stand is 21 ac (8.5 ha) in size, growing on land with northern exposures. The entire stand is located south of Cove Lick Road, and the growing sites are some of the best on the woodlot. A small portion of the stand was logged in a timber trespass event that occurred in 1998 prior to the completion of the boundary survey. The stand contains 14 different commercial tree species but is dominated by yellow-poplar. The understory is significantly different and is dominated by shade-tolerant species that include sugar maple 25%, black birch (*Betula nigra*) 18%, beech 13%, and red maple 15%. The average diameter of trees is the largest among the three principle forest types (Table 5.1). Overall, tree quality is very good, only 6% of the stand basal area is cull, and gypsy moths (*Lymantria dispar dispar*) are not a current threat to the stand. However, a forest improvement cut is a priority for the stand. During the recommended harvest, the best trees possible should be retained for continued growth and seed production. Because of the arrival of the emerald ash borer (*Agrilus planipennis*) in the local woodland, it is recommended that all white ash trees of commercial size be removed during the harvest, which was planned for the summer of 2013. In conjunction with the removal of all white ash from the stand, natural regeneration should be stimulated through the openings created, and as many high-quality red oak trees as possible should be retained for seed production.

Two areas make up the 91 ac (36.8 ha) of mixed oak and mixed hardwoods. These contain medium-sized sawtimber and were last logged in the middle of the 1950s. These sites are moderately fertile and the exposures are primarily to the northwest, west, and southwest. Cull beech and red maple are common in the stand; cull trees represent 12% of the basal area. Although nearly 20 different commercially valuable tree species are present in the stand, the overstory is dominated by just a few species, including chestnut oak, yellow-poplar, hickory, red maple, and other oaks. The poletimber portion of the inventory is significantly different and includes red maple (29%), hickory (20%), and chestnut oak (15%). Stocking in the stand is variable, with some areas covered with a low-value combination of red maple, hickory, and beech; however, most of the stand is fully-stocked. Gypsy moths are a potentially significant threat to the stand. Most of this stand is ready for an improvement-oriented commercial harvest. During any planned harvest in the stand, the highest quality trees possible should be retained for continued growth and seed production. The health of the white oak and chestnut oak in the stand should be monitored for an outbreak of a periodically destructive native insect pest of white oaks, the jumping oak gall wasp (*Neuroterus saltatorius*). Some of the healthiest beech and hickory culls should be retained for the benefit of wildlife, and the arrival of gypsy moths should be carefully monitored.

The 70 ac (28.3 ha) yellow-poplar stand contains small sawtimber growing on moderately sloping land that is dominated by eastern and northeastern exposures. This stand contains the majority of the merchantable sawtimber volume on the property (Figure 5.3). The stand was pastured and farmed as a part of a subsistence agriculture endeavor for more than 100 years prior to final abandonment in the middle part of the last century. The more gently sloping land near the valley bottom was tilled for generations, and several large piles of stones cleared from the fields remain in areas where tillage occurred. Grapevines, which were previously very dense in much of the stand, were treated in 2001; the growth response in the yellow-poplar trees that were released has been extremely good. Since 1999, the average diameter of this stand has increased by nearly 5 in (12.7 cm). Infestations of multifloral rose and autumn olive are a present and increasing problem in the stand's understory, and future regeneration success may require an herbicide treatment. The stand is ready for a forest improvement cut, and emphasis of the harvest will be to select the best possible trees for retention and continued growth. Although white ash represents a small proportion of the inventory, it is recommended that all white ash trees of commercial

TABLE 5.1 Summary Statistics for the Three Principle Forest Types on the Poole/Pianta Woodlot

Forest Type	Volume (Board ft per ac*)	Cull (%)	Average Diameter (inches)	Basal Area per ac (ft²)	Stocking (Trees per ac)
Mixed hardwoods	9,590	6	14.1	99	145
Mixed oak and mixed hardwoods	8,393	12	13.6	123	140
Yellow-poplar	9,965	0	13.4	113	190

*International 1/4-inch log rule.

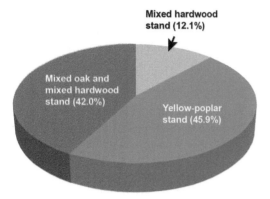

Mixed hardwood stand (12.1%)

Mixed oak and mixed hardwood stand (42.0%)

Yellow-poplar stand (45.9%)

FIGURE 5.3 Allocation of total volume within the three principle forest types.

size be removed during the initial harvest. Recent infestation of local woodland by the introduced forest pest, emerald ash borer, will result in the mortality of white ash within the next few years.

The wetland contains alder (*Alnus glutinosa*), witch hazel (*Hamamelis* spp.), American hornbeam (*Carpinus caroliniana*), and other bottomland hardwoods, and the surrounding hillsides have a heavy cover of autumn olive and multifloral rose. Use of the area by American woodcock (*Scolopax minor*) is heavy. The area previously flooded by beavers is about 2 ac (about 0.8 ha) in size. The wetland is important wildlife habitat, and improvement of the area for American woodcock is suggested. The site might be appropriate for one of the wildlife enhancement projects covered under the Wildlife Habitat Incentive Program administered by the U.S. Department of Agriculture, Natural Resources Conservation Service (2014).

Guiding Laws, Regulations, and Policies

The federal income tax laws are important considerations to the landowners when selling forest products. The U.S. government, through the West Virginia Division of Forestry, can provide up to 75% cost-sharing assistance for some of the silvicultural practices recommended. Some of these practices include timber stand improvement, cull tree removal, and grapevine treatment.

The Poole and Pianta woodlot is enrolled in the Managed Timberland Program (West Virginia Division of Forestry, 2009), which provides tax incentives for nonindustrial forest landowners who utilize sustainable forestry practices on private land. This program uses a tax-incentive approach to encourage landowners to actively and sustainably manage their forests, with the intent of increasing the amount and quality of the forest resource in West Virginia. If enrolled, private landowners might receive up to 75% cost-sharing assistance for the development of a forest plan. The plan developed must specify all of the foreseeable management activities and be developed in a way that ensures a continuous and perpetual yield of timber. In enrolling forest land in the Managed Timberland Program, landowners might then be taxed according to the land's potential for growing timber rather than a generalized market value. The program also requires landowners to maintain a sufficient number of commercially valuable tree species that represent at least 40% or more of the normal stocking of trees and that these are well-distributed across the forest.

West Virginia has also developed a set of best management practices for forestry activities (West Virginia Division of Forestry, 2005). Landowners understand that these practices exist, yet the liability for adhering to these practices falls upon the timber harvesters. One aspect of the best management practices that specifically applies to this woodlot is that either no harvest is allowed, or carefully planned minimal selection harvests can be prescribed, in and around wetland areas and riparian zones. At least a 50 ft (about 15 m) buffer zone should be maintained. All tree tops need to be pulled a minimum of 25 ft (about 7.5 m) from perennial and intermittent streams, and equipment should not be permitted within 100 ft (about 30 m) of these areas. A Logging Sediment Control Act (1992) also was enacted in an effort to protect and maintain forest water and site productivity by minimizing soil erosion from areas disturbed during timber harvesting operations. Further, both the best management practices and the Logging Sediment Control Act are included in a silvicultural nonpoint pollution management program (West Virginia Division of Environmental Protection, 2000).

OUTCOMES OF THE PLAN

Through the planning process, it was determined that two of the management areas in the woodlot will need a commercial harvest soon to remove mature sawtimber or enhance the growth of the developing forest. The proposed harvest cuts will leave adequate residual volume for another harvest in 10–12 years and will meet the silvicultural needs of the stand and the wildlife needs for the area, while also meeting the landowners' objectives and economic needs. The quality of the standing timber should also be improved by these harvest operations through the removal of a large portion of the undesirable species and trees of poor form or quality.

During sale preparation activities in 2012 and 2013, it was noticed that the emerald ash borer was present on the property, and the infestation was severe. Therefore, during the sale marking process, 100% of the white ash trees over 6 in (15.2 cm) in diameter were marked for removal. The residual basal area in most of the sale area should be around 65–75 ft^2/ac (14.9–17.2 m^2/ha), and very few of the yellow-poplar crop trees remaining will have less than 3.5 16-foot logs.

Other general recommendations for the woodlot that were contained in the plan include:

1. During harvesting activities, mature sawtimber, certain undesirable species, and most damaged trees should be harvested, leaving a good stocking of immature trees of desirable species for the future stand.
2. Diseased trees should be removed from the stand during cultural treatments.
3. Three to five standing dead snags or live den trees per ac (about 7–12/ha) should be left standing for cavity nesting birds and animals.
4. Work should be concentrated on the better sites first, where the largest increase in benefits for both timber production and wildlife can be obtained.
5. Water diversion measures and seeding should be undertaken on constructed log roads and landings to reduce the possibility of erosion and siltation and to create wildlife feeding areas.
6. Livestock and deer grazing of the woodlot should be restricted. Control efforts, primarily by managing deer hunting activities, should be considered to maintain population levels of deer compatible with the capacity of the habitat.
7. A good system of forest roads should be constructed and maintained on the property to provide easy access for future woodland management work and to serve as firebreaks and access lanes should fire suppression be needed in the future.
8. Open or understocked areas should be planted with desirable tree species, and competing vegetation should be controlled in established plantations for optimum growth and survival of the planted seedlings.
9. Various other silvicultural treatments should be considered. These include cleanings in sapling hardwood stands, thinnings in pole and light sawtimber stands to improve species composition and stocking of desirable crop trees, and crown release cuttings to release desirable regeneration from overtopping cull trees.
10. Climbing vines that are growing on merchantable trees should be cut. Vines can cause the formation of crooks and forks, thereby reducing the quality and value of future crop trees. However, vines (especially grapevines) may provide excellent wildlife food and cover opportunities. Grapevines in low-valued tree species should not be cut.
11. Plants that provide wildlife foraging opportunities, such as dogwood (*Cornus florida*), serviceberry (*Amelanchier* spp.), sumac (*Rhus* spp.), and viburnum, should be maintained, especially around the woodland edge. Border plantings of gray dogwood (*Cornus racemosa*), chestnut chinquapin (*Castanea pumila*), bear oak (*Quercus ilicifolia*), and other seedlings desirable for wildlife food should be established.
12. Buffer areas along all well-traveled roads and along streams should be maintained to contribute to the aesthetic appeal of the woodlot and to protect streams from siltation or dramatic temperature changes. However, such buffer areas can be harvested or otherwise treated if affected by windstorms, insects, diseases, over-maturity, or in cases where safety dictates.

DISCUSSION AND CONCLUSIONS

The activities suggested in the recently developed plan are meant to increase the scenic and aesthetic qualities of the woodlot and positively contribute to water quality goals. Given the changes in forest structure that should occur through the improvement cuts and other harvests, it is recommended that the plan be revised in no more than 10 years. The majority of forestry activities recommended should place the *quality* of the individual trees remaining after treatments of far greater importance than the *number* of trees remaining.

Sustainability Issues

The forest management plan was prepared through the West Virginia Forest Stewardship Program and was developed with long-term forest management in mind. The Poole and Pianta woodlot is a productive tract of property that, with careful and conscientious use, can provide quiet enjoyment, wildlife habitat, forest products, and income for generations to come.

Plan Development Challenges

There were no unusual or extraordinary problems associated with developing the plan for this property because it was an extension and continuation of an ongoing and previously developed forest management program for the property. However, since its inadvertent introduction into Massachusetts in 1869, the gypsy moth has spread naturally south and west at approximately 5–10 mi (8–16 km) per year. In the last 10 years, it has been spreading across the eastern panhandle counties of West Virginia. Predictions were that it would have eventually caused damage in every West Virginia County by 2000. According to the National Agricultural Pest Information System (2014), the gypsy moth has been found in Lewis County. As a result, the gypsy moth has arguably been the most destructive forest pest threatening West Virginia woodlands. In addition, the arrival and spread of other insects and diseases have been important to West Virginia forests. These include the hemlock woolly adelgid (*Adelges tsugae*), Asian longhorn beetle (*Anoplophora glabripennis*), emerald ash borer, thousand canker blight (caused by the combined activity of the walnut twig beetle (*Pityophthorus juglandis*) and a fungus (*Geosmithia morbida*)), and others that have all either arrived or spread to the point that they are permanently impacting forests. Therefore, the gypsy moth section of the standard West Virginia nonindustrial private forest plan will likely be changed in the future to something related to "biological threats" in order to more fully address the range of important issues facing the management of private forests.

Plan Implementation Challenges

Markets for sawtimber and pulpwood have recently been extremely weak, and a limited market for biomass exists in northern West Virginia at this time. Any future harvesting on the property will attempt to take advantage of biomass and carbon markets when they develop. One emphasis during the harvest planning process will be identification and selection of the best possible den trees to retain in the woodlot. Features that identify trees as having good den potential include a vigorous healthy crown, hollows that might become nesting or den sites for small game, and evidence of active use by other woodland creatures.

Currently, the Poole and Pianta woodlot is not infested with the gypsy moth, but the landowners will need to periodically identify where the severe impacts are likely to occur and when the defoliating populations are present. It is estimated that about 10% of the woodlot volume is moderately susceptible to mortality caused by the gypsy moth. Aerial application of pesticides may be considered as a control activity, as may silvicultural treatments. In advance of a gypsy moth infestation, silvicultural activities can be used to decrease the susceptibility to defoliation and to strengthen the stand against tree mortality. Thinning and improvement cuttings can increase the vigor of residual trees by increasing both crown and root growing space. Healthy, vigorous trees are more likely to survive and recover from gypsy moth defoliation and to resist attack by secondary organisms.

While most of the woodlot is accessible at the present time, several new roads will need to be constructed on the property during the scheduled harvest operations. These should be constructed in strategic locations to provide access to the majority of stands for future forest management work. Properly maintained access roads will also serve as excellent firebreaks and fire access roads, should the need for fire suppression develop in the future. These roads will also provide opportunities for recreation opportunities such as hunting, hiking, and horseback riding. These activities were considered some of the primary goals for the woodlot. Finally, deer are a serious problem in the Poole and Pianta woodlot, and it is recommended that hunting pressure increase to prevent these populations from getting any larger.

REFERENCES

National Agricultural Pest Information System, 2014. Survey status of gypsy moth (European)—*Lymantria dispar* (2011 to present). Purdue University, West Lafayette, IN. http://pest.ceris.purdue.edu/map.php?code=ITAXAIA&year=3year (Accessed February 2, 2014).

Richardson, R., 2013. Management recommendations, Denise Poole and Lisa Pianta. Appalachian Investments, Arnoldsburg, WV. 22 p.

U.S. Department of Agriculture, Natural Resources Conservation Service, 2014. Wildlife Habitat Incentive Program. U.S. Department of Agriculture, Natural Resources Conservation Service, Washington, DC. http://www.nrcs.usda.gov/wps/portal/nrcs/detail/national/programs/financial/whip/?&cid=nrcs143_008423 (Accessed July 16, 2014).

West Virginia Division of Environmental Protection, 2000. West Virginia's silvicultural non-point source management plan. West Virginia Division of Environmental Protection, Office of Water Resources Nonpoint Source Program, Charleston, WV, 176 p.

West Virginia Division of Forestry, 2005. West Virginia best management practices for controlling soil erosion and sedimentation from logging operations. West Virginia Division of Forestry, Charleston, WV. 31 p.

West Virginia Division of Forestry, 2009. Managed timberlands fact sheet. West Virginia Division of Forestry, Charleston, WV. 2 p.

Chapter 6

Rib Lake School Forest, Wisconsin, United States of America

Scott Mueller,[1] Gretchen Marshall,[2] Kirsten Held[1] and Jeremy Solin[3]

[1]*Wisconsin Department of Natural Resources, Madison, Wisconsin, USA,* [2]*LEAF Program, University of Wisconsin-Stevens Point, Wisconsin Center for Environmental Education, Wisconsin, USA,* [3]*University of Wisconsin-Stevens Point, Wisconsin Center for Environmental Education, Wisconsin, USA*

ABBREVIATIONS

K-12   Kindergarten Through 12th Grade
LEAF   Wisconsin K-12 Forestry Education Program (formerly Learning, Experiences & Activities in Forestry)

MANAGEMENT SETTING AND BACKGROUND

A school forest is an outdoor classroom. Officially, a school forest is a specialized type of community forest in which land is owned or controlled by a public or private school and used for environmental education and natural resource management. Wisconsin state statutes give school districts the authority to obtain school forest land and register it through the state community forest program. In 1928, Wisconsin established the first school forest program in the nation. In the 1920s, much of northern Wisconsin had been exposed to over-harvesting and forest fires. Even though the cutting and burning cleared the land for would-be farmers, it was too rocky and too far north to be suitable for farming. Abandoned farmlands became tax-delinquent. Any bright spot in the economy of northern Wisconsin depended either on the slow, natural forest regrowth or an aggressive reforestation program. Wakelin "Ranger Mac" McNeel, an early school forest visionary and state 4-H leader in the 1920s, sent students and teachers out across the state to reclaim cut-over, burned-over land with shovels and seedlings. And so, through sweat and dedication, Wisconsin school children became conservation stewards, or caretakers, as they replanted a Wisconsin their children and grandchildren could be proud of.

The idea of school forests was not a new one. It was borrowed from Australia and introduced to Wisconsin in 1925 by the late Dean Russell of the University of Wisconsin College of Agriculture. While visiting Australia, Russell watched schoolchildren planting trees on public tracts of land as an educational project. He thought it would be an idea that could be put to practical use in his home state. By 1927, Russell spearheaded legislation that permitted school districts to own land for forestry programs. Motivated by this legislation, Wisconsin adopted the idea of school forests to promote an urgent reforestation program. Within the year, three tracts of land were donated or purchased for the first school forests in Wisconsin. They were dedicated in the spring of 1928. Legislation was passed in 1935 mandating that conservation education be taught in all high schools, vocational schools, and universities or colleges. School forests provided great outdoor classrooms for this type of education.

The School and Community Forest Law (Section 28.20, Wis. Stat.), enacted in 1947, allows schools, villages, cities, and towns to own land and practise forestry. To enroll as a school forest, districts must intend to actively manage their forest and to provide sustainable forestry education as a component of their educational programs. School districts must also have an approved forest management plan. No timber can be harvested on any school forests unless marked or designated by a Wisconsin Department of Natural Resources (DNR) forester. School forest statutes indicate that school forests are eligible to receive free planting stock from state forest nurseries and use the services of state foresters for forest management plans.

School forests are exceptional outdoor education sites that are available to integrate environmental education into school curriculums, provide experiential learning, meet state education standards, demonstrate sustainable natural resources

Forest Plans of North America. http://dx.doi.org/10.1016/B978-0-12-799936-4.00006-0

management, strengthen school-community relations, and provide income for education activities. Today, there are 412 school forests in Wisconsin that are owned or used by 238 schools or school districts with some schools having more than one school forest. Forest management plans are on file for 137 school forests, and school forest education plans have been approved for 85 schools (LEAF, 2014).

The Wisconsin School Forest Program serves as a resource for all school forests in the state. The program is coordinated by the Wisconsin School Forest Coordinator, who is a member of the Wisconsin kindergarten through 12th grade (K-12) forestry education program, or LEAF. The school forest program and the Wisconsin School Forest Coordinator position are the result of a partnership between the Wisconsin Department of Natural Resources—Division of Forestry and the Wisconsin Center for Environmental Education, which is a center of the College of Natural Resources and UW Extension—Cooperative Extension at the University of Wisconsin-Stevens Point.

Sustainable management of the school forest lands allows students within local school districts to benefit from educational opportunities that can be incorporated into their classroom lessons. It provides an opportunity for outdoor forestry education through active sustainable management of the school forest property.

Rib Lake is a rural community in Taylor County in north-central Wisconsin with a population of about 900 people. The Rib Lake School District covers an area of approximately 225 square miles (583 km²) and has about 470 students in its three schools: an elementary school, a middle school, and a high school (City-Data.com, 2013).

In 1945, the Rib Lake School began a search for a suitable parcel of land for a school forest. An 80 acre (ac) (32.4 hectare (ha)) tract was found nearby that belonged to Taylor County. Rib Lake School was given permission by Taylor County to use that tract as its school forest. The history of this parcel of land is not well documented. An aerial photo from 1938 shows that most of the parcel was forested, although the quality and age of the forest cannot be determined; and within the parcel there was one open field, which was likely farmed at one time. It is likely, given the history of northern Wisconsin, that this parcel of land had been logged during the "cutover." Much of the land controlled by Wisconsin counties was acquired through tax delinquency following the Great Depression.

Between 1946 and 1951, students of the Rib Lake School district planted 8,500 red pine (*Pinus resinosa*) and eastern white pine (*Pinus strobus*) seedlings in what was a former field in the school forest. As the Rib Lake School utilized this tract, it felt the need to purchase the tract from Taylor County. In 1955, the county sold 200 ac (81 ha) to the Rib Lake School, which is now the Rib Lake School Forest (Figure 6.1). Located in the north-central forest ecological landscape of Wisconsin, the forest has a mix of traditional northern hardwoods, including sugar maple (*Acer saccharum*), American basswood (*Tilia americana*), white ash (*Fraxinus americana*), northern red oak (*Quercus rubra*), red pine (Figure 6.2), aspen (*Populus* spp.); swamp hardwoods, primarily black ash (*Fraxinus nigra*); and lowland herbaceous vegetation with

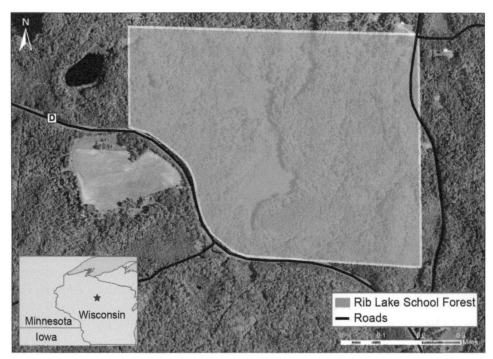

FIGURE 6.1 A map of the Rib Lake School Forest.

FIGURE 6.2 A road through a pine stand, in fall, within the Rib Lake School Forest.

lowland brush. The soils range from a very poorly drained organic to a well-drained loam. The topography is variable from relatively flat to hilly. Sheep Ranch Creek, a trout stream, flows through a portion of the school forest. The school forest is enrolled in the American Tree Farm System. The 2005 management plan identified five stand types (Table 6.1). A school forest committee consisting of teachers, an administrator, the DNR forester, and community members provides planning, oversight, and coordination of the school forest.

The Rib Lake School Forest is used extensively by the local community for recreational activities that include hiking, hunting, bird watching, snowshoeing, and cross country skiing. The Rib Lake Ski & Snowshoe Club actively manages the trails on the School Forest for recreational use. All trails in the Rib Lake School Forest are open to the public year-round for nonmotorized use.

PLANNING ENVIRONMENT AND METHODOLOGY

Forest Stewardship Plans for school forests follow the requirements indicated by the Wisconsin DNR's Private Forestry Handbook (Wisconsin Department of Natural Resources, 2013). Stewardship plans must include landowner objectives, a map of the property, soils information, stand information, silviculture objectives, threatened and endangered species information, archeological and historical structure information, and practices to be taken by the landowner to protect soil, water, recreation, timber, and fish and wildlife. These plans can be written by a Wisconsin DNR forester (at no charge), by a private forester (at owner's expense), or other interested parties (including students). Regardless of who writes the forest management plan, a Wisconsin DNR forester must approve it.

Stewardship plans must address Wisconsin's Forestry Best Management Practices (BMPs) for Water Quality (Wisconsin Department of Natural Resources, 2010). The BMPs are intended to provide methods for protecting and maintaining water quality before, during, and after forest management activities.

TABLE 6.1 Land Classes within the Rib Lake School Forest

Land Class	Acres	Hectares
Northern hardwoods	135	54.6
Lowland herbaceous vegetation with lowland brush	23	9.3
Red pine	7	2.8
Swamp hardwood	14	5.7
Aspen with northern hardwoods	21	8.5
Total	200	80.9

A Forest Stewardship Plan was written for the Rib Lake School Forest in 1980 and was later updated in 2005. Forest management plans are developed utilizing air photos to delineate the various stands prior to any field work. Field work consists of cruising each stand using a cruising stick, which includes a 10-factor angle gauge mounted on a Biltmore stick with a variable radius plot. Information that is collected includes current stand volume, species composition, age class distribution, and any unique stand characteristics (Figure 6.3). Each stand's cruise data is entered separately into a computer program, Kruzer (Wimme, 2003). The Kruzer program provides a detailed report for each stand. A stand age is obtained for any even-aged stand. Uneven-aged stands are not aged.

Objectives listed in the Rib Lake School Forest Stewardship Plan include:

1. Practise sustainable forestry.
2. Provide educational opportunities to the Rib Lake School District.
3. Provide periodic income to re-invest into the school forest.

Educational outcomes included in the Stewardship Plan include overviews of the following topics:

- Tree and shrub identification
- Timber harvesting techniques
- Invasive species identification and control measures
- Forest succession
- BMPs for water quality
- Basic ecology
- An understanding of natural versus artificial regeneration
- Tree measurements and tree measuring equipment
- Geology, landforms, and soil resources

Rib Lake also has an education plan developed for its forest that was written and adopted in 2008. The education plan describes how the school forest will be used for educational purposes. An education scope and sequence lays out the topics to be covered at each grade level and provides curricular recommendations for use by the teachers. The education plan connects to the Forest Stewardship Plan by indicating educational opportunities provided by forest management activities. The Rib Lake School Forest Education Plan identifies the school forest as a place that can enhance and enrich student, staff, and community learning. Learning that takes place at the school forest can help create environmentally responsible citizens and incorporates classroom work with hands-on experiences in the forest. The Rib Lake School Forest, as a working outdoor classroom, can be used to teach plant and animal identification, plant diseases, forest management, tree pruning, stream improvement, wildlife management, compass use and orienteering, elementary surveying, growth studies, and fire protection.

The School District of Rib Lake's Board of Education adopted a School Forest Resolution that states the management of the school forest properties shall be conducted in a manner that represents good stewardship of the resources and being good neighbors to adjacent landowners. The district administrator, Board of Education, and School Forest Committee are authorized within the resolution to make decisions regarding management of the forest. The resolution also states that income received from the sales of forest products, school forest properties, money received for rental of the school forest, and other revenues shall be placed in a segregated account and used to support development expenses and educational activities at the forest.

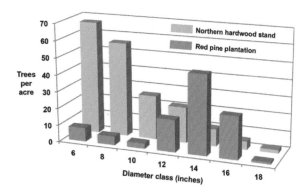

FIGURE 6.3 Diameter distribution of the red pine and northern hardwood stands within the Rib Lake School Forest.

An additional policy states that the school forest's primary use and goal is for environmental education. Recreational use by the public is encouraged if it does not disrupt the environment and is not in conflict with school use. Guidelines listed include:

1. The use of motorized recreational vehicles is not allowed.
2. Hunting is allowed, but is prohibited when school or rental groups are on the property.
3. Any trail clearance or modifications must be approved by the school forest committee.
4. All rules and regulations of the School District of Rib Lake apply at the School Forest.
5. The public is prohibited from harvesting timber and any other forest products.

OUTCOMES OF THE PLAN

In 2009, a timber sale was conducted on 50 ac (20 ha), with approximately 45,000 board feet (MBF) and 400 cords harvested (about 1,556 cubic meters (m³) in total). This harvest consisted primarily of a single tree selection in a portion of the northern hardwood stand and the thinning of a red pine plantation, which was planted by former students. All of Wisconsin's BMPs for water quality protection were followed, and the necessary water regulation permits were obtained for the sale. This included constructing an appropriate stream crossing for skidding. Another timber sale is tentatively scheduled for 2014–2015 on 30–40 ac (12–16 ha) of the forest.

The School Forest has been used consistently for educational purposes. During the timber sale in 2009, the sophomore class visited to observe and discuss the timber harvest. Field events were held in 2010 and 2012 for the Rib Lake middle school students. There were eight stations set up around the school forest featuring different outdoor educational themes. In 2012, an outdoor learning center was installed, including an outdoor amphitheater with a stage for the teacher and seating for up to 100 students. The center increases outdoor education opportunities for students and the community. Two times each year, approximately 230 Rib Lake elementary students visit the School Forest to hike, measure stream velocity, learn about their senses, study flora, explore animal habitats and adaptations, enhance writing and drawing skills, and participate in snowshoeing. A portion of teacher professional development time was spent at the school forest in 2013 to showcase the new learning station and encourage teacher exploration and use of this incredible learning environment.

DISCUSSION AND CONCLUSIONS

School forests are outdoor classrooms that integrate the twin goals of sustainable forest management and education. Forest management provides real-world learning opportunities for students to understand ecology, resource use and management, and resource decision making (among many other topics) to apply critical thinking skills and develop career interest. Engaging students and the community in forest management planning and implementation increases understanding of and support for management. Forest management plans that incorporate education and management objectives can be effective tools in accomplishing these goals. The Rib Lake School Forest provides an example of how this can be accomplished. However, the accomplishment of these twin goals can be difficult. Like all other forests, management can be complicated by diverse interests and understandings.

Learnings and Insights

The development and implementation of the Forest Stewardship Plan has enhanced the educational experience of the Rib Lake students. Students have been able to observe active management, and they have continued to learn about management concepts and outcomes through field experiences at the forest. The experiences have provided opportunities for students to develop an understanding and appreciation of forest management as well as to consider forestry-related careers. The school forest has created opportunities for intergenerational learning by bringing community members (e.g., DNR forester, biologists, and loggers) to the school forest to work with the students.

Like with most resource management, bringing people into the process early helps them to better understand management activities. Given the diverse interests connected to the Rib Lake School Forest (recreational users, students, teachers, school administrators, school board members, resource managers, and loggers) public outreach is essential to effective forest management. It can be difficult to find a balance between engaging and educating these diverse interests and moving forward within the constraints of the setting (e.g., seasonal restrictions, desire for income generation). Educating key stakeholders, in particular the school forest committee members, and involving them in forest management planning and implementation has been important to successful management. Multiple site visits were held with school forest committee members to educate them about the purpose of and approach to the forest management. In addition, the school forest policy that commits income generated from timber harvest and other revenue generating uses ensures ongoing management and educational utilization of the school forest.

Sustainability Issues

The sustainability issues for this parcel of land are less about the ecological health of the land and more about the social acceptance of, and support for, forest management. There is a need to continuously educate and engage stakeholders in the school forest so they understand the necessary silvicultural approaches being taken and the ability to meet the diverse interests and uses of the school forest. Given that school forests are public lands controlled by school districts, the management decisions are often being made by people (the school board and administration) with no forestry background. The support of the DNR forester is critical in managing these lands, and ongoing education and communication is necessary to ensure sustainable management.

Plan Development Challenges

The primary challenges of the plan development had to do with the diversity of interests involved in the school forests. Addressing educational opportunities, income generation potential to support the school forest program, and appropriate silvicultural practices took some thoughtful consideration. Ultimately, all interests were integrated into the plan.

Plan Implementation Challenges

Given the diverse interests in the management of the Rib Lake School Forest, it is not surprising that there were challenges in implementing the plan, although these were not substantial. As indicated earlier, there is considerable public recreation in the school forest. Most of the challenges with plan implementation came from esthetic concerns with timber harvesting. To address these concerns, buffers were maintained along primary trails. Harvesting was also planned to minimize conflict during high-use periods of time.

REFERENCES

City-Data.com, 2013. Rib Lake. Wisconsin. Advameg, Inc., Flossmoor, IL. http://www.city-data.com/city/Rib-Lake-Wisconsin.html (Accessed May 2, 2014).

LEAF, 2014. Wisconsin School Forest Program-2013 annual report. University of Wisconsin-Stevens Point, College of Natural Resources, Wisconsin Center for Environmental Education, Stevens Point, WI. http://www.uwsp.edu/cnr-ap/leaf/Documents/School%20Forest%20Annual%20Report%20 2013.pdf (Accessed May 2, 2014).

Wimme, C., 2003. Kruzer 2003, Version 1.6. Wisconsin Department of Natural Resources, Madison, WI.

Wisconsin Department of Natural Resources, 2010. Wisconsin's Forestry Best Management Practices for Water Quality. Wisconsin Department of Natural Resources, Division of Forestry, Madison, WI, FR-093. 162p.

Wisconsin Department of Natural Resources, 2013. Private Forestry Handbook 2470.5. Wisconsin Department of Natural Resources, Division of Forestry, Madison, WI. http://dnr.wi.gov/topic/ForestManagement/documents/24705.pdf (Accessed May 2, 2014).

Chapter 7

Shirley Town Forest, New Hampshire, United States of America

Peter Farrell

New England Forestry Consultants, Inc., Alton, New Hampshire, USA

MANAGEMENT SETTING AND BACKGROUND

The town of New Durham is a rural community located in south-central New Hampshire. The Shirley Town Forest (Figure 7.1), which totals 133.7 acres (ac) (54.1 hectares (ha)), was donated to the town in 1946 for the expressed use as timber management and conservation land. In 2011, after more than 15 years of inactivity, the town selectmen engaged New England Forestry Consultants, Inc. to prepare a forest management plan. Although previous forestry activity had occurred on the property, this plan was the first comprehensive evaluation supported by quantitative field data. The Shirley Town Forest is located in the Merrymeeting River watershed, which flows directly into Lake Winnipesaukee. The soils in this region are predominantly glaciated, being mixed tills with more clay on upper slopes and well-drained outwash soils in valley bottoms. This pattern is found on the Shirley forest, where the soil type varies from compacted tills on the upper slopes to pure sand and gravel at lower elevations.

The majority of the land in the New Durham region was cleared for settlement in the 1700s and 1800s. As agricultural activity declined in New Hampshire in the late 1800s and early 1900s, the fields were abandoned and land reverted to forest. The predominant forest species found on these former fields and pasture sites was eastern white pine (*Pinus strobus*). As the pine was subsequently harvested, the forest cover typically transitioned to mixed types with much higher percentages of hardwood species. Although this trend is evident on the Shirley Forest, the rate of transition was greatly slowed as a result of the forest stewardship activity by John Shirley, the previous owner and benefactor of the town. Current land uses are described as general forest (86.1 ac, 34.8 ha), which include timber management lands (84.3 ac, 34.1 ha) and upland forest reserves (<2.0 ac, <0.8 ha), nonforested wetlands (38.0 ac, 15.4 ha), fields (0.3 ac, 0.1 ha), and gravel pits (9.4 ac, 3.8 ha).

Although the area is predominantly forested (Figure 7.2), the landscape surrounding the property is well settled along the frontage of maintained town roads. There is relatively little industry in the region, and the local economy depends heavily on tourism and vacation-related trade. Many residents commute to work in more heavily developed southern and seacoast communities, and many retirees have chosen to live in this less densely settled town with the natural landscape it affords. These demographic factors are important to consider in planning activity on the town forest, particularly given the high visibility of the land from public roads, the town cemetery, and a public snowmobile and walking trail.

In 2011, the range of forest stand ages was determined to be 15–90 years, and the average timber volumes were 15.9 thousand board feet (MBF) per ac (15.7 cords per ac, 37.5 m^3 per ha). The approximate annual rate of growth was estimated to be about 2.5%, and the timber management land value was $2,400 per ac. No rare or endangered species were found to occur within, or adjacent to, the timber management area. Sawtimber volume is predominately composed of eastern white pine (85.2%), eastern hemlock (*Tsuga canadensis*, 9.7%), and red oak (*Quercus* spp., 4.0%). Pulpwood volume is predominately composed of eastern white pine (55.3%), eastern hemlock (29.9%), red oak (7.5%), and other hardwoods (7.4%). The contrast in sawtimber and pulp composition between species and species groups illustrates the lower quality of hemlock and other hardwoods.

Forest Plans of North America. http://dx.doi.org/10.1016/B978-0-12-799936-4.00007-2

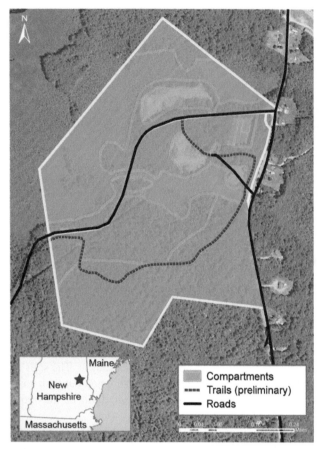

FIGURE 7.1 A map of the forest area management units, Shirley Town Forest, New Hampshire.

FIGURE 7.2 The landscape surrounding the Shirley Town Forest, within the Merrymeeting River Watershed.

PLANNING ENVIRONMENT AND METHODOLOGY

The management objectives for the Shirley Town Forest were stated in the owner's bequest to the town. His statements were:

1. Hold the land in perpetual trust, and periodically harvest wood and lumber as it reaches maturity.
2. Use care to prevent stripping the land of immature growth.
3. Forestry work should be done in accordance with the advice of the State Forestry Commission.

To this a statement was added that widened the focus of the management to include more than timber and which is compatible with current perspectives on forest stewardship.

> *In keeping with the spirit of the third objective, the work done on the forest will be conducted in a manner which is compliant with currently accepted standards regarding wildlife habitat, water quality protection, recreational use, aesthetics, and other values not specifically stated in the donors will.*

Other than the conditions set forth in the will of the donor, there were no formal or legal constraints on the management of the property. Informally, the greatest constraint is public attitudes that are resistant to forest management activities that involve harvesting or, more generally, change.

The development of the plan consisted of four phases. First, a solicitation of comments regarding the management of the land was requested by elected officials of the town (the selectmen and the Conservation Commission). These comments included concerns with planning and management costs, type of activities envisioned, potential income, and transparency in the conduct of forestry services. Next, field data was collected via a formal forest inventory. This included a confirmation of property boundaries using a Geographic Positioning System (GPS) and a Geographic Information System (GIS) and a variable plot timber cruise (20 basal area factor, 5×5 chain (100 meter (m) $\times$ 100 m) spacing of stations). Sample trees were measured by 2 inch (in) (5 centimeter (cm)) diameter classes, beginning with the 4 in (10.2 cm) diameter at breast height (dbh) class, and heights were recorded by product in 8 foot (ft) (2.43 m) segments. Wildlife trees, living and dead, were tallied, and a subjective observation of coarse woody debris was recorded. Furthermore, a subjective quantitative categorization of stand structure was made. Finally, a survey of the New Hampshire Natural Heritage Inventory database, made available through the statewide GIS clearinghouse (Complex Systems Research Center, 2014), was conducted in order to check for occurrences of rare and endangered species on the property as well as rare plant communities. The data and information was processed and analyzed, and a recommendation was formulated based on the findings. Finally, the plan and recommendations were presented at a public hearing. The public hearing resulted in questions but, finally, no objections to the proposed management schedule.

The regulatory environment in New Hampshire is highly compatible with the type of forest management activity encompassed in the stated management objectives. Forest management plans are required for properties enrolled in the stewardship category of the New Hampshire Current Use Assessment Program (New Hampshire Department of Revenue Administration, 2013), but otherwise no plan is required before harvesting can be conducted.

To get beyond the generalities, we have the advantage of using a comprehensive forest management practices reference that was produced by the joint effort of numerous individuals, agencies, and organizations. Titled *Good forestry in the Granite State: Recommended voluntary forest management practices for the State of New Hampshire* (Bennett, 2010), the publication fit the requirement of taking advice from the State Forestry Commission. The state's Best Management Practices (BMPs) for water quality (Moesswilde, 2005) are also compatible with the Clean Water Act (33 U.S.C. §§ 1251–1387).

The New Hampshire regulations that are relevant on the property are:

1. Wetland crossing regulations administered by the New Hampshire Department of Environmental Services (RSA 482-A:3, V). Temporary and permanent wetland and stream crossing must be mapped and meet minimum size criteria to qualify for an expedited notification rather than an in-depth permit process. All proposed activity fell within the notification classification.
2. Basal Area and Slash Laws (RSA 227-J:9 and J:10). Within the stands that are fronted on (adjacent to) town roads, the slash created by harvesting must be 50 ft (15.2 m) away from roads edge, and tops must be lopped to below 4 ft (1.2 m) within 150 ft (45.7 m) of the road. In the same zone, only 50% of the basal area may be harvested at a single time, unless prior approval is obtained by permit process from the New Hampshire Division of Forest and Lands.

There were no certifications related to the property at the time the plan was prepared, and the town has not expressed any interest in participation in any of the certification programs. However, American Tree Farm System and Forest Stewardship Council certification programs are readily available through the services provided by New England Forestry Consultants, Inc.

The town's organizational policies relate primarily to the process of obtaining services and for the selling of town assests (timber). These are generally bidding processes that are directly overseen by the town administrator, with all final contracts being signed by the selectmen.

OUTCOMES OF THE PLAN

Although the town had been entrusted with stewardship responsibilities for the land since 1946, it was 2011 before it engaged a consulting forester to prepare a comprehensive forest management plan. Previous plans and recommendations had been prepared or provided by state and extension foresters. With increasing budget and mission constraints within government agencies, the town was encouraged by these agencies to engage a privately employed forester for assistance with its management obligations. Forestry services, including timber sale administration and timber stand improvement, had been effectively and successfully provided by a private forester, but without a written plan in place. Anecdotal accounts of those activities, however successful, are often negative. By obtaining a plan that incorporated formal forest mensuration methods in its development, and clearly describes a long-term management strategy to meet the objectives, the town has created a document that can be used for rational decision making and as a baseline reference for evaluating performance. In short, the plan makes the selectmen's job of overseeing the property easier and helps fulfill their accountability obligations to the community.

For example, the issue of forest regeneration is often contentious, particularly in the public arena. *Selective cutting* is widely used by the public as a preferred method of harvesting; its perceived antithesis, *clear-cutting*, is generally scorned. Aside from the fact that the term *selective* can describe any range of forestry practice from the best to the worst, partial cutting of stands in New England almost inevitably results in the development of low-value understory regeneration. Left in place, these trees become the dominant timber type, degrading stand value and reducing multiple-use management options.

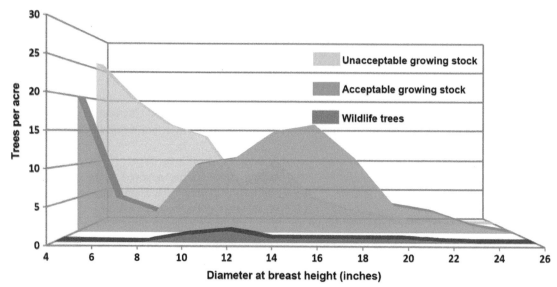

FIGURE 7.3 The development of a low-value understory stand in New Hampshire.

Within the plan, we were able to first indentify the problem of low-value species and trees in smaller diameter classes, and second, we provided a rational model for establishing an appropriate acreage of regeneration cutting (clear-cuts), however small. Figure 7.3, which was created from the data collected, simply and clearly shows the troubling condition of too many lower value trees in the smaller diameter classes. Acceptable growing stock includes stands of trees with sufficient vigor, quality, and/or species that have the potential to produce sawtimber. Unacceptable growing stock includes stands of trees of insufficient vigor, quality, and/or species to have the potential to produce sawtimber. This information acts as a powerful tool in making the argument for investment in silvicultural activity, despite the reduction in income. The chart shows the development of a low-value understory stand, and the need for vigilant protection of forest quality, particularly in precommercial harvest-diameter classes.

In presenting an opinion on when and how much silvicultural activity to undertake, particularly regeneration harvesting, it is useful to have a model as a reference. In the consultant's company management plan system, an *Area Regulation* guide (Table 7.1) was developed that allows the forester to present the case for various timber management activities and then compare the model to acreages of recommended, or actual, on-the-ground work. This tool, used in combination with timber inventory information, forms the basis for the recommendations for sustainable harvesting.

TABLE 7.1 Area Regulation Guidelines Table[a]

Timber Mgmnt Acres	Rotation Age	Operating or Age Class Interval	Planning Horizon
84.3	120	20	15
SUGGESTED TREATMENT COVERAGE GOALS BY AGE CLASS			
Ending Age	**Treatment Type**	**% of acres**	**# of acres**
20	Silvicultural investment—Release seedl/sap	12.5%	10.5
40	Silvicultural investment—PC thinning		
60	Intermediate harvest—1st thinning		
80	Intermediate harvest—2nd thinning		
100	Intermediate harvest—1st shelterwood		
120	Regeneration harvest—2nd shelterwood		

[a]As presented in the output of the Forest Pro Landbase© program.

Finally, the plan resulted in a valuation of the timber and its potential to economically support the ownership of the land, including its recreational and aesthetic values. No matter what surveys show, private landowners and New England towns are extremely cost-conscious owners of forest assets—the money does matter. Having an accurate financial estimate of total value, projected incomes, and costs greatly enhanced the town's ability to move forward with decisions regarding harvesting, silvicultural investment, and property maintenance.

The plan for the Shirley Town Forest was completed in 2011, but due to market conditions the first harvest was not prepared and sold until the fall of 2013. Harvesting commenced in the spring of 2014 and a combination of intermediate and regeneration harvesting was conducted on about 35 ac (14.2 ha).

DISCUSSION AND CONCLUSIONS

The preparation of a forest plan for the Shirley Town Forest in New Durham, New Hampshire required developing an understanding of the goals and constraints of the town's leaders, an assessment of the land and forest resources, an analysis of alternatives, and agreement on future direction.

Learnings and Insights

After 35 years of nonindustrial forest land management, it is hard to find anything approaching totally *new* while preparing a forest management plan. The challenges on these land holdings are not technical, but are social and economic. First, there is communicating the principle of forestry to a lay audience that has many preconceived ideas as to what is good and what is bad. Foresters have to balance simplifying the critical discussions without dumbing the information down to meaningless rhetoric. The discussion has to be detailed enough to be credible, but concise and clear enough to be digested in the limited time frame the client will allow. Furthermore, the plan has to be prepared within very limited budgets. The economic constraints often lead to the temptation of using a boilerplate model, which also has a high potential for undermining the goal of winning the audience. These challenges are faced with nearly every small-ownership client, but when the client includes a board of selectmen, a Conservation Commission, and an electorate, the scrutiny and challenges increase proportionally.

Sustainability Issues

The most common threat to sustainability in this region is the conversion of land to nonforest uses. Given the legal restrictions on the use of the town forest property, this issue has been taken off the table. The second most important challenge in sustainable forest management is having the data to keep a pulse on the forest condition and value. Obtaining this data requires periodic investments in woodland exams and inventories. The New England tradition of frugality is alive and well in New Durham, and selling the town selectmen on such services may be challenging as a result. Furthermore, time spans between data collection events of 5, 10, or 15 years is accompanied by changes in the elected selectmen who oversee the care of the forest, resulting in the need to resell these management concepts and responsibilities. Similarly, silvicultural investments needed to maintain forest quality are also difficult to sell. To meet these challenges, proposed schedules of work are developed that emphasize actions at shorter intervals and investments that are concurrent with income. This strategy keeps management concepts in near-term memory and takes some of the pinch out of the expenditures.

Plan Development Challenges

To meet these and other challenges, New England Forestry Consultants, Inc. has developed a forest management planning tool called Forest Pro Landbase©. This tool is a standardized forest management template built on a database chassis. As discussed above, the system include models for area and volume calculations related to sustainability, as well as a wildlife habitat model based on the work of DeGraaf et al. (2006). The database architecture allows for rapid reporting and searching of all plan information, maintaining forest histories, and prompting clients to notify them of recommended activities. Combined with cruising tools that easily allow for the creation of clear charts and tables, the forestry consulting company was able to overcome the traditional burden of handling data and focus more on delivering a meaningful discussion of the results.

Plan Implementation Challenges

As mentioned earlier, implementation of the plan was delayed for about a year due to changes in market conditions. In the fall of 2013, with an improvement in the white pine sawlog market, the selectmen were encouraged to undertake a harvest on roughly half of the forested area. This timber sale was put out to bid, conditional upon bucking at the stump (cut-to-length), and on a per-unit sale basis (mill scale). The harvest was successfully completed in the spring of 2014, meeting silvicultural, income, and aesthetic expectations.

Other Interesting Issues Related to the Plan

Overall, the plan was well received, and the subsequent harvesting has generated more curiosity and, so far, no objections. Limiting the harvest to about half the area of the forest and using a low-impact harvesting technology contributed to this favorable outcome.

REFERENCES

Bennett, K.P. (Ed.), 2010. Good forestry in the Granite State: Recommended voluntary forest management practices for New Hampshire. second ed. University of New Hampshire Cooperative Extension, Durham, NH. http://extension.unh.edu/goodforestry/toc.htm (Accessed March 24, 2014).

Complex Systems Research Center, 2014. New Hampshire's statewide GIS clearinghouse. Complex Systems Research Center, Institute for the Study of Earth, University of New Hampshire, Durham, NH. http://www.granit.unh.edu (Accessed March 24, 2014).

DeGraaf, R.M., Yamasaki, M., Leak, W.B., Lester, A.M., 2006. Technical guide to forest wildlife habitat management in New England. University Press of New England, London, NH.

Moesswilde, M., 2005. Best management practices for forestry: Protecting New Hampshire's water quality. University of New Hampshire Cooperative Extension, Durham, NH. http://extension.unh.edu/resources/files/Resource000248_Rep267.pdf (Accessed March 24, 2014).

New Hampshire Department of Revenue Administration, 2013. State of New Hampshire current use criteria booklet for April 1, 2012 to March 31, 2013. New Hampshire Department of Revenue Administration, Concord, NH.

Chapter 8

Arcata Community Forest, California, United States of America

Mark S. Andre

City of Arcata, Arcata, California, USA

MANAGEMENT SETTING AND BACKGROUND

The Arcata Community Forest (Figure 8.1) consists of 2,311 acres (ac) (935 ha) of mostly second-growth redwood (*Sequoia sempervirens*) forest (Figure 8.2) located near Humboldt Bay in north-coastal California. The city also owns more than 600 ac (243 ha) of wetland and riparian lowlands that are within the same small costal watersheds. Other conifer species found on the forest include Douglas-fir (*Pseudotsuga menziesii*), grand fir (*Abies grandis*), western hemlock (*Tsuga heterophylla*), western red cedar (*Thuja plicata*), and Sitka spruce (*Picea sitchensis*). The forest is a recreational attraction for the region due its extensive trail system that supports hiking, mountain biking, and horseback riding. Arcata is home to Humboldt State University and its School of Natural Resources. For many of these young people, the years that they spend at the university are formative ones, as they achieve or expand their ecological awareness. The Arcata Community Forest facilitates opportunities for these students to connect with a small but perhaps influential forestry program.

The forest forms the headwaters for five salmonid streams that flow to Humboldt Bay via state, federal, and local wildlife areas. The quality of those areas is influenced by management activities in the Community Forest. Revenue from timber harvests has been used to purchase wetlands, creek side conservation easements, and parkland that have benefited the Humboldt Bay area ecosystem and recreational uses around Humboldt Bay.

Prior to public ownership of the forest, lands within the Arcata Community Forest were claimed through land patents in the 1860s. Land patents are the legal documents that transferred land ownership from the U.S. government to individuals. Most of the Community Forest was logged during the 1880s. Trees were felled with axes, wedges, and crosscut saws. Large trees with defects and many smaller diameter trees were left following logging although they were usually consumed in the slash fires, which regularly occurred. Oxen teams were once used to skid massive redwood logs to Humboldt Bay. Most of the trees that remain today were naturally regenerated from the stumps and seeds of the original first-growth forest. Following the logging of this area, the Community Forest was used for grazing and for water supply. It was not until the 1930–1940s that the citizens of Arcata gained title to the Community Forest property for the purpose of providing water supplies to the town (Van Kirk, 1985). The Union Water Company collected and conveyed water from the Community Forest to Arcata's residents until 1963, when the Raney wells were constructed on the nearby Mad River.

The Community Forest was dedicated in 1955 as the first municipally owned forest in California, and was to be "managed for the benefit of all the citizens of the city, with attention to watershed, recreation, timber management and other values" (Humboldt Times, May 15, 1955). During the 1960s, much of the Community Forest was selectively logged. At that time, second-growth redwood was not a desired species and they were left in favor of Douglas-fir, grand fir, and Sitka spruce. The result of that logging episode was to create a simplified system of homogenous even-aged redwood stands (Table 8.1). Following voter approval of the Forest Management and Parkland Bond initiative of 1979, the Arcata Community Forest Multiple-Use Management plan was adopted and, subsequently, the plan was updated in 1994 and again in 2013. The plan directed the City of Arcata to manage the forest using ecological principles, with a portion of the net revenue to be used for parkland acquisitions and development. The Community Forest was the first municipal forest certified in the United States under the Forest Stewardship Council (FSC). Since 2003, the city has added more than 1,114 additional ac (451 ha) to the Community Forest. Most of the new acquisitions were funded from state, federal, and foundation grants that were leveraged with a portion of timber harvest revenue from the city's timber program.

Forest Plans of North America. http://dx.doi.org/10.1016/B978-0-12-799936-4.00008-4

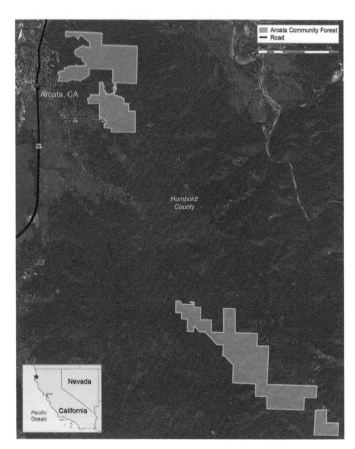

FIGURE 8.1   Arcata Community Forest lands.

FIGURE 8.2   Redwood forest within the Arcata Community Forest.

TABLE 8.1 Area (ac) of Forest Type by Structural Groups, Arcata Community Forest, California, United States of America

| Forest Type | Structural Group | | | | | |
	Open	Young (25–30 years)	Mature (41–70 years)	Older Mature (70–130 years)	Old Growth (150 years+)	Total
Redwood type		300	900	1,020	18	2,238
Riparian			10	20		30
Park meadow, roads, powerlines, rock pits	43					43
Total	43	300	910	1,040	18	2,311

PLANNING ENVIRONMENT AND METHODOLOGY

The current forest management plan was developed by Environmental Services staff with technical guidance provided by the Arcata Forest Management Committee. The operation of the Community Forest is tied to the approved forest management plans and a 1979 voter-approved initiative to manage the forests using "ecological principles." Much of the city's parks and open space lands were purchased under the park bond funded by timber harvest revenues. The 1979 management plan was updated in 1994, and that effort was greatly influenced by the federal Northwest Forest Plan, which was also adopted in 1994. This plan called for an extensive system of late-successional and riparian reserves along with some timber harvest on the intervening lands under a set of controls and safeguards. The current city forest management plan (2013) is an updated version of the 1994 plan framework and includes additional lands acquired since that date.

Currently, the Community Forest generates annual revenue of approximately $350,000–$400,000 U.S. dollars, which is more than is needed to be a self-supporting enterprise. No tax revenues are used for the forest management activity. Excess net revenue is used to purchase and maintain other city parkland and open space. Several parks and open space areas have been purchased with timber harvest revenue, including the city's main community park. The city pays timber yield tax to the state on timber harvested even though it does not pay property tax.

The forest is being managed to maximize habitat diversity, with an emphasis to move the forest toward an old-growth conditions. Management priorities include watershed, wildlife habitat, recreation, carbon sequestration, and timber harvest revenue. Approximately 30% of the land base is situated in reserves. The maximum allowable annual harvest is one-half of the annual growth increment on the "working landscape" portion (excluding the reserves); therefore, the forests are accruing volume and age over time.

The city directs management to be tiered to three elements of community-based forestry: ecological, social, and economic. The social component promotes engagement of all members of the community and builds local relationships of trust and reciprocity among diverse (and sometimes opposing) groups. It also enhances community knowledge and the skills necessary for planning and implementing sustainable forestry practices. The goal is not only more resilient forest ecosystems, but also more resilient communities, better equipped to respond to both challenges and opportunities. The ecological component involves the community in enhancing and restoring forested ecosystems, builds on local knowledge, and practises management and protection for the full range of social, ecological, and economic values. The economic strategy builds and sustains livelihoods based on natural resources. It often involves fostering small-scale, value-adding enterprises for timber. Economic benefits are often invested in the local community.

Guiding Laws, Regulations, and Policies

The Community Forest is owned by the city and managed by the city's Environmental Services Department. A volunteer Forest Management Committee advises staff and the city council on forest policy matters. The committee consists of seven members with backgrounds and expertise in botany, forest ecology, wildlife, fisheries, geology, recreation, and forestry. All committee meetings and field trips are public meetings, whereby the public is encouraged to attend and participate. The average tenure for committee members is 18 years. The committee members are respected natural resources professionals in the community, providing credibility to the city's forestry program. Current members include a retired U.S. Forest Service soil scientist, a geologist, fisheries biologist, forester, wildlife biologist from private sector firms, a University of California Extension forester and an engineering technician/watershed restorationist from the state of California. The forest management plan goals are

1. Maintain the health of the forest system, specifically, maintain the integrity of the watershed, wildlife, fisheries and plant resources, their relationships and the process through which they interact with their environment
2. Produce marketable forest products and income to the city in perpetuity, balancing timber harvest and growth
3. Provide forest recreational opportunities for the community
4. Serve as a model of managed redwood forests for demonstration purposes.

The goals were refined into a mission statement that was adopted following a public "visioning process." The Community Forest is managed whereby:

- Biological and physical elements of the forests, specifically wildlife, aquatic and plant species, plant and animal communities, and watershed processes are maintained
- Forest stewardship, including timber harvest, maintains forest integrity while generating public benefits
- Forest stewardship is fully supported by the community
- Community and visitors enjoy the forest setting and recreate in a respectful manner

- Public land ownership extends to include watersheds and headwater areas as well as corridors to neighboring communities
- Forests serve as outdoor laboratories for local schools and the university; research and other academic studies are fostered

Arcata has adopted the definition of its community-based forestry model as stated by The Aspen Institute's report on community-based forestry:

Community-based forestry (CBF) is a participatory approach to forest management that strengthens communities' capacity to build vibrant local economies while protecting and enhancing their local forest ecosystems. By integrating ecological, social, and economic components into cohesive approaches to forestry issues, community-based approaches give local residents both the opportunity and the responsibility to manage their natural resources effectively and to enjoy the benefits of that responsibility.

Wycoff-Baird (2005)

Elements of the management plan include restoring forested ecosystems and defining sustainability, building trust (social license) and relationships among diverse groups, and reinvesting in the community, as many livelihoods are based upon natural resources. Key features of Arcata's forest management program suggest that the community benefits from resources conserved, that community members support a conservation ethics and take pride in managing land for future generations, and that diverse viewpoints are respected and considered in the forest management planning process. A combination of working forests, special management areas, and ecological reserves creates a balanced approach to the management of the forest.

The plan developed for the Community Forest is consistent with the state of California Forest Practice Regulations that govern timber harvest on private lands under Title 14, California Code of Regulations. These regulations are used to implement the Forest Practice Act. They dictate minimum stocking levels and actions that govern the conduct of timber operations in the field. These regulations also determine that, for an entity the size of Arcata, the appropriate regulatory management plan for timber harvest and forest planning activities is a nonindustrial timber management plan. This type of plan is the formal environmental review document that must be prepared by a registered professional forester and approved by the California Department of Forestry and Fire Protection State and Federal Policies. Further applicable regulations include the California Timberland Productivity Act of 1982 (Gov. C. 51100 et seq.), the intent of which is to "encourage investment in timberlands based on reasonable expectation of harvest," and to "discourage premature or unnecessary conversion of timberland to urban and other uses." Other applicable regulations that govern activities within the forest relative to habitat and watercourse protection include the Porter-Cologne Water Quality Control Act (California Water Code, Division 7), the federal 1972 Clean Water Act, the California Endangered Species Act (Fish & G.C. 2050 et seq.), and the federal Endangered Species Act of 1973 (16 U.S.C. 1531-1544, 87 Stat. 884).

Forest Types and Current Silvicultural Practices

The Arcata forest is Site Class II (site index 167, 50-year base age) and Site Class III redwood forest type. Conifer species include redwood, Douglas-fir western hemlock, Sitka spruce, western red cedar, and grand fir (Figure 8.3). Inland portions support tanoak (*Lithocarpus densiflorus*), madrone (*Arbutus menziesii*), California bay (*Umbellularia californica*), and bigleaf maple (*Acer macrophyllum*). Red alder (*Alnus rubra*) is a common hardwood species found within the riparian zones. The primary natural disturbance in redwood forests, unlike mixed-conifer and most other California forests, is neither from fire nor insect damage, but rather blowdown from wind events. Other threats to the ecological integrity of the Arcata Community Forest include urbanization on the forest edge, invasive plants, and potential severing of ecological corridors that link the Community Forest to other intact forest areas to the east.

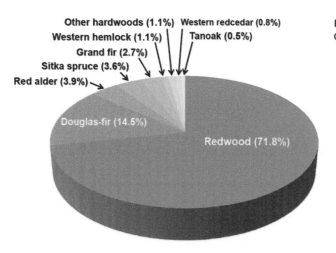

FIGURE 8.3 Current species composition by basal area, Arcata Community Forest.

The current merchantable inventory on the forest is 74 million board feet (MMBF) (174,640 m³) of wood. The projected inventory for 2025 is 94 MMBF (221,840 m³). The annual growth rate currently is about 1,000–3,000 board feet per ac (5.8–17.5 m³/ha) per year. The typical standing volume is 90–130 thousand board feet (MBF) per ac (525–758 m³/ha). Harvest levels are currently estimated to be 500–700 MBF (1,180–1,652 m³) per year. Reserve areas occupy about 35% of the land base.

Using tree spacing as a reference indicator, forests with tree densities of 120–200 trees per ac (297–494 trees per ha) would need to be slowly thinned over time to achieve a relative tree density of 20–35 trees per ac (49–86 trees per ha) as found on nearby old-growth stands. This could occur naturally through competition and mortality or be stimulated via mechanical thinning. Second-growth redwood has the ability to dramatically increase basal growth as a response to thinning (Jameson et al., 2005).

The Community Forest hosts an array of permanent continuous forest inventory ecosystem monitoring plots that were installed in 1985. The 0.2 ac (0.08 ha) plots are remeasured every 5 years. Measured ecosystem components include down logs, snags, seedling count, tree diameter, height, decay class, soil condition, live and dead carbon pools, lichens, fungi and bryophytes, wildlife use, and understory vascular plants. Plots in the old-growth reference stands are contrasted with the data in the Community Forest as part of a monitoring strategy.

Fortunately, there are reference stands of old-growth redwood in the region that can serve as blueprints for Arcata's Community Forest management trajectory. Many of the missing ecological qualities in the Community Forest can be found in nearby Redwood National and State Parks. Recruiting the structural elements commonly found in older forests is recognized as an important management objective in younger forest stands to address issues of biological diversity and forest integrity. Forest structure refers to species composition and the physical arrangement of trees, including the sizes, ages, spatial arrangement of gaps, and the sizes, heights, spatial arrangement of trees, snags, and dead and down material.

In an old-growth forest, natural disturbances (landslides, fire, and wind) create and maintain gaps in the canopy. The gaps, allowing light to reach the ground, give young seedlings and saplings the chance to grow and provide variety in the age and physical structure of a forest's trees. Thinning and group or "gap" cuts in a second-growth forests are an attempt to mimic natural disturbance. It relieves the forest's unnatural, uniform growth created by the initial clear-cut operation. Single-tree selection with a focus of thinning from below, and group selection harvests with green-tree retention are the main disturbance regimes used in the Community Forest. There has been a blend of restorative treatments that are considered adaptive management practices, in that city staff attempt to learn from new information and indicators when designing new interventions. The forest is a shifting mosaic of patches, thinnings, and gaps with the goal of allowing tree ages to exceed 100–300 years. The development of a "layering" effect in the tree canopy is one indicator of the shift to an uneven-aged condition. Efforts to increase species diversity have also included under-planting of shade-tolerant conifer species using transplanted stock from stands close in proximity to the Community Forest.

Other silvicultural methods are employed in order to increase stand complexity, and by structural objective, they include the following:

1. *Multi-layered canopies*: Modified single-tree selection timber harvests, release of advanced regeneration, establishment of new redwood tree cohorts, planting of shade-tolerant species (western hemlock, grand fir, etc.) in areas where they were eliminated years ago, and retention of some trees with complicated resprouted or reiterated tops.
2. *Elevated large snag densities*: Girdling of selected intermediate to codominate trees, usually Sitka spruce, grand fir, and Douglas-fir.
3. *Elevated downed woody debris densities and volumes*: Felling of trees and retention of large broken pieces to create large downed log material, and augmentation of debris by hauling in cull logs from nearby land clearing operations.
4. *Variable retention density harvests*: Thinning from below and harvest of stump sprout clusters with retention of dominant sprout trees, variable density harvests, reserve areas along watercourses, and modified group selection harvests (less than 2 ac (0.8 ha)) in size.
5. *Activities to re-allocate basal area to larger diameter classes*: High target basal area (96–157 ft² per ac (22–36 m²/ha)).

OUTCOMES OF THE PLAN

The updated plan is fundamentally designed to restore and move a relatively even-aged forest to a more structurally complex forest. The ultimate goal is to develop late-seral or old-growth forest characteristics. Tangible outcomes of management include:

- Accelerating the transition to an old forest stand structure through selective thinnings that promote light in the forest understory and stimulate recruitment of new tree age class
- Obtaining support from the community for management that includes timber harvests in close proximity to residential areas and recreational use areas

- Protecting and enhancing biological diversity and rare species, including maintenance of northern spotted owl (*Strix occidentalis caurina*) nesting pairs
- Contributing to the local economy by providing a source of wood products and jobs in the woods
- Providing an opportunity for residents to be involved in forest planning, as well as on-the-ground activities, with volunteer work days that amount to 5,000 volunteer hours per year
- Providing opportunities for nonmotorized recreation and contributing to the local tourism economy
- Testing different silvicultural practices and "no-cut" watercourse protection zones to protect and enhance water quality, as well as providing a network of connectivity of older seral forest habitat for species that require those conditions

DISCUSSION AND CONCLUSIONS

The Arcata Community Forest is a significant community asset. It is the responsibility for the current generation to convey this asset and pass it on to future generations as a healthy ecosystem. This makes it essential for citizens to understand what a healthy ecosystem is and why it is important. The role of local government as managers of such a community asset is to foster reciprocal relationships between forest and people.

The success of the ecosystem approach in managing Arcata's public forests depends on the community's interest and involvement as well as a degree of ecological awareness and understanding by the citizenry and elected officials. For the past 34 years, Arcata has been fortunate to have a high level of ecological knowledge and environmental ethics among its population and elected leaders. Future success will be measured in the ability to assist in the recovery of the forest structure and in the changes in composition and ecological processes that are necessary to more closely approximate reference conditions. Surrogate old-growth species such as the northern spotted owl, fisher (*Martes pennanti*), northern flying squirrel (*Glaucomys sabrinus*), and red tree vole (*Arborimus longicaudus*) are examples of indicators used in Arcata to demonstrate the positive trajectory.

The selective harvesting regime implemented over the past 34 years has visibly altered the forest. Larger trees that are more widely spaced and thus situated in a *park-like* condition, have more than anything else led to continued public support for the forest management program. By using a portion of the net timber revenue to leverage other funding sources, thus allowing the purchase of additional forest land and to enable the city to protect and restore urban streams and wetlands, tangible results of actively managing this public resource have been realized by the members of the public.

Learnings and Insights

Arcata's city forestry program is designed to document and share information with other landowners in the region as well as to provide assistance to other emerging community forest programs. The involvement of Humboldt State University faculty and students into the program has leveraged the city's ability to provide a functional demonstration forest for other landowners in the region. Additionally, other emerging community-based forestry programs have looked closely at Arcata's program as they attempt to develop a similar management approach.

Sustainability Issues

Sustainable harvest levels are a goal of Arcata's forest management plan, and it is a continuing challenge to adapt to new information from forestry research as well as to feedback from plan monitoring processes to ensure this goal is being achieved. By remeasuring permanent plots that include timber growth and other ecological habitat parameters, as well as by pursuing a new emphasis on managing for carbon sequestration, sustainability benchmarks will be reviewed so that corrections can be made as necessary.

Plan Development Challenges

In the development of the management plan, the city staff and the Forest Management Committee planners needed to set a high standard for public transparency and scientific rigor and to be consistent with state forest practice regulations as well as the federal Endangered Species Act. Development of a meaningful and cost-effective monitoring program on which to base future adaptive decision making was a particular challenge. The integration of third-party FSC certification and regular audits as part of the management regime helped to refine the plan and to set up a system for making periodic management (ecological and social) adjustments.

Plan Implementation Challenges

The Community Forest activity is funded almost exclusively by revenue derived from commercial timber harvesting. Maintaining sufficient funds for proper management of the forest is a challenge for the city during poor market conditions for timber. This can be minimized somewhat by attempting to time timber harvests with market peaks and by developing a reserve account for the forest that cannot be used for other purposes. Additionally, the regulatory environment for forestry and restoration in California is a particularly burdensome process even for management systems such as Arcata's relatively light touch harvest regime. More often than not, responding to a multi-agency permitting process diverts resources that may otherwise be directed toward restoration, recreation, or monitoring efforts. Many people involved in forestry in California, from industrial timberland managers to environmental groups, recognize that this problem is especially onerous on small-forest landowners and community-based forestry programs. Within the past 5 years, a market premium has been observed for larger diameter redwood logs, which has benefited Arcata economically. In addition, managing for larger logs has decreased logging and trucking costs. This has been especially important given the average haul distances to some of the regional sawmills.

Collaboration with conservation biologists and ecologists must be continually nurtured in order to provide information and maintain credibility to the restoration program. This can best be accomplished through a monitoring program that has clear goals and indicators of success. Given the small area of the land base, managing for species that require large landscapes is difficult. As urbanization continues to be a threat around the plan area, the city is actively engaged with industrial landowners, local land trusts, and the county of Humboldt to minimize parcelization or fragmentation so that the Community Forest is buffered from noncompatible uses such as residential uses.

Climate change also presents a particular challenge that requires a long-term approach to prepare the forest ecosystems to be as resilient as possible to stress caused by droughts, severe storms, and changes in species compositions.

Explaining to the public the complexity of forest management, especially using a tools such has a timber harvests to mimic episodic disturbances, is a continuing challenge that requires educational outreach. For example, our goal of thinning to accelerate the development of old-growth characteristics that have been simplified by past management is a difficult concept to explain to people, especially those who may have recently moved to the redwood region. There has been some skepticism that logging is a justifiable tool that can actually benefit the forest ecosystem. Involvement of community volunteers is a fundamental part of community-based forestry. Volunteer stewards engaged with land managers helps develop a constituency that will be better informed and prepared to accept the responsibilities associated with the privilege of owning a community forest.

The Forest Management Committee, whose average tenure has been 18 years, provides an important access point for public involvement. The members are well known and respected in the community, and this has helped the program weather political and economic shifts over the years. Certification by the FSC, annual audits, and transparent third-party monitoring of the forestry operations have also given the public and the city council a level of confidence that the forest resources are being managed in accordance with ecological principles.

ADDITIONAL READING AND RESOURCES

This chapter represents a synthesis of the 1994 Land and Resource Management Plan and subsequent amendments, and the 1999 nonindustrial timber management plan for the Arcata Forest in Arcata, California. To view the plans themselves, please visit this Internet site, which was available on May 13, 2014:

http://www.cityofarcata.org/departments/environmental-services/city-forests/forest-management-plan

If in the future the link to the site appears broken, search the Internet using the title of the plan and the keywords provided.

REFERENCES

Jameson, M.J., Reuter, E., Robards, T.A., 2005. Redwood and Douglas-fir leave trees Using Variable Retention. California Department of Forestry & Fire Protection, Sacramento, CA, California Forestry Note No. 119. 6 p.

Van Kirk, S., 1985. A history of the Arcata Community Forest. Unpublished Research paper.

Wycoff-Baird, B., 2005. Growth Rings: Communities and Trees. The Aspen Institute, Washington, DC. 225 p.

Chapter 9

Ejido Borbollones, Durango, Mexico

J. Javier Corral-Rivas[1], J. Ciro Hernández-Díaz[1], Carlos Antonio López Sánchez[1],
José Encarnación Luján Soto[2] and Klaus von Gadow[3]

[1]*Instituto de Silvicultura e Industria de la Madera, Universidad Juárez del Estado de Durango, Durango, Mexico,* [2]*Unidad de Prestación de Servicios Ejidales de El Salto Dgo. A. C., Durango, Mexico,* [3]*Burckhardt Institut, Georg-August University, Göttingen, Germany*

ABBREVIATIONS

FMP	Forest Management Plan
MDS	Método de Desarrollo Silvícola
MMFR	Mexican Method of Forest Regulation
MMIFR	Mexican Method of Irregular Forest Regulation
SEMARNAT	Ministry of Environment and Natural Resources (Secretaria del Medio Ambiente y Resursos Naturales)
SiPlaFor	Sistema de Planeación Forestal para Bosque Templado
UCODEFO	Unidad de Conservación y Desarrollo Forestal

MANAGEMENT SETTING AND BACKGROUND

Altogether, 64.8 million hectares (ha) (160.1 million acres (ac)) or 47% of the total national territory of Mexico is covered by temperate and tropical forests. These forests are very diverse and produce a wide range of economic benefits. As such, they are critical to the well-being of rural Mexican communities (World Wildlife Fund, 2008). Mexican pine and oak forests cover large areas in the states of Durango, Chihuahua, Michoacan, Jalisco, Guerrero, and Oaxaca. They cover 16% of the national territory, comprising 31.8 million ha (78.6 million ac) in total and occur throughout the major mountain ranges of the Sierra Madre Oriental, the Sierra Madre Occidental, the Sierra Madre del Sur, and the Transvolcanic Belt (Rzedowski, 1978).

Mexico is not only home to 50% of all the known pine species in the world, but also is a refuge to a remarkable 200 species of oak (World Wildlife Fund, 2014). Different sources cite different percentages, however it can be said that at least 60% of the forest areas of the country belongs to about 15,481 rural communities (*comunidades*) and *ejidos* (Madrid et al., 2009; Merino-Pérez and Martínez-Romero, 2014; Secretaría de Recursos Naturales y Medio Ambiente, 2002). *Ejidos* and *comunidades* are communal groups that live in rural areas and whose lands are managed with some level of governmental control. This arrangement resembles American tribal lands in the United States (Thoms and Betters, 1998; Velazquez et al., 2001). Between 20 and 35% of the forest area is privately owned, and the remainder belongs to the federal government, of which 3.7% is classified as protected areas (Reyes and D'Acosta, 2012). The majority of the rural communities living in the forests are indigenous peoples (Thoms and Betters, 1998).

About 21 million ha (51.9 million ac) of Mexico's forests have some commercial timber production potential out of which 15.6 million ha (36.6 million ac) belong to ejidos and communities. However, only 8.6 million ha (21.2 million ac) currently have timber harvest plans (Secretaría de Recursos Naturales y Medio Ambiente, 2002) approved by the Ministry of Environment and Natural Resources (SEMARNAT). The forest region of El Salto, Durango occupies an area of approximately 500,000 ha (1.24 million ac) and is very important for the country's timber production. In this region, the timber harvest amounts to about 544,700 cubic meters (m^3) (1.924 million cubic feet (ft^3)) annually, or 30% of the total production in Durango. The major economic activity is represented by the community forestry enterprises (*comunidades* and *ejidos*).

Harvesting of the pine-oak forests in this region began about 100 years ago, and harvest regulation relied on European forest management methods developed during the second half of the nineteenth century. During the past 100 years, a remarkable silvicultural and socioeconomic evolution has taken place in the El Salto region (Burgos and Villa, 1974).

From 1918 to 1943, the virgin natural forests, owned now by the *ejidos*, that once contained the best trees with the largest dimensions and located in readily accessible areas, were subject to intensive cutting by a U.S. company. The *ejidatarios* (members of *ejidos*) were treated not as owners but rather as tenants.

Between 1918 and 1927, there was practically no attention paid to systematic management, protection, regeneration, and research. The prevailing objective was to obtain the greatest possible gain with the minimum investment, without considering any social benefits for the local population. In 1928, the regulations of the first Mexican Forest Law of 1926 began to have a desirable effect starting with the formulation of the first forest management study that considered the impact of natural regeneration from seed trees. The prescribed silvicultural practices were, however, not applied correctly by the company, and the harvests were not sustainable. Until 1943, the ejidatarios could not participate in the management of their forests.

In 1943, a new forest management study was conducted with the task of formulating the Mexican Method of Forest Regulation (MMFR). This method was characterized by the application of selective harvesting of trees of a minimum diameter of 45 centimeter (cm) (17.7 inch (in)) and maximum harvest intensity between 35% and 50% of the standing volume per unit area. The cutting cycle was set arbitrarily and specifically adapted to incorporate a planned road expansion, with the condition that each year a high harvest level could be maintained. The area harvested was adjusted to be consistent with the number of years in the cutting cycle. The area of each cutting block was defined such that the harvested volume in each periodic block was more or less equal. This method was applied from 1944 to 1977 (Rodríguez et al., 1960; Secretaria de Agricultura y Recursos Hidraulicos Subsecretaria Forestal, 1985).

Between 1975 and 1984, several adjustments to the MMFR system were made. Subsequently, the name was changed to Mexican Method of Irregular Forests Regulation (MMIFR), which is still used in Durango. The new silvicultural and harvesting practices caused substantial changes in the structure of the forests, and the timber supply was also seriously affected. One problem, for example, was that only short and small-sized logs, which were easily harvested and transported, were sent to sawmills. Socioeconomic problems were also observed among the *ejidatarios*, *comuneros* (members of *comunidades*), and *pequeños propietarios* (small landowners), caused especially by limitations posed on the harvest intensity, which was between 35% and 50% of the standing volume per unit area.

Consequently, a new silvicultural management system was implemented in 1978 in all the *ejidos*, *comunidades*, and small private properties. The new system was known as the *Método de Desarrollo Silvícola* (MDS). The MDS system was established as a way of using silviculture to achieve sustainable forest management and is based on normal forest principles. The MDS system is characterized by a rotation of 60 years, subdivided into six 10-year cutting cycles. The incorporation of harvesting areas containing young trees contributed to a greater diversity of products and satisfied needs of the local sawmilling industry, which could also sell surplus volume to the mills in Durango City and to other markets.

The objectives of this chapter are to describe the silvicultural systems used in El Salto, Durango Forest Region, and the forest management plan (FMP) of the Ejido Borbollones, which was developed by the forestry department Unidad de Conservación y Desarrollo Forestal (UCODEFO) No. 6. The Ejido Borbollones is located in the Sierra Madre Occidental between 23°45′ to 23°49′ latitude and 105°41′ to 105°44′ longitude, 100 kilometers (62 miles) to the southwest of Durango City (Figure 9.1). The total land area of Ejido Borbollones is 2,488 ha (6,148 ac). Forests of Durango are rich in biodiversity with at least 27 coniferous tree species, including 20 species of pines (*Pinus* spp.) and 43 species of oaks (*Quercus* spp.). The predominant forest types are mixed and uneven-aged pine-oak forests, representing more than two-thirds of the total area (González et al., 2007), and they are often mixed with *Arbutus* and *Juniperus* species and other tree species (Zhao et al., 2014).

PLANNING ENVIRONMENT AND METHODOLOGY

Guiding Laws, Regulations, and Policies

The FMP of the Ejido Borbollones (Unidad de Conservacion y Desarrollo Forestal No. 6., 2012) was developed in accordance with the federal laws, following the Mexican Official Norm NOM-152-SEMARNAT-2006, which determines the guidelines, criteria, and specifications of the content of the FMPs for the utilization of timber forest resources in forest, jungles, and arid zones (SEMARNAT, 2008). In 2006, the Ejido was certified by SmartWood for both its forest management and chain of custody, following the principles and criteria of the Forest Stewardship Council (certificate: SW-FM/COC-1960). During the development and implementation of 2007–2017 FMP, several issues have been addressed by the forestry department. The main issue was a modification of FMP, which was performed in 2012, with the aim of improving land use and proposed silvicultural treatments. Results of the modification can be summarized as an incorporation of an area of 38 ha (94 ac) to the harvestable forest land base and an increase of 15% in the forest area managed under the MDS system.

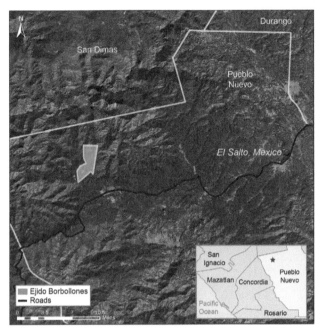

FIGURE 9.1   Location of the Ejido Borbollones forest in El Salto.

The implementation of the activities considered in the FMP is conducted by Forestry Unit UCODEFO No. 6, which acts as an advisor department of the *Comisariado Ejidal*, an executive council composed of three *ejidatarios*, including a president, a treasurer, and a secretary, who are all responsible for the administrative and management functions of the *ejido*. They are elected for 3-year periods. The president is the official representative of the *ejido* and is responsible for management-related decisions, including those concerning the *ejido's* natural resources. The secretary and treasurer, besides their assigned responsibilities, generally serve as advisors to the president (DeWalt, 1979). A Vigilance Council serves as a body of control for the Commissariat (*Comisariado*). The elected officers are typically recognized as leaders within the *ejido*.

Silvicultural and Analytical Systems

The first version of the FMP developed for Ejido Borbollones was generated with the Forest Planning System SPF-SUG-2001 (MARS SOFTWARE SA de CV), which was licensed for use by UCODEFO No. 6. In 2012, necessary adjustments to the long-term management direction and the planning were performed using the Sistema de Planeación Forestal para Bosque Templado (SiPlaFor), an open, Web-based software developed by the Faculty of Forestry of the Juarez University of Durango State. SiPlaFor is a tool to support foresters in planning, developing, and implementing FMPs (Comisión Nacional Forestal, 2014).

The forests of Ejido Borbollones (Figure 9.2) have been managed by local communities for more than 70 years, mainly using selective harvest for sustainable timber production, but also for the maintenance of biological diversity and uneven-aged stand structures. Currently, the Forest Rotation Management System (even-aged forests) is represented in Mexico by the MDS and the Continuous Cover Forestry System (uneven-aged forests) is represented by the MMIFR. Both systems are used to achieve management objectives and develop the desired future forest structures.

The Forest Rotation Management System (represented by MDS) relies on natural regeneration (first phase), resulting in fairly even-aged, regular stands. This is followed by a second phase in which four commercial thinnings are performed for tree improvement, and a third phase of final harvest when the rotation age is reached (60 years). The same silvicultural system is then successively applied.

The Continuous Cover Forestry System (represented by MMIFR) is characterized by the absence of a rotation age that defines the time of the harvest. The stand age is undefined because trees of all ages usually occur in close vicinity to each other. Commercial harvests are based on maintaining growing stock levels within an ideal and negative exponential diameter class distribution, which is also known in the literature as an inverse *J*-shaped distribution. In uneven-aged stands, the idealized diameter distribution is thought to be balanced, meaning that it can be maintained by applying a given harvest rate

FIGURE 9.2 A view of the Ejido Borbollones. *(Courtesy of M.C. Pedro Hernández Díaz).*

in perpetuity. In a balanced, inverse *J*-shaped distribution, a constant quotient exists that allows one to easily estimate stand density between successive diameter classes (Buongiorno and Mitchie, 1980; Rautiainen, 1998; Virgilietti and Buongiorno, 1997). This forest management system is preferred in Ejido Borbollones because it promotes mixed and irregular forest stands and high biodiversity.

For the development of a FMP in the El Salto, Durango forest region, a forest inventory must be conducted every 10 years. Here, a stratified systematic sampling design without replacement was performed. Circular plots of 1,000 m² were systematically distributed in each productive substand (minimum management unit that may vary between 1 and 40 ha) by means of SiPlaFor. A sample intensity of 4.2% on average was applied, verifying that there were at least two plots for each minimum management unit (substand). This sample intensity allowed for a sampling error lower than 10% at the land property level (8.2%). During this step, plot information and tree measurements are recorded on a special form (sheet), which contains several sections:

1. *Plot information*: Identification and ecological information along with plot number, date, location of the plot, and plot site attributes (slope, elevation, aspect, and recommended silvicultural treatment).
2. *Tree measurements*: Information about all living and dead trees equal or larger than 7.5 cm (3 in) along with tree number, species, tree class, breast height diameter, and total height.
3. *Understory information*: Within a 9 m² (96.75 ft²) subplot, diameter and height classes of woody species less than 7.5 cm (3 in) in breast height diameter.
4. *Increment cores*: A determination of age and growth of dominant standing trees.

Additional recommendations to promote, protect, and conserve associated resources are noted by the field workers.

The development of this 4-year operational management strategy was based on the use of the SiPlaFor, an integrated timberland management planning tool that covers forest management process from potential to operational planning for the existing wood resources. It provides a tool to facilitate decision making associated with the sustainable utilization of wood resources. SiPlaFor is a Web-based application that can easily be used over the Internet (Comisión Nacional Forestal, 2014). It contains modules for:

- Sample plot inventory and stand data management, including map data
- Data entry and validation
- Estimates of stem volume, timber product distribution, increment rates, site index, available harvest volume, tree diversity, and stand structure indices based on inventory data
- Reports on the results of the above mentioned parameters at substand level, including silvicultural treatments

SiPlaFor uses several biometric equations developed recently for the most important pine and oak tree species in the forest region of El Salto. These equations define species-specific biometric equations that are used for the estimation of species attributes (standing volume, product distribution, site index, etc.) and for the calculation of the utilization potential of each of the subcompartments.

OUTCOMES OF THE PLAN

A process developed through the use of SiplaFor, and supported by the field data, facilitated the development of a land classification. Harvestable forest areas represented about 86% of the land area, conservation and controlled utilization areas represented about 9%, and areas for other uses represented about 4% (Figure 9.3). Therefore, as a result of this classification, about 86% of the total area is considered available for commercial logging, of which 265 ha (655 ac) are considered in the current 4-year management period. About 10% of the area is used to provide habitat for wildlife mainly because it is very inaccessible.

A summary of the mean stand variables estimated from the forest inventory data indicates that 80.9% of the trees are pines, 8.4% are oaks, and 10.7% are other types of trees, mainly *Juniperus* and *Arbutus*. Thus, timber production areas of the *ejido* are generally covered with pine-oak forests. The average growing stock of pine is estimated in about 202 m³ of standing timber (m³rta) per ha (2,887 ft³ per ac), which includes commercial branches in the volume. The total volume of the allowable cut for the proposed 4-year management strategy is 23,060 m³rta (814,479 ft³), of which 78% are pine, 12% are oak, and 10% are other conifers and hardwoods. Harvestable timber per hectare is estimated in 68.4 m³ (2,413.4 ft³), 76% of which is pine.

The average current annual increment for pine is estimated to be 7.4 m³ per ha (105.1 ft³ per ac) per year, while the average mean annual increment is about 5.1 m³ per ha (72.6 ft³ per ac) per year. This growth rate is considered high for the commercial forests of Durango, where the reported average is about 2 m³ (28.6 ft³ per ac) per year (Secretaría de Recursos Naturales y Medio Ambiente, 2002). Twelve pine species (*Pinus strobiformis, P. chihuahuana, P. cooperi, P. douglasiana, P. durangensis, P. engelmannii, P. herrerae, P. leiophylla, P. lumholtzii, P. michoacana, P. oocarpa,* and *P. teocote*), and eight oak species (*Quercus candicans, Q. crassifolia, Q. durifolia, Q. eduardii, Q. obtusata, Q. rugosa, Q. scytophylla,* and *Q. sideroxila*) were selected for harvesting. However, *Pinus cooperi, Pinus durangensis,* and *Quercus sideroxila* compose 77% of the total allowed harvest volume.

FIGURE 9.3   Land classification of the Ejido Borbollones forest.

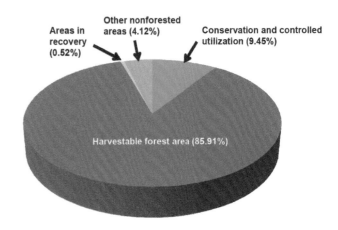

DISCUSSION AND CONCLUSIONS

The management plan for Ejido Borbollones is based on current forest inventory and the application of the MDS and MMIFR. These systems involved both even-aged and uneven-aged management of forests where appropriate.

Learnings and Insights

Durango´s forests exhibit a great diversity in plant and animal communities (González et al., 2007). However, individual *ejidos* are not specifically managed for diversity but primarily for timber production. This mixed forest management approach used in the Ejido Borbollones is largely the result of the community´s dependence on timber products for income; the usual main objective is to produce high volumes of commercial tree species on the most productive sites. On the other hand, the MMIFR method, with selective harvesting, aims to maintain stable the ecosystems that provide timber as well as other important products and services. These include the continuity of structural and compositional diversity, with benefits for tree-dwelling plants, wood-inhabiting insects, and fungi (Dedrick et al., 2007). In the FMP of the Ejido Borbollones, a total of 23 tree species are considered for timber production. However, the timber harvest consists almost totally of pine and oak species. Other hardwood and softwood tree species are harvested at very low rates. The commercially valuable timber species are mainly pines, followed by different species of oak. Extensive areas of low scrub forest are not under management, because they are mostly inaccessible and of high environmental value.

Sustainability Issues

Currently in Mexico, as in many other countries, it is accepted policy to practice sustainable forest management that is environmentally responsible, socially beneficial, and economically viable. The current FMP also considers the implementation of some complementary treatments to both the soil and the vegetation, with the aim of improving the forest conditions and ensuring the establishment of tree seedlings. They include logging residue control, where branches and tips of trees typically left on site following harvest are chopped and dispersed perpendicular to the slope. This helps to impede water runoff and to promote higher rainwater infiltration. Also pruning is considered to improve the timber quality.

Plan Development Challenges

Many forest management planning processes include a planning phase, a decision-making phase, and a monitoring phase. These phases may refer to different scales: the region, the ownership (e.g., *ejidos* in Mexico), the stand, the subcompartment, and the individual tree. Each scale represents a particular scope and level of available information (Diéguez-Aranda et al., 2009; Valsta, 1993). The stand and subcompartment levels are those with a real meaning in forest planning, decision making, and monitoring.

A monitoring program, through the establishment of permanent observational studies, will allow forest managers and researchers to observe different silvicultural, economic, ecological, social, and culturally relevant variables because they provide objective evidence in terms of base information. This information is important for evaluating the planned management objectives regarding conservation and use of biodiversity, the maintenance of essential ecological processes, and the maintenance and improvement of relevant economic outputs. The information obtained from such observational studies is also essential to the detection of negative impacts caused by the application of silvicultural treatments, and for the development of key indicators for sustainable forest management. Once the negative impacts are identified, they can be reduced or eliminated, as necessary, through modifications to the FMP until sustainable forest management and utilization is attained (Corral-Rivas et al., 2009).

Information is the key to successful management of the property, and some information is still being collected. During 2012, two long-term silvicultural trials were established for *P. cooperi* in the Ejido Borbollones, with the goal of providing important information on tree response to thinning treatments. Each trial has five plots that are 25 m² (0.154 ac) in size. All trees were tagged, and individual tree measurements have been carefully kept. Four thinning intensities (10%, 20%, 40%, and 60% of standing basal area) were applied in different plots, and one plot was left in its natural state for comparison. These trials will permit examination of the response to thinning of these forest stands, managed with the MDS-System in their mid- to late-rotation. The same trials will also facilitate validation of models to forecast forest stand development, growth and yield, and timber supply. Results from these long-term field trials will allow the *ejido* to assess the potential for increases in productivity resulting from commercial thinnings.

Plan Implementation Challenges

The forests of El Salto, Durango represent an important national asset and, in a global perspective, a unique socioecological system that has survived all kinds of pressures over 100 years thanks to the knowledge and responsibility of the local population and the support of the state. A special mix of management methods has evolved during the years. Some of these are well developed while others, especially in uneven-aged multispecies communities, could be improved. The continuous support of local scientists is essential in the critical analysis of existing forest management systems, with the ultimate aim of providing a broad mix of products and much-needed environmental services. In the past, forest management practice and theory have focused on the development of plantation monocultures to maximize the supply of timber at a low cost. Societal expectations are changing, however, and uneven-aged multispecies ecosystems are often believed to be superior to monocultures in addressing a wide range of expectations, which makes forest ecosystem management in El Salto, Durango relevant.

Other Interesting Issues Related to the Plan

In the Forestry Unit UCODEFO No. 6, professional foresters and forest technicians work and operate as a separate department or consulting agency for the *ejido*. Forest department officials are under the authority of the SEMARNAT, but their salaries are paid by the *ejidos*. The UCODEFO No. 6 prepares the annual timber management plans and budgets while overseeing forest operations, all of which must be approved by the commissariat. At the same time, the federal government instructs the forest department to ensure that the *ejido* manages its lands in accordance with federal law and regulations.

The UCODEFO No. 6 is also responsible for plan modifications, extension work, technical studies, and the use of non-timber products. Within the frame of the forest management of the *ejidos*, this organization also carries out other activities related to the timber harvest and its control, advising the *ejidos* in the administration of their forest enterprises.

REFERENCES

Buongiorno, J., Mitchie, B.R., 1980. A matrix model of uneven-aged forest management. Forest Science 26 (4), 609–625.

Burgos, M.F., Villa, S.A.B. 1974. La silvicultura en el desarrollo económico-social de México. D.F. Bol. Núm. 5. U.I.E.F. San Rafael. S.A.G.-S.F.F. México, 144p.

Comisión Nacional Forestal, 2014. Sistema de planeación Forestal para bosque templado. Comisión Nacional Forestal, Zapopan, Jalisco, México. http://fcfposgrado.ujed.mx (Accessed June 13, 2014).

Corral-Rivas, J., Vargas, B., Wehenkel, C., Aguirre, O., Álvarez, J., Rojo, A., 2009. Guía para el Establecimiento de Sitios de Inventario Periódico Forestal y de Suelos del Estado de Durango. Universidad Juárez del Estado de Durango, Facultad de Ciencias Forestales, Mexico, 89 p.

Dedrick, S., Spiecker, H., Orazio, C., Tomé, M., Martinez, I., 2007. Plantation or conversion—the debate! European Forest Institute, Joensuu, Finland, Discussion Paper 13. 98 p.

DeWalt, B.R., 1979. Modernization in a Mexican ejido: A Study in Economic Adaptation. Cambridge University Press, Cambridge, UK.

Diéguez-Aranda, U., Rojo Alboreca, A., Castedo-Dorado, F., Álvarez González, J.G., Barrio-Anta, M., Crecente-Campo, F., González González, J.M., Pérez-Cruzado, C., Rodríguez Soalleiro, R., López-Sánchez, C.A., Balboa-Murias, M.A., Gorgoso Varela, J.J., Sánchez Rodríguez, F., 2009. Herramientas selvícolas para la gestión forestal sostenible en Galicia. Xunta de Galicia, Mexico.

González, E.M.S., González, E.M., Márquez, L.M.A., 2007. Vegetación y Ecorregiones de Durango. CIIDIR-IPN. Plaza y Valdés, S.A. de C.V. México, D.F. 219 p.

Madrid, L., Núñez, J.M., Quiroz, G., Rodríguez, Y., 2009. La propriedad social forestal en México. Investigación Ambiental. 1 (2), 179–196.

Merino-Pérez, L., Martínez-Romero, A.E., 2014. A vuelo de pájaro. Las condiciones de las comunidades con bosques templados en México. Comisión Nacional para el Conocimiento y Uso de la Biodiversidad, México.

Rautiainen, O., 1998. Modelling the yield and growth of uneven-aged Shorea robusta stands. Ph.D. thesis, University of Joensuu.

Reyes, J.A., D'Acosta, S. (Eds.), 2012. Memorias del Seminario Propiedad Social y Servicios Ambientales 8 de noviembre de 2011, Ciudad de México. Proyecto de Cooperación Registro Agrario Nacional (RAN)—Instituto Interamericano de Cooperación para la Agricultura (IICA), en Coordinación con el Consejo Civil Mexicano para la Silvicultura Sostenible (CCMSS) y la Comisión Nacional Forestal (CONAFOR). México, D.F. 85 p.

Rodríguez, C.R., Mendoza, M.R., Barrena, G.R., 1960. El Método Mexicano de Ordenación de Montes. In Algunas Prácticas de Ordenación de Montes. Comisión Forestal del Estado de Michoacán, Morelia, Michoacán, pp. 9–74.

Rzedowski, J., 1978. Vegetación de México. Limusa, México, 432 p.

Secretaria de Agricultura y Recursos Hidraulicos Subsecretaria Forestal, 1985. Normas mínimas de calidad para la formulación de estudios dasonómicos en bosques. Secretaría de Agricultura y Recursos Hidraulicos, Mexico, D.F, 298 p.

Secretaría de Recursos Naturales y Medio Ambiente, 2002. Programa Estratégico Forestal 2030. Secretaría de Recursos Naturales y Medio Ambiente del Estado de Durango, Durango, Mexico, 242 p.

SEMARNAT, 2008. Norma Oficial Mexicana NOM-EM-152-SEMARNAT-2006: que establece los lineamientos, criterios y especificaciones de los contenidos de los programas de manejo forestal para el aprovechamiento de recursos forestales maderables en bosques, selvas y vegetación de zonas áridas. Viernes 17 de octubre de 2008. Diario Oficial, México D.F.

Thoms, C.A., Betters, D.R., 1998. The potential for ecosystem management in Mexico's forest ejidos. Forest Ecology and Management 103, 149–157.

Unidad de Conservacion y Desarrollo Forestal No. 6., 2012. Modificación de Manejo Forestal para las Áreas de Corta 2012–2016 del programa de manejo del Ejido Borbollones. El Salto, Durango, Mexico. 108 p.

Valsta, L.T., 1993. Stand management optimization based on growth simulators. The Finnish Forest Research Institute, Joensuu, Finland, Research Paper 453.

Velazquez, A., Bocco, G., Torres, A., 2001. Turning scientific approaches into practical conservation actions: The case of Cominudad Indigena de Nuevo San Juan de Parangaricutiro, Mexico. Environmental Management 27, 655.

Virgilietti, P., Buongiorno, J., 1997. Modeling forest growth with management data: a matrix approach for the Italian Alps. Silva Fennica 31 (1), 27–42.

World Wildlife Fund, 2008. WWW en México. World Wildlife Fund, Gland, Vaud, Switzerland. http://awsassets.panda.org/downloads/2008_fs_wwf.pdf (Accessed May 10, 2014).

World Wildlife Fund, 2014. Southern North America: Western Mexico into the southwestern United States. World Wildlife Fund, Gland, Vaud, Switzerland. https://worldwildlife.org/ecoregions/na0302 (Accessed May 10, 2014).

Zhao, X., Corral-Rivas, J.J., Zhang, C., Temesgen, H., Gadow, K.v., 2014. Forest observational studies—an essential infrastructure for sustainable use of natural resources. Forest Ecosystems 1 (8), 1–10.

Chapter 10

Harvard University Forest, Massachusetts, United States of America

Audrey Barker Plotkin, John O'Keefe and David Foster

Harvard Forest, Harvard University, Petersham, Massachusetts, USA

ABBREVIATIONS

CRs	Conservation Restrictions
ForestGEO	Forest Global Earth Observatory of the Smithsonian Institution
LTER	Long-Term Ecological Research Program of the National Science Foundation
NEON	National Ecological Observatory Network of the National Science Foundation

MANAGEMENT SETTING AND BACKGROUND

The Harvard Forest was founded in 1907, as Harvard University's outdoor classroom, laboratory, and forest demonstration area. It is a department within the Faculty of Arts and Sciences of Harvard University. From a center comprising 3,650 ac, (1,477 ha) of land, research facilities, and the Fisher Museum, the scientists, students, and collaborators at the Harvard Forest explore topics ranging from conservation and environmental change to land-use history and the ways in which physical, biological, and human systems interact to change our earth. Over the past century, major research themes have encompassed forest-site relationships, silviculture, forest policy and economics, botany, ecology, and atmospheric science. The Harvard Forest has a long history of partnerships with national networks of ecological research sites, including the Long-Term Ecological Research (LTER) program and National Ecological Observatory Network (NEON) of the National Science Foundation, the National Institute for Global Environmental Change of the Department of Energy, and the Forest Global Earth Observatory (ForestGEO) network coordinated by the Smithsonian Institute.

Located within the New England Upland physiographic region in north-central Massachusetts (42.5° North; 72° West), the Harvard Forest's rolling hills and valleys range from 722 ft (220 m) to 1,345 ft (410 m) above sea level. The bedrock underlying the terrain is a mixture of metamorphic rocks formed through continental collisions during the middle Devonian Period. Local site conditions are driven by the stony glacial tills deposited during the Wisconsin Ice Age. These are interspersed with local glacial outwash deposits and wetland peats. Soils are stony and acidic with a wide range of depth and moisture. The very few calcareous, open land, wetland, and aquatic sites contain a disproportionate amount of the floral diversity of the forest. The cool, moist, temperate climate is becoming warmer and wetter regionwide (Melillo et al., 2014). Based on meteorological records from 1961 to 1990, July mean temperature was 68.2 °F (20.1 °C), January mean temperature was 19.8 °F (−6.8 °C), with 42 in (1,067 mm) average annual precipitation distributed evenly throughout the year (Greenland and Kittel, 1997). Records from 2002 to 2013, however, show higher temperatures, especially in the winter, and possibly higher precipitation: July mean 69.3 °F (20.7 °C), January mean 23.0 °F (−5.0 °C) and annual precipitation averaged 48.7 in (1,237 mm) (Boose, 2001).

Almost all of the forests in the region are second-growth, following extensive agricultural clearing and logging that peaked in the mid-1800s (Foster and Aber, 2004). The primary forests that remain (i.e., those forests that were never cleared for agriculture) were typically utilized as woodlots. This land-use history is a primary driver of current forest structure and ongoing dynamics (Figure 10.1). The long-term trends in forest cover and human population in the six New England states show that even as the population grew, forest cover increased between 1850 and the late 1900s. In recent years, conversion from forest to developed land has begun to reverse this trend. Eastern white pine (*Pinus strobus*) declined after the 1938

Forest Plans of North America. http://dx.doi.org/10.1016/B978-0-12-799936-4.00010-2

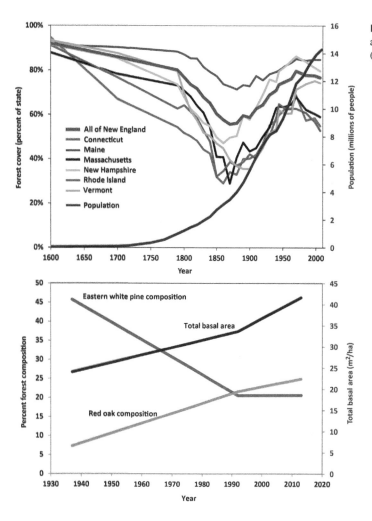

FIGURE 10.1 The regional context for the Harvard Forest (top) and forest growth and composition change at the Harvard Forest (bottom).

hurricane, whereas red oak (*Quercus rubra*) has become an increasingly important component of forest growth, based on data from 60 permanent plots at the Harvard Forest. Wind is the major natural disturbance agent, from isolated windthrown trees to extensive hurricane damage such as occurred in 1938. Pests and pathogens tend to be species-specific. The 1910s-era chestnut blight, caused by the fungus *Cryphonectria parasitica*, and the current hemlock woolly adelgid (*Adelges tsugae*) outbreak have caused long-term tree species declines, whereas periodic outbreaks of pests and pathogens, such as gypsy moth (*Lymantria dispar dispar*), have led to short-term pulses of mortality and decreased tree growth. Wildlife populations have changed substantially with reforestation and changes in hunting activity; many forest species have increased in recent decades. White-tailed deer (*Odocoileus virginianus*), moose (*Alces alces*), North American beaver (*Castor canadensis*), and porcupine (*Erethizon dorsatum*) exert subtle to significant effects on forest ecosystems (Bernardos et al., 2004).

The Harvard Forest comprises five tracts in the towns of Petersham and Phillipston, Massachusetts (Figure 10.2). Three smaller tracts are located in nearby Royalston (the Tall Timbers Tract), eastern Massachusetts (the Matthews Plantation), and southwestern New Hampshire (the old-growth Pisgah Tract; see Foster, 2014). The Prospect Hill Tract is the hub of research and educational activity, with offices, laboratories and greenhouses, the archives building, a maintenance garage and shop, housing for short-term visitors, and the Fisher Museum, which houses the Harvard Forest Dioramas. Research infrastructure across the tract includes seven research towers (Figure 10.3), two gaged headwater streams, a wireless and electrical loop, the Shaler meteorological station, many long-term plots and manipulative experiments, and an 86.5 ac (35 ha) ForestGEO plot.

The Harvard Forest is situated within the transition hardwood/eastern white pine/eastern hemlock (*Tsuga canadensis*) region of the northeastern United States, and more than 90% of the land base is forested. Major forest types include oak-maple (*Quercus* spp., *Acer* spp.), red maple (*Acer rubrum*) swales, hemlock (often mixed with hardwoods and pine), oak-pine (*Pinus* spp.), and remnant conifer plantations. Stands 75 to 125 years old dominate the forest age structure and originate from agricultural abandonment, the logging of old-field white pine, and areas with forests damaged by the historic

FIGURE 10.2 A map of the Harvard Forest properties showing land-use zones. All areas zoned as Managed Woodlands have specific forest stewardship management plans.

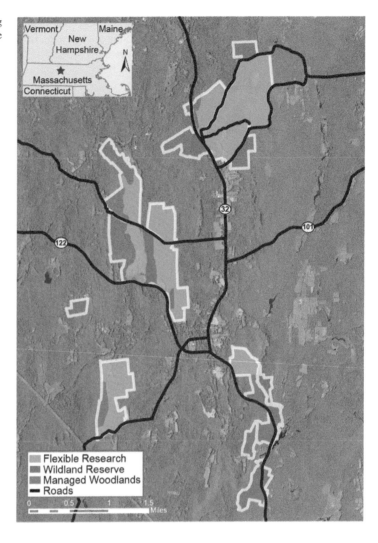

FIGURE 10.3 A canopy-view of the Harvard Forest, looking from one research tower to another.

1938 hurricane. The oldest trees tend to be hemlocks (greater than 200-years old) but scattered black gum (*Nyssa sylvatica*) trees in wetlands exceed 300 years in age. The 20 ac (8.1 ha) Pisgah Tract was never logged, but the oldest and largest trees in this magnificent old-growth forest were blown down by the 1938 hurricane (Foster, 2014). Young forests cover less than 5% of the Harvard Forest, in stands ranging from 1 to 35 ac (0.4–14.2 ha) that were created by plantation removal harvests in the early 1990s and between 2008 and 2010. Forested and open wetlands cover about 4% of the land base. Most streams are intermittent, and headwater brooks flow into the Swift River and Miller's River watersheds. One major stream, the East Branch of the Swift River, flows through the Slab City Tract. Harvard Pond is a 70 ac (28.3 ha) dammed pond within the Tom Swamp Tract, and the Harvard Forest borders two other ponds.

Cultural features include ample evidence of the agricultural past: extensive stone walls, and foundations of historical homes, mills, outbuildings, a tavern, and a tannery. The Harvard Forest maintains approximately 100 ac (40 ha) as open pasture. Many of the current gravel woods roads (over 30 mi (48.3 km)) are formerly farm or town roads. A gate system limits access by motorized vehicles. A number of sites with limited Native American artifacts have been located.

The Harvard Forest is embedded in a heavily forested landscape. Most of the forested land in Massachusetts is owned by nonindustrial private landowners, in parcels averaging less than 10 ac (4 ha) (Butler, 2013). However, the 83,615 ac (33,839 ha) state-owned Quabbin Watershed Forest is nearby. Much of the forested land in the North Quabbin region is enrolled in current-use forest management programs or is permanently protected from development.

Current field research at the Harvard Forest encompasses more than 100 distinct projects led by more than 60 investigators from the Harvard Forest, other departments of Harvard University, NEON, co-investigators with the LTER program, and universities and research institutes from New England and worldwide. Many of the ongoing major experiments were initiated between 1988 and 1990 and include a nitrogen saturation experiment, two soil warming experiments that use buried heating cable, a simulated hurricane experiment, and chronic manipulations of above- and below-ground litter inputs. The first eddy-flux tower began monitoring forest-atmosphere carbon exchange in 1991 and is now complemented by eddy-flux measurements in an old hemlock forest and in a recently harvested forest. Associated with these towers are complementary studies of tree physiology, plot-based estimates of carbon flux in soils and vegetation, and monitoring of forest-atmosphere exchange of ozone, volatile organic compounds, and nitrogen. In addition to the recently established 86.5 ac (35 ha) mapped forest (ForestGEO) plot, there are more than 50 ac (20.2 ha) of permanent forest plots that were installed between 1937 and 2007 for a variety of purposes. They provide long-term information on forest development. Many of these long-term studies incorporate the well-documented land-use history of the Harvard Forest into their study design and interpretation of how this history influences current forest structure and function. Prototype remote sensing equipment is often field-tested at the Harvard Forest, and the field wireless network supports collection of sensor-based data streams. There are also ongoing studies of microbial ecology, small mammal and amphibian diversity and function, ant biodiversity and ecosystem function, and invasive plant populations.

Education at the Harvard Forest encompasses: (a) programs for K-12 schoolchildren, undergraduate, and graduate courses; (b) tours and workshops for professional groups; (c) self-guided trails; and (d) the Fisher Museum. Two trails that begin at the Fisher Museum are the most frequently visited, but the entire forest is utilized as an outdoor laboratory and classroom. Each summer, a group of 20–30 undergraduates from across the United States work with researchers located at the Harvard Forest to learn and collaborate on research projects.

PLANNING ENVIRONMENT AND METHODOLOGY

As the Harvard Forest enters its second century, its mission is to develop and implement interdisciplinary research and education programs investigating the ways in which physical, biological, and human systems interact to change our earth. The central focus on research and education has been unchanged since the forest's founding in 1907. The Land Use Master Plan supports this overall mission, as the overarching goal of the plan is to allow for flexibility of research and educational uses while protecting current and future research opportunities and natural and cultural resources.

Although the Harvard Forest has maintained a consistent mission and long-standing management approaches for some of its notable forest areas, much of this information is in scattered written sources originating with senior staff and directors (Fisher, 1920, 1921, 1931; Gould, 1960). As research and educational activity at the Harvard Forest increases over time, a formal Land Use Master Plan was developed to describe long-standing activities and to help guide future collaborators and leaders. The plan focuses on identifying broad land-use zones, mapping these zones across the Harvard Forest land base, and articulating guidelines for acceptable research, educational, recreational, and forest management activities for each zone. This will allow better guidance for siting new activities on the Harvard Forest, protecting sensitive areas, and providing a broad framework for our local and worldwide research collaborators.

The museum coordinator and the site coordinator led the planning process from 2004 to 2008, in close consultation with the Harvard Forest director, guided by historical documents describing the mission and goals for management and a series of discussions with the Harvard Forest research group. Early discussions helped to clarify major types of land use at the Harvard Forest and zone types. True to its mission, the primary land uses are for research and education activities. Recreational use, protection of fragile sites, and forest management are supporting uses. Informed by the Harvard Forest, emphasis on understanding ecological dynamics in the New England region resulting from natural disturbances and environmental change, management response to forest disturbances was also considered. Based on the land-use types identified, a parallel effort was completed to compile and synthesize historical and current assessments of information about the Harvard Forest lands. These supporting documents include base maps and history of the forest, a summary of its natural resources, and current and historical policies and activities for research, education, recreation, land protection, and forest management.

At first, zones defined by use restrictions based on long-standing Harvard Forest traditions, site fragility, state and local regulations, the occurrence of long-term studies, and unique habitat values, were favored. Further discussions shaped an approach of designating reserves and active management areas within a matrix of land available primarily for research and educational use. The matrix of land zoned for Flexible Research reflects the unique mission of the Harvard Forest. As an *ecological* research forest, current major studies mentioned above suggest an emphasis that is somewhat different from a research forest focused on silviculture and forest management strategies, although this is an important part of the history of research at the Harvard Forest.

Guiding Laws, Regulations, and Policies

The Wildland Reserves and Managed Woodlands are closely aligned with the history of forest management and the regionwide Wildlands and Woodlands vision published by the Harvard Forest. Concurrent with the Harvard Forest planning process, the original Wildlands and Woodlands initiative was developed for Massachusetts, involving many researchers at the Harvard Forest (Foster et al., 2005). Since then, the vision was expanded by a large consortium of academics to encompass the entire New England region (Foster et al., 2010). The Wildlands and Woodlands vision calls for a 50-year effort to conserve 70% of New England as forest permanently free from development. Through the leadership and commitment of landowners, these conserved lands will continue to power the region's traditional land-based economy and provide environmental and social benefits for current and future generations. The Wildlands and Woodlands vision strikes a balance between active, long-term forest management and preservation. Ninety percent of forests would be expansive woodlands that are voluntarily protected from development and managed for a multiple use of forest products, water supply, wildlife habitat, recreation, aesthetics, and other objectives. Ten percent of the forestland, or seven percent of the region, would be wildlands that are established as large landscape reserves subject to minimal human impact and shaped by natural processes. Designating about 40% of the Harvard Forest land base as wildlands or woodlands reflects this larger vision and allows the Harvard Forest to serve as a forest demonstration area, although in a different form than envisioned in 1907.

Within the broad Flexible Research, Wildland Reserve, and Managed Woodland land-use zones, activities and projects are subject to state and federal regulations and internal management policies. The Massachusetts Rivers Protection Act and Wetlands Protection Act regulate activities near water bodies and wetlands. The Massachusetts Natural Heritage Program enforces state and federal endangered species protection. Timber harvesting is regulated by the Massachusetts Forest Cutting Practices Act, which includes review by The Natural Heritage and Endangered Species Program, and adherence to water quality Best Management Practices.

In addition, more than 700 ac (283 ha) of the Harvard Forest are subject to parcel-specific Conservation Restrictions (CRs; called conservation easements in most other states) held by the Massachusetts Department of Conservation and Recreation, the Massachusetts Division of Fisheries and Wildlife, Mount Grace Land Conservation Trust, and the Town of Petersham. The reduced cost of parcels with these conservation restrictions has allowed the Harvard Forest to finance the purchase of significant additional lands in the past 20 years that buffer research sites, while providing permanent protection from development. Within these areas, research and educational activities are reserved rights, as is timber harvesting if part of a long-term approved forest management plan. Much effort was devoted to crafting unique CRs that protected these rights, and these may represent the first agreements in which research-focused easements, written to allow novel and even currently unthought-of future research activities, had been developed.

Management of the Harvard Forest is also subject to memoranda of understanding with agencies utilitizing the site. The Commonwealth of Massachusetts and the Harvard Forest have an agreement that allows the state to maintain a working fire tower on Prospect Hill, and access to that tower. The NEON and the Harvard Forest have developed a memorandum of understanding that secures NEON's infrastructure and research at the Harvard Forest for its 30-year program.

The Harvard Forest has working policies on invasive species management and research and response to forest disturbances including wind, fire, North American beaver encroachment, and pest and pathogen outbreaks. Discussions to clarify and put in writing such policies is one ongoing outcome of the planning process. Finally, major land transactions, including sale or purchase of land, placing conservation easements on land, and appointment of the Harvard Forest director, are decisions that ultimately lie with the President and Fellows of Harvard College.

OUTCOMES OF THE PLAN

A land-use matrix is at the heart of the plan. Land-use zones include Flexible Research (about 60% of the land base, or 2,155 ac (873 ha)), Wildland Reserves (about 20%, or 702 ac (284 ha)), and Managed Woodlands (about 20%, or 804 ac (325 ha)). The forest management and disturbance management activities allowed in the three land-use zones are illustrated in Table 10.1. Flexible Research areas include the major experiments and research infrastructure on the Prospect Hill Tract, much of the Tom Swamp and Simes Tracts, and the northern portion of the Slab City Tract. Rugged and less-studied native forests form the major Wildland Reserves along the east side of Harvard Pond and most of the Slab City Tract, which is bordered by other conservation lands. Managed Woodlands include parcels recently acquired to buffer research areas from possible land conversion. These are suitable for active timber management, as they lack the history of intensive land-use documentation and existing research infrastructure. The 45 ac (18.2 ha) Schwarz Tract is considered a Managed Woodland, in fulfillment of the donor's intent that it "serve as an experimental area in the study of aesthetic or landscape forestry." The western portion of the Tom Swamp Tract includes large conifer plantations and 1920s-era forest management studies, and it forms the largest (470 ac, 190 ha) Managed Woodland. Where possible, Managed Woodlands and Wildland Reserves are paired; for example, the Managed Woodland on the west side of Harvard Pond complements the Wildland Reserve on the east.

For each zone, acceptable research, education, recreation, and forest management activities are defined. For example, research projects in the Flexible Research zone may include a range of intensities, from observational studies to major installations and manipulative experiments. In Managed Woodlands, research activities are allowed that are compatible with the site-specific stewardship plans and, in the spirit of long-term planning, may run for many years. In Wildland Reserves, research is limited to observational studies and low-impact sampling, although plot markers can be used to document the locations of long-term plots. The use guidelines also specify that a written research project application must be approved and updated annually for all studies. This illustrates how the land-use zones provide a broad framework, whereas specific projects are considered and sited on a case-by-case basis. Educational activities for each zone include interpretative signs

TABLE 10.1 Forest Management and Disturbance Management Activities Allowed in Land-Use Zones on the Harvard Forest

	Land-Use Zone		
Activities	**Wildland Reserves**	**Flexible Research**	**Managed Woodlands**
Forest management	• Fell hazard trees and maintain existing trails and roads	• Maintain roads and trails • Remove hazard trees • Create access and infrastructure if required by approved research and if compatible with existing research • Plantation harvest and other management as compatible with research	• Forest management as described in approved Stewardship Plan (including Best Management Practices)
Disturbance management	• Clear existing trails and roads • Precautionary/abatement measures for invasive species • Invasive plant removal	• Road maintenance and plowing if needed • Protect infrastructure • Other measures as compatible or required by research • Precautionary/abatement measures for invasive species • Eradication of all invasive species manipulations at end of experiment	• Exisiting road maintenance • Skid roads as compatible with research • Precautionary/abatement measures for invasive species • Invasive plant removal

and trails, and guided group visits are allowed. Passive recreational activities, including hiking, cross-country skiing, and snowshoeing are allowed in all zones. Hunting is allowed except in posted research areas. Limited woods roads are open to horseback riding or mountain biking.

In addition to the three major zones, a Development Envelope and a Minimum Impact area are identified. These are sub-zones of the Wildland Reserve (Minimum Impact areas) and Flexible Research (Development Envelope) land-use zones. No new buildings are anticipated in the near future, but identifying a suitable area where additional buildings would be placed is helpful in planning land-use and conservation activities. The one Minimum Impact area is a fragile, steep slope with erodible soils and a rare plant population.

The plan is a useful tool for siting new research activities and for communicating with research users of the Harvard Forest. In addition, the planning process identified activities and policies that needed further work. Further discussions were held to develop policies for invasive plant management and guidelines for invasive plant research, beaver management, and recreational uses of the Harvard Forest internal roads and trails. Management of open lands was not considered in the 2008 plan, but management and research planning for pasture lands is now active.

Once the Managed Woodlands were identified by the plan, detailed stand inventory and management recommendations were developed in a suite of six Massachusetts Forest Stewardship Plans. Some of these were already in place, as required for parcels subject to Conservation Restrictions. Once the suite of site-specific plans was in place, the Forest Manager then developed a 5-year projection of timber and fuelwood needs for the Harvard Forest and a recommended harvest schedule.

DISCUSSION AND CONCLUSIONS

If the Harvard Forest founding director, R. T. Fisher, were to see the current Land Use Master Plan, it would not likely be what he expected. However, he would likely appreciate its overall directions and the extent to which it is informed by historical decisions and activities. Fisher's (1920) report on the Harvard Forest emphasized practical forest production issues, so the laboratory was "for forest research and the training of advanced students in the operation of timberlands." At that time, the Harvard Forest was funded through its timber revenues, and so research, demonstration and education focused on planting non-forested lands, harvesting (mainly old-field white pine), improvement cutting, and increasing the timber volume. In 1931, Fisher noted the success of the Harvard Forest in increasing its timber volume and managing the site as a regulated forest. He also emphasized that forest management must be consistent with natural forest development and site conditions. In particular, he noted better success of managing for mixed hardwood-pine or hardwood forests after harvest of old-field white pine, rather than attempting to perpetuate the white pine type by planting. These early insights to the value of learning from natural forest dynamics are reflected in the Flexible Research and Wildland Reserve land-use zones. The choice to use broad zones in this plan originates with Foster's (2002) call for a strategy for forest conservation that recognizes change over time and takes a broad-scale approach.

Learnings and Insights

The Harvard Forest is one of the most thoroughly documented forests in the world, yet finding useful summaries of site information can be daunting. The summary maps and information compiled for the site's land-use history, natural resources, research, education programs, recreation use and policies, land protection history and efforts, and 100 years of forest management are useful starting points to share with many site users. Concurrently, many of the core stand records for the Harvard Forest were digitized and indexed, so it has become easier to use the summaries in the plan to delve into the details of a particular site. The Land Use Master Plan is also strongly complemented by the *Harvard Forest Flora* book (Jenkins et al., 2008). The compilation not only provides species accounts and location maps for all vascular plant taxa found at the Harvard Forest, but also includes a compelling narrative of the Harvard Forest's history and natural resources.

Most of the Harvard Forest has a long history of human management, from clearing for agriculture to harvesting old-field pine, and the interactions between humans and forests is the hallmark of research at the Forest. Yet, the basic policy for managing disturbances that damage or kill trees—the hemlock woolly adelgid, damage from ice-storms, windthrown stands from the next great hurricane—is to observe change and allow recovery to unfold in the absence of further intervention. This is not always a comfortable option. Witnessing the decline of beloved hemlock groves is painful (Foster, 2014), yet this studied management decision provides a valuable contrast to more active management responses often practised on other land in the region. The managers have taken an active approach in some cases, such as beaver encroachment on roads and study plots, or invasive plant populations, but these continue to be considered and debated.

The Harvard Forest management response to disturbance will be reviewed and discussed over the next decades, as the Land Use Master Plan is updated each decade. The second century of Harvard's ecological research forest will surely experience many surprises, engaging many researchers and students in the quest to observe and understand both continuity and change. The Harvard Forest Plan provides a framework for this continued exploration. The Land Use Master Plan was developed at a critical time for the Harvard Forest. As it enters its second century, the forest's research infrastructure and detailed stand and land-use history records attract a growing community of researchers and scholars who utilize the land base for research studies and educational projects. At the same time that research and educational use has increased, recreational use of the Harvard Forest has increased.

Sustainability Issues

Timber supply on the subset of woodland-zoned lands is adequate for projected needs, which are mainly for on-site use. The main ongoing need is firewood for a new, efficient, thermal biomass system, which provides heating to the five main buildings at the Harvard Forest. The system, installed in 2013, encompasses a forwarder, firewood processor, three efficient wood-burners, and a 2,500 gallons (9,464 L) hot water tank. The system is projected to use less than 100 cords (362 m³) per year, some of which is supplied simply as a by-product of woods road maintenance with the rest from planned woodland harvest. In keeping with the research and education mission of the Harvard Forest, wood and labor inputs, and system performance and heat outputs, are carefully documented.

In addition, small amounts of sawtimber are milled on-site for construction projects at the Harvard Forest. At times, timber is harvested as part of an experimental manipulation. For example, a study implemented in 2005 to study loss of hemlock by harvesting or girdling provided the siding for a new maintenance garage. Most wood sales from the Harvard Forest are small-scale; revenues from the large plantation harvest in 2008–2009 were dedicated to funding land protection around the Harvard Forest's core research areas. The Harvard Forest land base is an important asset to Harvard University's 2008 goal to to reduce greenhouse gas emissions, including those associated with prospective growth, by 30% as measured from a 2006 baseline through calendar year 2016. Using wood harvested on-site for heat and building material, as part of a long-term forest management plan, reduces the University's carbon footprint. Enrolling Harvard Forest lands in the carbon marketplace for improved forest management is another option under consideration.

Plan Development Challenges

The Harvard Forest has a rich and deep knowledge of the forest, but lacks current, forestwide forest inventory data. The last full inventory was completed more than 20 years ago. This poses a challenge to assessing the consequences of the plan. The main difficulty is discerning what, if any, comprehensive suite of information would be most useful to the research and educational uses of the Harvard Forest. The 804 ac (325 ha) of Managed Woodlands have current timber inventories, as part of their parcel-specific Forest Stewardship Plans, but these data are less relevant to Flexible Research and Wildland Reserve zones. Traditional timber-oriented inventory data is of limited use to the broad range of ecological studies that span scales from individual organisms to regions. Ecological mapping systems, such as the U.S. National Vegetation Classification, may be more relevant but are unfamiliar to many of the researchers who use the site. The *Harvard Forest Flora* (Jenkins et al., 2008) does provide one aspect of an ecological inventory. Funding is also a barrier to conducting a comprehensive forest inventory, as most of the work of the Harvard Forest is funded by grants that are based on specific research questions.

Plan Implementation Challenges

Making the Land Use Master Plan a living, relevant guide for all users of the Harvard Forest is a common challenge to any plan implementation. Beyond making the plan available on the Harvard Forest web site, products have been developed for specific user groups. For example, maps showing suitability of the internal roads and trails for research vehicle use are helpful for researchers planning access to field sites and have resulted in less driving and fewer stuck vehicles in the forest. A different set of trail maps and signage communicate with recreation users which areas are open for hiking, equestrian, and mountain biking use. Perhaps most importantly, ensuring that the Land Use Master Plan is used by future Harvard Forest managers is critical to continuity and the sense of place developed over decades. The current director acquired most of his knowledge conversationally and through reading dispersed archival materials. A major impetus for creating the master plan was to codify that knowledge and history and to make guidelines readily accessible.

Balancing the benefits and impacts of these uses is an increasing challenge. For example, the research hub at the Prospect Hill Tract now includes long-term experiments and monitoring infrastructure that is part of the LTER program,

ForestGEO forest dynamics plot, and NEON infrastructure—including a tower, soil array, and permanent plots. These major research sites are co-located to maximize synergies among the studies, but at the same time, site impacts must be monitored and carefully controlled. Another example is increasing enthusiasm in the North Quabbin region for recreational trail networks. The Harvard Forest is open to the public, but new policies limiting equestrian and mountain bike use became necessary to protect research, especially as recreational use has increased over the past 15 years.

Other Interesting Issues Related to the Plan

The planning process has prompted us to think proactively about decision making in the face of prospective and ongoing impacts. These include, for example, the hemlock woolly adelgid's arrival and future impacts such as the next major hurricane. A somewhat unusual category in the zone guidelines is one called *Disturbance Management*. The Harvard Forest has long studied how human management response alters forest response to disturbances. For example, the regionwide response to the 1938 hurricane was to salvage downed trees, and in the process, remove surviving trees, pile and burn slash, and scarify the ground. The salvage operation following the hurricane in some ways had a larger effect on ecosystem function than did the storm itself (Foster and Aber, 2004). Salvage harvesting occurred across the Harvard Forest as well, except in one place. Against the regionwide trend and admonishments from federal and state agencies, the great windthrown pines of the old-growth Pisgah Tract were left in place (Foster, 2014). Seventy-five years later, many of these fallen giants remain, mossy but remarkably intact, providing an unparalleled glimpse into how unmanaged forests function. We hope that the Land Use Master Plan will provide a lasting framework in which long-term experiments and studies, sustained woodland management, and wildland reserves can yield insights for many decades.

REFERENCES

Bernardos, D., Foster, D.R., Motzkin, G., Cardoza, J., 2004. Wildlife dynamics in the changing New England landscape. In: Foster, D.R., Aber, J.D. (Eds.), Forests in time: the environmental consequences of 1000 years of change in New England. Yale University Press, New Haven, CT.

Boose, E., 2001. Fisher Meteorological Station at Harvard Forest since 2001. Harvard Forest, Petersham, MA. Harvard Forest Data Archive: HF001, http://harvardforest.fas.harvard.edu:8080/exist/xquery/data.xq?id=hf001 (Accessed March 30, 2014).

Butler, B.J., 2013. Massachusetts' Forest Resources, 2012. U.S. Department of Agriculture, Forest Service, Northern Research Station, Newtown Square, PA, Research Note NRS-189, 3 p.

Fisher, R.T., 1920. The Harvard Forest at Petersham. Harvard Alumni Bulletin 22, 829–835.

Fisher, R.T., 1921. The management of the Harvard Forest, 1909–1919. Harvard Forest, Petersham, MA, Harvard Forest Bulletin 1, 27 p.

Fisher, R.T., 1931. The Harvard Forest as a demonstration tract. Quarterly Journal of Forestry 25, 1–12.

Foster, D.R., 2002. Thoreau's country: a historical–ecological perspective to conservation in the New England landscape. Journal of Biogeography 29, 1537–1555.

Foster, D.R. (Ed.), 2014. Hemlock: A Forest Giant on the Edge. Yale University Press, New Haven, CT.

Foster, D.R., Aber, J.D. (Eds.), 2004. Forests in Time: The Environmental Consequences of 1000 Years of Change in New England. Yale University Press, New Haven, CT.

Foster, D., Kittredge, D., Donahue, B., Motzkin, G., Orwig, D., Ellison, A., Hall, B., Colburn, B., D'Amato, A., 2005. Wildlands and Woodlands: A Vision for the Forests of Massachusetts. Harvard Forest, Petersham, MA.

Foster, D.R., Donahue, B.M., Kittredge, D.B., Lambert, K.F., Hunter, M.L., Hall, B.R., Irland, L.C., Lilieholm, R.J., Orwig, D.A., D'Amato, A.W., Colburn, E.A., Thompson, J.R., Levitt, J.N., Ellison, A.M., Keeton, W.S., Aber, J.D., Cogbill, C.V., Driscoll, C.T., Fahey, T.J., Hart, C.M., 2010. Wildlands and Woodlands: A Vision for the New England Landscape. Harvard Forest, Petersham, MA.

Gould Jr., E.M., 1960. Fifty years of management at the Harvard Forest. Harvard Forest, Petersham, MA, Harvard Forest Bulletin 29, 30 p.

Greenland, D., Kittel, T., 1997. A Climatic Analysis of Long Term Ecological Research Sites. http://intranet2.lternet.edu/sites/intranet2.lternet.edu/files/documents/Scientific%20Reports/Climate%20and%20Hydrology%20Database%20Projects/CLIMDES.pdf (Accessed January 15, 2014).

Jenkins, J., Motzkin, G., Ward, K., 2008. The Harvard Forest flora. An Inventory, Analysis and Ecological History. Harvard Forest, Petersham, MA, Harvard Forest Paper 28. 266 p.

Melillo, J.M., Richmond, T.C., Yohe, G.W. (Eds.), 2014. Climate Change Impacts in the United States: the Third National Climate Assessment. U.S. Global Change Research Program, Washington, DC, pp. 841. doi:10.7930/J0Z31WJ2.

Chapter 11

Dubuar Memorial Forest, New York, United States of America

James Savage

SUNY-ESF Ranger School, Wanakena, New York, USA

ABBREVIATION

SUNY-ESF   State University of New York College of Environmental Science and Forestry

MANAGEMENT SETTING AND BACKGROUND

The James F. Dubuar Memorial Forest (Dubuar Forest) has been home to the State University of New York College of Environmental Science and Forestry (SUNY-ESF) Ranger School since 1912. The property was donated to the college (then known as the New York State College of Forestry at Syracuse University) by the Rich Lumber Company of Wanakena, New York for the purposes of establishing a forest experiment station. Much of the companies' lands, including those to be donated, had been thoughtlessly cut over and subsequently subjected to wildfires. For the Rich Lumber Company it was time to move on, but not without a responsible hope that the nascent College of Forestry could begin to restore the land and conduct experiments in scientific forestry. The College enthusiastically accepted the challenge, and quickly received permission from the donors to establish a Ranger School rather than an experiment station. Such a school would, and still does, provide an ideal place for students to engage in field-based, experiential learning related to forestry, natural resources management, and environmental conservation. In more recent decades, the property also serves as a place to conduct practical research and to demonstrate traditional and experimental forest practices. In these capacities, the Dubuar Forest benefits undergraduate and graduate students, faculty, forestry and natural resources professionals, forest landowners, public officials, and the interested public.

Importantly, the Dubuar Forest also provides outdoor recreation opportunities for the students, faculty, staff, local community, and the general public. Hiking is a common activity, with the most popular trail leading to a restored fire tower that offers panoramic views of the region's forests, mountains, and waterways. Three self-guided nature trails are available three seasons of the year, and a major snowmobile corridor traverses the forest in winter.

The Dubuar Forest is located in northern New York State, in the southeast corner of St. Lawrence County. It is currently one contiguous 2,732-acre (ac) (1,106 hectare (ha)) tract, bordered or split only by state, county, and town roads. These roads provide access to the property, via locked gates, and viewing opportunities for passing motorists. About 13 miles (mi) (21 kilometers (km)) of graveled forest roads and 10 mi (16 km) of hiking trails provide additional access to nearly all parts of the property. Many of the forest roads are built atop logging railroads dating back to the Rich Lumber Company days. Elevations on the Dubuar Forest range from 1,400 feet (ft) (427 meters (m)) to 1,900 ft (579 m). Importantly, the Dubuar Forest lies entirely within the nearly 6 million ac (2.43 million ha) Adirondack Park. The park is a unique, balanced mix of private and public land. The public land, known as the Forest Preserve, is owned and managed by the State of New York. The Forest Preserve, managed to be "forever wild" according to the New York State constitution, includes land designated both as Wild Forest and Wilderness. Timber harvesting is not permitted, and the lands are managed mainly for recreation and wildlife. The Dubuar Forest adjoins Forest Preserve land to the west, south, and east, and private, undeveloped land to the north.

Forest Plans of North America. http://dx.doi.org/10.1016/B978-0-12-799936-4.00011-4

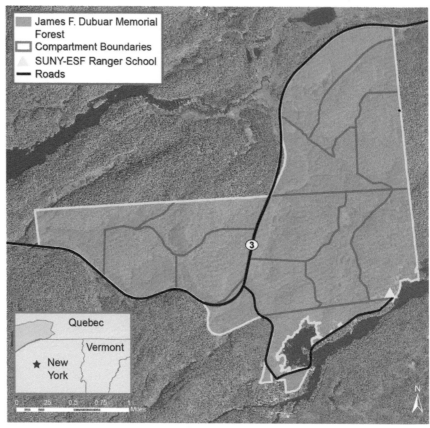

FIGURE 11.1 The Dubuar Forest is grouped for convenience of the academic programs into 13 compartments of about 200 ac (80.9 ha) each.

The forest (Figure 11.1) is composed of 119 stands, ranging in size from 5 to 175 ac (2 to 71 ha), and delineated primarily based on forest type changes, natural boundaries, and property boundaries. The management plan (Savage et al., 2007) contains a one-page, detailed stand description for each stand. The stands are grouped for convenience of the academic programs into 13 compartments, each about 200 ac (81 ha) in size. Following heavy logging in the early 1900s (essentially clear-cuts), 50% of the forest has regenerated naturally to northern hardwoods: sugar maple (*Acer saccharum*), red maple (*Acer rubrum*), yellow birch (*Betula alleghaniensis*), and American beech (*Fagus grandifolia*), occasionally mixed with black cherry (*Prunus serotina*) or white ash (*Fraxinus americana*). Conifer plantations—dominated by eastern white pine (*Pinus strobus*), red pine (*Pinus resinosa*), Norway spruce (*Picea abies*), and Scots pine (*Pinus sylvestris*)—now comprise 30% of the forest (Figure 11.2). Most of the plantations have been established and maintained by students as part of their coursework at the Ranger School. Various types of wetlands—wooded and non-wooded—and a small bog comprise 13%

FIGURE 11.2 A 78-year old pine plantation established, tended, and now being regenerated by students and faculty at the SUNY-ESF Ranger School.

FIGURE 11.3 Approximate age class structure of the Dubuar Forest's timber resource.

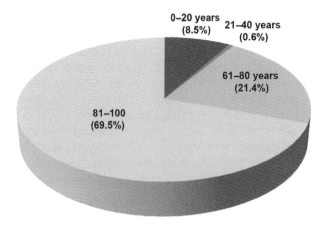

of the forest. Natural softwood stands, open areas, and administrative areas comprise the remaining parts of the forest. No threatened or endangered species exist on the property according to the New York Natural Heritage program (New York State Department of Conservation, 2014).

As expected, the Dubuar Forest has been subjected to natural disturbance from time to time, mainly from wind and insects. Damage has typically been inconsequential, with a few notable exceptions. A 1950 hurricane affected numerous areas of the Dubuar Forest and surrounding state lands. White pine blister rust (*Cronartium ribicola*) was a fairly common problem among white pine plantations from the 1930s to the 1960s. In 1995, a derecho (straight-line wind storm) brought winds in excess of 100 mi per hour (161 km per hour) that flattened about 350 ac (142 ha) of forest, especially those oc-curring on northwest aspects. Between 2004 and 2006, a major outbreak of the native forest tent caterpillar (*Malocasoma disstria*) killed or severely weakened trees, mostly sugar maple, on 979 ac (396 ha) of the Dubuar Forest. In the latter two cases, salvage operations facilitated the establishment of several new stands, either through natural regeneration or hand planting. Age-class diversity (Figure 11.3) of the forest increased as a result.

PLANNING ENVIRONMENT AND METHODOLOGY

Although it has been managed since 1912, the first, formal management plan for the Dubuar Forest appears to have been written in 1985, shortly after the nascent Forest Properties department took over responsibility for managing the Dubuar Forest from the Ranger School faculty. After 20 years without revision, a major effort began in 2005 to revise, augment, and modernize that plan, especially in light of the developing forest certification movement. Equally important, by 2005, implementation of the management plan had evolved to become the responsibility of three distinct departments within SUNY-ESF: Department of Forest and Natural Resources Management (academic staff), Forest Properties (forest manage-ment staff), and Physical Plant (maintenance staff). However, the respective roles of each department were becoming less clear, especially when it came to fiscal responsibilities. It was also recognized by 2005 that additional stakeholders needed to be included in the planning process, most notably representatives of the summer programs that now actively use the property. In response, a "Dubuar Forest Management Review Committee" was established to conduct all planning related to management of the Dubuar Forest. The committee met several times in an effort to formulate the current management plan and now meets at least once per year (usually in late winter) to review all plan components, to review and approve requests for use of the Dubuar Forest, and to prepare an annual operation plan for the coming field season.

Implementation of the plan is the responsibility of the three departments mentioned above and as follows. Day-to-day management of the Dubuar Forest is conducted by the Forest Properties department. Typical duties include inventory, regu-latory compliance, and timber sale administration. Physical Plant staff are responsible for the construction and maintenance of forest roads. The Ranger School academic staff, with considerable assistance from the students, focuses on construction and maintenance of recreational resources like trails and picnic areas. They also conduct tree plantings, forest thinnings, and other cultural activities as a part of classes, such as silviculture and timber harvesting. To facilitate communication, understanding, and effective cooperation moving forward, the current plan for the Dubuar Forest outlines the common roles and typical management responsibilities of each department.

The ultimate challenge to managing the Dubuar Forest is that there are three administrative units responsible for man-aging the forest, with no one having more authority over the other. Furthermore, annual budget allocations specifically targeted toward forest management activities are limited to non-existent. The Ranger School academic staff, for example, receives no specific allocation to support its management role. In general, the cost of conducting work that is atypical or beyond routine maintenance requires advance coordination and planning. Examples of the latter include repairing roads after natural disasters, or replacing durable equipment, machinery and vehicles.

The costs of managing the Dubuar Forest are high and only increase with attempts to manage it in a more sustainable and publicly acceptable way. Some revenue is generated from the sale of natural resources and forest products like timber, firewood, maple syrup, and gravel, but these are generally not enough to cover even routine costs. There is increasing interest and budget-related pressure to improve the annual revenue stream from this and other ESF-owned forests, but the College still maintains that the primary purpose for the Dubuar Forest is to generate educational opportunities.

The plan for the Dubuar Forest was thoughtfully structured into three components: strategic, tactical, and operational. The Strategic Forest Management Plan (Savage et al., 2007) is intended to provide a long-range, sustainable management direction for the forest. It is intended to be flexible, dynamic, and reviewed at least once every five years. It describes the purpose of management, a vision for the forest, and 13 long-term management goals set to meet the same.

(1) The Dubuar Forest will be widely recognized and publicly-noted as a healthy, sustainably-managed, critically-important resource for the Ranger School, the College, the local community, and the forestry and natural resource professions as a whole.

(2) Biological diversity, ecological integrity, and general forest health will be maintained or enhanced over time.

(3) Opportunities for experiential learning will be maintained. For example, there should be regular opportunities for students and/or workshop participants to help establish and maintain forest plantations, conduct commercial harvesting operations, or assist with timber stand improvement (TSI) work.

(4) The Dubuar Forest will be a showcase for both traditional and experimental forest practices. Examples of even-aged, two-aged, and uneven-aged silviculture will be found on the Forest, as will examples of sugarbush management and plantation forestry for Christmas trees and timber.

(5) In support of other long-term goals, the age diversity of the Dubuar Forest will increase. Ultimately, no one age class will occur on over 50% of the total area of the forest.

(6) The Dubuar Forest will comprise a balance of cover types, including both natural forests and plantations, but plantations will not exceed 50% of the total area of the forest.

(7) The road and trail system on the Dubuar Forest will allow students and faculty reasonable access to the property throughout the year. Every compartment will contain or be bordered by a well-maintained forest road, and every compartment will contain or be accessible by at least one well-maintained foot trail.

(8) The Dubuar Forest will be a place where students, landowners, forestry and natural resource professionals, and the public at large can learn directly and indirectly about forests and the methods and benefits of science-based, sustainable forestry. The Forest will contain interpretive trails, exhibits, signs, and demonstration areas with this goal in mind.

(9) All management of the Dubuar Forest will attempt to minimize stand fragmentation and, where possible, aggregate existing stands or sub-stands.

(10) The volume and dollar value of the timber on the Dubuar Forest will be optimized. The production of high-quality, high-value timber will be encouraged when and where possible.

(11) To help off-set the high costs of managing the Dubuar Forest, some amount of revenue will be generated from the forest each year.

(12) The Dubuar Forest will be a place where people can go to experience peace, beauty, relaxation, and spiritual and physical renewal. There will be roads and trails for hiking, biking, skiing and the like, and there will be nature trails, picnic areas and places to hunt, climb, and enjoy scenic beauty. Public recreation will be allowed but controlled so that it does not interfere with the primary purpose for the Forest and the goals of this management plan.

(13) The quality and use of the recreation and interpretive resources on the Dubuar Forest will increase over time.

These goals support the stated intent to manage the forest primarily for education, secondly for extension and research, and thirdly for forest recreation.

Forest certification had an important influence on the development and content of the Strategic Plan. Although the Dubuar Forest is a certified Tree Farm through the American Tree Farm System (2014), efforts to certify through the Sustainable Forestry Initiative (SFI), Forest Stewardship Council (FSC), or similar programs was deemed neither cost-effective nor necessary at this time in light of the unique mission and purpose for the forest. That said, existing certification programs that promote stewardship and sustainable forestry provide a useful framework for developing a management plan and, indeed, managing a forest like the Dubuar Forest. Recognizing this, management of the Dubuar Forest is voluntarily guided by a set of stewardship and sustainability principles that has been adapted from both SFI and FSC. Use of these widely accepted principles is intended to demonstrate SUNY-ESF's commitment to environmentally responsible, socially beneficial, and economically viable forest management. Such a strategy also facilitates the process of becoming an official participant in one or more certification programs should the College ever desire to do so.

As mentioned, the plan also contains a tactical component and an operational component. The Tactical Plan (Savage, 2007) outlines 12 short-term management objectives.

(1) Improve the forest road system by day-lighting and maintaining or repairing the highest priority road sections. Minimum standards, including Best Management Practices (New York State Department of Conservation, 2011), will be provided by Forest Properties. The highest priorities currently are: (1) the North Road—1.2 mi (1.9 km), (2) the Cathedral Rock Road—1.1 mi (1.8 km), and (3) the Sugarbush Road—2.4 mi (3.9 km).

(2) Develop an accurate, up-to-date trail map for the Dubuar Forest and add it to the GIS database for the Forest.

(3) Evaluate the merit and feasibility of improving, modernizing or relocating the Ranger School's sugarbush area.

(4) Utilize the existing continuous forest inventory sample-plot network to establish and monitor plant diversity (species richness and evenness) on the Dubuar Forest.

(5) Regenerate 5 to 10 ac (2 to 4 ha) of the 80–100 year old stands per year for the next 5 years using the clearcut system of silviculture. This will help augment the under-stocked 0–20 year old age class, mitigate the over-stocked 80–100 year old age class, and provide a place for Ranger School students to plant trees every year. If possible, harvest non-native Scots pine stands first and replace with native species of conifers or hardwoods (or both).

(6) Develop 100 additional ac (40 ha) of uneven-aged stands for teaching, demonstration, and research purposes.

(7) Develop and maintain a demonstration stand containing the two-aged silviculture system. The demonstration stand should be accessible by road or short walk, and at least 25 ac (10 ha) in size.

(8) Document, control and, if at all possible, eradicate non-native, invasive species growing on the Dubuar Forest.

(9) Add or enhance directional and informational signage on the forest, especially at trail and road intersections, and especially along nature trails and the drive-through tour of the property. Use high-quality, durable signs with a consistent color and design scheme throughout the forest.

(10) Develop a checklist of known wildlife and fish species that exist on the Dubuar Forest, using undergraduate and graduate student assistants. Determine average white-tailed deer (*Odocoileus virginianus*) density on Dubuar Forest.

(11) Plan, design, and layout a new section of road to extend the Spur Road between Compartments 10 and 12 and perhaps join it into Route 3 to the north.

(12) Determine what the carbon dioxide equivalent is of the timber currently growing on the Dubuar Forest, and at what rate carbon is being sequestered, as a first step to determining whether SUNY-ESF would benefit from entering the carbon-offset market, and as a demonstration of non-timber forest values.

Each objective describes a specific project or activity that should receive priority in the ensuing five years. Listed with each objective is a budget estimate and a suggested performance measure. Unlike the goals of the Strategic Plan, each short-term objective is specific and measurable. Equally important, each objective supports the vision for the Dubuar Forest, one or more long-term goals, and is consistent with the stated principles of stewardship and sustainability. Operational Plans (i.e., work plans) are prepared annually, consistent with the Strategic Plan and Tactical Plan. They are prepared by the Forest Properties staff in consultation with the Dubuar Forest Management Review Committee.

OUTCOMES OF THE PLAN

In an attempt to relate the Strategic Plan to on-the-ground forest management activities, a table was prepared to describe and emphasize the relationship of management purposes and goals to the individual stands comprising the Dubuar Forest. An abridged version of that table demonstrates the relationships that exist for the 13 stands of Compartment I (Table 11.1). Analysis of such a table describes the management priorities that guide day-to-day management decisions and activities. For example, one could identify those stands having only an educational purpose (purpose #1), a long-term management goal of high-quality or high-value timber (long-term management goal #10), and having no special considerations or restrictions noted as places to manage specifically for timber or revenue production. Further, stands with a listed purpose of "extension and research" (purpose #2) or "forest recreation" (purpose #3) should receive cultural treatments only after consultation with the Dubuar Forest Management Review Committee. Such stands are easily identified by reviewing the table. Regarding recreation, the unabridged table shows that most opportunities are confined to 5 of the 13 compartments that comprise the Dubuar Forest. It is important information for the absentee Forest Properties staff, prospective researchers, and other stakeholders who are less familiar with day-to-day use of the property. Further summarization of the unabridged table both clarifies and validates the purposes for owning and managing the Dubuar Forest: 100% of the stands are managed for education, 45% of the stands are managed for extension and research, and 40% of the stands are managed for recreation.

TABLE 11.1 The Relationship of Management Purposes and Goals to Individual Stands Comprising the Dubuar Forest—Compartment I Example

Stand	Management Purposes[a]	Long-Term Management Goals[b]	Special Considerations or Restrictions
0107	1,2,3	1,2,3,4,8,10,12,13	Microburst interpretive trail; driving tour route
0121	1,3	1,2,3,4,10,12,13	Latham trail—high use
0122	1,3	1,2,3,4,10,12,13	Latham trail—high use
0123A	1,2,3	1,2,3,4,10,12,13	Latham trail and Reservoir trail—high use
0123B	1	1,2,3,4,9,10	
0124	1	1,2,3,4,9,10	
0125	1,3	1,2,3,4,10,12,13	Reservoir trail and Trail of Wimps—high use
0126	1,3	1,2,3,4,10,12,13	
0127	1,3	1,2,3,4,10,12,13	Reservoir trail—high use
0128	1,2,3	1,2,3,4,8,10,12,13	Microburst interpretive trail; Christmas trees; scenic view; driving tour route
0129	1,3	1,2,3,8,12,13	Reservoir buffer; driving tour route
0130	1,2	1,2,3,4,8,10,12,13	Fire training area; scenic view
0131	1,3	1,2,3,4,10,12,13	Use by the Ranger School and the Summer Program in Field Forestry; Peavine Swamp trail / Cranberry Lake 50 trail; river shoreline

[a] 1, education; 2, extension and research; 3, forest recreation.
[b] listed by number in the chapter.

Revising and updating the management plan for the Dubuar Forest brought key players to the table to focus on management priorities and strategies. It was a valuable process for current stakeholders, and a critical process for the forest in light of the College's and the Ranger School's missions. And it was written with a thoroughness that should facilitate more efficient and consistent management in the future. Several of the short-term objectives, actually 6 out of 12, have already been accomplished as per the five-year Tactical Plan. Each of those, in turn, has contributed to the vision and one or more of the long-term goals for the forest. The Dubuar Forest Management Review Committee has met as intended nearly every year since the plan was written, forcing stakeholders to revisit, review, and prioritize management goals and objectives. The plan has also been used on several occasions by those wanting more information about the forest, or by those wanting to clarify a process or a policy. The plan is an invaluable resource for new faculty and staff as an orientation piece and for new and existing faculty as they plan for outdoor laboratory exercises and projects. Finally, the plan and the process used to develop it have served as a discussion point and learning tool for numerous ESF students in several different courses, but most notably the Natural Resources Management course.

DISCUSSION AND CONCLUSIONS

The Dubuar Forest was not poorly managed before the current management plan was developed, but the new plan has improved efforts and outcomes, and it has also better-connected ongoing management to SUNY-ESF's mission and ownership goals. The plan itself accomplishes nothing: the forest is not better managed simply because the plan exists. Nor is there a guarantee that anyone now or in the future will utilize the plan. But the plan is a lighthouse that can guide those obligated or interested toward common purpose and institutionally agreed-upon goals and objectives. The current plan also does an effective job of explaining current conditions and activities as they relate to, or are determined by, historical efforts and circumstances. In this way, it is the agent that binds generations of Dubuar Forest users together, which builds appreciation for tradition and acceptance of change. At a 100-plus-year-old institution like the SUNY-ESF Ranger School, the latter is and will continue to be very important.

Sustainability Issues

There are three integrated plans (strategic, tactical, operational) that guide the management of the Dubuar Forest, and each objective of the tactical plan supports the vision for the Dubuar Forest and one or more of its long-term goals. The tactical plan is consistent with general principles of stewardship and sustainability held by the College.

Plan Development Challenges

On a more tangible note, the plan has explicitly stated the typical management responsibilities of the three departments tasked with operating the Dubuar Forest. The management responsibilities were discussed, agreed upon, and then accepted when the plan was adopted. Recognizing the challenges and constraints to such a commitment, the Tactical Plan reads "*A continuous attempt to understand and meet these responsibilities will facilitate the implementation of the strategic, tactical and operational management plans and the achievement of their stated goals and objectives.*" This aspect of the current plan, not found in previous versions, has helped to foster mutual understanding, coordination, and motivation. More importantly, it has resulted in real progress on the ground for the benefit of all stakeholders. The forest road system, for example, long neglected and randomly maintained, has vastly improved and is consistent with the current plan. As a consequence, satisfied stakeholders positively reinforce the progress, which in turn, fosters more efforts and more progress. As expected, key personnel changes have occurred since the plan's adoption, but the plan remains, and thus new employees and department heads can quickly align themselves and their efforts with the plan's vision, goals, and objectives.

As mentioned previously, one of the ongoing challenges to shared management responsibility for the Dubuar Forest relates to budgets and revenue. There are real costs associated with managing the forest, and there are occasional revenues generated by management. How much of cost can or should any one department bear, and how should revenues be shared among the three departments? These are questions that were vigorously debated during the planning process and remain only partially resolved. Also, with the College feeling pressure to increase revenues, the questions become more pertinent. The current plan reflects much that was agreed upon, and hopefully the next round of planning can more satisfactorily, or at least more explicitly, address the pertinent questions related to revenues and expenses associated with management. In this way, they will no longer be distractions to the process. Indeed, another resolved issue only serves to strengthen the plan's *lighthouse* effect.

The enduring contribution of the plan and the enabling planning effort is the Dubuar Forest Management Review Committee. The intention and the reality so far is that the committee will meet at least once per year to review requests to conduct research or make significant modifications to the forest and to formulate a work plan that is consistent with long-term goals and short-term objectives. Such an effort keeps the plan alive and encourages, if not forces, a regular review of the plan elements. These meetings also serve to remind stakeholders of their respective roles in management and to catalyze action. However, as much as these are positive results that have resulted in real achievement, not all of the short-term objectives set for the first five-year period were met. It seems more realistic, now that the plan and its management process have been tested a bit, to reduce the short-term, tactical objectives from 12 to about 6 or 8. This change would force the review committee to further refine and prioritize short-term objectives, facilitating success and, in turn, continued acceptance of and support for the plan. Without that acceptance and support, the guiding light of the plan dims, and management becomes uncoordinated and less likely to meet the needs of current and future stakeholders.

Plan Implementation Challenges

As was suggested, the main challenge in managing the Dubuar Forest involves the three administrative units that are responsible for the Forest and the coordination required in meeting the management goals. In addition, annual budget allocations specific to forest management are either limited or non-existent. In general, accommodating the cost of conducting work can require significant advance notice and patient creativity. Although some revenue is generated from the forest through the sale of commodity and non-timber forest products, those revenues are generally not sufficient to cover the total cost of management of the forest. With due respect and concern for such realities, the College continues to support the forest as an educational asset pertinent to its mission.

REFERENCES

American Tree Farm System, 2014. About American Tree Farm System. American Forest Foundation, Washington, D.C. https://www.treefarmsystem.org/about-tree-farm-system (Accessed April 17, 2014).

New York State Department of Conservation, 2011. New York State forestry best management practices for water quality. New York Department of Environmental Conservation, Albany, NY, 81 p.

New York State Department of Conservation, 2014. New York Natural Heritage Program. New York State Department of Conservation, Albany, NY. http://www.dec.ny.gov/animals/29338.html (Accessed May 28, 2014).

Savage, J., 2007. A Tactical Forest Management Plan for the James F. Dubuar Memorial Forest. SUNY College of Environmental Science and Forestry, Syracuse, NY.

Savage, J., Breitmeyer, B., Gooden, M., Drew, A., Dawson, C., 2007. A Strategic Forest Management Plan for the James F. Dubuar Memorial Forest. SUNY College of Environmental Science and Forestry, Syracuse, NY.

Chapter 12

McPhail Tree Farm, South Carolina, United States of America

Thomas J. Straka[1] and Tamara L. Cushing[2]

[1]School of Agricultural, Forest, and Environmental Sciences, Clemson University, Clemson, South Carolina, USA, [2]Department of Forest Engineering, Resources and Management, Oregon State University, Corvallis, Oregon, USA

MANAGEMENT SETTING AND BACKGROUND

The McPhail Tree Farm is located in Anderson County, South Carolina, which is situated in the Piedmont region of the southern United States. This is the western part of the state, where the borders of Georgia, North Carolina, and South Carolina meet. It includes the foothills of the Blue Ridge Mountains. The original inhabitants were Cherokee Indians and after a series of "Cherokee Wars," plantations became established in the region. By the late nineteenth and early twentieth centuries the entire Upstate was managed predominantly for cotton cultivation, resulting in poor land management practices and highly eroded soils. New Deal (1930s era) conservation programs took much of the land out of cultivation and put it into timber production. Large forested areas were subsequently acquired by timber companies throughout the region. By the 1980s, these same timber companies began to divest of their timberland. Brunswick Pulp and Paper was one of these timber companies.

Timberland was not just an investment for corporations; many individuals and families saw timberland as an attractive financial investment. Walter McPhail, Sr. was one of those individuals. He started purchasing working farms and small forested tracts in the 1940s and 1950s. His family (Figure 12.1) has always seen timber as being a sound foundation for the family finances. In the mid-1980s, his sons, Walt and James McPhail, acquired 1,005 ac (407 ha) of timberland from the Brunswick Pulp and Paper Company as a continuation of that family financial philosophy. The land (Figure 12.2) was partially clear-cut, and required a substantial investment in forest management planning and forest regeneration. The family land is in the American Tree Farm System (ATFS), which requires a commitment to sustainable forest management. The ATFS has 82,000 family forest owners who control 24 million ac (9.71 million ha) of forest land (American Tree Farm System, 2014). It is the oldest sustainable forest management certification/recognition systems in the United States. Members are expected to conform to eight standards of sustainability, which were developed for the size, scale, and management intensity of family-owned small woodlands in the United States. While fiber production and profitability are not necessarily primary objectives of tree farmers, it is fair to say that many owners look upon their tree farms as investments and expect reasonable financial returns from the forest assets. The McPhail Tree Farm is a crucial part of the family's investment portfolio, part of the family retirement plan, and perhaps the family's primary asset. This is not an unusual situation for a tree farmer. Walt McPhail is a practicing veterinarian and "considers the tree farm to be his 401(k) retirement plan" (Walt McPhail, personal communication, March 7, 2014).

The McPhail Tree Farm is divided into four sections: a North section of 437 ac (177 ha), a South section of 256 ac (104 ha), an East section of 169 ac (68 ha), and a West section of 143 ac (58 ha). Prior to selling the property to the McPhails in 1985 and 1991, the timber company harvested and regenerated 456 ac (185 ha) with loblolly pine (Pinus taeda), leaving 172 ac (70 ha) of hardwood along the drains or creeks and another 377 ac (153 ha) of cut-over stands. The hardwood stands are in streamside management zones (SMZs) along major creeks and intermittent streams to protect water quality, to enhance wildlife habitat and diversity, and to provide wildlife corridors. The hardwood areas had been harvested in about 1970 and naturally regenerated. Today, they represent well-stocked stands of yellow poplar (Liriodendron tulipifera), sweetgum (Liquidambar styraciflua), and red maple (Acer rubrum). Minor tree species are willow oak (Quercus phellos), water oak (Quercus nigra), southern red oak (Quercus falcata), wild cherry (Prunus avium), sycamore (Platanus occidentalis), and river birch (Betula nigra). Unfortunately, an invasive species, Chinese privet (Ligustrum sinense), is also present. The site index on the hardwood stands is 90 (expressed as the height in feet of the dominant and codominant trees at a base age,

Forest Plans of North America. http://dx.doi.org/10.1016/B978-0-12-799936-4.00012-6

FIGURE 12.1 A tree farm is a multigenerational endeavor.

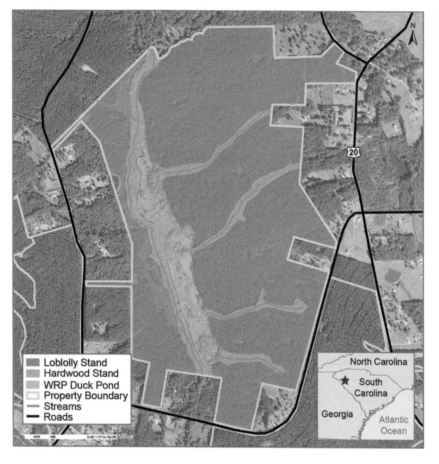

Loblolly Stand
Hardwood Stand
WRP Duck Pond
Property Boundary
Streams
Roads

North Carolina

South Carolina

Georgia

Atlantic Ocean

FIGURE 12.2 Representative map of the North Section of the McPhail property.

which is 50 years in this case). Tree ages range from 40 to 60 years. The McPhails have also planted "Gobbler" sawtooth oak (*Quercus acutissima*), swamp oak (*Quercus bicolor*), and cherrybark oak (*Quercus pogoda*) in wildlife food plots and 6 ac (2.4 ha) of baldcypress (*Taxodium distichum*) in a wetlands protection area. Native persimmon (*Diospyros virginiana*) trees are protected. There are no operations planned for the hardwood stand, and it will be left to follow its natural course for the next 50 years. Possibly, while not part of the current plan, sweetgum and maple species may be removed during future/adjacent harvesting operations.

The soil productivity on the pine lands is representative of a site index of 75 (base age 25) for loblolly pine. Eight hundred and thirty-three acres (337 ha), or 83% of the tract, has been regenerated into productive even-aged stands of loblolly pine. The timing of the regeneration provides pine plantations with staggered ages; meaning that capital requirements for establishment costs were spread out over time and, more importantly, future cash flows from the forest will be better timed to support retirement. However, the land may go unstocked for several years following harvest.

A rudimentary road system existed when the McPhails purchased the property; however, there were major road problems after 10–15 years of post-harvest neglect. Three-to-five-feet (1–1.5 m) deep gullies, numerous large pot holes, and ruts developed. Previous access roads became inaccessible as adjacent parcels with road frontage were sold off as building lots. The McPhails, with a family heritage of soil and water conservation, developed a new forest road system with the aid of technical assistance and cost-share funding from the Natural Resources Conservation Service (NRCS).

State-developed voluntary best management practices (BMPs) that protect water quality were a part of prior forest management. The productive pine timberland was converted to plantations, and the streamside hardwood areas were left as buffers between pine stands to enhance wildlife habitat values. Watershed issues are a consideration on the property, as only about 30% is flat, half is gently sloping, and 20% is steeply sloping. A portion of the hardwood drains is being developed as a wetlands restoration project. This also enhances wildlife values. Preferred wildlife species include white-tailed deer (*Odocoileus virginianus*), turkey (*Meleagris gallapavo*), and quail (*Colinus virginianus*). The McPhails are enthusiastic tree farmers and encourage forestry education activities on their property, and they have developed a trail system to provide better access and enhance recreational opportunities.

PLANNING ENVIRONMENT AND METHODOLOGY

Objectives for the Analyzed Area

Tree farmers usually tend to have financial objectives for their forests, which are often considered a financial asset and part of the owner's investment portfolio. Legacy issues—keeping the forest intact and in the family for the next generation— also tend to be dominant objectives. Tree farmers belong to a program that is based on forest sustainability, so management objectives focus on future forest conditions and multiple forest resources. The Tree Farm forest management template (American Forest Foundation, 2011) is accepted as meeting the requirements of the U.S. Department of Agriculture, Forest Service Forest Stewardship Program and the NRCS guidelines. Forest stewardship or forest sustainability is the framework of this management plan type and must be reflected in the management objectives.

The McPhail family defines its primary management objectives as a desire to practise sustainable forest management with an emphasis on commercial timber production and income generation. A secondary objective is to promote wildlife habitat. The family has a "hands-on" approach to forest management on the property. Dr. Walt McPhail takes the lead on management issues and states, "Our primary objective is to keep the Farm in the family—without being a burden on our children—by being profitable. We want our successful Tree Farm to stay in the family from generation to generation versus selling to housing developers. Plus this is our Individual Retirement Arrangement, 401(k) account (a tax-qualified, defined-contribution pension account), and nursing home protection" (Walt McPhail, personal communication, March 7, 2014). McPhail clearly takes a financial approach to the Tree Farm and added that financial performance depends on selecting the best properties at the right price. "We look for timber production and land appreciation potential. We research many tracts. For every one we inspect, very few are selected for purchase. We purchased nine tracts and all are profitable" (Walt McPhail, personal communication, March 7, 2014). The formal management objectives include the forest sustainability requirements and show that profitability is just one of the major management objectives. Formally listed, the management objectives for the McPhail Tree Farm are

- Quality and quantity of wood production
- Increased and enhanced wildlife diversity and habitat
- Soil and water conservation
- Esthetics
- Recreational activities
- Keeping the tree farm active and in the family

Constraints for the Analyzed Area

There are no unusual constraints on forest management for this property. Every property has area and land suitability constraints. Except for the streamside hardwood areas, all of the property is suitable for loblolly pine plantations. Caution must be used on portions of the property located on steep slopes, but no slope is so steep as to exclude intensive forest management. BMPs are required under ATFS forest certification and address problems created by steep slopes. While the forest owner has a primary management objective of financial profitability, he is willing to forgo some profit to maintain secondary goals like enhanced wildlife habitat, recreational opportunities, and soil and water conservation. Neighbors are located near the property boundaries and may be impacted by some management activities, especially smoke from controlled burns. A portion of the hardwood drain, 41 ac (16 ha), consists of a wetlands restoration project that includes four waterfowl impoundments and planted corn and millet. Wildlife plantings include bald cypress, sawtooth oak, and fruit trees. The wetlands restoration project is funded under the NRCS Wetlands Reserve Program (U.S. Department of Agriculture, 2014) and must follow program requirements. Since the goal of forest management is to produce retirement and other income, there are cash-flow constraints related to periodic revenue generation.

Guiding Laws, Regulations, and Policies

South Carolina has a set of BMPs that pertain to forest operations. BMPs are voluntary in South Carolina, but the ATFS forest certification standards make them mandatory. BMPs are considered the foundation of good forest stewardship and are designed, interpreted, monitored, and updated by the South Carolina Forestry Commission. They emphasize protection of lakes, streams, and ponds and were designed to meet the requirements of Section 404 (dredge and fill) of the federal Clean Water Act and to address jeopardizing endangered species as well as modification or destruction of critical habitat as defined in the Endangered Species Act. BMPs address SMZs, stream crossings, forest road construction, timber harvesting, site preparation, reforestation, prescribed burning, pesticides, fertilization, minor drainage, and wildlife management (South Carolina Forestry Commission, 1994). Since 2009, South Carolina has had a Right to Practice Forestry Act (H3651, R109, A48) that limits the authority of counties and municipalities to restrict or regulate certain forest activities. Such laws are common in the southern United States and protect forest owners' right to practice standard forest management on their property.

Federal and state income taxes impact the cash flows expected from the forest. The timing of cash flows can significantly affect the magnitude of taxes on ordinary income and capital gains income. Federal tax law has included different incentives for forestry, including deductions for reforestation, amortization of excess reforestation expenses, capital gains treatment of timber sale income, and potential exclusion of cost-share payments.

Federal and state laws also impact the forest as estate and inheritance taxes can be a problem for family forest owners. The timber and land are included in the value of the estate. Over time, the estate tax exclusion has fluctuated, but is now indexed for inflation. Family forest owners will need to have not only a management plan, but a succession plan discussing how the property is to be distributed among heirs. This plan will make the ownership transition easier for all involved and will allow management on the property to continue.

Certification Requirements

There are at least three major forest certification systems in the United States; the ATFS standards of sustainability were designed specifically for tree farmers or owners of small family forests. The standards were developed taking into consideration the scale of management utilized on these small forests. ATFS forest management plans satisfy the requirements to participate in U.S. Department of Agriculture conservation incentive and cost-share programs. They specifically meet both Forest Service Stewardship Program (U.S. Forest Service, 2013) and NRCS incentive program requirements. Thus, certain elements are required to be in all plans: (1) landowner objectives, (2) forest condition and health, (3) management activities/prescriptions, (4) tract maps, (5) soils and water resources, (6) wood and fiber production, (7) threatened and endangered species, (8) high-conservation value forests and other special sites, (9) invasive species and integrated pest management, (10) special cultural and environmental sites, (11) periodic monitoring to address changes that could require modifications of management objectives, and (12) practical efforts to eradicate or control invasive species (American Tree Farm System, 2011).

Forest and natural resource enhancement and protection efforts must be a strong component of these forest management plans. This includes protection of special sites and consideration of social impacts. Special sites encompass archeological, cultural, historical, geological, biological, ecologically valuable, and high-conservation value areas of the forest. Forest certification programs, including ATFS, describe high-conservation value areas as forests of exceptional value "due to their environmental, social, biodiversity, or landscape values" (American Tree Farm System, 2011). Special sites must be delineated and protected. Issues with adjacent stands and ownership concerns must also be addressed. This includes esthetics, wildfire concerns, privacy, wildlife habitat and movement patterns, invasive species, and urban encroachment. The visual impact of forestry practices is considered (public view) and could impact logging deck locations, timber sale layout, road locations, scale or timing of operations, and logging debris. Access and boundary issues should be addressed. Potential impacts of forestry operations on recreation and wildlife resources are also important. Furthermore, air, water, and soil protection must be considered in the management plan, according to the South Carolina BMPs.

An ATFS management plan addresses forest resources first, in a broad sense, around general management issues that affect the entire property. Later, stand-level information is developed to address specific management activities within stands. Broad issues include reforestation and afforestation (natural versus artificial regeneration and related site preparation impacts); prescribed fire or controlled burns as a management tool and fuel reduction technique; home and community *Firewise Safety Programs*; insects and diseases (including integrated pest management); invasive species control; inventory systems; and control, monitoring, and prevention guidelines to be used. All forest resource management plans have implementation constraints (e.g., available markets for wood products, forest owner capital and time limitations, land-use ordinances, operability problems due to issues like seasonal access problems, steep slopes, or wetlands) and these must be addressed.

The ATFS forest certification program is recognized by the Programme for the Endorsement of Forest Certification systems. Like the other major certification systems, the Sustainable Forestry Initiative and the Forest Stewardship Council, the ATFS has a set of rigorous standards of sustainability for forest certification intended to "promote the vitality of renewable forest resources while protecting environmental, economic, and social benefits of sustainable forestry" (American Tree Farm System, 2011). Each standard has various performance measures and indicators. The following eight standards, performance measures, and indicators are crucial when certified forests are audited to ensure compliance with the requirements for certification (American Tree Farm System, 2011):

1. A commitment to practising sustainable forestry. The performance measure is the existence of a written forest management plan suitable for the forest in terms of size, scale, and intensity of management.
2. A requirement to comply with applicable laws. This includes federal, state, county, and local laws and regulations.
3. A requirement of timely reforestation and afforestation. Regeneration of desired species, adequate stocking levels, and compliance with the forest owner's objectives are required.
4. A requirement of air, soil, site quality, and water protection. BMPs must be met or exceeded. Integrated pest management must be considered as an option for forest health problems. Pesticide use is addressed and prescribed fire must only be used in proper situations.
5. A requirement of the plan to contribute to the conservation of fish, wildlife, and biodiversity. Threatened or endangered species must be maintained or enhanced. Desired species are encouraged, while efforts must be made to control invasive species.
6. A recognition of forest esthetics. Visual impacts of forest management activities must be managed.
7. A requirement of the protection of special sites. Special sites must be maintained.
8. A requirement of forest products harvests and other activities to conform to the management plan and to consider other forest values. The potential of the forest to produce future forest products and other benefits must be maintained.

Methods for Developing the Plan

The planning process begins with the forester, or other natural resource professional, and the forest owner jointly taking active roles in the plan development. Preliminary information is collected prior to the first meeting of the forester and forest owner. This can include the owner's contact information, property description, property history, and maps. The property description should contain sufficient description for a forester to easily locate the tract, topography, access, slope, road locations and condition, and watershed data. Three forest area descriptions are fundamental to the management planning process: total ownership area, total forested area, and total area covered by the plan. Property history can provide crucial background information that impacts the management potential of the property. This includes length of current ownership, past management activities, and impacts of surrounding properties. Property maps are fundamental to all management plans. They include location maps, boundary maps, topographic maps, soils maps, stand maps, and aerial photographs. These maps should include locations of special sites, threatened and endangered species, water resources, roads, current management activity areas, cultural features, and conservation areas. The maps for the property are most easily utilized on a section basis, and we included the North Section map that represents the entire property well and illustrates typical stands.

Forest goals and objectives are the crux of the management plan. These represent the most important part of the information gathered in the process. They only come from the forest owner and reflect his or her expectations about the future forest conditions, personal values, and management potential for the forest. Goals are broad and epitomize the forest owner's vision for the property, while objectives are specific and can be measured in terms of being satisfied or not. Hunting for white-tailed deer, turkey, and rabbits (*Sylvilagus floridanus*) accounts for the highest level of recreational activity on the property. Youth hunts are promoted. Camping is common on the property. The McPhails consider esthetics a primary concern and desire "curbside appeal" for the property. Logging decks are kept to a minimum and placed at an angle well off the public roads. Logging debris is distributed throughout the forest and the logging decks are planted. The forest is clear-cut, but it is also a "clean cut," with no debris left on the ground. Chemical hardwood control and prescribed burning combine to control unwanted vegetation, creating a beautiful savannah effect. Forest roads are planted with a mixture of fescue (*Festuca* spp.), browntop millet (*Panicum ramosum*), Korean lespedeza (*Kummerowia stipulacea*), sericea lespedeza (*Lespedeza cuneata*), bahiagrass (*Paspalum notatum*), red clover (*Trifolium pratense*), and white clover (*Trifolium repens*) to prevent erosion and provide wildlife benefits for game species including quail, turkey, doves (*Zenaida macroura*), rabbits, and white-tailed deer.

The planning process behind the ATFS management plan template follows the standard steps commonly used in any planning process. In the beginning, the forester, or natural resource professional, and the forest owner are co-creators of the management plan as the forester is preparing the plan to meet the objectives of the landowner. The first step is a meeting of the forester and forest owner to gather the preliminary information, like maps, property descriptions, special sites, and other background. The primary piece of information supplied by the forest owner is the management goals and objectives. These can only come from the forest owner(s) and can only be prioritized by him or her. Often the goals of future owners are considered and a time frame must be established.

The second step is an assessment of resource condition and history. Some of this can be completed in the office using aerial photographs and other information. Much of it must be completed in the field. Perhaps existing information may be suitable for preliminary plan development of summary tables, charts, and maps. Prior ownership and history is important and must be considered, especially if there was a prior planning system. Data on the resources must be obtained for all physical resource attributes. External constraints need to be identified, like legal and regulatory issues.

The third step in the planning process is to develop plan alternatives. The forest owner is the decision maker and will evaluate all alternatives to determine which ones meet management objectives. The forest owner may formulate alternatives not considered in the proposed management plan. Forest resource management plans for public lands may require some sort of public assessment or hearing. Private sector plans sometimes also come under public scrutiny. However, most tree farms are relatively small enterprises and usually don't generate issues with the public.

The fourth step involves a decision (choice of alternative) by the forest owner. This is the alternative that will be implemented, subject to limitations set by the forest owner (budgetary limits, revenue expectations, and management limits). The forester is not the decision maker; all authority is derived from the forest owner.

The fifth step is preparation of plan documents and implementation of the plan strategy. The forest owner's goals and the resource assessment form the beginning of the plan. Stands, maps, charts, tables, and schedules are developed for the selected alternative. Current conditions and expected future conditions are reported. Expected outcomes and cash flows are projected. A complete, written, operational management plan is produced.

The last step is a feedback loop and may lead to changes in the plan and repeating the steps. Once the plan exists the forest owner(s) will evaluate it over time. He or she may direct changes. Once implementation begins, the need for a change of direction may become apparent (Davis et al., 2001).

Stand-level information is the foundation of a forest management plan (Bettinger et al., 2009). Stand objectives should relate to overall forest owner's goals. Current stand conditions are described in terms of past history, site index, topography, slope, stand quality and health, and timber inventory (age, species, volume, growth rate, size classes, height, stocking level). Desired future stand condition is also described. Both current and desired future stand condition are primarily described in terms of forest type and current age, then in terms of desired forest type and expected longevity. Natural regeneration or planting is discussed. The ATFS provides a template (American Forest Foundation, 2011) to help develop a management plan following its requirements. The template utilizes a "bird's eye view" of current and future stand conditions, which illustrates the typical spacing and arrangement of trees within wild stands, evenly spaced stands, and variable density stands. The intention is to better clarify for the forest owner, by graphical means, where the plan is headed silviculturally. That is, where the forest is now, in terms of condition, and what the future forest is expected to look like. Because most of the McPhail Tree Farm is managed using even-aged pine plantations, the diagram of future stand conditions, using the template provided by the ATFS, illustrates a fairly simple situation (evenly spaced trees). However, the diagram would also prove very useful in explaining silviculture in more complex situations (variable density spaced with openings). In sum, the template uses the diagram twice; once for current conditions and once for desired future conditions.

Forest management activities are outlined for each stand, including forest health management activities, like pruning, precommercial thinning, prescribed fire, and salvage cuts. Timber harvesting activities are specified by type of treatment as even-aged (clear-cut or thinning) or uneven-aged (group selection, single-tree selection, overstory removal, understory removal), and type of harvesting system to be used. Post-harvest activities must also be addressed and include slash management, log deck location, and road condition assessment. Relevant BMPs are discussed for each stand, and the monitoring of management activities planned should be identified. Obviously, these stand-level requirements are repeated for each stand.

Finally, management activity and an implementation schedule are discussed. The final section of the plan is a schedule of management activities on the forest, stand-by-stand and year-by-year. This allows the forest owner to track when each activity was completed, what revenue or cost was generated, and what incentive programs might have been utilized.

OUTCOMES OF THE PLAN

All ATFS management plans include multiple non-timber objectives. Many of these plans have a strong timber management focus as the forest owners are counting on predictable cash flows from their forests to fund retirement, college educations, or other activities. The McPhails represent an example of a timber focus, but one that also recognizes non-timber objectives. Seventeen percent of their forest holdings are in hardwood "reserves" and a wetland protection area, providing for wildlife, recreation, esthetic, water, and soil conservation objectives. The remainder of the property, while still contributing to non-timber goals, is being managed for intensive sawtimber production.

Dr. McPhail gives a presentation on his forest property with the title: *Does Money Grow on Trees*? He stresses any forest area will grow the same volume of wood: 1,000 tiny trees, 400 small trees, 200 medium trees, or 100 large trees. The difference in value in his example ranges from $3,245 U.S. dollars (USD) per ac ($8,019 per ha) for pulpwood sized trees to $7,265 USD per ac ($17,952 per ha) for sawtimber and pole sized trees. Quality and size determine harvest values and are improved with intensive forest management. High-valued sawtimber and poles are the final products and over 75% of total rotation revenues. Management regime protocols include chemical site preparation, plantings using the best available genetically improved seedlings, herbaceous weed control, hardwood control, frequent thinning, and final harvest. The target rotation on site index 70 (25 base age) is 35 years, producing an 18 in (46 cm) diameter at breast height, 3–4 knot, clear 16 ft (4.9 m) logs with 4–6 uniform annual growth rings per inch. This results in a high percentage of pole products. Current stand ages vary from 1 to 34 years. Only the oldest age classes contain sawtimber and pole products. Figure 12.3 illustrates the distribution of age classes of the loblolly pine plantations. Only the three oldest age classes contained high-quality sawtimber and poles. Of course, younger stands contained pulpwood and small sawtimber. Total timber volume of high-quality sawtimber and poles on the property is 5,644 thousand board feet (MBF) (13,300 m³). This is distributed as 2,372 MBF (5,600 m³) on the 34-year-old plantations; 3,142 MBF (7,400 m³) on the 30-year-old plantations, and 130 MBF (300 m³) on the 27-year-old plantations. Mature plantations are producing 14–15 MBF per ac (81–88 m³ per ha).

The pine stands are thinned as early and as often as possible to encourage high-density late wood (dark, dense rings) and diameter growth on crop trees. Thinning is controlled using residual basal area rules (generally thinnings) triggered when basal area reaches about 120 ft² (11.2 m²) and basal area is reduced to about 70 ft² (6.5 m²) with a goal of maintaining 35–40% crown ratio to ensure optimal growth throughout the rotation. Higher stand residual basal area promotes wood quality, through quicker natural pruning and higher form class, whereas lower stand residual basal area allows more sunlight to reach the forest floor, enhancing understory plant production and thus wildlife habitat. Stand improvement through periodic thinnings, the selling of products in good markets, and proper merchandizing of trees by loggers into high-valued products all enhance the profitability of the tree farm. The timber management regime for loblolly pine plantations on the McPhail Tree Farm is shown in Table 12.1. It is definitely intensive forest management. Significant costs are incurred, and, likewise, significant timber revenues are realized.

FIGURE 12.3 Forest area of loblolly pine plantations by age class on the McPhail Tree Farm.

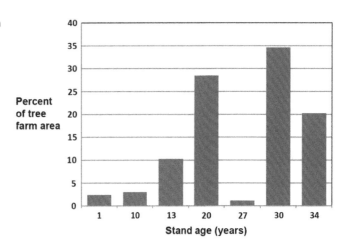

TABLE 12.1 McPhail Tree Farm Timber Management Regime for Loblolly Pine Stands

Year	Forest Management Practice	Cost/Revenue (US Dollars per ac)[a]
0	Chemical site preparation	(100)
1	Plant about 500 genetically improved seedlings per ac (1,235 per ha) machine planting preferred, on a 9×10ft (2.74×3.05m) spacing, between December and February	(150)
2	Herbaceous weed control in April or May	(35)
14	First thinning, with a residual density of 250 well-spaced crop trees per ac (618 per ha)	300
15	Chemical hardwood control	(60)
16	Fertilization	(150)
21	Second thinning, with a residual density of 150 crop trees per ac (371 per ha). The objective is to remove all pulpwood trees, and leave quality pole and sawtimber crop trees. This creates proper spacing to promote growth, nutrient utilization, and healthy trees	200
28	Third thinning, with a residual density of about 90–120 well-spaced quality pole and sawtimber trees per ac (222–297 per ha). Pole thinning is considered depending on stand quality	400
35	Final harvest timed to maximize profitability, based on forest growth and market conditions. Final harvest can be delayed to maximize returns (while thinnings should occur as scheduled despite market conditions in order to maximize final outputs)	3,500

[a] Costs are in parentheses, revenues are not.

A central focus of this management plan is placed on financial results. At the same time, the McPhails have an equal focus on forest sustainability. All resources on the forest are protected, particularly the forest soil as the basis of forest productivity. Major efforts have been extended toward wildlife and recreation goals at the same time. The large amount of protected hardwood area and wetlands speaks to that. The fact that this plan has a strong financial base does not distract from the strong forest sustainability framework.

DISCUSSION AND CONCLUSIONS

Because this is an ATFS management plan, outcomes should cross the various natural resources and include a broad forest sustainability focus. While timber was stressed as the engine that drives the financial outcome (profitability) and is an important revenue stream expected from the forest, all forest resources were addressed in the management plan and each one has important outcomes that impact forest sustainability. Soil and water conservation is a crucial part of this management plan. Trees planted on slopes are planted on top of raised beds on contours to catch and hold water. Water bars are situated on forest roads at 45° angles to divert water runoff into the forest and prevent erosion. Broad-based dips are also utilized to divert water when practical. Water bars, broad-based dips and areas where there is soil disturbance by heavy equipment are planted with a mixture of seeds as soon as practical, often before equipment leaves the site, to establish protective vegetation. Logging decks are stabilized to prevent erosion. Culverts are installed and portable bridges are used over larger streams in harvesting operations. SMZs are maintained along perennial and intermediate streams preventing runoff and providing wildlife corridors and habitat. Roads and areas that might not be conducive to heavy traffic in wet weather are protected with a layer of geotextile fiber and gravel.

Learnings and Insights

This forest management plan has the advantage of having a forest owner who was National Tree Farmer of the Year. The forest owner aggressively pursues forest sustainability objectives, and the plan represents forest certification at its best. At the same time, the forest owner aggressively pursues maximum timber productivity on a portion of this holding (while leaving a significant portion in a protected status). The timber management regime illustrates forest practices and investment levels to achieve that maximum forest productivity. It also illustrates simultaneous management for all non-timber objectives. The forest owner enthusiastically works toward a showcase for forest sustainability.

One problem that the management plan highlights is the need for timber harvest scheduling in order for the forest owner to achieve the family goals of developing a retirement cash flow from the property. Multi-aged pine plantations are ideal for this situation; however, the McPhail plantations are concentrated in a few age classes, mostly older ones. This management plan illustrates the tradeoff the forest owner will have to make in terms of forest age-class structure, stable cash flows for the future generation, and harvesting current forest stands at the optimal age of about 35 years. This represents an opportunity cost, or a cost of forgoing optimal timber harvesting today in order to structure the forest age classes for a more even cash flow over time. For example, in a fully regulated forest with an optimum rotation age of 35 years that produced a perfectly uniform annual cash flow, 1/35 of the total timber would be harvested annually. This forest has 456 ac (185 ha) of pine plantations aged 30 years and above. This means that if perfect cash flow is desired, 1/35 of the oldest plantation area would be harvested annually (or perhaps 5/35 every 5 years). It would take about 15 years to cut through the oldest plantations, and the harvest age of the oldest stands would be about 45 years. To maximize the cash flow, all timber should be cut at age 35 years, so the extra 15 years of "slow" growth would represent that opportunity cost. The forest owner would be the one to evaluate that tradeoff between future cash-flow stability and current sub-optimal timber harvesting. This is an ideal problem to explain the optimization, alternatives, and temporal accommodation involved in timber harvest scheduling.

Sustainability Issues

The ATFS forest management plan is an example of a plan type that is well-suited to timber management, but strongly supports forest sustainability objectives at the same time. It also illustrates the impact of forest certification on management plan development. The McPhail family has a strong timber production orientation. This is not rare for tree farm management plans. Timber production and financial profit are therefore important parts of the management plan, and more than 75% of that profit comes from high-valued sawtimber and poles. Dr. McPhail describes "growing trees as like growing your garden." He adds, "Plant in good soil, prepare a good seed bed, plant the best seedlings, control undesirable competition, provide nutrients and abundant growing space for a bountiful harvest" (Walt McPhail, personal communication, March 7, 2014). The management regime for loblolly pine (Table 12.1) describes exactly that same approach. A key management philosophy is not to delay thinning due to bad markets, as this impacts forest health.

Plan Development Challenges

Traditional forest regulation sometimes uses the concept of area control to stabilize cash or wood flows over time. While this may not have been an explicit goal of forest management, the staggered age classes of loblolly pine plantations on the McPhail property are actually well-suited to provide a stable cash flow as the family retirement fund. This is easily seen by closely observing the age classes in Figure 12.3. Notice there are 833 ac (337 ha) of loblolly pine plantations, ranging in age from 1 to 34 years. Applying simple area control, and a 35-year optimal timber rotation age that produces 14–15 MBF per ac (81–88 m³ per ha), approximately 1/35 of the pine plantations could be harvested annually over 35 years. If each clear-cut was promptly replanted, a series of stands would develop over time, each one 23.8 ac (9.6 ha) in size. A steady stream of wood and revenue would develop from the roughly 350 MBF (826 m³) harvested annually. However, the oldest plantation is currently 34 years old and would take 7 years to harvest (168 ac (68 ha) or 23.8 ac (9.7 ha) per year). The second oldest age class is currently 30 years old and would be 37 when the oldest age class was completely cut. It would take about 12 years to harvest the second age class at the rate suggested. Obviously, the plantations would need to be harvested on an accelerated schedule or some harvested timber would be 59 years old when scheduled. However, the two age-class harvests could be timed to develop more stable cash flows for the grandchildren in 35 years. The current 20-year-old stands will take just over 10 years to harvest and would be 39 years old when the older stands are completely harvested. So the same problem exists there. Cash flows and area harvested will have to be planned if more stable cash flows are desired for the next generation.

Plan Implementation Challenges

The property contains no endangered or threatened species. The quail population in the state is on the decline, mainly due to a loss of habitat. To improve quail habitat, pine thinning protocols include thinning every 5–7 years, burning, disking, and mowing. This results in soil disturbances that encourage early successional vegetation that improve quail and turkey habitat. The property includes protected Native American campsites with artifacts. Hanna Hall Cemetery, with tombstones dating back to the late 1700s, is also protected. It is a small knoll surrounded by old oaks and persimmon trees.

Outbreaks of the Southern Pine Beetle (*Dendroctonus frontalis*) affected two newly acquired tracts of 15-year-old heavily stocked loblolly pine plantations that were under stress from recent droughts. Salvage cuts with wide edges (twice the average tree height) limited the amount of spread. Some unmerchantable areas were cut and left on the ground to prevent spread. Areas of beetle infestation on neighboring unthinned pine plantations were discovered. The infestation actually stopped at the McPhail boundary line. The McPhail plantation had been thinned, fertilized, and hardwood-controlled, producing a healthy forest that vigorously fought off the pine beetles.

Other Interesting Issues Related to the Plan

ATFS members tend to be enthusiastic forest owners, dedicated to sustainable forest management. Forest certification requirements often are welcome road maps, rather than obstacles that others may see. ATFS members post large signs on the roadway, "bragging" about their membership. Usually, the entire family is involved, and the tree farm becomes part of the family heritage. This pride of achievement and example of forest sustainability often lead to tree farm owners using their forests as educational tools. Often, tree farms are used as demonstration areas or for forestry tours. The McPhail Tree Farm is one of those and has hosted educational programs for forestry associations, Boy Scout groups, university forestry students, school groups, legislative tours, and university short courses. More than 1,200 visitors have toured the forest property. The McPhails actually include this as an objective of the management plan.

Family forests are the focus of federal and state forestry incentive programs (Butler, 2008). ATFS management plans encourage the use of these incentives to foster environmental and conservation goals. The plan described here benefited from past use of federal conservation incentive programs like the Conservation Reserve Program, the Wetlands Reserve Program, the Environmental Quality Incentives Program, and the Southern Pine Beetle Prevention and Restoration Program (South Carolina Forestry Commission, 2010). These conservation programs were used to partially fund plantations on erodible cropland, replant harvested pine plantations, repair gullied roads and add water bars to meet BMP requirements, wetlands restoration, prescribed burning, and pine plantation thinnings to help prevent southern pine beetle infestations. This forest management plan provides insights into how forestry incentive programs can be used to improve forest management on family forests.

REFERENCES

American Forest Foundation, 2011. Managing Your Woodlands: A Template for Your Plans for the Future. American Forest Foundation, Washington, DC, 14 p.

American Tree Farm System, 2011. 2010–2015 Standards of Sustainability for Forest Certification. American Forest Foundation, Washington, DC, 8 p.

American Tree Farm System, 2014. About American Tree Farm System. American Forest Foundation, Washington, DC. https://www.treefarmsystem.org/about-tree-farm-system (Accessed April 17, 2014).

Bettinger, P., Boston, K., Siry, J.P., Grebner, D.L., 2009. Forest Management and Planning. Academic Press, Burlington, MA, 331 p.

Butler, B.J., 2008. Family Forest Owners of the United States, 2006. U.S. Department of Agriculture, Forest Service, Northern Research Station, Newtown Square, PA General Technical Report NRS-27, 72 p.

Davis, L.S., Johnson, K.N., Bettinger, P.S., Howard, T.E., 2001. Forest Management: To Sustain Ecological, Economic, and Social Values, fourth ed. McGraw Hill, Boston, MA, 804 p.

South Carolina Forestry Commission, 1994. Best Management Practices for Forestry. South Carolina Forestry Commission, Columbia, SC, 65 p.

South Carolina Forestry Commission, 2010. Southern Pine Beetle Prevention and Restoration Cost Share Program. South Carolina Forestry Commission, Columbia, SC, 20 p.

U.S. Department of Agriculture, 2014. Wetlands Reserve Program. U.S. Department of Agriculture, Natural Resources Conservation Service, Washington, DC. http://www.nrcs.usda.gov/wps/portal/nrcs/main/national/programs/easements/wetlands/ (Accessed April 17, 2014).

U.S. Forest Service, 2013. Forest Stewardship Program. U.S. Department of Agriculture, Forest Service, Washington, D.C. http://www.fs.fed.us/spf/coop/programs/loa/fsp.shtml (Accessed April 17, 2014).

Chapter 13

Molinillos Private Forest Estate, Durango, Mexico

Gustavo Perez-Verdin[1], Juan Manuel Cassian-Santos[2], Klaus von Gadow[3] and Jose Carlos Monarrez-Gonzalez[1]

[1]Instituto Politécnico Nacional, CIIDIR Durango, Durango, Mexico, [2]Col. Periferia, Durango, Mexico, [3]Burckhardt Institut, Georg-August University, Göttingen, Germany

ABBREVIATIONS

CCF	Continuous Cover Forestry
CONAFOR	National Forestry Commission
FMP	Forest Management Plan
MDS	Silvicultural Development Method
MMOBI	Mexican Forest Management Method
UMA	Wildlife Management Unit

MANAGEMENT SETTING AND BACKGROUND

Molinillos is a 2,866 ha (7,082 ac) private forest estate composed of five adjoining lots owned by five families. The estate is located in the southeastern part of the state of Durango (Figure 13.1), in a region where access is restricted and most properties are communal. The main business in Molinillos is forestry and ecotourism, with agricultural crops and livestock providing additional income in some areas. The current five owners acquired the land in 1994 from different sellers and decided to consolidate the properties to operate as one estate. Mexican agrarian law does not allow private ownerships to exceed an area of 800 ha (1,977 ac), but does allow the unification of land by several owners. The pine-oak forests of the region, where the ownership is located, have been subject to timber management since the 1920s with the introduction of the Durango-El Salto railroad (Aserraderos Station). At that time, forest studies concentrated on estimating timber volumes based on an inventory of trees in 20 m by 50 m (66 ft by 164 ft) transects, oriented from east to west or from north to south. Volumes per ha were estimated for 5 cm (2 in) diameter classes, using a volume tariff of a single input, derived from a sample of trees that included diameter and height data. Harvest volumes were estimated using a simple area-volume method and associated cutting cycle. In these early years, commercial interests prevailed where only the best and largest trees were harvested, with minimal investment in silvicultural treatments (such as thinning or pruning) or protection and conservation activities. Eventually, these interests led to high-grading of the forests.

The management operations were mostly concentrated in forest areas with easy access, often using only the most valuable butt log and leaving the remainder of the tree bole on the ground. The result of employing these management practices was a significant reduction of trees with larger diameters, the establishment of natural regeneration in open spaces, and the interspersed presence of older residual trees, presenting an appearance of forest structures with one or two stories (height classes). For many years, large areas with these structures were regarded as "irregular forests."

The Molinillos ownership is situated in the Sierra Madre Occidental and has demonstrated a leading role in the fields of silviculture, biodiversity conservation, and ecotourism. Therefore, various important research and educational institutions like the College of Postgraduate, the National Institute of Forest, Agricultural and Livestock Research, and the National Polytechnic Institute (the senior author's institution) are active in the area. The ownership provides a rich diversity of forest ecosystems and management types for these institutions. They also ensure security so that the institutions can leave

Forest Plans of North America. http://dx.doi.org/10.1016/B978-0-12-799936-4.00013-8

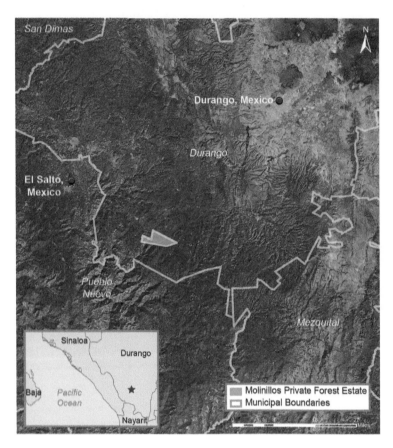

FIGURE 13.1 Location of the forest ownership Molinillos.

their instruments and equipment working all year long. The research topics addressed by these institutions include forest nutrition, growth and yield, site preparation and reforestation, environmental services, the monitoring of microclimatic conditions (there are two automated climate stations), the establishment of monitoring plots to evaluate the impact of growth dynamics, and assessments of wildlife habitats.

Almost 75% of the Molinillos estate is allocated to commercial timber production (Table 13.1). The predominant vegetation consists of pine-oak forests. About 57% of the area is stocked with several species of pine, 39% with different oak species, 1% with other hardwoods, and 3% with other conifers. The most prominent species are Durango pine (*Pinus durangensis*), Apache pine (*Pinus engelmannii*), Chihuahua pine (*Pinus leiophylla*), Lumholtz pine (*Pinus lumholtzii*), teocote pine (*Pinus teocote*), alligator juniper (*Juniperus deppeana*), madrones (*Arbutus* spp.), Durango Emory oak (*Quercus durifolia*), and a number of other oaks, including *Quercus coccolobifoli*, *Quercus eduardi*, *Quercus laeta*, *Quercus magnolifolia*, *Quercus obtusata*, *Quercus rugosa*, and *Quercus sideroxyla*.

According to the current forest management plan (FMP) (Consultoria Forestal, 2013), the average age of the pines is 44 years, while the periodic and mean annual increments are estimated at 4.6 and 1.9 m³ per ha per year (65.9 and 26.6 ft³ per ac per year), respectively. The average site index is 16 m (52 ft) representing the dominant tree height at a base age of 50 years. The estimated average timber growing stock of pines is rather low (75 m³ per ha (1,072 ft³ per ac)), resulting in an estimated total stock of 162,000 m³ (about 5.7 million ft³).

The climate in the forest area has been described as mostly "semi-cold" and "wet." The temperature range in the coldest month is between −3 °C (26.6 °F) and 18 °C (64.4 °F), while the annual average fluctuates between 5 °C (41.0 °F) and 12 °C (53.6 °F) (González-Elizondo et al., 2012). Precipitation occurs from July to September with an annual average rainfall of 800 mm (31.5 in) and occasional hailstorms. Winter rainfall is between 5% and 10% of the annual total, while the *R/T* ratio (representing total annual rainfall in mm/average annual temperature in °C) exceeds 55. Plateaus separated by streams represent the most common topographic feature in Molinillos. Steep and inaccessible canyons also exist. The typical slope ranges between 21% and 40%. Runoff flows toward the hydrographic system of the Acaponeta River basin and eventually to the Pacific Ocean. Total annual runoff in the ownership is estimated at 10 million m³ (over 353 million ft³).

Ecotourism began to grow in 2006 and has been expanding ever since. The ecotourism project includes the construction and operation of a 20-room hotel, a restaurant, a convention center, cabins, bridges, a 10 km (6.2 mi) network of hiking

TABLE 13.1 Land Use Categories and Associated Areas in the Molinillos Private Forest Estate

Land Use	Hectares	Acres	Percent of Total
Conservation			
Wildlife habitat	213	526.3	7.43
Riparian buffers	147	363.2	5.13
Road buffers	22	54.4	0.77
High conservation value forests	17	42.0	0.59
Production			
Timber production	2,049	5,063.1	71.49
Restricted timber production	99	244.6	3.45
Experimental plots	<1	2.5	0.03
Restoration	10	24.7	0.36
Other uses (grasslands, roads, human settlements)	308	761.1	10.75
Total	2,866	7,081.9	100.00

trails, viewpoints, an observatory, and booths for wildlife watching, which offer a great variety of services for nature tourism. Also, as part of the project, the ownership offers tourist packages for hunters that include transportation, food, and hotel accommodation. The National Forestry Commission (CONAFOR) subsidized part of the developed infrastructure, but the landowners carried a large part of the investment.

PLANNING ENVIRONMENT AND METHODOLOGY

From 1940 to 1980, the federal government approved several regulations related to the application of the Mexican forest management method (MMOBI, for its Spanish acronym), which basically stated that irregular forests (uneven-aged stands) with a diameter distribution of trees as described by de Liocourt (1898) were to be managed irregularly (Moreno-Sanchez and Torres-Rojo, 2010). Volume estimation was based on a compound interest formula, in which the value of the interest was equivalent to the value of the periodic annual increment expressed as a percentage of standing volume. Subsequently, the general practice was to apply a maximum harvest intensity of 35% of the existing growing stock. Once again, the formulae and calculations were largely theoretical since timber harvesting concentrated on the removal of the best and biggest trees on site. The effect of this management was the establishment of natural regeneration resulting from irregular timber removal or fires, which created new stands with irregular features. Another problem was the lack of attention to areas with dense natural regeneration, which required thinning. These stands suffered from severe competition for light, nutrients, and water, and from high mortality rates.

The first FMP for Molinillos was compiled in 1987 in which the total pine and oak volumes were estimated to amount to 37,816 and 14,340 m^3 (1,335,661 and 506,489 ft^3), respectively, which translates to 18.4 and 6.9 m^3 per ha (263 and 99 ft^3 per ac). This forest plan, based on the government-promoted silvicultural development method (MDS), was intended to create even-aged, regular forests. The idea of converting all stands to even-aged forest management was to apparently maximize the use of the site with more intensive management. However, the method ignored site-specific conditions such as slope, aspect, and soil nutrients, and was the main cause of changes in the structure and species composition in many areas.

From 1996 to 1999, the ownership did not register any timber production. In 1999, a new FMP for the 1999–2008 cycle was completed. This new forest plan was developed following a combination of the MMOBI and MDS, incorporating irregular and regular forest management principles. The forest plan estimated a total timber volume of 26,814 m^3 (947,070 ft^3) of pine (13.1 m^3 per ha (187 ft^3 per ac)) and 11,107 m^3 (392,299 ft^3) of oak (5.4 m^3 per ha (77 ft^3 per ac)), a rotation of 50 years plus a final "liberation" period of 10 years after which the seed trees were removed.

The latest and currently active FMP is the so-called *2009 Advanced Level Forest Management Program*. This plan considers two types of management: one for irregular forest areas where the focus is on selective harvesting, and the other for areas with even-aged or regular features. The annual average timber removal in both types for the remaining years of the forest plan is about 5,355 m³ (189,139 ft³) of pine (29 m³ per ha (415 ft³ per ac)) and 3,426 m³ (121,006 ft³) of oak (18.5 m³ per ha (264 ft³ per ac)). It is important to emphasize that this program also considered the application of best management practices for the conservation of biodiversity—the identification of habitats and sites with attributes of high conservation value.

There is a noticeable increase in timber production since the start of the first forest plan for this ownership. This is explained by the steady incorporation of additional areas into regular forest management and the application of more intensive silvicultural treatments, such as the use of seed trees and clear-cuts, which both feature prominently in the current management plan. An additional "twin" management plan also has been compiled that considers habitat requirements for the Wildlife Management Unit (UMA), which incorporates hunting. Ecotourism is also an important part of estate planning and includes a range of activities such as horseback riding, bird watching, sightseeing, and mountain biking. The ownership has developed special infrastructure features (some of which is in progress) such as an on-site restaurant, a hotel with neighboring cabins, a convention center, and other facilities. Molinillos has been developing one of the most extensive ecotourism facilities in the state of Durango and possibly in the entire Sierra Madre Occidental.

Through the existing forestry program, the owners wish to promote the sustainable use of forest resources through intensive silvicultural techniques designed to optimize productivity and increase timber production. At the same time, restoration activities and protection of the ecosystem should take place, promoting and adding to the conservation of biodiversity.

Previous forest plans have included a diversity of silvicultural options such as selective harvesting, harvesting with seed tree retention, and clear-cuts, as well as intermediate treatments like thinning and pruning. This diversity has created situations where stands have been overexploited to the extent that many of them lack harvestable diameters, particularly above 40 cm (15.7 in). The current management plan includes the application of non-intensive methods (selective harvesting) and intensive methods (seed tree retention or clear-cuts). Attempts are made to re-establish a forest condition that satisfies many objectives, including higher biodiversity, improved wildlife habitat, lower operational costs, and superior net economic benefits. The need for this new multi-objective approach arose quite naturally from the existing diversity of species composition, forest structure, soil quality, and topography. The current forest plan is valid for one harvesting cycle of 12 years and, based on the Forest Law (Diario Oficial de la Federación, 2003), the owners are required to present a new plan at the end of the cycle. The quantitative, silvicultural, and ecological flow of information is supported by a geographic information system.

The MMOBI silvicultural method maintains existing irregular forest structures through the removal of single trees or groups of trees. The method has been widely applied in multispecies ecosystems, favoring shade-tolerant species whose new regeneration can establish itself and survive under the conditions created by the gaps resulting from the extraction of superior trees (Moreno-Sanchez and Torres-Rojo, 2010; Pukkala and von Gadow, 2012). The objective of the selection treatment is to generate a continuous forest cover, which is often believed to be superior to single-species planted forests in addressing a wide range of objectives (Wehenkel et al., 2014). In this system, the dominant trees of merchantable size are harvested. There is no need for costly thinning or planting. Originally, this method ignored the treatment of young, dense forests in which products were not easily commercialized. Now, as new markets for smaller size trees emerge, the method is being applied to all types of irregular forests. A forester marks the trees to be removed. Special harvesting equipment is used to extract the logs, which may include less valuable timber used for cellulose, poles, or small-dimensions lumber. To determine the application of the selection method, the areas are evaluated considering some physical and biological criteria, such as slope (>60%), site index (<11 m), the Reineke stand density index (<247), and the proportion of oak trees (>30%). Cutting intensity ranges from 25% to 35% of the total volume in the stand.

MDS is based on the retention of seed trees, which remain after the final harvest to produce natural regeneration. The aim is to establish a regular forest, preferably with one commercially desirable pine species (Moreno-Sanchez and Torres-Rojo, 2010). The MDS approach involves natural or artificial regeneration, depending on site conditions and includes a precommercial "release" thinning and intermediate commercial thinnings, which vary with stand age and site. The objective of the precommercial release thinning is to reduce competition in the newly established stand while the commercial thinnings provide some small-size products and concentrate the growth on the residual trees. The cutting intensities of these treatments vary from 15% to 100% and depend on the diameter structure and growth.

Clear-cuts, which may be followed by natural or artificial regeneration, are also included in the forest plan. To ensure the establishment of an artificial regeneration, the following site preparation activities are performed: subsoil ripping using heavy machinery (preferably perpendicular to the slope), retention of part of harvest residues (the other part is chipped and dispersed within the treatment area), fencing against livestock and during the rainy season (June–August), and reforestation with native trees. The clear-cut areas need to provide connectivity in the landscape with minimum visual impact. This is generally achieved by retaining a 30 m (98.4 ft) wide corridor that connects the clear-felled sections with natural areas like riparian communities or stands with a high proportion of oak.

Small group fellings are also applied, but the harvested area should not exceed 20% of the total stand area. In this case, the clear-felling treatment is referred to as a group selection harvest followed by immediate replanting. The idea is to continue maintaining uneven-aged forests, although this is done using larger clearings instead of a single-tree selection. The clear-felled area is usually reforested immediately after the harvest. The remaining area (80%) is then managed by individual selection of trees. The process is repeated in the following cutting cycles until all five-fifths are clear-felled. Indicators of the success of a well-established regeneration include a minimum of 1200 trees per ha (486 trees per ac) and no areas without surviving trees greater than 25 m² (269 ft²) (Consultoria Forestal, 2013). To determine the best silvicultural treatment, a rule-based decision approach is used, which starts by discriminating poor stands from high-quality stands. If any of the following factors are true, a selection system is employed, otherwise a seed tree or clear-cut system is employed:

1. The stand is stocked with less than 40 m³ per ha (572 ft³ per ac) of pine.
2. The ground slope is greater than 60%.
3. The site has a degradation index greater than 15.
4. The site has a Reineke stand density index (pine) less than 247, or a site index less than 15.
5. The stand has more than 30% of trees in oak species.
6. The median tree age is greater than 100 years.
7. The stand is not a candidate for even-aged forest management.

The degradation index is the result of an estimate of the propensity of a forest soil to be damaged by erosion. It considers the slope, land use, soil type, and level of compaction (Carmona, 1986). The higher the index, the more degraded the stand. Within the ownership, complementary treatments are applied such as:

- Control and arrangement of harvest residues (chipped and scattered, or arranged perpendicularly to the slope)
- Stand improvement with reforestation of native plants in places where natural regeneration is not established
- Prevention and control of forest fires with a paid firefighting crew, construction of fire breaks, and permanent surveillance in the property
- Weed control in regeneration areas containing highly valued tree species
- Establishment of rock or wood-based dams to reduce sediment transportation and erosion
- Pruning, noncommercial thinning, and construction of burrows and artificial nests to improve wildlife habitat

Some of the environmental restrictions are primarily normative but others are self-imposed. These restrictions ensure compliance with the objectives of the management plan, but at the same time aim to safeguard the sustainability of forest resources, which is the ultimate goal of the owners. For example, stands that are to be excluded from timber production are identified using the decision tree illustrated in Figure 13.2. Other measures to prevent and mitigate environmental impacts are established during the various planning stages, and include:

- 20 m (65.6 ft) unmanaged buffers to protect streams, water bodies, as well as primary and secondary roads
- 10–20 snags per ha (4–8 trees per ac) for nesting and wildlife habitat
- Adequate disposal of organic and inorganic waste generated by harvest operations
- Road closure in some areas
- Minimization of soil erosion in procurement operations (Figure 13.3)

The owners receive about 30 permits per year to hunt white-tailed deer (*Odocoileus virginianus*) and wild turkey (*Meleagris gallopavo*). The permits are managed under the UMA and the corresponding wildlife management plan. Because the population of white-tailed deer is still recovering from intensive hunting, permits are being used for wild turkey only. However, the owners pay all registration fees and comply with all required procedures to keep active the original UMA plan, which includes strategies and activities to increase deer population. Like the forest plan, the wildlife management plan needs approval from the environmental office, which issues the tags for each individual. The wildlife management plan is based on the General Law of Ecological Equilibrium and Environmental Protection (Diario Oficial de la Federación, 1988), which also oversees compliance of harvest operations such as the correct application of silvicultural treatments, the connectivity of stands, nesting areas, refuge areas, and food for wildlife, among others.

The Forest Stewardship Council (FSC) initially certified the Molinillos ownership in June 2002 and granted a certificate for being a well-managed source of timber products with compliance to strict environmental and socioeconomic forestry standards. The certification, however, was discontinued 3 years later due to the high costs associated with the monitoring process. In April 2012, CONAFOR awarded the ownership with a certificate for "Adequate Compliance to Forest Management" through a technical audit procedure in compliance with the Sustainable Forest Development Law (Diario

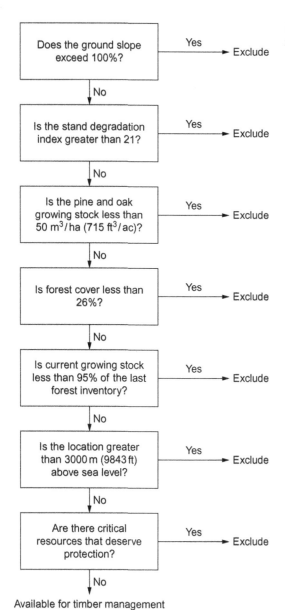

FIGURE 13.2 A decision tree for determining the areas available for timber management.

Available for timber management

FIGURE 13.3 Cattle log loading is a protection measure to minimize soil erosion during forest extraction in the private ownership Molinillos.

Oficial de la Federación, 2003). To compensate for the lack of a FSC certification, the owners are turning to a national system of certification, which involves the fulfillment of the Mexican norm NOM-AA-143-SCFI-2008143 for sustainable forest management.

By law, both the owners and the forest manager are responsible for the implementation of the FMP. To ensure proper planning, owners get support from two external consultants besides the forest manager (who executes the management plan). One of the consultants is a moderator and is responsible for the management of resources in government programs. The other serves as head of the monitoring activities and makes sure activities are carried out within the limits of the permits. In addition, the monitoring head is a specialized advisor and technical link between the forest plan and non-timber management, such as ecotourism, research, or actions related to the social and governmental sectors. To execute the forest plan, the ownership hired permanent field staff dedicated to surveillance, protection, conservation, and harvesting operations. The staff receives frequent training, particularly those involved with ecotourism. They also have a modern radio communication system and recently established a system for personal satellite-based phone communication.

OUTCOMES OF THE PLAN

Through the implementation of current and previous FMPs, the ownership features two distinctive areas. One is intended for timber production and the other for ecotourism activities aimed at landscape and biodiversity conservation. Based on the application of the forest plan, a number of results can be expected. First, the estate has two areas defined by the type of desired target structures, one involving intensive, even-aged management (seed trees and clear-cuts) and the other extensive, irregular management (selection cuts). The proposed rotation length in area one is 60 years and the cutting cycle in the other is 12 years, to ensure the recovery of growing stock volumes in the area, given the diversity of structures to be generated. Given also the application of complementary treatments to residual trees, such as pruning, thinning, reforestation, and forestation with quality trees, the forest plan ensures the continued supply of desirable products, both in quantity and quality.

The application of best management practices improves soil conservation and water management. The forest plan includes the establishment of protection buffers, fencing of areas to reduce overgrazing, encouragement of natural or artificial (induced) regeneration, soil conservation activities (construction of filter dams), and chipping and spreading forest residues to promote infiltration, among other actions. The goal is to reduce the environmental impact caused by timber operations. The wildlife management plan, through the UMA, which oversees the management of white-tailed deer and wild turkey populations, is compatible with the FMP. Both plans are applied simultaneously throughout the ownership. Given the background of certification for sustainable forest management, the owners have shown interest in moving toward environmental, social, and economic sustainability within a national and international perspective. The existing bond between the ownership and several research institutions has also paved the way to improve current technologies related to the multiple-use management of temperate forest ecosystems in northern Mexico.

Given the conditions of the venue in forestry, Molinillos serves as an agent of change for the people who are living in surrounding areas. The majority of these areas are communal lands, which closely follow what the ownership is doing. Both the Molinillos estate and surrounding communities often benefit from regional financial support involving road improvement and construction of hospitals and schools. The ecotourism project may turn out to be one of the most productive activities of the ownership, and may change, in the medium-term, the balance of how the environmental services are valued by generating additional income and creating jobs.

Timber production has been stimulating other economic activities. The owners have invested in equipment and infrastructure for logging operations and are operating as forest contractors in other areas. The equipment includes truck loaders, a skidder, a bulldozer, and cattle for log loading, trucks, an administration building, and a lumber yard which stores over 10,000 m³ of wood (353,200 ft³). While they do not own a sawmill, the owners pay for the processing of roundwood into lumber. Thus, they sell both roundwood and lumber, as required by particular customer. The infrastructure, equipment, and technical know-how available in Molinillos has positioned the estate as a regional "model forest" regarding provision of environmental services, creating jobs, and generating tax income.

DISCUSSION AND CONCLUSIONS

According to the owners, the application of the FMP is a serious endeavor and a commitment to their descendants and future generations in general. The forest plan has evolved from previous owners whose only purpose was to maximize net benefits from timber sales to a current, holistic management of natural resources.

Learnings and Insights

The owners of Molinillos want their descendants to see timber and non-timber resources as a worthwhile alternative for economic livelihood. The forest resources are renewable, economically profitable, and safe to produce. The planning experience has made them see the usefulness of the forest resource, and they hope to see this transmitted to future generations. Practices such as noncommercial thinning, construction of small dams, and fencing of areas with regeneration, for example, are now not only seen as a cost item, but also as a long-term investment. This current attitude of the owners represents a big change when compared to the way their forest resources have been managed since the 1920s.

Sustainability Issues

The owners learned that forest resources are important not only because of the various ecosystem services they provide but also because of the jobs that are created. According to the owners, 15 families depend directly on the benefits obtained from the resources, including the 5 owners, and about 30 families indirectly depend on the forest and wildlife. To preserve these opportunities, the owners expect to comply with local and federal legislation as well as with the application of best management practices suggested by an eventual forest certification. They further expect to reduce intensive treatments, such as clear-cuts, and replace them with less-intensive treatments (single-tree selection). The reason is the compatibility of the latter with ecotourism and the high costs associated with the former (namely soil preparation, fencing, continuous sapling replacement). Finally, they expect to diversify forest management through hunting, ecotourism, timber, and non-timber product opportunities, and federal programs related to the payment for environmental services.

Plan Development Challenges

The Molinillos ownership has been subject to forest management since the 1920s. At the beginning, the landowners were reluctant to change the traditional view of single-objective forest management to a more integrated planning approach involving all natural resources. They rejected the application of basic silvicultural treatments such as thinning of dense, young forest stands, which eventually resulted in the formation of irregular, unbalanced forest structures. During the 1970s, a new way of timber management (MDS) acknowledged the need to treat forests in their earlier stages, but it failed to recognize the uneven structure of some areas and enforced all stands to be incorrectly treated through even-aged forest management. It also failed to address the environmental impacts of timber management on other resources like water, soil, and wildlife. During the early 2000s, the owners, along with their forest manager, changed this perspective and allowed the inclusion of both even-aged and uneven-aged forest management, depending upon the species composition, site productivity, and the owners' own decisions. They also started addressing environmental impacts. Now, recognizing the advantages of irregular forest management, their plan is to reduce the application of intensive silvicultural treatments (i.e., clear-cuts) and increase the application of continuous cover forestry (CCF) by individual tree or group selection.

A particularly challenging objective of CCF silviculture is to derive economic benefits without modifying the key features of the natural ecosystem. A range of practical silvicultural rules and "recipes" has been developed in various regions of Europe, North America, and in some tropical and subtropical forests of South Africa, Asia, and South America. In all these applications, forest development oscillates around some level of stocking, which is assumed to be near-natural (Schütz et al., 2012). In theory, the degree of naturalness of an ecosystem is the difference between its current observed state relative to some assumed natural state. Unfortunately, however, the *natural state* is not known. Naturalness is a moving target because ecosystems are subject to continual change even when they are protected from human use. The natural state is something that cannot be defined. We cannot measure and objectively evaluate the degree of naturalness. However, it is possible to describe some aspects of the ecosystem that appear to be natural, such as the distribution of tree species, and to relate these to the effects of human interventions. The CCF retention strategies that were recently proposed for multispecies forests may turn out to be more suitable for Molinillos than the simple de Liocourt curves (von Gadow et al., 2013).

Plan Implementation Challenges

As mentioned before, the property became FSC certified in 2002, but the certification was discontinued due to the high transaction costs. Now, with the support of the federal government and of the United Nations Development Program, the possibility of a recertification by an official Mexican norm (mentioned earlier) is being negotiated. The certification might not be the ultimate goal, but it does help to ensure the commitment of the owners to comply with high-quality forest management standards.

The estate has positioned itself as an example of economic diversification and use of natural resources that include the sale of tourist packages, hunting, and resource protection. Included in the ecotourism plan is a convention center, a restaurant, and a hotel (in addition to five existing log cabins). This development implies decreasing net profits, at least initially. The challenge also requires the improvement of existing roads, including the pavement of a 10 km (6.2 mi) segment and the introduction of electricity to the ownership.

Other Relevant Issues Related to the Plan

The owners are facing many challenges, uncertainties, and even threats in the quest for an effective and holistic approach to sustainable forest management. One important aspect to consider is that, despite the numerous challenges that have already been overcome, the owners feel that they still need to adapt to changing markets as well as to political and socioeconomic changes. For instance, during the past 5 years, the revenues generated by ecotourism were negligible because of the violence caused by drug trafficking, which seriously reduced the number of visitors to the ownership. Yet, using revenues from timber sales, the owners sacrificed part of their income to establish the hotel and a convention center in the middle of the Sierra Madre Occidental. In addition, support from the federal and state governments has been essential to achieve current objectives. During the last 10 years, more than US $150,000 have been awarded to the owners for various purposes, including payment for ecosystem services, ecotourism, and application of best management practices. The support for forest development varies as a function of the nation's economic priorities and the commitment of politicians to forest stakeholders.

ACKNOWLEDGMENTS

We would like to thank the owners of Molinillos, particularly Antonio Mancinas and Eliseo Sariñana, for their support and willingness to share important information with us. Also, thanks to Roberto Trujillo, a forest consultant, for allowing us to review information of previous forest management plans. CONACYT, IPN, and COFAA provided financial support. The editors of the book and Celina Perez-Herrera provided helpful comments in an earlier version of the chapter.

REFERENCES

Carmona, A.H., 1986. Aproximaciones al concepto de manejo integral de los espacios físicos — bióticos de las cuencas hidrográficas. In: Memoria del ciclo de conferencia bajo el convenio S.A.R.H. – FAO. Durango, Dgo, Mexico, 30 p.

Consultoria Forestal, 2013. Modificacion al programa de manejo del P.P. Molinillos. Documento de Trabajo enviado a la SEMARNAT. Consultoria Forestal, Durango, Mexico, 116 p.

de Liocourt, F., 1898. De l'aménagement des Sapinières. Bul. de la Sociétié Forestière de Franch-Conté et Belfort, Besancon.

Diario Oficial de la Federación, 1988. Ley General del Equilibrio Ecológico y la Protección al Ambiente (LGEPA). Ultima reforma publicada 16-01-2014. Diario Oficial de la Federación, Available at http://www.diputados.gob.mx/LeyesBiblio/pdf/148.pdf (Accessed May 15, 2014).

Diario Oficial de la Federación, 2003. Ley Forestal y Desarrollo Sustentable (LFDS). Últimas reformas aplicadas, 07-06-2013. Diario Oficial de la Federación. http://www.diputados.gob.mx/LeyesBiblio/pdf/259.pdf (Accessed May 15, 2014).

González-Elizondo, M.S., González-Elizondo, M., Tena-Flores, J.A., Ruacho-Gonzálezy, L., López-Enríquez, L., 2012. Vegetación de la Sierra Madre Occidental, México: Una síntesis. Acta Botánica Mexicana 100, 351–403.

Moreno-Sanchez, R., Torres-Rojo, J.M., 2010. Forest management in Mexico: their characteristics and context for their creation and evolution. In: Manos, B., Paparrizos, K., Matsatsinis, N., Papathanasiou, J. (Eds.), Decision Support Systems in Agriculture, Food, and the Environment. Information Science Reference, Hershey, PA, pp. 74–100.

Pukkala, T., von Gadow, K. (Eds.), 2012. Continuous Cover Forestry. second ed. Managing Forest Ecosystems, vol. 23. Springer, New York, 296 p.

Schütz, J.-P., Pukkala, T., Donoso, P., Gadow, K.V., 2012. Historical emergence and current application of CCF. In: Continuous Cover Forestry. second ed.. Managing Forest Ecosystems, vol. 23. Springer, New York, pp. 1–28.

von Gadow, K., Zhao, X.H., Corral Rivas, J.J., 2013. Retention strategies for multi-species forests. In: Proceedings of the International Symposium for the 50th Anniversary of the Forestry Sector Planning in Turkey. Orman Genel Müdürlüğü, Ankara, Turkey, Orman İdaresi ve Planlama Dairesi Başkanlığı Yayın. No. 107.

Wehenkel, C., Corral-Rivas, J.J., von Gadow, K., 2014. Quantifying differences between ecosystems with particular reference to selection forests in Durango/Mexico. Forest Ecology and Management 316, 117–124.

Chapter 14

Ross Forests, Tennessee, United States of America

Kerry Livengood[1] and John Ross[2]

[1]Tennessee Division of Forestry, Nashville, Tennessee, USA, [2]John Ross Tree Farm, Savannah, Tennessee, USA

ABBREVIATIONS

EQIP   Environmental Quality Incentive Program
TDEC   Tennessee Department of Environment and Conservation
TDF   Tennessee Division of Forestry

MANAGEMENT SETTING AND BACKGROUND

In 1881, I.W. Ross, an American Civil War veteran, county sheriff, and farmer and H.R. Hinkle bought a small tract of land. Ross and sons, and eventually grandsons, added 316 acres (ac) (128 hectares (ha)) in 1883, 800 ac (324 ha) in 1901, 177 ac (72 ha) in 1912, and 2,011 ac (814 ha) in 1936. Other additions in following years brought the total to more than 7,750 ac (3,136 ha) (Figure 14.1). The first pine plantation was planted in 1972 under a publicly funded program, the Forestry Incentive Program, through the U.S. Department of Agriculture's Soil Conservation Service, which was later renamed the Natural Resources Conservation Service (NRCS). In 1975, a long-term management agreement was entered into with a local paper mill, and the Ross family began to grow pine forests primarily on short rotations to supply fiber to a local kraft linerboard mill. More than 900 ac (364 ha) of hardwood, streamside areas, and natural areas were reserved from conversion to pine plantations, and some areas remained as farmland. The first forest management plan for this property was developed by a private consulting company in 1976; the landowner joined the American Tree Farm System in 1979.

In 1988, the partnership began offering hunting leases, and currently there are 10 leases covering the entire ownership. The partnership became a limited liability company (LLC) in 1998, and finally a general partnership in 1999 because of taxation levied by the State of Tennessee on LLCs. The remaining cropland of about 400 ac (162 ha) was planted as wildlife food plots in 1999, the same year that prescribed burning was included as a management tool to restore fire-dependent ecosystems. In 2002, the owner began a process to terminate the long-term management agreement with the paper mill. This allowed the owner to include broader management goals that involve recreation, wildlife, non-game species, and biodiversity objectives. The forest includes 20 miles (mi) (32 kilometers (km)) of mountain bike and hiking trails that were created in 2002 and 2004. In that same period, a management plan was developed for the protection of biodiversity, which included an inventory of mussels and fish that took 5 years to complete (2003–2007). Many other projects were completed, including:

- The development of rock jetties to protect stream banks
- Wetland restoration projects led by the NRCS
- The replanting of pine after southern pine beetle (*Dendroctonus frontalis*) outbreaks
- The creation of a State Natural Area
- The maintenance of U.S. Department of Agriculture Conservation Reserve Program contracts for upland row crop areas, which began in 1985 and will continue through 2021

Forest Plans of North America. http://dx.doi.org/10.1016/B978-0-12-799936-4.00014-X

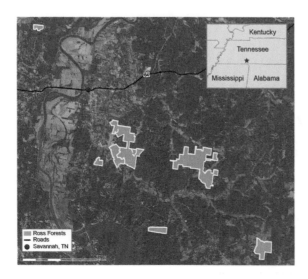

FIGURE 14.1 Map of the Ross Forests properties in Tennessee.

FIGURE 14.2 Pine plantation, Horse Creek, and riparian protected areas on the Ross Forests properties.

In 2009, a management plan was developed through the Environmental Quality Incentives Program (EQIP), which is administered through the NRCS. EQIP was created by the 1996 Farm Bill to provide cost-share assistance to farm and forest landowners in order to plan and implement conservation practices that produce environmental benefits. The state of Tennessee developed a sub-category of EQIP plans called the Forest Habitat Fund, to enhance habitat for non-game wildlife and plant species of greatest conservation need. The Tennessee Division of Forestry (TDF) and a private forest management firm developed the plan to emphasize silvicultural options that are compatible with habitat requirements of non-game species while maintaining the landowner's opportunities for sustainable timber production.

Figure 14.2 illustrates the variety of habitat found in the Ross Forests. The steam plume in the background is the Packaging Corporation of America paper mill located on the Tennessee River in Counce, Tennessee. In the foreground is a 7-year-old loblolly pine (*Pinus taeda*) plantation that will eventually be thinned and harvested for production of paper at the Counce mill. The stream, Horse Creek, contains many threatened and endangered species and is protected from silvicultural activities by a streamside management zone (SMZ).

PLANNING ENVIRONMENT AND METHODOLOGY

In the United States, forest management plans are often a requirement or are encouraged for participation in a number of public and private programs that include the U.S. Forest Service's Forest Stewardship Program (FSP), programs administered by the Farm Services Administration (FSA) and NRCS, state programs that were designed to reduce property taxes for conserving forests (Greenbelt), and certification programs such as the American Forest Foundation (Tree Farm System) and the Forest Stewardship Council (FSC). The TDF develops hundreds of plans each year, and in the past it has been a time-consuming process that provided no procedure for monitoring whether the suggested practices were implemented.

Plans may produce no landowner or public benefits if the recommended practices, such as erosion control, protection of cultural and environmental resources, or forest improvement, are not followed.

Forest Management Program

In order to improve the consistency of plans, to allow tracking of practices, and to decrease the time it takes to develop plans, the TDF developed a personal geo-database application for Environmental Systems Research Institute's ArcGIS software to enable the creation and editing of feature datasets and tables. Nonspatial data is entered through a program developed in Visual Basic, and the data is stored in a Microsoft Access database. The Access program contains forms for entering data and resource attributes. The program as a whole guides the planning procedure, specifies what data is required, allows plan implementation monitoring, and writes the management plan based on selections made by the user. The outcomes include checklists, activities, narratives, and practices. The program meets the standards of EQIP, Tree Farm, Greenbelt, and Stewardship plans.

There are 15 main parts (tabs) to the program (Figure 14.3), which were designed to lead a forester through the planning process.

1. Client: Owner contact information, objectives, location, other contact information, forester
2. Tracts: Selector for choosing tract, description, adjoining landowner list, location, acres
3. Health checklist
4. Erosion/water quality checklist
5. Habitat checklist
6. Cultural resources checklist
7. Invasive/exotics checklist
8. Wildlife/recreation checklist
9. Fire risk checklist
10. Timber inventory
11. Land-use areas
12. Practices checklist
13. Tract maps
14. Attachments checklist
15. Plan preview and export

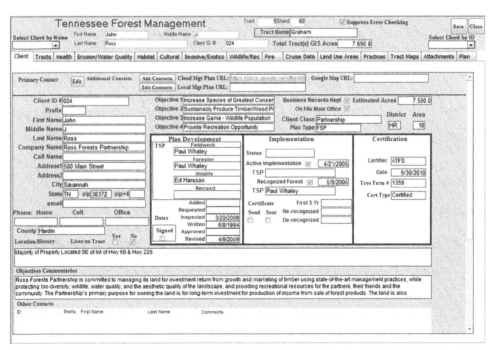

FIGURE 14.3   Forest Management Program main screen.

The final tab ("Plan preview and export") allows a forester to monitor plan progress, preview each plan section, and export the plan to an Adobe Reader (pdf) format. Completed plans are stored on a shared computer resource available to all users, accessed by clicking on a hyperlink field.

After gathering contact information, a forester and the landowner will select their objectives from a list of 13, and will add comments within the "Client" tab. The information is stored in the database and a cover sheet with contact information and the following section of the plan is produced. These are the objectives and the comments for the Ross Forest:

"In discussions during your property inspection, you indicated the following management goals apply to your property:

Goal 1: Increase Species of Greatest Conservation Need

Goal 2: Sustainably Produce Timber/Wood Products

Goal 3: Increase Game—Wildlife Population

Goal 4: Provide Recreation Opportunity

Ross Forests Partnership is committed to managing its land for investment return from growth and marketing of timber using state-of-the-art management practices, while protecting bio-diversity, wildlife, water quality, and the esthetic quality of the landscape, and providing recreational resources for the partners, their friends and the community. The Partnership's primary purpose for owning the land is for long-term investment for production of income from sale of forest products. The land is also valued by the members for personal recreational activities and is leased for hunting and other recreational activities, as demand exists. While extraordinary opportunities for sale of land might be considered, the lands are not for sale. If any lands are sold, the proceeds are to be used for reinvestment in additional land, preferably adjoining property, or invested in improvements on remaining property. New lands will only be acquired if they adjoin and significantly enhance the value of existing property."

The second tab ("Tracts") allows the user of the program to add multiple tracts. Only landowners with scattered, large properties will have multiple tracts, since contiguous tracts are delineated into land-use areas (e.g., pastures, wildlife food plots) or stands (for forests). The Ross Forests are divided into nine tracts, based on past ownership, location, or topographical features, and these vary in size from 86 to more than 3,000 ac (34 to 1,214 ha) as shown in Figure 14.1. Major topographic features are available as layers in the geographic information system (GIS). Stream centerlines are provided by the National Hydrography Dataset (U.S. Geological Survey, 2013), SMZs are obtained by buffering the stream centerlines, and topography was obtained from contours developed by the U.S. Geological Survey.

A checklist is included in each tab, so that a forester developing the plan will know what concerns to note or inventory while inspecting the property. For example, the "Health" tab lists insect pests and diseases found on the property during field surveys. A forester fills out the form with a checklist of 30 health problems commonly found in the forests of Tennessee. The table includes the common name, the scientific name, a narrative explaining the problems caused by the insect or disease, and the recommended control practice. The "Tract Maps" tab displays maps that have been exported from ArcGIS and displayed based on a hyperlink field. Land-use areas (stands) are also shown.

On the "Health" tab, a forester can note insect or disease problems found when inspecting the property. For example, the eastern tent caterpillar (*Malacosoma americanum*) is commonly found on the Factory Hollow tract in black cherry (*Prunus serotina*) forests. When checked in the form, the following information is provided in the management plan:

Forest Health

A diverse mix of insects and diseases exist within every forest; it is part of normal forest ecology. Typical insect and disease presence and activity is referred to as "endemic." It is common for trees to occasionally become weakened or die in any forest due to regular environmental stresses such as weather, competition with other trees, old age and other factors including endemic insect or disease populations. However, forest owners should be concerned when an endemic insect or disease population becomes numerous enough (epidemic levels) to cause significant tree damage or death. Your property has been examined for areas that are at risk of or have current harmful insect and disease activity. Practices should be applied to these areas as described below in order to protect and enhance forest health.

Factory Hollow

Resource Threat Eastern Tent Caterpillar

Severity Light

Minimal risks—continue to monitor for problems

Description

The forest tent caterpillar may be found throughout the United States and Canada wherever hardwoods grow. It is a native insect that has attracted attention since colonial times. Region wide outbreaks have occurred at intervals varying from 6 to 16 years in northern areas. Southern gum forests in southwest Alabama and southern Louisiana have had continuous infestations, especially in water tupelo "ponds," since 1948. Here, varying degrees of defoliation occur annually in 3.5 million acres (1.4 million hectares) of gum forests.

Recommendation

In some years hatch is low. High mortality of larvae in the egg is associated with temperatures below −42 °F (−41 °C). Freezing weather just prior to, during, and following hatching kill many of the young caterpillars. When trees are completely stripped of leaves, larvae starve. In the North, temperatures above 100 °F (38 °C) in the shade during moth emergence and egg laying have caused death of adults and low viability of eggs. Several species of flies and wasps parasitize the eggs, larvae, and pupae of the forest tent caterpillar. Most important are large gray flies, Sarcophaga aldrichi Parker, in the North; and S. houghi Aldrich in the South. Female flies deposit maggots on cocoons. The maggots penetrate the silk and move into the prepupae or pupae, killing them as well as any other parasites that may be present. S. Aldrichi becomes extremely abundant and contributes greatly to the termination of out breaks of the caterpillars in aspen forests. Although the flies do not bite, they annoy people by lighting on them and regurgitating on clothing and laundry hung outdoors. In southern gum forests, S. houghi is much less conspicuous and usually goes unnoticed. Nevertheless, it can be an important control agent. Itopletis conquisitor (Say) is an important ichneumonid wasp parasitoid of the pupal stage. Parasitization up to 20 percent by five species of egg wasps has been recorded in Alabama. Small trees can be protected by collecting and destroying egg masses, destroying colonies of young larvae at the end of branches, or killing larvae clustered on the trunks of branches during molting and resting periods. Several chemical insecticides and a microbial insecticide, Bacillus thuringiensis are registered for control of this insect.

On the "Erosion/Water Quality" tab, the user of the program can locate water quality problems by finding problems grouped on a tree structure. The first branch on the tree, best management practices (BMPs), has concerns grouped under haul roads, skid trails, log decks, SMZs, stream crossings, debris and hazardous materials, site preparation and planting, and wetlands. This structure was based on Tennessee's BMP manual, *Guide to Forestry Best Management Practices in Tennessee* (Tennessee Department of Agriculture, Division of Forestry, 2003). In the Rogers Creek tract, it was noted that there was moderate erosion associated with some of the haul roads. By checking this problem in the BMP tree under haul roads, the following narrative is produced:

Erosion/Water Quality

An abundant supply of clean water has long been one of the main benefits derived from forested watersheds. Protecting water quality and preventing soil erosion associated with all land uses is an important responsibility for forest managers and forest landowners. Any threats to water quality identified in this plan must be addressed to maintain Stewardship recognition and Tree Farm certification.

Your property has been examined for areas of active or potential soil erosion. Soil stabilizing practices should be applied to these areas as described below in order to protect and enhance water quality.

Rogers Creek

Resource Threat Logging Road Erosion

Severity Moderate

Forest roads are one of the single greatest sources of sediment. Construction of logging roads also causes soil compaction. Forest roads should be designed, constructed and retired in a way that prevents sediment from entering waterways. Construct roads well in advance of logging, vary the slope of the road, add outsloping, wing ditches, broadbased dips, water bars and/or cross-drain culverts. Follow all best management practices for locating, constructing, maintaining and retiring logging roads.

For threatened and endangered species, the approach is to inventory habitat rather than individual species, the choices on the "Habitat" tab include types such as karst or sinkholes, caves, mines, bottomland swamp, and other types. In the Factory Hollow tract, there are a number of springs and seeps. Checking this option produces a report, such as this:

Habitat Features

Special habitat features include unique plants, trees and forest communities, geological formations, and aquatic resources. Quite often these features exist as relatively small areas embedded within the surrounding forest. Such areas provide specialized habitats or habitat elements that are utilized by wildlife. These areas also sustain elements of ecosystem functions that help maintain or increase forest community diversity.

Your property has been examined for special habitat features. Such features could be recommended for establishment or protection. The features recommended for establishment are not required but implementation will significantly improve wildlife habitat. Features recommended for protection, at a minimum, should not be destroyed during management activities. Protecting to the level recommended is not required but gives an indication of how these features can be further developed to enhance wildlife habitat. These levels of implementation and protection are especially well suited for those landowners who have a strong interest in providing habitat for non-game wildlife.

Factory Hollow

Feature spring/seep

Narrative

Springs and seeps are small wetlands typically found in sloping, forested terrain. They are often found at or near the headwaters of streams. Springs flow from a clearly defined opening, and seeps form a saturated area where water percolates slowly through the soil. Small springs and seeps are so similar that both are usually called spring seeps. They provide critical habitat for salamanders and frogs and also food and a year-round water supply for many terrestrial birds and mammals. These systems are often small, isolated and embedded within a larger habitat matrix. The surrounding forest is important in maintaining the micro-habitat conditions associated with springs and seeps.

Practice

Establish a minimum 50' use exclusion/no timber harvest buffer surrounding springs and seeps; for maximum protection extend buffer to 150' (with outer 100' limited to 50% tree canopy removal)

The Ross Forests were extensively surveyed for threatened and endangered species in a separate report, but a forester would normally use a spatial layer to determine the location and habitat for threatened and endangered species. This is because surveys for locating threatened and endangered species are expensive, time consuming, and often outside the expertise of foresters. The alternative to surveys is to check for species already found in the area and look for habitats that match. Spatial data regarding existing threatened and endangered species are obtained by querying the Natural Heritage Inventory Program (Tennessee Department of Environment and Conservation, 2014). A tool was developed for ArcGIS to simplify the process of adding information on management for endangered species. By selecting a tract in ArcGIS, the tool will buffer the tract polygon by a distance determined by the user. After setting the buffer polygon, a query is performed to determine which species are within the buffer area. These species are added to a table and the plan links to the table to produce a report in the plan listing each and describing the most common habitat where each species is found. An example of this report for only a few of the species for Factory Hollow is shown below:

Buffer Distance: 4 mile(s)

The following threatened or endangered species have been identified within the specified buffer distance of the property based on data kept in the Tennessee Division of Environment and Conservation, Natural Heritage Inventory Database.

Factory Hollow

COMMON NAME	TYPE	HABITAT
Cracking Pearly mussel	*Mollusc*	*Medium-sized rivers of mod current, deeply buried in mud, sand, gravel, and cobble substrates; Tennessee & Cumberland river systems*
Gray bat	*Mammal*	*Cave obligate year-round; esp. forested areas; migratory.*

In addition, the landowner manages four natural areas identified in a report entitled, *Floral Survey of the John Ross Property, Hardin County*, by Claude Bailey, May 29, 2005 and in a report entitled, *Biodiversity on Ross Forests Partnership Lands, Hardin County, Tennessee*, prepared by Conservation Southeast, Inc., Andalusia, Alabama. The voluntary agreement with the Tennessee Department of Conservation and Environment (TDEC), Division of Natural Areas, was developed to formalize the cooperative efforts between the State and Ross Forests to protect populations of rare species. The agreement excluded these areas from forest management practices. Three of the four sites are significant primarily due to the presence of the beaked trout lily (*Erythronium rostratum*). Clack Springs is a cold-water spring that supports a diversity of ferns not observed in other similar sites. Factory Hollow is an example of a hardwood community made up of rock chestnut oak (*Quercus montana*) and shortleaf pine (*Pinus echinata*) that is not observed elsewhere on the Ross Property. Sugar Camp Hollow includes scenic waterfalls as well as rare state plants. The last area, Horse Creek Mesic Forests and Bluffs, has steep limestone bluffs above the narrow Horse Creek floodplain with rare plants such as American ginseng (*Panax quinquefolius*) and goldenseal (*Hydrastis canadensis*). These areas are open for recreation but timber harvesting is excluded.

The Cultural Resources Tab includes options with check boxes for cemeteries, archaeological and historical sites, burial mounds, scenic waterfalls, and sections of state and national wild and scenic rivers. The Blevins tract contains a cemetery. Selecting the cemetery check-box indicates that a cemetery is present on the tract and will produce the following section in the plan:

Tennessee contains a wealth of historical resources including prehistoric Middle Woodland mounds, Mississippian mounds, settlements, trails, forts, civil war battlefields, cemeteries, and other historic sites. Many sites and their artifacts have not been located or protected. The Tennessee Historical Commission is the state agency responsible for historical sites while the Division of Archaeology handles pre-historic finds. Please contact these agencies if you are interested in obtaining more information relating to cultural resources. Your property has been examined for cultural features that are worthy of protection. Any future activities planned on your property should take into account these areas as described below and protect them accordingly.

Blevins

Feature Cemetery

Description

Cemeteries are important features for a number of reasons; they are storehouses of family and community history, genealogists use the inscriptions and record to build family trees and are held as places of quiet and solitude. Tennessee State Law states that "No person shall willfully destroy, deface, or injure any monument, tomb, gravestone, or other structure placed in the cemetery, or any roadway, walk, fence or enclosure in or around the same, or injure any tree, plant or shrub therein ..." (Tenn. Code Ann. § 46-1-313) Violation of the statute is a Class E felony.

Recommendation

Unfenced cemeteries should be marked and/or flagged with a 50' buffer zone if silvicultural activities are planned in the same area.

The "Invasive/Exotics" tab provides a checklist of 24 non-native plants that in some way interfere with the reproduction or survival of native species. The list includes kudzu (*Pueraria lobata*), which was introduced in the 1920s to control erosion. Kudzu shades out plants in the understory, and can form dense thickets, preventing normal reproduction of even shade-tolerant species such as sugar maple (*Acer saccharum*). The Graham tract was the original Ross family home place and early settlers often planted various privets (*Ligustrum* spp.), which, much like bush honeysuckle (*Lonicera* spp.), have escaped from yards and moved into the forest. Checking "privets" in the Invasive/Exotics checklist produces the following report and recommendations.

Some of the insects, diseases, and plants found in our forests today are not native to the area and are here because they were accidentally introduced into our forests. These invasive/exotic species have the potential to interfere with native ecosystems because they lack natural predators and can rapidly exploit their environment. In some situations, invasive/exotic species alter ecosystem functions to the point management objectives are difficult to achieve.

Your property has been examined for areas that are at risk of or have current harmful levels of invasive/exotic species activities. Practices should be applied to these areas as described below in order to protect and enhance your ability to achieve your management objectives.

Graham

Resource Threat Privets

Severity Severe

Description

There are several nonnative privets. Chinese privet and European privet are difficult to distinguish except at flowering, both are evergreen to semi-evergreen. Both are thicket forming shrubs to 30 feet (9 m) in height that are soft woody, multiple stemmed with long leafy branches and opposite leaves less than 2 inches long. Showy clusters of small white flowers in spring yield clusters of small ovoid, dark-purple berries during fall and winter.

Recommendation

Thoroughly wet all leaves with one of the following herbicides in water with a surfactant (August to December): a glyphosate herbicide as a 3-percent solution (12 ounces per 3-gallon mix) or Arsenal AC* as a 1- percent solution (4 ounces per 3-gallon mix). For stems too tall for foliar sprays, apply Garlon 4 as a 20-percent solution in commercially available basal oil, diesel fuel, or kerosene (2.5 quarts per 3-gallon mix) with a penetrant (check with herbicide distributor) to young bark as a basal spray. Or, cut large stems and immediately treat the stumps with Arsenal AC* or Velpar L* as a 10-percent solution in water (1 quart per 3-gallon mix) with a surfactant. When safety to surrounding vegetation is desired, immediately treat stumps and cut stems with Garlon 3A or a glyphosate herbicide as a 20-percent solution in water (2.5 quarts per 3-gallon mix) with a surfactant. *Nontarget plants may be killed or injured by root uptake.

Information on game wildlife and recreation resources is collected during field visits and recorded by checking the appropriate box on the "Wildlife/recreation" section of the program.

Wildlife and recreation features include trails, vistas, waterfalls, streams with game fish, and all the game wildlife resources such as deer, quail, wild turkey, rabbit, and squirrel. Many species that were once scarce in the state have been restored to abundance in many areas. This return is generally due to a decrease in hunting pressure, changes in farming practices and an increase in food supplies and cover. Habitat is the general term used to describe wildlife food supply and cover. There are a number of techniques that can be utilized to improve wildlife habitat on private lands. The map accompanying this document details the habitat types existing on your property. Generally, they are cropland, grassland, forestland, wetlands, riparian areas, and inhabited areas. The number and type of wildlife will vary in each of these areas but all will depend on food, water and cover.

Your property has been examined for wildlife habitat and the presence of the important game species.

General guidelines for managing those observed are given below.

Rogers Creek

Feature Wild turkey

Description

The wild turkey is an omnivore with its annual diet consisting of 90% plant and 10% animal matter. Mast, fruits, seeds, greens and agricultural crops are the principal plant food groups consumed. Acorns make up about one third of their diet. Soft-mast producing shrubs like wild grape, dogwood, black gum, wild cherry, hackberry and similar species are also important foods, particularly when hard mast crops fail. Grasses and seeds are important winter and spring foods, while insects comprise the majority of the summer diet for young turkey poults. A mixture of forested and open lands provides the best turkey habitat. Open lands should comprise 10–50% of the area. The size and distribution of open areas is important with a system of well dispersed smaller clearings being most favorable.

Recommendation

Timber lands should be managed to optimize hard and soft mast production and to provide a dispersed system of permanent forest openings. The even-aged harvest method is recommended to maintain oak regeneration, to create open understory conditions and to provide stand diversity. Long timber rotations are recommended to provide a high percentage of trees of mast producing age. Because white oaks live longer, longer rotation ages for this group are recommended. At least 60 % of the trees on your property should be in mast producing age (50+ years). Rotations from 120–200 years are recommended for wild turkey, depending on the forest type.

Fire risk is, at present, the least advanced of the approaches to forest resource planning in this program. Currently, the program design is simply a checklist of potential hazards to structures with non-fire resistant roofs or chimneys lacking a spark arrestor. The goal is therefore to incorporate the spatial data being developed by the Southern Wildfire Risk Assessment (SWRA). The project will provide spatial data on features, such as surface fuels, canopy bulk density, fire occurrence areas, and a suppression-difficulty rating, and will be able to assign a risk assessment rating to a given property.

Using aerial photographs in the GIS system, a forester delineates the various land uses, such as pine plantation, upland hardwood forests, SMZs, and wildlife food plots. GPS receivers may be used to determine size and location of some features, such as wildlife food plots. Attributes of the areas are collected in field surveys and recorded in the database. Attributes include site quality, stocking, product class, species, and cover type. Cover types are based on *Cover Types of the United States and Canada* (Eyre, 1980), and the plan automatically includes information on associated species, midstory, and herbaceous species found on similar sites.

Practices are also selected and grouped by landowner objectives. Silvicultural practices are grouped under *Timber & Wood Products*, and broken into subclasses, such as *Pine Plantation, Hardwood Plantation, Natural Hardwood*, and other broad forest management types. More than 200 practice narratives have been developed. On the Factory Hollow tract, when the practice "Burning—site preparation" is selected, the following practice narrative is added to the plan:

Fire reduces the levels of slash, debris and litter, while releasing nutrients back into the soil. A properly timed fire will kill vegetation that initially invades a harvested stand and will also increase the ease of planting seedlings. Controlled fire should be used only under ideal conditions. If the site is too wet, the application will be useless. If the site is too dry or wind and weather conditions are not ideal, the fire can burn too hot and create potential short-term nutrient and erosion problems in the area to be planted. Smoke management should be a priority when applying fire. Considering the liability and safety issues surrounding the application of fire, landowners should always work with a professional forester and double check to ensure that all the proper permits and regulations have been filed and followed. Research has shown that the better a site is prepared, the more likely that a release treatment will not be needed two or three years later. Investments made prior to the establishment of the stand will result in higher returns when the harvest is conducted.

One goal of the project was to standardize the components making up a forest management plan. With more than 30 foresters across the state writing plans, there was considerable variation in topics covered even though the Stewardship Program has fairly specific national guidelines. To correct this problem, each plan is automatically evaluated based on a checklist of required and optional components (Table 14.1). A grade is assigned to the plan, and points are received when components have been completed in the database. In the table, required components receive 5 points, and optional components receive 1 point. A plan needs 100 points to receive a passing grade, and all required components result in a 100 point deduction when not present. The Ross Plan meets the Stewardship Standards and has earned more than 100 points.

OUTCOMES OF THE PLAN

A checklist of scheduled practices for the 10-year life of the plan is produced by the program described here for tracking plan accomplishments. A few of these past activities included:

1. *Graham Tract*: Mowing/weeding within the hardwood plantations in land-use area 2 in 2008; regeneration cutting in land-use area 1 in 2010; thinning in land-use area 2 in 2011.
2. *Blevins Tract*: Crop tree release in land-use area 22 in 2008; creation of a vernal pool in land-use area 1 in 2010.
3. *Factory Hollow Tract*: Creation of a vernal pool in land-use area 1 in 2010.
4. *Rogers Creek Tract*: Row thinning in land-use area 1 in 2010; herbicide applications in land-use areas 12 and 13 in year 2012.
5. *Horse Creek Tract*: Creation of wildlife food plots in land-use area 71 in 2008; prescribed burning in land-use area 2 in 2010.

There were 46 recommended practices over the 10-year planning period in 65 stands and land-use areas. Recommended practices include 10 vernal pools, 3 crop tree releases, one food plot, 4 herbicide applications, 18 growth and mortality checks, one mowing and weeding, 5 prescribed burns, 1 regeneration cutting, and 3 row thinnings. With an area as large as the Ross Forest, spatial tracking of practices that change over time due to unexpected problems and disturbances greatly reduces the job of management. For example, stand boundaries will change due to disturbances, such as fire, wind damage, and tornados. Further, often a harvest is begun but cannot be completed due to inclement weather, access problems, or mill quotas. The practices and the stand boundaries can be easily changed in both the spatial layer and the attribute tables of the database described, and they immediately show up when plan documents are printed.

DISCUSSION AND CONCLUSIONS

Improvements are planned to the forest management plan development program described here and applied to the Ross Forests. These include the incorporation of forest inventory data from commercial cruising programs, such as TCruise, growth and yield modules to predict future stand volumes, and a financial component to estimate returns to investments. Managing a forest where events such as wind storms, wildfires, and insect outbreaks occur several times a year requires changes in plans and makes static planning documents obsolete. Therefore, it is our contention that a management plan should be a compendium of many actions and information stored in databases and in GIS that can be exported as hardcopy products whenever the entire document is requested.

Learnings and Insights

Forest management planning often requires a team; management decisions are made by a landowner with input from foresters, wildlife biologists, farm managers, timber buyers, and loggers. Each should have some access to planning documents

TABLE 14.1 Checklist of Important Pieces of a Forest Management Plan

Number	Item Name	Required?	Points Earned
1	Client name	Yes	5
2	Client address	Yes	5
3	Client phone number	Yes	5
4	Plan preparation data	Yes	5
5	Name of technical service provider	Yes	5
6	Address of technical service provider	Yes	5
7	Legal description or directions to tract		1
8	Tract acres		1
9	Stewardship acres	Yes	5
10	Landowner objectives	Yes	5
11	General property description	Yes	5
12	Adjacent landowners		1
13	Map with tract boundary, stands, labels	Yes	5
14	Heritage database check for threatened and endangered species	Yes	5
15	Soils information	Yes	5
16	Forest habitat survey		1
17	Forest health survey		1
18	Invasive/exotic survey		1
19	Erosion/water quality survey	Yes	5
20	Wildlife/recreation survey		
21	Stand objectives	Yes	5
22	Stands by cover type	Yes	5
23	Stand acres	Yes	5
24	Stand dominant vegetation		1
25	Stand tree species	Yes	5
26	Stand health		
27	Stand site quality		1
28	Stand stocking	Yes	5
29	Stand volume		
30	Stand timber quality		1
31	Stand growth rate		1
32	Stand age		1
33	Stand site index		1
34	Stand basal area		1
35	Stand history		1
36	Activities schedule		1
37	Implementation schedule		1
Total			106

to make suggestions or to change planned activities. If a harvest operation is begun but is interrupted by rainfall events then the extent of the work needs to be gathered and incorporated into GIS, plans need to be made for the remaining stand, and work needs to be coordinated between site preparation and planting crews. Areas where prescribed fires have been used in the past must also be recorded. Information on the extent and timing of implemented practices needs to be gathered and recorded in the database. A plan, represented as a database, can allow a landowner to monitor activities and to determine whether they fit together and match the landowner's objectives. For example, when John Ross became more concerned with non-market objectives over time, this required a broader approach to planning and involved many more professionals and agencies. The plan is also a broad-brush approach to land management and does not specify, for example, the exact way vernal pools are to be built. This information may depend on more detailed, operational plans with specifications provided by the agency or organization funding the practice. These issues are important because the Ross Forests does not employ a full-time forester nor a land manager, but utilizes the advice of consultants and foresters employed by state and federal agencies.

Sustainability Issues

From the perspective of a private landowner in Tennessee, there are many different levels associated with the issue of sustainable forest management. For example, productive, sustainable forests can be converted to agricultural or urban uses. Over time, family farms, especially dairy farms, are disappearing in the United States as children leave family farms for higher paying, less demanding professions. Family forests also incur property taxes and are often pressured to be converted to other uses by nearby urban development. Greenbelt tax exemptions are one possibility to reduce the property tax burden, but in Tennessee these are limited to 1,500 ac (607 ha) by state law, so the Ross family is paying full-rate property taxes on the remaining area of forest. In Tennessee's hardwood forests, the issue is not one of deforestation due to overharvesting; as the U.S. Department of Agriculture Forest Inventory and Analysis data indicates, much more volume is growing than is being harvested each year. Hardwoods in Tennessee naturally regenerate quickly if land is not paved over, tilled, burned, or grazed, but the quality of the timber in forests has dropped greatly, and eventually hardwood forests may only produce low-valued products, such as pulpwood, railroad ties, or firewood. The issue with sustaining hardwood forests is that they can require at least twice as much time as a pine plantation to produce an acceptable, positive return to the landowner. A pine plantation can provide income from thinnings in 12 years and from a final harvest at age 25, while a hardwood forest would need more than 50 years. Hardwood forest area in the Ross Forests has increased, but pine plantations are still required to make the endeavor a sound economic investment. Overall, the Ross Forests are considered to be managed sustainably for timber production and economic purposes, but investments in site preparation, planting, and herbicide treatments can raise the productivity of the forest and the rate of return on investment. There is no one level of sustained yield for the Ross Forests because it depends on management decisions and investments in increasing the productivity. Planting genetically improved loblolly pine has more than doubled the productivity of pine stands over other forests regenerated from natural seed of wild trees. The increase in productivity in some of the pine plantations frees up areas that can be devoted to recreation and natural areas without reducing the volume being harvested over time by the Ross family.

Plan Development Challenges

Information and suggestions come from many sources when managing a forest as large as the Ross Forest. There are yearly changes in government programs coming from multiple agencies, many with different objectives. Some focus on environmental protection, others on game wildlife. The short planning horizon precludes use of economic analysis, such as present net worth and rates of return, over the entire development cycle of the forest. This makes it nearly impossible to determine if recommended practices are truly successful from an economic point of view. Because this area of the forest management planning program is under development, revenues from hunting leases and recreation are not currently integrated in to the Ross Forest plan. Finally, many mountain biking trails have been developed on the logging roads and were made available to the public. Yet, by opening the forest to recreational use, the risks of fire and illegal dumping of trash have increased, and these risks are difficult to quantify and to incorporate into a plan.

Plan Implementation Challenges

Every organization needs a memory, or documentation of what decisions were made, when the activities were begun, and whether the planned activities were completed. The database described in this chapter does not directly fulfill all of these needs. For example, changes in the plan can be made, and information regarding the original, planned practice(s) can be lost. Ultimately, the forest landowner is responsible for plan implementation and all decision-making processes. The various forms of advice provided by public agencies can only serve as recommended practices. Since public agencies do not bear the costs of making wrong decisions on private lands, the landowner needs to ensure that the correct practice is suggested.

REFERENCES

Eyre, F.H. (Ed.), 1980. Forest cover types of the United States and Canada. Society of American Foresters, Bethesda, MD. 148 p.

Tennessee Department of Agriculture, Division of Forestry, 2003. Guide to forestry best management practices in Tennessee. Tennessee Department of Agriculture, Division of Forestry, Nashville, TN. 50 p.

Tennessee Department of Environment and Conservation, 2014. Natural Heritage Inventory Program. Tennessee Department of Environment and Conservation, Nashville, TN. http://www.tennessee.gov/environment/natural-areas/natural-heritage-inventory-program.shtm (Accessed May 22, 2014).

U.S. Geological Survey, 2013. Hydrography. U.S. Department of the Interior, U.S. Geological Survey, Reston, VA. http://nhd.usgs.gov/data.html (Accessed May 22, 2014). TNMap, Tennessee's Enterprise GIS PPortal, NMap_Data_Library.DBO.Contours_LOFT. http://tnmap.tn.gov/ (Accessed May 22, 2014). Southern Group of State Foresters, Wildlife Risk Assessment Portal. http://www.southernwildfirerisk.com.

Chapter 15

Willow Break LLC, Mississippi, United States of America

Ian Munn,[1] William C. Wright,[1] William Hunter[2] and Greg Bentley[3]

[1]Department of Forestry, Mississippi State University, Mississippi, USA, [2]U.S. Marine Corps, Hampstead, North Carolina, USA, [3]Campbell Global, Forest and Natural Resources Investments, Mississippi, USA

ABBREVIATIONS

DFC	Desired Forest Condition
DMAP	Deer Management Assistance Program
LLC	Limited Liability Corporation
LMVJV	Lower Mississippi Valley Joint Venture
MAV	Mississippi Alluvial Valley
MDWFP	Mississippi Department of Wildlife Fisheries and Parks

MANAGEMENT SETTING AND BACKGROUND

The 3,370 acre (ac) (1,364 hectare (ha)) Willow Break LLC property is located in Warren County, Mississippi and lies entirely within the historic flood plain of the Yazoo River in the Mississippi Alluvial Valley (MAV). The Yazoo River forms its western boundary. The Yazoo River Levee and Collins Creek are two physical features that dominate the landscape. Both dramatically influence management through their impact on the hydrology, the levee—by containing the floodwaters of the Yazoo River, and Collins Creek—by alternately draining and flooding much of the land outside the levee. Prior to 1995, the property was dedicated to agriculture; however, its propensity to flood made agriculture a marginal operation. Consequently, the bulk of the property (2,906 ac, 1,176 ha) was enrolled in the Wetlands Reserve Program (WRP) (U.S. Department of Agriculture, 2014). Water control structures were installed to help restore the original hydrology and to create functional wetlands. Hardwoods were planted on roughly 1,950 ac (789 ha). In 2001, a group of investors established Willow Break LLC and purchased the property for use as a private hunting ground.

Willow Break LLC (Figure 15.1) consists of timberland (2,555 ac, 1,034 ha) 28 duck holes (528 ac, or 214 ha), 42 food plots (103 ac, or 42 ha), a river levee (114 ac, or 46 ha), and a camp compound (69 ac, or 28 ha). The timberland includes four timber types: mature bottomland hardwoods (317 ac, 128 ha), hardwood plantations (1,950 ac, 789 ha), natural hardwood regeneration (91 ac, 37 ha), and streamside management zones (SMZs) (197 ac, 80 ha). A well-maintained road system provides access throughout the property, making traveling to and from recreational sites relatively easy. There are more than 15 miles (mi) (24.1 kilometers (km)) of improved roads and a maze of secondary roads and trails.

The mature bottomland hardwoods (Figure 15.2) include five stands inside the Yazoo River Levee and one adjacent to the compound. Prominent timber species include southern red oak (*Quercus falcata*), water oak (*Quercus nigra*), white oak (*Quercus alba*), green ash *(Fraxinus pennsylvanica)*, shagbark hickory (*Carya ovata*), black willow (*Salix nigra*), sweetgum (*Liquidambar styraciflua*), bitter pecan (*Carya aquatica*), American sycamore (*Platanus occidentalis*), and sugarberry (*Celtis laevigata*). The timber is generally poor quality. The stands are relatively open with a basal area of around 65 ft^2 per ac (14.9 m^2 per ha), and advanced reproduction, where present, consists primarily of shade-tolerant tree species. The only merchantable timber volumes on the property are located in these stands, averaging less than 4.0 thousand board feet (MBF) per ac (23.3 m^3 per ha) (Table 15.1).

Forest Plans of North America. http://dx.doi.org/10.1016/B978-0-12-799936-4.00015-1

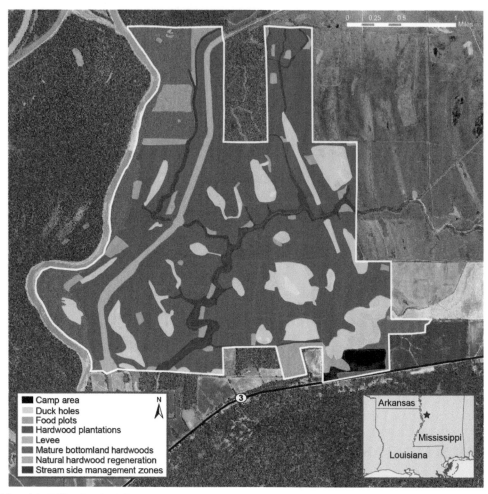

FIGURE 15.1   Map of the Willow Break LLC property.

The hardwood plantations were initially established in 1997, shortly after the property was enrolled in the WRP. Initially, the agriculture fields were direct-seeded with acorns but germination and establishment was generally poor. Next, hardwood seedlings were hand planted in 1998. Areas of poor survival were replanted in 2000. Thus, tree ages range from 14 to 17 years. Species planted included Nuttall oak (*Quercus texana*), green ash, willow oak (*Quercus phellos*), overcup oak *(Quercus lyrata)*, sawtooth oak (*Quercus acutissima*), and baldcypress (*Taxodium distichum*). Other species that established naturally within the plantation include sweetgum, sycamore, and black willow. Stocking averages 490 trees per ac (1,211 trees per ha) but widely ranges across the plantation. Eastern baccharis (*Baccharis halimifolia*), common buttonbush (*Cephalanthus occidentalis*), trumpet creeper (*Campsis radicans*), and redvine (*Brunnichia ovate*) are prevalent in the understory, and in the case of the vines, often reach well up into the overstory. Forbs and grasses are abundant. In areas where stocking exceeds 300 trees per ac (741 trees per ha), the canopy is beginning to close and the understory is dying out. This has serious implications for the white-tailed deer (*Odocoileus virginianus)* herd. As the canopy continues to close throughout the plantation, the available browse on almost two-thirds of the land area of Willow Break will decrease dramatically.

There are four naturally regenerated, immature hardwood stands, and the predominant species are sweetgum, green ash, sycamore, and black willow. Stocking averages 2,100 trees per ac (5,189 trees per ha). Due to the high stocking levels, there is very little understory and browse present. Streamside management zones (SMZs) border Collins Creek and its tributaries, ranging in width from 66 to 198 feet (ft) (20–60 meters (m)) and encompassing 197 ac (80 ha). Bitter pecan, overcup oak, black willow, and sweetgum are the most common timber species. The Yazoo River Levee is a dominant feature on the landscape. Maintained by the Yazoo River Levee Board, it traverses the property from north to south, 3.6 mi (5.8 km) in length and encompassing 114 ac (46 ha). The levee is planted with Bermuda grass (*Cynodon dactylon*) and is mowed biannually, providing excellent quality forage.

FIGURE 15.2 Mature bottomland hardwood forest area within the Willow Break LLC property.

TABLE 15.1 Merchantable Timber Volumes on the Willow Break LLC Property

Product	Species	Trees per ac	Basal Area per ac	Volume per ac	Total Volume
Pulpwood	All	37	15	3 cords	806 cords
Sawtimber	Red oak	9	20	1.8 MBF	579 MBF
	White oak	3	3	0.3 MBF	85 MBF
	Ash	2	4	0.4 MBF	127 MBF
	Hickory	3	4	0.3 MBF	91 MBF
	Willow	3	3	0.2 MBF	70 MBF
	Other	15	16	0.9 MBF	291 MBF
Total sawtimber		35	50	3.9 MBF	1,243 MBF

The soils on the property are primarily heavy, poorly drained clay soils such as Sharkey-Tunica-Dowling clays and Alligator clay. The depth to the water table ranges from 0 to 24 inches (in) (0–51 centimeters (cm)). These characteristics severely limit the operability of these soils. Representative site indices (base age 50) for Sharkey-Tunica-Dowling clays are 90 for green ash, 96 for overcup oak, 105 for eastern cottonwood (*Populus deltoides*), and 100 for sugarberry. Site indices (base age 50) for Alligator clay are 70 for green ash, 80 for water locust (*Gleditsia aquatica*), 90 for sugarberry, 90 for sweetgum, 90 for water oak, and 95 for willow oak. These are reasonably good site indices for hardwood management; however, much of the MAV has site indices over 100.

There are 28 duck holes on the property, the majority of which were established by the Natural Resources Conservation Service (NRCS) as part of a WRP easement. Levees and water control structures were installed to help restore the natural hydrology of the area. Their primary purpose is to provide wetland habitat for migrating waterfowl and shorebirds. Cover varies considerably and includes flooded timber, hemi-marsh (a mix of emergent vegetation or vegetation with floating leaves that are interspersed with a plant community that is otherwise submersed in water), moist soil complexes, and areas managed for agricultural crops. The food plots are managed to provide year-round and seasonal forage, primarily for white-tailed deer. The camp compound features a lodge, a bunk house that sleeps 16 people, private condominiums, and a utility building equipped with a skinning rack, game processing equipment, and a walk-in cooler for storing game. A tractor shed is available to store all equipment used for maintenance of the grounds and for habitat manipulation. The camp compound includes a 20 ac (8.1 ha) lake that is stocked with black bass (*Micropterus* spp.) and bream (sunfish).

Wildlife is abundant throughout the Willow Break property. Game species include white-tailed deer, eastern wild turkey (*Meleagris gallopavo silvestris*), fox squirrel (*Sciurus niger*), gray squirrel (*Sciurus carolinensis*), eastern cottontail rabbit (*Sylvilagus floridanus*), swamp rabbit (*Sylvilagus aquaticus*), and feral hogs (*Sus scrofa*). Furbearers include beaver (*Castor canadensis*), bobcat (*Lynx rufus*), otter (*Lotra Canadensis*), nutria (*Myocastor coypus*), and raccoon (*Procyon lotor*). Coyote (*Canis latrans*), opposum (*Didelphis virginiana*), and nine banded armadillo (*Dasypus novemcinctus*) are also common. Other wildlife species of note include a nesting pair of bald eagles (*Haliaeetus leucocephalus*), a resurging population of American alligator (*Alligator mississippiensis*), and a Louisiana black bear (*Ursus americanus luteolus*) that periodically visits the property. American wood storks (*Mycteria americana*), listed as threatened under the Endangered Species Act, are common during the spring and summer months. Migratory waterfowl are abundant from the late fall to early spring.

PLANNING ENVIRONMENT AND METHODOLOGY

The primary objective for Willow Break LLC is to provide family-oriented outdoor recreation. The favorite activity is hunting. White-tailed deer and waterfowl are the predominant species of choice for hunting, but feral hog, squirrel, and rabbits are also hunted. Other activities include fishing, trapping, ATV riding, camping, and wildlife viewing. A secondary objective is to provide revenue through provision of environmental services. Management efforts focus on providing optimal wildlife habitat to sustain a trophy deer herd and attract abundant waterfowl during hunting season.

Management is constrained by both legal and environmental factors.

1. The majority of Willow Break LLC is enrolled in a permanent WRP easement. All timber harvests are strictly regulated by the NRCS and are permitted only to improve wildlife habitat.
2. Although the bald eagle has been delisted from the Endangered Species Act, the ownership has voluntarily restricted management activities within 660 ft (201 m) of the bald eagles' nest as recommended by the U.S. Fish and Wildlife Service (2007). No disturbance is allowed during the nesting season and only low-impact activities (e.g., those that can be conducted on foot or by small all-terrain vehicles) are permitted during the remainder of the year.
3. The property frequently floods in the winter and spring, limiting access and delaying habitat management.
4. There is a large feral hog population on the property. These animals do substantial damage to roads and duck hole levees through their rooting and wallowing behavior. They also limit the suite of agricultural crops that can be planted for wildlife. Any corn planted, for example, is entirely consumed by the hogs.

Guiding Laws, Regulations, and Policies

The 2,906 ac (1,176 ha) of land enrolled under the WRP easement must comply with the terms of that agreement. The easement restricts the types of activities that can be conducted. For example, no permanent structures such as fences and buildings can be constructed. The most limiting constraint is that timber can only be harvested to benefit wildlife and prior approval from the NRCS is required.

Following the guidelines set by the Yazoo River Levee Board, no hunting is allowed within 200 ft (61 m) of the toe of the levee on either side. Additionally, access to the levee is controlled by the Levee Board and is currently not open to the public. It is in the best interest of the LLC (the "club") to keep the levee closed because poaching and trespassing are major problems. The club assists the Levee Board by mowing the levee twice annually, thereby saving the Levee Board substantial expense. In exchange, the Levee Board keeps the levee closed to the public.

Wildlife management is subject to both state and federal regulations. Migratory species and those protected under the Endangered Species Act are under federal jurisdiction, and all others are under the state's jurisdiction. This primarily impacts deer herd management and control of feral hogs, a nuisance species. Because of the presence of the Louisiana black bear, a protected species, baiting and trapping of feral hogs must be modified to eliminate the possibility of harm to the bear. Deer populations are manipulated through harvest. Because the harvest must comply with statewide harvest regulations, the ability to manipulate the herd can be limited.

Willow Break LLC is also in the process of applying to the American Tree Farm System to become a Certified Tree Farm.

Willow Break LLC is a limited liability company wholly owned by its shareholders. A managing member is responsible for the day-to-day operations of the company and is elected by the membership. An executive committee, consisting of six committee members also elected by the membership, oversees the managing member and sets the direction for the company's activities. The managing member has authority to approve all routine expenses necessary for the day-to-day operation of the company up to a specified cap. The executive committee has the authority to approve all routine expenses plus capital purchases up to a specified cap. Capital expenditures above the cap specified for the executive committee must be approved by a majority vote of the shareholders. A detailed report of the financial activities of the company is submitted annually to the shareholders for review. There are two standing committees to implement the routine habitat operations. The deer committee develops and implements those activities benefitting white-tailed deer such as planting food plots and roadsides, and enforcing the harvest plan. The duck committee develops and implements activities to enhance waterfowl habitat. This includes manipulation of water levels, planting agricultural crops, and moist soil management. Ad hoc committees are formed as necessary for a variety of nonroutine activities. Membership on these committees is voluntary.

The 42 food plots, encompassing 103 ac (41.7 ha), are managed to provide year-round and seasonal forage, primarily for white-tailed deer. On roughly one-third of these plots, ladino clover (*Trifolium repens*) has been established as a year-round, high-protein forage. The remaining two-thirds are intensively managed as winter food plots. The predominant crop planted is winter wheat (*Triticum aestivum*) but other cold, hardy-forage crops are sometimes included. Roadsides are either planted with ladino clover or managed for natural browse by annual light disking to stimulate new growth. In addition, the duck holes are managed to maintain early successional moist soil vegetation, such as grasses, wild millets, sedges, smartweed (*Polygonum* spp.), and toothcup (*Ammania coccinea*). This requires that the vegetation be set back approximately once every three years to maintain the early successional plant communities. Disking or herbicides are used to accomplish this. One-third to one-half of the area is treated each year depending on weather conditions. The areas treated each year are rotated among the duck holes so that all areas are treated at least once every three years. Typically, one of two types of treatments is applied to the areas designated for manipulation that year: agricultural cropping or light disking. Fifty to 100 ac (20.2 to 40.5 ha) are planted in agricultural crops each year. These areas are typically disked in preparation for planting and subsequent herbicide treatments are applied where necessary. Crops include sorghum (*Sorghum* spp.), browntop (*Agrostis capillaris*), Japanese millet (*Echinochloa frumentacea*), Chiwapa millet (*Echinochloa frumentacea var.*), soybeans (*Glycine max*), and rice (*Oryza* spp.). Light disking is applied in the late fall to the remaining areas targeted that year. Due to weather and ground conditions, many wetter areas cannot be treated in most years. Therefore, the areas designated for treatment are flexible to permit opportunistic treatments of areas that only dry up sufficiently for treatment in drought years. Other treatments include planting Japanese millet on exposed mudflats, spraying broadleaf herbicides such as 2,4-D and Dicamba to foster natural grasses, manipulating water levels, and supplemental irrigation.

The primary game management objective is to generate quality trophy white-tailed deer bucks. To achieve this goal, Willow Break LLC participates in the Deer Management Assistance Program (DMAP) offered by the Mississippi Department of Wildlife Fisheries and Parks (MDWFP). In this program, MDWFP helps the club manage for a healthy deer herd. The club members collect data on all harvested deer including sex, weight, age (from jawbones), lactation (does), and antler size (bucks). Based on this data, a habitat evaluation, and the club objectives, the MDWFP biologist suggests appropriate harvest strategies, which the members then implement. In addition, the biologist also provides a harvest report that includes an in-depth analysis of the previous year's harvest and identifies progress toward goals. As a result of participating in the DMAP program, the state has the option of easing harvest restrictions should the condition of the deer herd warrant it.

Willow Break stocks three impoundments, two with bream and black bass and the third with channel catfish (*Ictalurus punctatus*). In addition, Alligator Lake (a natural lake), Alligator Slough (a permanent slough), and Collins Creek sustain fishable populations of native species. The compound lake is professionally managed. Stocking levels of desirable and nondesirable species are monitored closely, and harvest rates and stocking rates are adjusted as appropriate.

The primary concern facing the ownership of the LLC is that almost two-thirds of the forested part of the property is essentially one age class consisting of hardwood plantations and natural stands that regenerated at essentially the same time that the plantations were established. These stands currently provide excellent cover and browse for the primary game species of interest (white-tailed deer). The canopy, however, is fast approaching closure throughout the property. Once this happens, the amount of available cover and browse will plummet. This will significantly reduce the property's carrying capacity, reducing both the quantity and quality of deer available for harvest. Timber management is necessary to provide a diversity of habitat conditions amenable to white-tailed deer and other game species. Management activities in the hardwood plantation, however, are governed by the terms of the WRP easement. Only activities benefiting wildlife are permitted and must first be approved by the NRCS, in consultation with the U.S. Fish and Wildlife Service (FWS). The first step in the planning process was to identify management practices that would be amenable to both the NRCS and the FWS. The Lower Mississippi Valley Joint Venture (LMVJV) has developed a set of silvicultural guidelines for managing bottomland hardwoods in the MAV to create desired forest conditions (DFCs) for white-tailed deer, wild turkey, and other wildlife. The NRCS and the FWS are both members of the LMVJV; therefore, the DFC guidelines were adopted as the basis for all timber management activities.

At the landscape level, the LMVJV recommends that 70–95% of the forested area be managed to create desired habitat conditions and the remaining portions be set aside to serve as "wilderness areas" or unmanaged forests. Within the managed area, 5% or less should be shrub-scrub habitat but no more than 10% should be in large (>7 ac, or 2.8 ha) regenerating forests. Regenerating forests are defined as even-aged stands that have achieved less than one-third of their typical height at maturity. Thirty-five to 50% of the landscape should meet desired forest conditions (Lower Mississippi Valley Joint Venture Forest Resource Conservation Working Group, 2007).

At the stand level, the desired forest structure consists of basal areas in the 60–70 ft^2 per ac (13.8–16.1 m^2 per ha) range, with 60–70% overstory canopy cover, and midstory cover in the 25–40% range. Secondary criteria include the presence of two or more dominant trees, tree cavities, both large and small, coarse woody debris, and standing dead trees. Regeneration should cover 30–40% of the area (Lower Mississippi Valley Joint Venture Forest Resource Conservation Working Group, 2007). The LMVJV recommends a combination of single-tree and group selection to develop DFCs within stands (Lower Mississippi Valley Joint Venture Forest Resource Conservation Working Group, 2011). The two techniques serve different purposes. Single-tree selection is employed to achieve target basal areas, while group selection serves to establish or release regeneration of shade-intolerant species (Lamson and Leak, 2000) such as oaks—a highly desirable species group for our purposes. While the LMVJV recommends group selections not exceeding 2 ac (0.8 ha) in size, regenerating Nuttall oak requires substantially larger openings (3.6+ ac, or 1.5+ ha) (Emile Gardiner, Research Forester, Center for Bottomland Research, USDA Forest Service, personal communication, April 9, 2014). Openings of this size are more appropriately called patch clear-cuts. Nuttall oak is the predominant oak species planted in the WRP plantation.

Based on these guidelines, the timber management plan was developed. The intent of the plan was to establish a forest that provides optimal habitat for white-tailed deer and is consistent with DFC guidelines. At the end of the planning horizon (40 years), the existing, essentially single-aged forest will be transformed to include multi-age classes and a range of density conditions, with multiple canopy layers developing within specified stands. Cover, browse, and hard mast should be in abundant supply.

The forested portions of the property are delineated into six management units. Management Unit 1 includes the mature timber stands. These five stands constitute roughly 12% of the forested area and currently meet the basal areal guidelines for DFCs at 65 ft^2 per ac (14.9 m^2 per ha). These stands will constitute the set-aside areas recommended at the landscape level. The SMZs constitute Management Unit 2. The primary purpose of this unit is to protect water quality in compliance with Mississippi's Best Management Practices (BMPs) (Mississippi Forestry Commission, 2008). No management will take place within the SMZs and thus, to some degree, they will serve a similar function as Management Unit 1. Due to their geographic configuration, however, they represent a significantly different habitat niche. In aggregate, these two management units represent 20% of the forested area, still within the recommended guidelines for unmanaged areas.

Management Units 3–6 consist of the hardwood plantation and the naturally regenerated hardwood stands. This area is delineated into four contiguous blocks, roughly equal in size, about 500 ac (202 ha). Each block constitutes one management unit. Because the timber is not yet merchantable size, no harvesting will be scheduled for at least five years so that prescribed timber harvests will be financially viable. Once the timber is large enough to sustain a commercial harvest, these blocks will be treated on 5-year intervals over a 20-year cycle, with Unit 3 treated in the first interval, Unit 4 the next, and so on until all four have been treated. The cycle will then be repeated, although the treatments will vary. Thus, the planning horizon is 45 years, consisting of two, 20-year treatment cycles initiating after a delay of approximately 5 years until the trees reach suitable size for a financial timber harvest. Progress toward plan goals will be assessed every 5 years and supplemental activities may be planned if warranted. Each unit will be treated at least once every 20 years. In the first entry into a unit, 60% of the block will be harvested in the following manner. Twenty percent of the unit (about 100 ac, or 40.5 ha) will be cut employing patch clear-cut methods. Approximately twenty 5 ac (2 ha) patch clear-cuts will be designated for harvest. These openings should be distributed relatively uniformly across the unit; however, existing openings should be incorporated and augmented whenever possible. Every attempt to coordinate the timing of the harvest within the 5-year interval to coincide with the presence of adequate shade-intolerant regeneration will be made. Forty percent of the unit (about 200 ac, or 81 ha) will be thinned using the single-tree selection method to approximately 50 ft per ac (11.5 m^2 per ha), targeting DFCs. Cutting to this basal area will allow the basal area to straddle the target basal area of 65–70 ft per ac (14.9–16.1 m^2 per ha) over the 20-year evaluation cycle. The remaining 40% will be reserved until the efficacy of the single-tree selection cut and patch clear-cuts can be assessed. This area may be thinned, if needed, to approximately 100 ft^2 per ac (23 m^2 per ha), to reduce the likelihood of stand stagnation and to increase crown size to stimulate mast production. In the second cycle, the area previously managed for DFCs will be re-evaluated and, if necessary, thinned back again to 50 ft^2 per ac (11.5 m^2 per ha). The patch clear-cuts will be evaluated for satisfactory regeneration. Mast bearing tree species should represent roughly 40% of the species composition. Assuming satisfactory regeneration in the patch clear-cuts, another 20 5 ac (2 ha) patch clear-cuts will be installed across the area managed for DFCs and the reserve. The remaining reserve may be lightly thinned, if needed, again to no less than 100 ft^2 per ac (23 m^2 per ha). After the second cycle, each block will consist of about 100 ac (40.5 ha) of newly naturally regenerated, mixed species, even-aged stands in the 0–20 age class, about 100 ac (40.5 ha) of naturally regenerated, mixed species, even-aged stands in the 21–40 age class, about 150 ac (60.7 ha) of low basal area stands whose overstory consists of predominantly oaks and other mixed hardwoods from the original plantation with substantial vertical structure and browse created by the two single-tree selection cuts. The remaining 150 ac or so (60.7 ha) will be composed of an even-aged hardwood plantation.

The condition of the forested portion of the property will be assessed every 5 years. Factors to be evaluated include the amount of available browse and mast, quantity, and species composition of regeneration throughout, presence of advanced regeneration in areas scheduled for regeneration cuts, basal area (as a measure of competition and stress), canopy structure in the DFC managed areas, and development of secondary DFC criteria such as cavities, coarse woody debris, and standing dead trees. Should desirable conditions not develop as expected, changes in the timing and intensity of the timber harvests, supplemental planting of desired species, and chemical injection of trees to create standing dead trees or to increase light penetration to the forest floor (thereby stimulating regeneration and browse) are all possible corrective actions.

OUTCOMES OF THE PLAN

If the prescriptions suggested in the management plan are followed, at the end of the planning horizon the property will contain 800 ac (324 ha) of naturally regenerated, even-aged, mixed-species stands, 600 ac (243 ha) of low basal area stands managed for DFCs, 600 ac (243 ha) of even-aged hardwood plantation, and 514 ac (208 ha) of old growth "reserve" timber. The naturally regenerated, even-aged stands will range in age from 5 to 40, and each age class will consist of about 100 ac (40.5 ha) (Figure 15.3). In the stands are managed by single-tree selection, basal areas will range from 50 ft^2 per ac (11.5 m^2 per ha) to 85–90 ft^2 per ac (19.5–20.7 m^2 per ha) in four condition classes of 150 ac (60.7 ha) each. These condition classes differ only in the time since last thinned with those most recently thinned having basal areas at the 50 ft^2 per ac (11.5 m^2 per ha) at the end of the planning horizon.

With regard to the secondary objectives (alternative revenue sources), to generate revenue to offset its operating costs, Willow Break is active in environmental services markets. In the past, it has generated revenues through the provision of shorebird and waterfowl habitat, rights-of-way, and carbon credits. The proposed timber harvests, if approved by the NRCS, may in time generate income. Because of the relatively small size of the timber and the complexity of the proposed harvests, initial sales are likely to be revenue neutral. Sales toward the end of the planning horizon are likely to generate substantial income.

The stand prescriptions and harvest schedule for the next nine 5-year time periods are as follows:

2014–2020: No harvests are planned during this period, as the timber will not yet be merchantable. Optional treatments may include reducing the number of trees in the naturally regenerated stands via chemical injection or "hack and squirt" methods.

2021–2025: 200 ac (80.9 ha) of single-tree selection harvests and 100 ac (40.5 ha) of 5 ac (2 ha) patch clear-cuts will be conducted in Unit 3. The patch clear-cuts will be located in areas of desirable advanced regeneration where feasible. Every fifth row may be cut as needed to provide access throughout the unit.

2026–2030: 200 ac (80.9 ha) of single-tree selection harvests and 100 ac (40.5 ha) of 5 ac (2 ha) patch clear-cuts will be conducted in Unit 4. The patch clear-cuts will be located in areas of desirable advanced regeneration where feasible. Every fifth row may be cut as needed to provide access throughout the unit. Unit 3 will be inventoried to assess the quantity and quality of browse and regeneration in the harvested areas.

2031–2035: 200 ac (80.9 ha) of single-tree selection harvests and 100 ac (40.5 ha) of 5 ac (2 ha) patch clear-cuts will be conducted in Unit 5. The patch clear-cuts will be located in areas of desirable advanced regeneration where feasible. Every

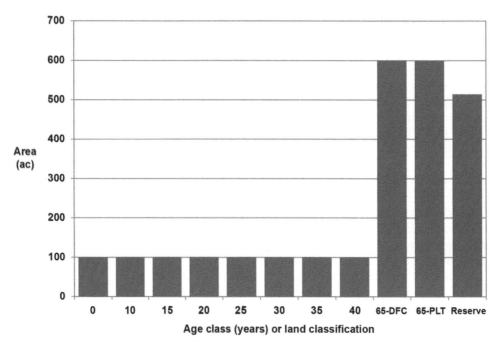

FIGURE 15.3 Age-class distribution of the forests on the Willow Break LLC property at the end of the planning horizon.

fifth row may be cut as needed to provide access throughout the unit. Units 3 and 4 will be inventoried to assess the quantity and quality of browse and regeneration in the harvested areas.

2036–2040: 200 ac (80.9 ha) of single-tree selection harvests and 100 ac (40.5 ha) of 5 ac (2 ha) patch clear-cuts will be conducted in Unit 6. The patch clear-cuts will be located in areas of desirable advanced regeneration where feasible. Every fifth row may be cut as needed to provide access throughout the unit. Units 3, 4 and 5 will be inventoried to assess the quantity and quality of browse and regeneration in the harvested areas.

2041–2045: This initiates the second cycle through the original plantation area. 100 ac (40.5 ha) of 5 ac (2 ha) patch clear-cuts will be implemented in Unit 3. The patch clear-cuts will be systematically distributed across the area previously thinned using single-tree selection and the residual unmanaged plantation, avoiding previous patch clear-cuts and located in areas of desirable advanced regeneration where feasible. Basal areas in the DFC managed area will be cut back to 50 ft^2 per ac (11.5 m^2 per ha) where needed using single-tree selection. In this cycle, greater attention will be paid to enhancing the secondary criteria, such as providing den and cavity trees. Units 4, 5, and 6 will be inventoried to assess the quantity and quality of browse and regeneration in the harvested areas.

2046–2050: 100 ac (40.5 ha) of 5 ac (2 ha) patch clear-cuts will be implemented in Unit 4. The patch clear-cuts will be systematically distributed across the area previously thinned using single-tree selection and the residual unmanaged planta-tion, avoiding previous patch clear-cuts and located in areas of desirable advanced regeneration where feasible. Basal areas in the DFC managed area will be cut back to 50 ft^2 per ac (11.5 m^2 per ha) where needed using single-tree selection. In this cycle, greater attention will be paid to enhancing the secondary criteria, such as providing den and cavity trees. Units 3, 5, and 6 will be inventoried to assess the quantity and quality of browse and regeneration in the harvested areas.

2051–2055: 100 ac (40.5 ha) of 5 ac (2 ha) patch clear-cuts will be implemented in Unit 5. The patch clear-cuts will be systematically distributed across the area previously thinned using single-tree selection and the residual unmanaged planta-tion, avoiding previous patch clear-cuts and located in areas of desirable advanced regeneration where feasible. Basal areas in the DFC managed area will be cut back to 50 ft^2 per ac (11.5 m^2 per ha) where needed using single-tree selection. In this cycle, greater attention will be paid to enhancing the secondary criteria, such as providing den and cavity trees. Units 3, 4, and 6 will be inventoried to assess the quantity and quality of browse and regeneration in the harvested areas.

2056–2060: 100 ac (40.5 ha) of 5 ac (2 ha) patch clear-cuts will be implemented in Unit 6. The patch clear-cuts will be systematically distributed across the area previously thinned using single-tree selection and the residual unmanaged planta-tion, avoiding previous patch clear-cuts and located in areas of desirable advanced regeneration where feasible. Basal areas in the DFC managed area will be cut back to 50 ft^2 per ac (11.5 m^2 per ha) where needed using single-tree selection. In this cycle, greater attention will be paid to enhancing the secondary criteria, such as providing den and cavity trees. Units 3, 4, and 5 will be inventoried to assess the quantity and quality of browse and regeneration in the harvested areas.

DISCUSSION AND CONCLUSIONS

Willow Break LLC is committed to providing family-oriented recreation. With hunting as the primary recreational activity enjoyed on the property, members also engage in a broad range of outdoor activities such as fishing, trapping, camping, trail riding, bird watching, and wildlife photography. The diverse habitats created by this plan will enhance all of these activities. Over the life span of the plan, the forest will evolve, and the reserve areas will continue to mature and take on more and more characteristics of old growth forests. The area of the hardwood plantation managed to maintain its current even-aged structure will develop into a mature hardwood forest, providing ample amounts of hard mast, open vistas due to the lack of an understory, but little cover or browse. The area managed by single-tree selection will mimic old growth conditions with multiple canopy layers, multiple age classes, and habitat niches not currently found on the property. The patch clear-cuts will provide stand level age-class diversity across the landscape, with the younger age classes providing both cover and browse. In combination, these treatments will create a forest to be enjoyed by all.

Sustainability Issues

Maintaining the carrying capacity of the property is critical to the club's primary objective of providing optimal hunting opportunities. The impending crown closure over two-thirds of the property will drastically reduce the carrying capacity unless action is taken. The proposed plan offers a pathway to maintaining and possibly even improving the carrying capacity of the property. Buy-in by, and cooperation with, the NRCS is critical to the success of the plan. By adopting a plan that complies with timber management recommendations endorsed by NRCS, Willow Break has taken an approach likely to facilitate buy-in by the NRCS.

The plan was a conservative approach. First, harvesting is scheduled on less than 12% of the forested property in any 5-year time period. Second, clear-cutting is scheduled on less than 4% of the forested property in any 5-year period. Furthermore, clear-cutting is restricted to 5 ac (2 ha) patch cuts, thereby complying with the guidelines to minimize clearcuts over 7 ac (2.8 ha). Third, 20% of the forested area is reserved from any type of harvesting. Fourth, 30% of the original plantation will not be converted. Cuts in this area will be constrained to the minimum required to maintain the health and vigor of the stand. Fifth, conversion is slow, occurring over a 40-plus year time horizon. Although a more aggressive approach would enhance habitat sooner, this conservative approach allows ample time to evaluate the efficacy of the single-tree selection harvests and their ability to create the desired canopy structure, cover, and browse and to evaluate the patch clear-cuts as a means of successfully regenerating a desirable species mix.

Plan Development Challenges

Developing the plan posed two unique challenges. First, all timber harvests are strictly regulated by the NRCS and are permitted only to improve wildlife habitat. Traditional timber harvests to improve growth and quality of the residual stand, maintain stand health, salvage dead or dying timber, or provide financial returns to the landowner do not qualify. Providing wildlife justifications for what would otherwise be standard silvicultural practices has not proven to be a successful tactic with NRCS. It was necessary to propose a timber harvesting system that had clear wildlife benefits and would not be seen as a thinly veiled attempt to conduct a traditional harvest. Although the LMVJV provides guidelines for managing bottomland hardwoods for wildlife, its system was developed with natural stands in mind. Thus, the challenge was to adapt these guidelines for even-aged plantations. A second challenge was the membership. A plan of this magnitude required a consensus of the membership. Most of the members, however, have no forestry or wildlife expertise and many have biases against timber harvesting in any form. The challenge was to present the plan to the membership in such a way as to garner the necessary support. In addition to a formal presentation to the club, many informal discussions in the duck blinds, bunkhouse, and around the lodge fireplace accomplished this goal.

Plan Implementation Challenges

Implementation faces three challenges. The first is a consequence of climate. The property routinely floods during the winter and spring and much of the area is inoperable until early-to-mid-summer. The second is that the LLC membership will not permit timber harvesting during hunting seasons. Hunting begins in September with mourning dove (*Zenaida macroura*) season and ends in mid-March with turkey season. Deer, small game, and waterfowl seasons occur within this window of time, effectively closing harvesting down for the duration. The third is a silvicultural restriction. Advanced oak regeneration must be present prior to harvesting the patch clear-cuts if successful establishment of shade-intolerant species is to occur. Thus, all harvesting operations are restricted to the late summer and, ideally, should be targeted to areas where oak regeneration is present. Clearly, these restrictions severely limit when and where harvesting can occur.

REFERENCES

Lamson, N.I., Leak, W.B., 2000. Guidelines for Applying Group Selection Harvesting. U.S. Department of Agriculture, Northeastern Area, State and Private Forestry, Newtown, PA, NA-TP-02-00. 8 p.

Lower Mississippi Valley Joint Venture Forest Resource Conservation Working Group, 2007. In: Wilson, R., Ribbeck, K., King, S., Twedt, D. (Eds.), Restoration, Management, and Monitoring of Forest Resources in the Mississippi Alluvial Valley: Recommendations for Enhancing Wildlife Habitat. http://www.lmvjv.org/library/Mgt_Board_June_2007/Tab7/MAV_Desired_Forest_Conditions_Final_Report_2007.pdf (Accessed April 27, 2014).

Lower Mississippi Valley Joint Venture Forest Resource Conservation Working Group, 2011. Wildlife Forestry in Bottomland Hardwoods: Desired Forest Conditions for Wild Turkey, White-tailed Deer, and Other Wildlife. http://www.lmvjv.org/library/DFC_Private_Landowner_Brochure_2011.pdf (Accessed April 9, 2014).

Mississippi Forestry Commission, 2008. Mississippi's BMPs. Mississippi Forestry Commission, Jackson, MS. MFC Publication 107, http://www.mfc.ms.gov/pdf/Mgt/WQ/Entire_bmp_2008-7-24.pdf (Accessed April 27, 2014).

U.S. Department of Agriculture, 2014. Wetland Reserve Program. U.S. Department of Agriculture, Natural Resources Conservation Service, Washington, DC. http://www.nrcs.usda.gov/wps/portal/nrcs/main/national/programs/easements/wetlands/ (Accessed April 27, 2014).

U.S. Fish and Wildlife Service, 2007. National Bald Eagle Management Guidelines. U.S. Fish and Wildlife Service, Washington, DC. http://www.fws.gov/midwest/eagle/pdf/NationalBaldEagleManagementGuidelines.pdf (Accessed April 9, 2014).

Chapter 16

Blue Ridge Parkway, Virginia and North Carolina, United States of America

Gary W. Johnson
Blue Ridge Parkway, Asheville, North Carolina, USA

ABBREVIATIONS

BRP Blue Ridge Parkway
GMP General Management Plan
NPS National Park Service

MANAGEMENT SETTING AND BACKGROUND

Construction of the Blue Ridge Parkway (BRP) motor road (Figure 16.1) began in 1933 under authority of the National Industrial Recovery Act (48 Stat. 195, Public Law 73-67). Congress then authorized the National Park Service (NPS) to administer and manage the BRP on June 30, 1936 (49 Stat. 2041, Public Law 74-848) as amended on June 8, 1940 (54 Stat. 249, Public Law 76-566). The Blue Ridge Parkway was established as the first national rural parkway in the United States. It was designed as a recreational destination-oriented driving experience with limited access, free from commercial traffic. Comprised of a scenic recreational motor road corridor and 15 major recreation areas, the park includes approximately 82,000 acres (ac) (33,185 hectares (ha)) of federally owned and managed land. Averaging 800 feet (ft) (244 meters (m)) in width, the motor road corridor traverses some 469 miles (mi) (755 kilometers (km)) through a variety of scenic ridge, mountainside, and pastoral farm landscapes in the central and southern Appalachian Mountains in Virginia and North Carolina. Fifteen major recreation areas were established along the length of the BRP at 30 mi (48 km) intervals (Figure 16.2).

The Parkway corridor encompasses the crests, ridges, and valleys of three major mountain ranges with 16 peaks above 5,000 ft (1,524 m) in elevation, and traverses several geographic and vegetative zones ranging from 600 to over 6,000 ft (183 to 1,829 m) above sea level. It is one of the most biologically diverse units in the NPS. Within the park, there are 600 mi (965 km) of streams, 2,074 plant species, and a wide variety of animal species including 43 amphibian, 99 fish, 60 mammal, 225 bird and 31 reptile species. There are 75 distinct plant communities documented, including 24 considered globally rare (7 of these are considered globally imperiled) (U.S. Department of the Interior, National Park Service, 2012).

As an example of pre- and post-World War II era automotive rural parkway design, the BRP retains the highest degree of designed landscape integrity of any parkway in the United States creating a unique driving experience that brings nature to the automobile. As such, it has been determined eligible for listing as a national historic landmark on the National Register of Historic Places. The Parkway's historic designed landscape includes 20 cultural landscapes, 26 tunnels (36% of NPS road tunnels), 91 historic buildings, 176 bridges (10% of NPS bridges), 382 overlook and parking areas, and 910 maintained roadside vistas. In addition, more than 200 archeological sites have been identified, and there are 690,000 objects and papers in the Parkway's museum and archival collection.

In support of the 15 million visitors that experience the BRP annually, managers and staff are responsible for operating and maintaining 9 campgrounds, 15 picnic areas, 3 concession lodges, 14 visitor centers, approximately 370 mi (595 km) of hiking trails, numerous cultural and natural exhibits, 90 waste water systems, and 50 water systems. More than 1,000 mi (1,609 km) of roadside vegetation (turf road shoulders and bays, wildflower bays, flowering trees and shrubs, and forest edges and vistas) are maintained on a cyclical basis. One important outcome of roadside vegetation maintenance is that more than 1,200 views are made available to showcase the beauty of the mountain region, ranging from intimate views of

Forest Plans of North America. http://dx.doi.org/10.1016/B978-0-12-799936-4.00016-3

FIGURE 16.1 A vista of the Blue Ridge Parkway in western North Carolina *(courtesy Blue Ridge Parkway).*

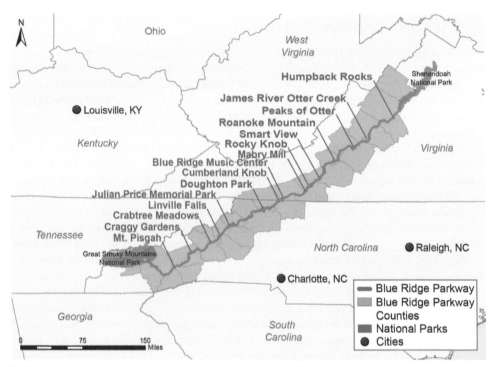

FIGURE 16.2 Vicinity map, segment planning units, and recreation areas of the Blue Ridge Parkway.

farms to lofty panoramas. The BRP includes more than 500,000 ac (202,347 ha) of scenic views within 1.0 mi (1.6 km) of its boundary for visitors to enjoy. These views are presented from 264 overlooks and hundreds of managed roadside vistas and open agricultural land views.

Administrative complexity is a suitable descriptor for the Parkway's corridor of overlapping jurisdictions, interests, and responsibilities. The BRP travels through 2 different U.S. Fish and Wildlife Service regions, 6 U.S. Army Corps of Engineers districts, 7 congressional districts, 12 planning regions, and 29 counties (12 in Virginia and 17 in North Carolina).

BRP staff interacts with 29 Natural Resource Conservation Service districts and consults with 8 associated federally recognized American Indian tribal governments. The BRP is also directly affected by four cities, four U.S. national forests, three state parks, and the reservation of the Eastern Band of the Cherokee Indians.

Land management is an everyday endeavor for park staff. The BRP shares over 1,000 mi (1,600 km) of its boundary with some 4,000 adjacent landowners who have more than 1200 individually held deed reservations on Parkway land. More than 100 of these reservations are for private road accesses. The Virginia and North Carolina Departments of Transportation have deed reservations for 39 primary access roads and almost 200 secondary roads that intersect the parkway motor road or cross park land.

PLANNING ENVIRONMENT AND METHODOLOGY

The NPS developed visitor facilities and operated the Parkway since its establishment using master plans that were developed in the late 1930s and early 1940s. Parkway land-use maps were also used for guidance, as well as applicable laws and policies. With completion of the Parkway motor road construction in 1987, management's focus began to shift to ensure that the Parkway's scenic, natural, cultural, and recreational qualities were protected. The early master plans only focused on facility development and provided no guidance for addressing current complex resource and visitor-use management issues confronting BRP managers and staff.

The National Parks and Recreation Act of 1978 requires that each unit (park) of the national park system must have an up-to-date general management plan. The NPS began preparing a general management plan for the BRP in the early 1990s, but that effort was placed on hold until 2000 when the project was refunded. The second General Management Plan (GMP) planning project for the Blue Ridge Parkway spanned almost 13 years incurring a cost of about $1.8 million U.S. dollars (USD). Communication with agencies and the public was accomplished through five NPS newsletters between 2002 and 2008 during the preparation of the draft general management plan. A *Draft General Management Plan (GMP) and Environmental Impact Statement* (EIS) was distributed in October 2011 to other agencies and interested organizations and individuals for their review and comment.

The *Final GMP and EIS* for the BRP was based on input from the NPS, other agencies, American Indian tribes, and the public. Consultation and coordination among these groups was vitally important throughout the planning process. The public had three primary venues for participation during development of the plan: participation in public meetings, responses to newsletters, and comments submitted by way of the National Park Service Planning Environmental Public Comment (PEPC) web site and regular mail. Changes and clarifications were made to the plan in response to comments received. Following the distribution of the final plan and a 30-day no-action period, a *Record of Decision* was signed on April 13, 2013.

General management planning for the BRP was accomplished using the NPS 11-step process (U.S. Department of the Interior, National Park Service, 2004; U.S. Department of the Interior, National Park Service, 2009) (Figure 16.3). The first three steps addressed factors that must be considered in the plan: the park's purpose and significance, special mandates, and NPS law and policies. Steps 4–11 included factors where NPS and BRP staff planners and managers worked in conjunction with other agencies and the public to determine how they wanted to see the park managed.

For Step 1, the BRP's enabling legislation, as well as laws and policies governing management of all national park system units, provided the basis for determining management purposes for the BRP. Park purpose statements establish the foundation for understanding why a park was established and its importance. During the GMP process the BRP's purposes where reaffirmed. They are to:

- Connect Shenandoah and Great Smoky Mountains National Parks by way of a national rural parkway, or a destination and recreational road that passes through a variety of scenic ridge, mountainside, and pastoral farm landscapes.
- Conserve the scenery and preserve the natural and cultural resources of the Parkway's designed and natural areas.
- Provide for public enjoyment and understanding of the natural resources and cultural heritage of the central and southern Appalachian Mountains.
- Provide opportunities for high-quality scenic and recreational experiences along the Parkway and in the corridor through which it passes.

During implementation of Steps 2 and 3, staff planners acknowledged all special mandates, commitments and laws, and NPS policies that had a direct bearing on the management planning. Next, during Step 4, scoping and information gathering meetings were employed to identify issues and concerns of the general public, NPS staff, county, state, and other federal agency representatives, parkway partners, resource experts, and representatives from various other organizations. The summary of the comments demonstrated the importance that public placed on the purpose for which the Parkway was

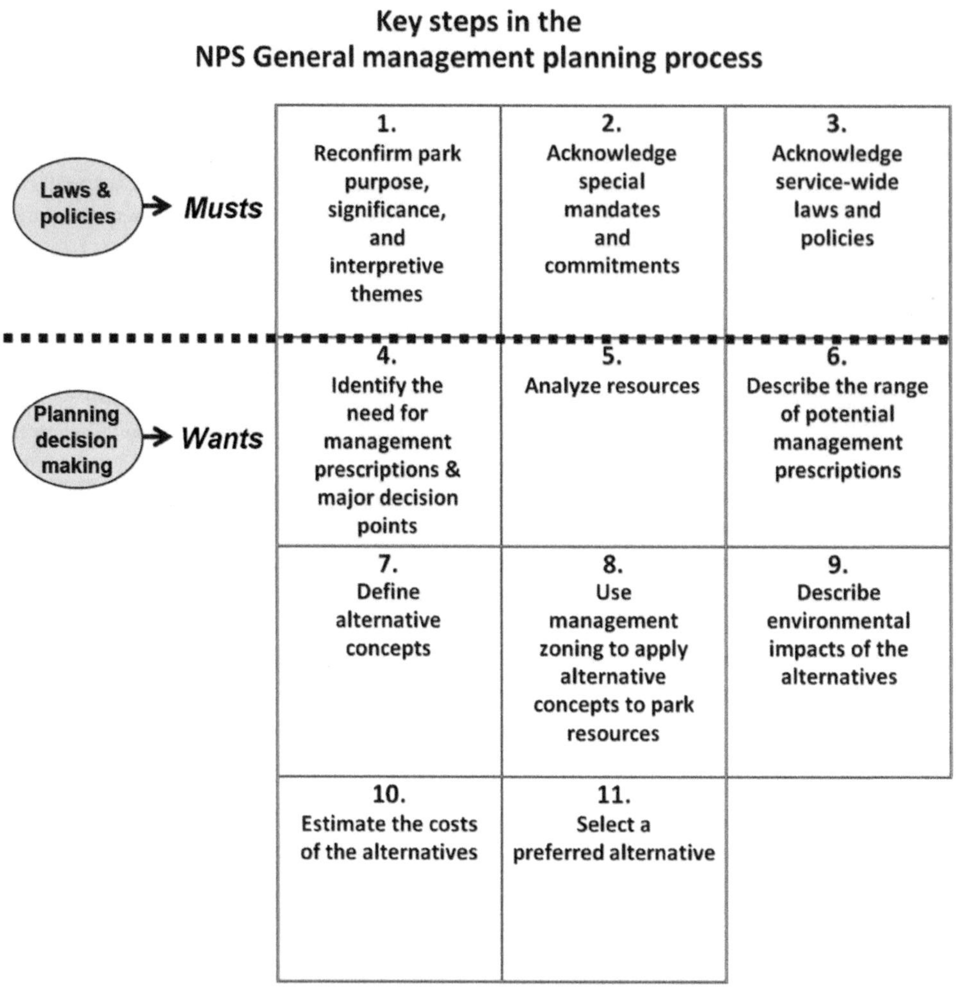

**Key steps in the
NPS General management planning process**

Laws & policies → *Musts*	1. Reconfirm park purpose, significance, and interpretive themes	2. Acknowledge special mandates and commitments	3. Acknowledge service-wide laws and policies
Planning decision making → *Wants*	4. Identify the need for management prescriptions & major decision points	5. Analyze resources	6. Describe the range of potential management prescriptions
	7. Define alternative concepts	8. Use management zoning to apply alternative concepts to park resources	9. Describe environmental impacts of the alternatives
	10. Estimate the costs of the alternatives	11. Select a preferred alternative	

FIGURE 16.3 Key steps in the National Park Service general management planning process.

created, especially the protection of scenery and high-elevation habitats, the local history and heritage, access to trails, the opportunity that the Parkway offers for escape from commercialism and traffic, and the beauty of the road and parkway architecture.

All of the issues and concerns identified in Step 4 were translated into six major decision points that were crafted as decision-point questions (U.S. Department of the Interior, National Park Service, 2012). These were used repeatedly by NPS planners and managers in Steps 5–8, and included the following:

1. To what extent should the original design of the Parkway be preserved, or under what circumstances might some design elements be modified for purposes of visitor convenience and safety, management of special resources, or fiscal or operational efficiency?
2. Are current visitation patterns and activities appropriate and sustainable, or are changes needed to protect special and/or fragile resources or the range of visitor opportunities?
3. Is the present range and mix of car, recreational vehicle, bicycle, motorcycle, and pedestrian use of the Parkway road appropriate and sustainable, or are changes needed for visitor experience and safety or for resource protection?
4. What are the desired conditions for park natural and cultural resources and what management strategies need to be implemented to ensure long-term sustainability of those conditions?
5. Can the park protect scenic views, cultural resources, and natural habitats important to the park via partnerships and agreements with park neighbors or to what extent are other approaches needed?
6. What criteria should the National Park Service use to determine whether or not and how secondary local or regional roads should be allowed to intersect or cross the Parkway?

Steps 5–8 create the primary building blocks (management zones and the management alternatives) for developing the approved management plan for a NPS unit. Building blocks must be joined, and that was the role for the six questions. The questions focused the effort of Step 5, the analysis of park resources, to determine the range of potential management zone prescriptions (Step 6). For Step 7, they placed limits on the definition of alternative concepts, and in Step 8 they guided applying management zoning to the alternative concepts. Products developed in Steps 5–8 were within the scope of the park's purpose, significance, mandates, and legislation described in Steps 1–3.

No comprehensive management zoning had been applied to BRP lands prior to starting the GMP planning in 2000. Development of management zoning was based upon the Parkway's existing natural, cultural, and scenic resources and visitor experiences and types of uses. Zoning defined land areas by type of zone and described how they are to be managed. The prescriptions were designed to provide a range of desired resource conditions and visitor experiences for the park and included statements about the appropriate kinds and levels of management, use, and development appropriate for each zone.

Eight management zones and their associated prescriptions were defined for the BRP (U.S. Department of the Interior, National Park Service, 2012). Two zones (Special Natural Resources and Natural zones) emphasized managing the Parkway's natural resources. The Special Natural Resources zone represents areas that would emphasize the highest level of protection of sensitive habitats. Natural resources and processes would be preserved to maintain their pristine conditions and ecological integrity. Visitor opportunities would be limited to avoid human-caused impacts on these sensitive or rare ecosystems. The Natural zone represents areas that would support the broader ecological integrity of the Parkway where natural processes predominate. Only low-impact recreational activities would be allowed. Visitors would be immersed in nature with opportunities to experience solitude and tranquility.

Cultural resources were emphasized in two zones: Special Cultural Resources and Historic Parkway zones. The Special Cultural Resources zone represents areas that would emphasize protection of cultural landscapes and historic structures not associated with the design and development of the Blue Ridge Parkway. The Historic Parkway zone represents areas that emphasize protection and interpretation of the historic parkway corridor, which includes the road prism and its original supporting structures and constructed landforms.

Landscape architects designed the Parkway for the provision of scenic driving experiences. Their goal has been realized, as 95% of BRP visitors give viewing scenery as their primary reason for visiting the Parkway. To manage the scenic attributes of the Parkway, a Scenic Character zone was developed. It represents areas of the Parkway that emphasize protection of scenery and viewing opportunities of the scenic landscapes along with the natural and cultural settings of the central and southern Appalachian highlands.

Parkway land areas where visitor use is emphasized were either zoned as a Recreation or Visitor Services zones. The Recreation zone represents areas that would support moderate levels of visitor use to accommodate a wide range of recreational, educational, and interpretive opportunities. Although some resource modifications could occur, natural and cultural resources would remain largely intact. The Visitor Services zone represents areas of the Parkway that would support moderate to high levels of development and visitor services in order to accommodate concentrated visitor use and diverse recreational, educational, and interpretive opportunities.

An eighth zone is for Park Support, and this zone represents areas of the Parkway that support administrative facilities for operations and maintenance. There are 14 maintenance and ranger office compounds along the Parkway.

A *Draft Blue Ridge Parkway GMP and EIS* presented three alternatives for the future management of the Parkway (U.S. Department of the Interior, National Park Service, 2012). Alternative A was the "no-action" alternative and proposed a continuation of current management direction. The development and discussion of a no-action alternative is required by law. The current management direction for the BRP, prior to the development of the plan described in this chapter, was not based on management zones, as there were none in place. The no-action alternative served as a baseline for comparing the two "action" alternatives, Alternatives B and C. The three alternatives encompassed a reasonable range of what the public and the NPS desired to see accomplished with regard to natural and cultural resource conditions, scenery conservation, land protection, visitor opportunities and experiences, traffic and transportation, concessions, and other services.

The two action alternatives presented approaches for managing the BRP in a more proactive manner. Both of the action alternatives presented an overall management concept and proposals about how different Parkway programs and areas could be managed through the application of the management zones and other strategies. In formulating the two action alternatives, management zones were placed in different locations on a map of the BRP according to each alternative's overall management concept. For example, the overall concept for Alternative C proposed a more ecosystem-based approach to natural resource management, and as a result, the Natural zone covered more Parkway lands than under Alternative B. This is because the Natural zone includes management prescriptions that emphasize an integrated natural resource management approach; management prescriptions define the kinds of resource conditions and visitor experiences that should be achieved and maintained in a management zone.

Given the geographical length of the BRP, information and proposed actions in the alternatives were organized into three levels of detail: (1) parkway-wide, (2) parkway segments, and (3) major recreation areas. The parkway-wide discussion presented the overall concept for each alternative followed by proposals that affect parkway-wide programs, activities, or resources. For the second level of detail, seven parkway segments were identified based primarily on physiographic and landscape characteristics. The segment organization also separated out the two major urban centers along the parkway—Roanoke, Virginia and Asheville, North Carolina. Separate planning actions and zoning maps were developed for the 15 major recreation areas. Parkway segment limits and recreation area locations are shown on Figure 16.2.

Under Alternative A, the "no-action" alternative no-management zoning had been implemented. A comparison of the land area encumbered by each management zone for Alternatives B and C is shown in Table 16.1.

Identification of the Preferred Alternative and Analysis of Impacts

The three alternatives were evaluated using an objective analysis process called "Choosing by Advantages" developed by the U.S. Forest Service and adopted in the 1990s by the NPS (Suhr, 2007). A three-day workshop was held in which 19 staff members, representing all management divisions of the Blue Ridge Parkway, worked together to develop the preferred alternative. Through this process the relative advantages of each alternative was compared using a set of factors. These factors were selected based on the benefits or advantages of each alternative to fulfill the purpose of the plan, while addressing the planning issues. Factors included the following:

- *Factor 1*—Maintains or enhances natural conditions and processes
- *Factor 2*—Preserves cultural resources
- *Factor 3*—Provides for an appropriate range of visitor services and recreational opportunities
- *Factor 4*—Provides a traditional parkway and scenic driving experiences
- *Factor 5*—Improves operational effectiveness and sustainability
- *Factor 6*—Provides other advantages to the Blue Ridge Parkway, regional communities, partners, and stakeholders

The "Choosing by Advantages" workshop identified Alternative B as the NPS's preferred alternative (U.S. Department of the Interior, National Park Service, 2010). This alternative provided the best combination of strategies to protect the park's unique natural and cultural resources and visitor experience, while improving the park's operational effectiveness and sustainability. It also demonstrated other advantages to the Parkway, regional communities, partners, and stakeholders. The significant advantage to cultural resources was one of the largest determining factors in identifying Alternative B as the agency's preferred management alternative.

TABLE 16.1 Management Zone Areas for Alternatives B and C of the Blue Ridge Parkway Management Plan

Management Zone	Alternative B (Preferred)		Alternative C	
	(ac)	(ha)	(ac)	(ha)
Scenic character	34,322	13,890	33,997	13,758
Natural	19,491	7,888	24,584	9,949
Special natural resource	10,068	4,074	10,074	4,077
Historic parkway	9,623	3,894	9,349	3,784
Recreation	7,751	3,137	2,946	1,192
Special cultural resource	388	157	388	157
Visitor services	356	144	662	268
Park support	193	78	193	78
Total	82,192	33,263	82,192	33,263

Key components of the NPS preferred alternative included the following:

1. Focusing resources on the traditional Parkway experience, including management based upon the original Parkway land-use maps as closely as possible.
2. Embracing a regional, ecosystemwide approach to natural resource management.
3. Enhancing outdoor recreational opportunities on Parkway lands, including regional trail connection through collaborative planning.
4. Emphasizing strategic planning and partnerships to address land and viewshed protection issues, education, and interpretation.
5. Recognizing that concessions are a vital part of the parkway experience and seeking to invest in those structures and businesses to make them more viable.
6. Allowing for moderate upgrades to campgrounds, rather than wholesale redesign.

An important part of NPS management planning is seeking to understand the consequences of making one decision over another (U.S. Department of the Interior, National Park Service, 2011). To that end, a park general management plan is accompanied by an EIS, which identifies the anticipated impacts of possible actions on resources and on park visitors and neighbors. Impacts were organized by topic, such as "impacts on the visitor experience" or "impacts on vegetation and wildlife." Impact topics serve to focus the environmental analysis and to ensure the relevance of the impact evaluation.

Actions described in the GMP alternatives were conceptual in nature and the impacts of those actions were analyzed in general qualitative terms, thus, the EIS analysis was completed at a programmatic level. The programmatic impact analysis compared the beneficial and adverse effects of implementing each of the alternatives to determine which alternative would create the most desirable combination of benefits with the fewest adverse effects on the park. The impact analyses were based on the planning team's professional judgment, research of existing studies and literature, opinions from experts within the NPS and other agencies, and the study of previous projects that had similar effects. Several impact parameters were analyzed for each alternative. Potential impacts of the three alternatives were described in terms of four criteria: type, intensity, duration, and context.

1. *Type of impact* was determined to be either beneficial or adverse. The beneficial and adverse impacts on resources and values were assessed by comparing the anticipated changes that would result from implementing each action alternative to the results of continuing current management. Once it was determined if an impact was beneficial or adverse, the three other impact measurement criteria were assessed.
2. *Intensity of impact* refers to the degree, level, or strength of the impact on the respective resource or value. The impact intensities for beneficial and adverse effects were quantified as negligible, minor, moderate, and major. Because the definitions of intensity vary by resource topic, separate intensity definitions are for each impact topic.
3. *Duration of impact* refers to the length of time the impact affects the resource or value. In this analysis, impact durations were defined as short-term if impacts would last less than five years and long-term if impacts would persist for five or more years, or may be permanent.
4. *Context of impact* refers to the setting or geographic scope of the impact on the particular resource or value. In this analysis, impacts were measured relative to two context levels: local, if impacts would be limited to a specific site or relatively small area within the Parkway boundaries; or regional, if impacts would occur over a large, widespread area within or beyond the Parkway boundaries, or in several areas along the Parkway.

OUTCOMES OF THE PLAN

The most important outcome of the GMP process is having an approved comprehensive Parkway-wide approach for resource and visitor-use management with the associated management zoning in place. Specific management prescriptions for each of the eight zones were applied to Parkway lands for each of the action alternatives. Prescriptions detail acceptable resource conditions, visitor experience and use levels, and appropriate activities and development. Both the comprehensive management approach and zoning provide the necessary guidance for resolving conflicts among natural and cultural resource issues and visitor-use demands and for setting the scope of projects, determining project priorities, and funding requests.

The approved management plan (U.S. Department of the Interior, National Park Service, 2013) takes the long view and creates a realistic vision for the future, setting a direction for the park that takes into consideration the environmental and financial impact of proposed facilities and programs and ensures that the final plan is achievable and sustainable. This management approach is more proactive in blending newer law and policy requirements and operational constraints with the traditional parkway self-contained, scenic recreational driving experience and the designed landscape developed during the Parkway's historic period of significance, 1935 to 1955. Under the approved GMP, Parkway recreation areas will provide enhanced opportunities for dispersed outdoor recreation activities. Opportunities to enhance resource protection, to facilitate regional natural resource connectivity, and to build stronger connections with adjacent communities will be more proactively pursued as directed in the approved GMP.

Land protection and scenery conservation activities now have a more integrated and directed approach. The Parkway's land protection process was updated to provide a strategy that does not identify specific tracts of land to be acquired. The approved process now incorporates resource and visitor-use management criteria, park management zoning and land-use compatibility factors, and other protection goals that will be used to evaluate the merits of acquiring an interest in adjacent property when it becomes available from willing sellers. Parkway staff will actively collaborate with adjacent landowners, county officials, and developers on a site-specific project basis and on regional efforts to conserve priority off-Parkway scenery through long-term strategies for conserving views seen from Parkway overlooks and vistas.

NPS management of natural resources will be a more proactive, long-term, and multi-year approach to inventory and management of natural resources, and will strive to advance regional ecosystem health through active partnerships with public and private entities. Some landscape areas traditionally managed for scenery, such as roadsides, vista clearings, and agricultural leases will be modified to actively protect natural resources. A Class I air-quality classification will be pursued in accordance with Clean Air Act, 1970 as amended 1977 and 1990. The Class 1 classification would provide funding and support for monitoring air quality of 500,000 to 1,000,000 ac (202,347 to 404,694 ha) of viewshed adjacent to the Parkway. Improving air quality is critical for maintaining visual quality of views that millions of tourists come to the Parkway and adjacent communities to experience each year.

Designation of the designed Parkway corridor as a national historic landmark will be sought while continuing to manage it as an eligible resource. Priority for preservation of historic structures will be provided for those that are directly associated with the Parkway's original design intent. Vistas above 4,000 ft (1,219 m) in elevation will be maintained, but their size and configuration will be determined by best practices for managing the potential habitat of sensitive species. The historic Parkway land-use maps will be updated to protect the Parkway's historic integrity while accommodating newer law and policy requirements and operational constraints.

Interpretation and visitor services operations at selected locations will be expanded to provide services for a nine-month visitor season. Visitor education will be increased using publications and waysides and emerging technology. The BRP will continue to maintain 20 recreation areas along the length of the Parkway with traditional visitor services that support a recreational and scenic driving experience, including camping, lodging, restaurants, camp stores, and picnic sites. The BRP will also continue to implement curriculum-based school outreach programs using current staffing levels at schools and in the Parkway, as available, during the school year. Operation of the existing campgrounds will be continued, including future repairs and rehabilitations focused on meeting backlog maintenance needs. Selected campground comfort stations will be upgraded to provide showers and all will be upgraded to be universally accessible. Existing recreational vehicle sites will be upgraded with water and electrical hookups except at Roanoke Mountain. Recreational vehicle access will also be improved to portions of the Peaks of Otter and Julian Price campgrounds. Finally, park management will continue to provide viable concession services at all existing locations to ensure the long-term availability of in-Parkway lodging, food, and other services.

The moratorium on secondary road improvement projects in both Virginia and North Carolina will be continued until a comprehensive corridor access management plan and environmental impact statement are completed.

DISCUSSION AND CONCLUSIONS

The approval and implementation of the general management plan represents a pivotal milestone for management of the Blue Ridge Parkway. It marks the significant shift from an infrastructure development and maintenance-oriented management strategy to a comprehensive resources and visitor-use management approach.

Learnings and Insights

Throughout the GMP process, project planning staff worked with adjacent county officials, local residents, travel and tourism organizations, and other stakeholders. From those conversations, many issues were identified, and people had definite ideas about how "their" Parkway was to be managed. But what came across most clearly for planning staff was just how personally important the Parkway is to the numerous individuals and communities whose lives are touched every day by the Parkway. For folks living in the Parkway's geographic region, it is a source of renewal and recreation, a place to enjoy scenic beauty and, for many it has economic benefit from the visitors who spend money in adjacent communities. While the Parkway is a unit of the National Park System and is visited every year by some 15 million people from around the world, it is those local park neighbors who feel most passionately about the Parkway and what should happen to it in the future.

The approved GMP management strategy responded to local interest and calls for the NPS to enhance resource protection, to act regionally in supporting natural resource connectivity, and to build stronger connections with adjacent communities.

Sustainability Issues

Sustaining the park's diversity of plant and animal species, historic resources, clean water, and scenic quality requires park staff to move toward a regional-centric style of management and away from the traditional parkway-centric style. The approved GMP sets forth just that kind of management strategy where park management and staff will need to think and act on a scale that extends beyond the park boundary to manage regional natural resources, protect cultural resources, and to conserve scenic resources. The plan emphasizes collaboration with park partners, other federal and state agencies, and surrounding communities.

Plan Development Challenges

The Parkway's general management planning process was a relatively slow and expensive one as it required 13 years and cost of $1.8 million USD. Four factors contributed to the length of time and expense required to complete the plan. First, the linear nature of the BRP traveling 469 mi (755 km) through 29 counties—12 in Virginia and 17 in North Carolina—presented a logistical challenge for planning staff to adequately and efficiently involve all of the various stakeholders during project scoping and throughout the planning process. Next, contractors were hired for four years to gather the necessary resource and visitor-use data that planning staff required for alternatives development and impact analysis. The third factor resulted from the political reality of how federal agencies are affected by changing administrations in Washington, DC. The BRP general management planning process extended through two presidential administrations, two Republican Party terms, and one Democratic Party term. There were internal Department of the Interior and National Park Service changes in leadership and associated shifts in park management and general management planning philosophy. As a result, midway through the planning process and after alternatives were fully developed, the planning team was directed to re-think the array of proposed actions within each of the alternatives and to conduct an extensive campground analysis. A fourth factor, project funding, became the most critical obstacle to completing the GMP. Decreases in available funding for the NPS GMP program and the need to complete other park GMP's made finding additional funding for the BRP GMP almost impossible, thus slowing project completion.

Plan Implementation Challenges

The approval of the GMP does not guarantee that the funding and staffing needed to implement the plan will be available. For example, implementation of the plan's management actions requires an increase of $4 million USD in the park's base operations budget to pay for additional NPS park staff, supplies, and materials. Current nondefense federal agency funding trends that flatten or reduce operations cost make receiving a budget increase unlikely. So implementation of the plan will be slow at best.

Other Interesting Issues Related to the Plan

During public review of the Draft GMP and EIS, unanticipated reactions came from the bicycling community. While the GMP had no proposed action to limit, beyond existing law and policy, or close the Parkway motor road to bicycle use in the plan, information contrary to what was in the plan appeared in an article in an outdoors magazine and on several local and national cycling group web sites. The misinformation then spread quickly through blogs, Twitter, and Facebook. None of these groups called the park to fact-check before printing information. Bicycling groups also went on record to oppose the Parkway's nomination as a National Historic Landmark because they believed that designation would preclude bicycle accommodations (i.e., a bike lane along the road). Numerous individuals and bicycling organizations communicated their concerns to the NPS at the park, the southeast regional office, and the Washington office. Over a several-month period, park planning staff conducted numerous local, regional, and national interviews, including radio, television, and print media, and worked with various bicycling organizations to correct the misinformation those organizations had provided the public.

ADDITIONAL READING AND RESOURCES

This chapter represents a summary of the Blue Ridge Parkway General Management Plan and planning process. To download a copy or view the final General Management Plan Environmental Impact Statement, please visit this Internet site, which was available on May 13, 2014:

http://www.nps.gov/blri/parkmgmt/index.htm

Once you have accessed the site, select *Final General Management Plan/Environmental Impact Statement FGMP/EIS*, which takes you to another page where you can download the Final GMP/EIS (25.2 MB, PDF file).

REFERENCES

Suhr, J., 2007. Choosing by Advantages: Sound Decisionmaking. The Institute for Decision Innovations, Inc., North Ogden, UT.

U.S. Department of the Interior, National Park Service, 2004. Program Standards Park Planning. U.S. Department of the Interior, National Park Service, Washington Office, Washington, DC.

U.S. Department of the Interior, National Park Service, 2009. General Management Planning Dynamic Sourcebook. Version 2.2, U.S. Department of the Interior, National Park Service, Washington, DC.

U.S. Department of the Interior, National Park Service, 2010. Blue Ridge Parkway, Virginia and North Carolina, general management plan choosing by advantages draft workshop meeting report. U.S. Department of the Interior, National Park Service, Denver Service Center, Denver, CO.

U.S. Department of the Interior, National Park Service, 2011. Director's Order 12 Handbook. U.S. Department of the Interior, National Park Service, Washington, DC.

U.S. Department of the Interior, National Park Service, 2012. Blue Ridge Parkway, Virginia and North Carolina, Draft General Management Plan/Environmental Impact Statement. U.S. Department of the Interior, National Park Service, Denver Service Center, Denver, CO.

U.S. Department of the Interior, National Park Service, 2013. Blue Ridge Parkway, Virginia and North Carolina, Summary of the Final General Management Plan. U.S. Department of the Interior, National Park Service, Denver Service Center, Denver, CO.

Chapter 17

Chesapeake Forest Lands, Maryland, United States of America

Kip Powers,[1] Anne Hairston-Strang,[2] Steven Koehn[3] and Kenneth Jolly[2]

[1]Maryland Department of Natural Resources, Forest Service, Salisbury, Maryland, USA, [2]Maryland Department of Natural Resources, Forest Service, Annapolis, Maryland, USA, [3]USDA Forest Service, Washington, District of Columbia, USA

ABBREVIATIONS

DFS	Delmarva Fox Squirrel
ESA	Ecologically Significant Area
FIDS	Forest Interior Dwelling Species (bird species)
HCVF	High Conservation Value Forest
MD DNR	Maryland Department of Natural Resources
TCF	The Conservation Fund

MANAGEMENT SETTING AND BACKGROUND

Chesapeake Forest was part of the national trend of integrated forest products companies selling off timberlands starting in the late 1990s (Butler and Wear, 2013). On September 10, 1999, approximately 76,000 acres (ac) (about 30,757 hectares (ha)) of forested land on the Delmarva Peninsula were sold by the Chesapeake Corporation to an innovative partnership between The Conservation Fund (TCF), the Richard King Mellon Foundation, and the State of Maryland. In addition, both Hancock Timber Resources Group and the Chesapeake Bay Foundation played significant roles in developing and supporting the acquisition by their partnership. Of the 58,400 ac (23,634 ha) purchased in Maryland, approximately half of the land was acquired by the Maryland Department of Natural Resources (MD DNR). TCF purchased the remaining half of the land on behalf of the Richard King Mellon foundation. It, in turn, then gifted these lands to the State of Maryland in December 2000. The division of tracts between the state and the TCF was based on a MD DNR review that resulted in State purchase of the most environmentally sensitive tracts as well as those adjoining existing MD DNR properties. Since the original purchase, several new acquisitions have added 9,372 ac (3,793 ha) to the forest, bringing the total now under the scope of the Chesapeake Forest Lands Sustainable Plan to 67,772 ac (27,427 ha) (Figure 17.1).

The geography of the Chesapeake Forest Lands reflects its industrial forest heritage, with more scattered, smaller parcels than is typical for state or federal lands in this area. The forest is divided among 238 parcels currently in 187 land management units, which are spread across a six-county area on the middle and lower eastern shore of Maryland. The abundance of parcels, extensive boundaries, and young age distribution of the forest at the time of acquisition posed significant challenges for sustainable forest management. Therefore, one of the management challenges inherent in the land base is that, in spite of the attempts to create the most manageable units, there are many small, isolated properties. For example, 7 management units are less than 30 ac (12 ha), and 27 are less than 50 ac (20 ha) in size. Ninety of the management units on the forest are less than 150 ac (61 ha) in size. Most of these areas adjoin or are surrounded by agricultural or developed land (Figure 17.2). The number of small parcels and their interrelationship with adjacent private landowners make their management very comparable to that of many nonindustrial private landowners in the region. The MD DNR must weigh the effects of various management activities that affect adjoining properties and seek to maintain good community relations with neighbors, who can be important eyes and ears over such an extensive and dispersed ownership.

Forest Plans of North America. http://dx.doi.org/10.1016/B978-0-12-799936-4.00017-5

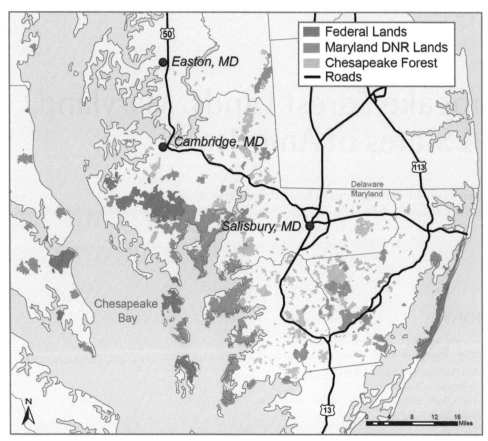

FIGURE 17.1 Chesapeake Forest Lands on the lower eastern shore of Maryland.

FIGURE 17.2 Aerial view showing the flat terrain typical of Chesapeake Forest Lands and adjoining agriculture lands on the Delmarva Peninsula.

The Chesapeake Forest Lands provide habitat for species such as the Delmarva fox squirrel (DFS) (*Sciurus niger cinereus*) and esthetic diversity, as well as the opportunity for a variety of ecological restoration projects. The typical physical constraints to management, goals for enhanced environmental protection of rare species, water quality improvement, and expanded public hunting opportunities all have arisen during a period of heightened public interest in state land management. The public-private partnership and clear commitment to economic as well as environmental goals have proved instrumental for effective management. However, adjoining land uses, such as agriculture or development, constrain the set of potential management activities on state lands. Smoke hazards are a particular concern near public roads and poultry farms, found throughout the Delmarva Peninsula, that limit the use of prescribed fire.

Young loblolly pine (*Pinus taeda*) forests, mostly established since the early 1980s, characterize a high proportion of the Chesapeake Forest Lands (Table 17.1). Naturally regenerated mixed pine and hardwood forests still occupy some of the lands, and many riparian areas and flood plains contain stands of mixed hardwoods. In general, the mixed pine-hardwood and hardwood stands are older, mature forests. Atlantic white-cedar (*Chamaecyparis thyoides*) and marshes, fields, and power lines occupy a small percentage of the forest but are important areas for meeting biodiversity goals. Atlantic white-cedar occupies only a fraction of its former range in Maryland, so opportunities are sought to plant seedlings where soil and hydrologic conditions permit. The non-forested lands include some light-loving rare species and habitats, some of which are related to the historic frequent fire intervals and others to distinctive wetland habitats like Delmarva Bays.

The current inventory approach for the forest is stratified by pine, hardwood, and mixed types because of their different uses by rare species. These species include the Delmarva fox squirrel, listed as endangered in 1967, as well as forest-interior-dwelling species (FIDS) that are in decline, especially neotropical migratory birds. The Chesapeake Forest Lands and other state and federal lands play a significant role in protecting these habitats and in tailoring management to support an appropriate matrix of habitats over time. Having the surety of suitable habitats on public lands may help relieve management constraints on private forests, depending on species needs and the requirements in an approved recovery plan. The latest 5-year review for DFS considered the population no longer in danger of extinction due to population and habitat expansion, creating a potential opportunity to modify recovery plan requirements.

PLANNING ENVIRONMENT AND METHODOLOGY

The original sustainable forest management plan was written for the TCF half of Chesapeake Forest as part of the two-step acquisition partnership described in the introduction. Work began on this plan in 1999 and was completed in December of 2000. It was developed for TCF by a planning team assembled by The Sampson Group, Inc. (2000), under the direction of Neil Sampson, Planning Coordinator. Guidance and planning decisions were provided by a steering committee under the leadership of David Sutherland of TCF, and included participation by representatives of the Chesapeake Bay Foundation, MD DNR, and the Smurfit-Stone forest products corporation. Two teams and a steering committee were created and reviewed and provided input on the original 2000 TCF plan. This plan was adopted as the interim guiding management

TABLE 17.1 Area (ac) of Forest Type by Structural Groups, Chesapeake Forest Lands

Forest Type	Structural Group							
	Open (0–5 yr)	Sapling (6–15 yr)	Growing (16–25 yr)	Maturing (26–40 yr)	Mature (41–60 yr)	Big Trees (61+ yr)	Uneven-aged	Total
Atlantic white cedar	4	2	1	0	0	0	0	7
Loblolly pine	635	7,322	20,141	14,812	5,177	702	555	49,344
Mixed pine/ hardwood	591	447	304	748	1,147	4,611	14	7,862
Mixed hardwoods	189	290	290	109	2,247	5,171	12	8,308
Marsh/fields/ powerlines	2,251	0	0	0	0	0	0	2,251
Total	3,670	8,061	20,736	15,669	8,571	10,484	581	67,772

document for the MD DNR half of the Chesapeake Forest Lands until a new sustainable plan could be developed for the entire 58,000 ac (23,472 ha) forest.

The 2004 sustainable forest management plan for the combined area (Maryland Department of Natural Resources, Forest Service, 2004) was reviewed by an advisory committee (a group consisting of various resource professionals), private citizens, industry representatives, and local political leaders. Committee review was conducted at monthly open public meetings. Based on the input provided by the advisory committee, along with updated resource information provided by MD DNR resource professionals, several sections of the 2000 TCF plan were revised, a few new chapters were added, and several were deleted. Following the completion of this draft document, additional input was received at a public meeting and during a 30-day public comment period. In 2004, dual certification of the forest by the Sustainable Forestry Initiative (SFI) and the Forest Stewardship Council (FSC) was obtained.

A 2007 Sustainable Forest Management Plan (Maryland Department of Natural Resources, Forest Service, 2007) now guides the management of the forest. Since receiving dual certification, the plan has become a "living document," undergoing periodic revisions since 2007 in order to keep track with certification standards. As of 2013, there have been seven revisions to the 2007 plan.

Guiding Laws, Regulations, and Policies

The plan developed for the Chesapeake Forest Lands needed to be consistent with Maryland's rules that govern private and public land, such as the Maryland Critical Areas Law (Maryland Title 27), Sediment and Erosion Control, Non-tidal Wetlands, Air Quality for Open Burning Regulations (Maryland Title 26), and the State Endangered Species Program (Maryland Title 08.03.08). Some state policies go beyond the minimum, such as requiring on-site inspections of all harvests and using Master Loggers for timber sales. State funding for the land purchase also came with some goals for public access and public hunting, a layer of complexity beyond typical private land constraints.

MD DNR intended Chesapeake Forest to represent a national model of certified sustainable forestry. To meet that objective for the forest, third-party certification was sought under both the SFI standard and the FSC standard. The dual certification for the gifted portion of the forest achieved in 2004 was followed a year later with dual certification for the entire forest. Compliance with these certification programs is monitored through annual audits conducted by a combined SFI/FSC audit team. Since the 2004 plan was completed, there have been a number of revisions, mostly driven by the SFI/FSC certification process. Changes have included revised chapters dealing with sensitive species management, high-conservation-value forest (HCVF), and forest modeling. In addition, several new tracts of land have since been added to the forest. All of these revisions and changes are now covered under the current (2007) plan, which includes major revisions to the chapter on forest modeling guidelines along with a completely revised chapter on ecologically significant areas (ESA). Several other chapters in the 2007 plan underwent minor revisions to update sections with current data. The 2007 plan went through a process of reviews and approvals similar to the 2004 plan. Additional management issues have been raised and addressed through the annual certification reviews, further tailoring management to better address sustainability principles. Issues have included vernal pool protection, management of DFS habitat, off-road vehicle trails, rutting and compaction policies, and invasive species.

The State of Maryland and TCF publicly committed that these forests will continue to play their role in a viable forest-based economy on the eastern shore. As stated in the 2007 plan, the primary goal of the forest is to demonstrate that an environmentally sound, sustainably managed forest can contribute to local and regional economies. This goal will be pursued subject to the following five broad constraints:

1. The quality of the water flowing through the properties will not be impaired due to any actions on the land, and in many cases will be improved. Where feasible, wetlands, riparian areas, and ditches will be the site of watershed improvement practices specifically aimed at improving the quality of water entering the Chesapeake Bay.
2. The management policies and actions will be consistent with state and federal requirements for protecting and managing rare, threatened, and endangered species of plants and animals. The MD DNR will identify locations of rare, threatened, and endangered species habitat and forest conditions associated with the habitat requirements of these species. Management actions will consider opportunities to enhance existing habitats and provide for corridors. Abundance and distribution goals for common species will be periodically updated through MD DNR resource assessments. Habitat goals for common species will be reflected in forest management activities.
3. Forest harvest levels will comply with targets established by a long-term sustainable harvest plan. To the extent possible, harvest and thinning activity levels will produce reasonably uniform flows of products and contractor activities year-to-year. Short-term deviations due to natural disturbances, operational logistics, or unusual events are anticipated,

but exceptions for an extended period will require re-evaluation of the sustainable harvest level. Spatial and timing constraints will prevent thinning or harvesting operations from concentrating impacts in any watershed or visual scene in violation of water quality goals, habitat diversity and connectivity goals, or the green-up requirements imposed by the SFI.

4. The MD DNR will make use of the best available data to determine what activity levels are consistent with the sustainability of the forest ecosystems so that harvests will not decrease the ability of the forests to continue to produce an average level of yield. Ecosystem sustainability means, in addition to the factors listed above under the first two points, no net loss in soil fertility and no loss of nontarget species due to forestry practices. Past and present data are limited, so future harvests will be based on adaptive response to monitoring, forecasting, and revision.

5. Forest recreational opportunities will be accommodated as appropriate and will be consistent with other goals for each site. Public use of the forest will be accommodated through a combination of revenue-generating hunting leases and public access recreation. The MD DNR will determine the appropriate level of public use for each tract as part of its ongoing evaluation and monitoring process.

Forest Types and Current Silvicultural Practices

The forest types and silvicultural practices pursued provide the technical backbone for developing the forest management plan within a multitude of constraints, which include everything from access limitations to meeting multiple goals for habitat, water quality, and recreation. The following information is a breakdown of forest types that occur on the forest and the recommended regeneration and harvesting methods based on information contained in annual work plans. These plans contain proposed management activities to be carried out each fiscal year. Such activities are reviewed by an interdisciplinary team of resource managers and specialists, presented to the citizens advisory committee, and posted online for public comment before being approved for use.

Nonforested land types include 1,145 ac (463 ha) of open marshes and swamps and 636 ac (257 ha) of power lines, with additional areas in agriculture fields and rights-of-way maintained by public drainage associations. Ecosystem restoration harvests that are maintained as brush or open-cover conditions comprise 237 ac (96 ha) of the Chesapeake Forest Lands. These are commonly designed to better meet habitat requirements for light-loving rare plants. The forest road system comprises 360 miles (mi) (579 kilometers (km)) of main access roads and side feeder roads, which amount to approximately 45 ac (18 ha) of open land. Road edges and roadside ditches are maintained with herbaceous vegetation. Mowing and control of invasive species occurs on a two- to three-year interval. Approximately 3.1% of the forest is of this land type.

Forested swamps with mixed hardwoods, baldcypress (*Taxodium distichum*), and Atlantic white cedar tend to retain surface water all year. Therefore, the management prescriptions are designed to protect their wetland functions. Where possible through restoration activities, some of these sites will be restored through the planting of native wetland forest species such as Atlantic white cedar. There are 2,528 (1,023 ha) of this forest type, which accounts for about 3.7% of the forest area.

Mixed pine-hardwood, hardwood-pine, and mixed hardwood forests are managed toward mature stands of mixed hardwoods and pine. Common practices include commercial thinning, selection harvesting, and small-opening harvests designed to encourage regeneration of desired native species, such as oaks, loblolly pine, and shortleaf pine (*Pinus echinata*). A minimum post-harvest basal area of 70 feet^2 (ft^2) per ac (16 meters^2 (m^2) per ha) usually is the stocking target. Herbicides are limited to ground applications to achieve specific goals in improving species balance or to control invasive species. Prescribed burning may be used to manage for a specific tree species. Natural regeneration is the preferred method used to reestablish forests after harvesting. The 13,715 ac (5,550 ha) in this forest type cover 20.2% of the forest area. This forest type will increase over time as the expanded 300 foot (ft) (91 meters (m)) HCVF water quality buffers are established along riparian areas in pine plantations.

There are both naturally regenerated and planted loblolly pine forests within the Chesapeake Forest Lands. Other tree species mixed in this forest type include a variety of gums, maples, oaks, Virginia pine (*Pinus virginiana*), and shortleaf pine. There are about 49,466 ac (20,019 ha) in this forest type, or about 73% of the forest. The loblolly pine plantations are intensively managed to maintain an annual flow of forest products. Silvicultural activities involve commercial thinning operations followed by a clear-cut regeneration harvest or a shelterwood regeneration harvest. A year after harvesting, reforestation needs are determined and either addressed through hand planting or natural regeneration. The naturally occurring loblolly pine and mixed pine stands are managed to maintain the naturally occurring species mix. Silvicultural activities involve commercial thinning operations followed by regeneration harvesting either by the seed tree, shelterwood, or clear-cut method. In order to maintain and/or reestablish certain pine species such as shortleaf pine, prescribed burning may be used to manage for this species. In most cases, natural regeneration will be the preferred method to reestablish the stand; some hand planting of shortleaf pine may be done to ensure the species mix in this forest type is maintained.

Precommercial thinning in a 6- to 10-yr-old naturally regenerated loblolly pine stand is a form of density control that is useful for concentrating growth on larger stems and maintaining an even distribution of trees across a site. The practice usually is accomplished by hand crews. As management activity shifts away from intensive site preparation and more toward natural regeneration, precommercial thinning will play a more important role. Commercial thinning of loblolly pine is performed several times during the life of a stand to capture value at an earlier date while concentrating growth on more desirable, larger diameter stems. Typically, a first thinning between the ages of 15 to 18 yrs removes every fifth row in a plantation and smaller trees in residual rows. A first thinning produces pulpwood-sized material. A second thinning, which typically occurs between the ages of 25 to 30 yrs, again removes smaller diameter trees but also produces merchantable sawtimber. Based on management prescriptions for a particular site, any subsequent thinning produces higher quality merchantable sawtimber.

The current age distribution of pine plantations (Figure 17.3) shows that most areas contain trees less than 30 years old. As a consequence, only modest areas of loblolly pine are available for final harvest over the next 15 years. First thinning of pine plantations is usually scheduled around age 14 to 18. Figure 17.3 indicates that there still are several hundred acres eligible for a first thinning treatment. This trend should continue into the near future as younger stands move into this age class. A majority of the pine stands will now be held for longer rotations. With second thinning generally occurring in the 25- to 35-yr age range, a fairly significantly increase in this practice will occur over the next decade.

Loblolly pine is intolerant of shade, and regeneration is best on sites with exposed mineral soils and full sunlight. Clear-cut harvesting provides the optimum conditions for subsequent stand establishment. Clear-cut harvesting on upland loblolly pine forests that are properly planned and follow best management practices can be expected to have little or no impact on water quality. The goal is to maintain a maximum regeneration harvest area of 40 ac (16 ha) per FSC standards. Cutting boundaries follow natural features to encourage irregular shapes that help diversify wildlife habitats and improve aesthetic appearance. Clear-cut harvests are not conducted until adjacent stands have reached the age of 3 years or an average tree height of 6 ft (about 2 m), in keeping with the SFI standard.

Either natural regeneration (seeding from remaining seed trees or adjacent stands) or artificial regeneration is used to reestablish loblolly pine stands in accordance with Maryland's pine tree reforestation law. However, one goal is to continue to reduce and eventually phase out the use of artificial regeneration. Both methods of regeneration will seek to reduce soil disturbance associated with site preparation practices, such as shearing, piling, bedding, ditching, and so forth. Careful harvest planning may be required to achieve natural regeneration wherever possible as well as to test new techniques and equipment that promise to achieve desired regeneration results with acceptable costs and reduced soil disturbance.

In addition to the forest types and silvicultural systems described above, there are a number of sensitive species and special habitat concerns. These can greatly impact how silvicultural systems are implemented and how the forest is managed over the long term. As previously described, the main objective for the Chesapeake Forest Lands is to maintain a

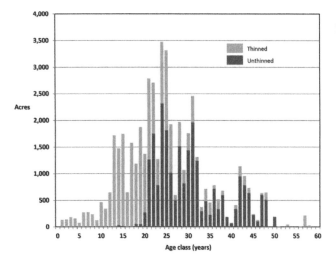

FIGURE 17.3 Age distribution of pine plantations on Chesapeake Forest Lands.

sustainable and economically self-sufficient forest while providing for or accommodating clean water, soil stabilization, populations of native plants and animals, critical habitats, and scenic, recreational, and educational values. As a result, a determination had to be made on the various levels of forest management activities that could occur across the entire forest. To accomplish this goal, the MD DNR Forest Service in partnership with the Wildlife and Heritage Service completed a land classification process in January 2006 detailing all the special habitat areas on the forest. This process identified several management areas:

- Ecologically significant areas (20.2%)
- Riparian forest buffers, which are 300 ft stream buffers (5.3%)
- Core FIDS habitat areas, including core DFS habitat areas (30.0%)
- DFS future core and translocation sites (16.6%)
- General management areas (27.9%)

Following the land classification process, specific management recommendations and silvicultural practices were developed that could be carried out in each of the management areas. For example, core FIDS and core DFS habitat areas are being managed for mature forest conditions, yet thinning can be used to improve the tree species mix or stand structure for these rare or declining species in some cases. Future core areas do not currently provide mature forest conditions, but thinning operations can be designed to create desired tree species and structural conditions in the future. The ability to carry out forestry operations compatible with habitat objectives is critical, since the emphasis of areas other than general forest management cover more than 70% of the forest.

To help guide general decision making, the original 2000 plan utilized a modern forest management model (HABPLAN) that allowed different management options to be compared in a wide variety of ways, and recognized costs and economic returns, production of wildlife habitat values, and the types of forest structures and diversity that will likely result. Since then, the MD DNR Forest Service has studied several available forest modeling systems and ultimately chose the Remsoft Spatial Woodstock model for the development of long-term timber projections.

OUTCOMES OF THE PLAN

This plan is designed to ease a transition between the former industrial forest management focus and the current state management focus. Some of the changes will become important over time, and they will likely include:

- Maintaining or enhancing water quality through restoration projects and patterns of green tree retention.
- Protecting a wide range of natural resources, including biological diversity and habitat for rare species.
- Contributing to the local resource-based economy through the generation of a variety of high- and low-value wood products.
- Providing opportunities for appropriate low-impact, resource-based public use, a regional draw for recreation.
- Widening of riparian forest and wetland buffers to protect and enhance water quality, as well as to provide mature forest habitat for species that need such conditions.
- Accommodating more mixed hardwoods and hardwood/pine forests associated with the buffers, in which timber harvesting maintains a mature forest stand after it is achieved.

In addition, the plan acknowledges that longer pine plantation rotations will be necessary than under the previous management, particularly in areas where wildlife habitat relies on large pine trees. These forests will eventually be harvested, however, when the trees are older and larger, which has implications for the future timber industry on the eastern shore. Further, the plan includes less intensive methods of forest regeneration and emphasizes the use of natural pine regeneration whenever and wherever it can succeed. This has been shown to result in somewhat slower tree growth for the first 2–4 years compared to the more intensive methods of soil preparation and the use of planted seedlings, but those early differences disappear later in the rotation. As a result, when forests are being managed for longer rotations, the less intensive regeneration methods should not result in a loss of productivity. They do, however, reduce up-front costs significantly as well as produce less soil and site disturbance.

Some of the changes required during the transition from the prior management focus to the new focus described in the plan will take years to emerge and may be almost imperceptible for a long time. Those include:

- The plan's shift to longer rotations in the loblolly pine and mixed pine/hardwood stand types (and the additional production of sawlogs) will emerge slowly as today's young stands reach larger sizes. The emphasis on thinning will produce significant amounts of pulpwood and forest-based jobs.

- The development of riparian forest buffers in areas now containing young pine plantations will take time. These areas must grow into buffers, so for the near future, there may be more pine pulpwood produced from buffer zones than from outside them as additional pines are removed to create openings for hardwood reestablishment.
- Measurable improvements in stream water quality may come slowly. Much of the water flowing across these forests comes from agricultural and developed areas. Efforts will be made to create areas that can trap nutrients, but groundwater lag times average more than 20 years on the Delmarva Peninsula (Sanford and Pope, 2013), so measured progress is likely to be slow to emerge.
- Major impacts on wildlife habitat will depend on the development of large trees, which will not occur until today's young forests have had time to grow. Improved DFS habitat in the future core areas will emerge more rapidly after about 20 years but not before.
- A different mix of public and private recreational opportunities will require time for the MD DNR to assess all the tracts, ensure public safety, and maintain good landowner relationships.

DISCUSSION AND CONCLUSIONS

The vision for Chesapeake Forest Lands is one that demonstrates a wide variety of management conditions and approaches that will likely result in sustainable forest management of the area. Public interaction and interest will likely continue to be intense. These include the occasional roadside and streamside viewers, other visitors, hunters, logging contractors, local businesses, industry personnel, and government leaders. Expectations will be diverse, often conflicting and changing. Forest industry skepticism exists about the ability of the MD DNR to maintain timber outputs from this forest. The gap in timber outputs caused by the current age-class distribution of the forest and the MD DNR's adherence to the management of endangered species habitat may be misread by the forest industry as not being able to meet economic goals on the forest.

Learnings and Insights

One of the most challenging aspects of sustainable forest management is the need to balance environmental, social, and economic goals to achieve the vision of a truly sustainable future. This plan addresses that need with guidelines based on the character of the land itself. For general planning purposes, the plan brought together the known conditions of forest vegetation and wildlife habitat with available information on the soil and water resources. This has allowed identification of key areas for water quality, wildlife habitat, and other values. At the same time, it also identified those areas where the production of economic timber harvests is most economically rewarding and environmentally sound. In addition to those general guidelines, the plan calls for intensive and ongoing fieldwork to identify and manage specific areas. Some important areas such as wetlands, Delmarva Bays, bald eagle (*Haliaeetus leucocephalus*) nest sites, and historic cemeteries are too small to be located on large-scale maps, but they still must be managed where they exist. The forest managers are tasked with precisely locating these special areas with GPS equipment, then crafting management approaches that maintain the special values that these areas possess.

With appropriate care of key environmental values, this land can produce timber for local industry and jobs for local workers, as well as opportunities for public recreation and enjoyment. This includes a hunting program that provides public hunting opportunities on about half, and leased opportunities on the other half, of the forest. The combination of timber and hunting lease revenues provides a critical balance of income for implementing this plan. The use of private contractors also expands contribution to the local economy and so far has been a successful management model. Five-year contracts are used to build in some efficiency and stability for management and allow periodic competitive bidding among qualified contractors, an important mechanism to constrain management costs. With some of the routine forestry and harvest oversight work contracted, the state staff can address the variety of other less predictable management tasks, from restoration projects, research requests, and boundary issues to managing hunting, public access, and citizen input.

Sustainability Issues

To accurately measure or assess the probable sustainability of a plan for the Chesapeake Forest Lands is a challenge for forest managers. First, one needs to recognize that there is never just one way to manage a forest toward a particular vision and set of goals. There are often many alternative management options for which rational arguments are made. The task in developing this plan has been to select management options that seem to offer the best balance of opportunities within the constraints and conditions that exist on the land. Understanding everything necessary to manage these lands sustainably is an incredible challenge, because lands with forests, waters, and wildlife are complex systems that are full of surprises. Perhaps the only thing one can be absolutely sure of with forested landscapes is that the unexpected is to be expected, whether it is a new pest, a major storm, new rules, or changing prices. The current forest plan is designed to increase diversity, foster the elements of the natural forests, build resilience, and hedge bets for future change. While this may not guarantee sustainability, it provides the best set of indicators we currently know.

Sustainable harvest levels are a goal of the Chesapeake Forest plan, but the current forest age distribution and goals for future stand condition mean that product mixes and harvest area differ from harvests levels under previous management. Since the median age of the pine stands was less than 15 years at the time of purchase, the first decade saw greater volumes of pulpwood from first thinnings. Products are now shifting to a mix of pulpwood and sawlogs as more stands reach second thinning or an early regeneration harvest to help balance the age distribution. The Chesapeake Forest will not contribute to local economies at levels consistent with prior management for another decade or so, given the young age of the pine plantations, the lack of merchantable timber in the mixed stands, and the need to maintain existing large trees in streamside management zones and special management areas.

Plan Development Challenges

In the development of the management plan, the planners needed to be consistent with the physical facts, biological potentials, economic constraints, and environmental conditions affecting the forests. The outcomes of the plan needed to contribute to a set of public expectations that are reasonable for the situation at hand. During the process, they needed to be open and transparent about likely results from various management options, what tradeoffs existed, and what has resulted from prior activities implemented. Of course, the plan also needed to be in compliance with current laws and regulations, such as the Maryland Critical Areas Law and the State Endangered Species Program. To meet these needs, the planning process involved developing and maintaining the best resource assessment possible under the limits of time and funds and assembling and updating a broad, interdisciplinary base of scientific knowledge and theory to support management decisions. Further, the planning team needed to create an integrated system of field data gathering, monitoring, information feedback, and data analysis that could be used to learn from research and field experience to support constant improvements in resource assessments, scientific understandings, and management techniques. The creation of an adaptive management process enables managers to more effectively respond to unforeseen disturbances. Finally, by involving third-party certification as part of the regular management regime, the environmental performance of field activities is evaluated annually and management adjustments are made as necessary.

Plan Implementation Challenges

One of the main challenges relates to the silvicultural goals and certification restrictions for harvest sizes. Loblolly pine, the dominant commercial species in the area, requires adequate light for regeneration and therefore needs sufficiently large openings. Small clear-cuts, while visually more acceptable than large ones, create habitat fragmentation, and are not recommended by many wildlife scientists. It is suggested that openings of 50–100 ac (20–40 ha) in size are more in keeping with the natural disturbance regimes needed by many species (Smith, 2007). However, clear-cut harvest sizes in excess of 40 ac (16 ha) are in conflict with the FSC certification standards, which the MD DNR has committed to follow. The FSC Southeast Region Indicator 6.3.g.1.a allows for exemptions to the 40-ac standard provided the harvest is for environmental or ecologic necessity. Nonetheless, the additional burden of proof adds to management complexity, increases time spent documenting rather than implementing, and lengthens the planning and approval time frame that already has multiple layers of review.

Another challenge relates to the interaction of silviculture, water quality, and habitat goals. Pine plantation management practices such as bedding, chemical hardwood suppression, and fertilization may be inconsistent with watershed and wildlife habitat enhancement goals, creating difficult tradeoff choices. Different management options are available in some situations, but many management methods exist today because little else worked in the past to regenerate pine forests. In the final decision, maintaining sustainable forest health may depend on doing what works best for the tree species and sites involved. Further, prescribed fire goals and burning restrictions may be more difficult due to air quality and liability issues. The timely creation of mature forests featuring large trees with some mature hardwood component and open understory for DFS habitat depends on aggressive use of practices like thinning and prescribed fire and the prudent use of herbicides. The use of prescribed fire at the desired scale may be an issue, in light of local objections to smoke, and the number of acceptable burning days that meet burn plan requirements.

Finally, landscape-scale goals and land ownership patterns can create challenges for plan implementation. The implementation of an ecosystem management approach that addresses landscape-level issues over a variety of unit sizes has inherent constraints. For example, restoration of habitat for species that need large areas of diverse conditions is feasible on some of the larger management units, but may not be feasible on many of the smaller units. However, many of the best water quality improvement projects are located on small- or medium-sized units because of their connectivity to other lands such as farms. Extensive mapping of land use, ownership, ecosystem characteristics, and accurate boundaries have proved essential to developing well-planned annual work plans.

In conclusion, after 13 years of active management by the MD DNR Forest Service, the Chesapeake Forest Lands have shown success as expressed by several measures. More than 1,000 acres of wetland and wildlife restoration projects have been completed or are in process. A local forestry consulting firm, several logging crews, and other local contractors are regularly employed to carry out various forest management activities. Local sawmills and chip mills are supplied on average with more than 60,000 tons of forest products annually. The State of Maryland receives almost a million dollars annually in revenue from timber sales and hunting leases, which is then used to fund restoration projects, road maintenance, recreational enhancements, and salaries of personnel. As the MD DNR moves forward with implementing the forest plan, it will continue to strive to maintain a sustainable and economically self-sufficient forest while at the same time protecting and restoring the many significant environmental qualities located on the Chesapeake Forest Lands.

ADDITIONAL READING AND RESOURCES

This chapter represents a synthesis of the 2007 Sustainable Forest Management Plan for the Chesapeake Forest Lands in Maryland. To view the plan itself, please visit this Internet site, which was available on January 28, 2014:

http://www.dnr.maryland.gov/forests/pdfs/CFL_SFMP_110513.pdf

If in the future the link to the site appears broken, search the Internet using the title of the plan and the keywords provided.

REFERENCES

Butler, B.J., Wear, D.N., 2013. Forest ownership dynamics of southern forests. In: Wear, D.N., Greis, J.G. (Eds.), The Southern Forest Futures Project: Technical Report. U.S. Department of Agriculture, Forest Service, Southern Research Station, Asheville, NC, pp. 103–121, General Technical Report SRS-GTR-178.

Maryland Department of Natural Resources, Forest Service, 2004. Sustainable forest management plan for Chesapeake Forest Lands. Maryland Department of Natural Resources, Salisbury, MD.

Maryland Department of Natural Resources, Forest Service, 2007. Sustainable Forest Management Plan for Chesapeake Forest Lands. Maryland Department of Natural Resources, Salisbury, MD. 216 p, http://www.dnr.maryland.gov/forests/pdfs/CFL_SFMP_110513.pdf (Accessed January 2, 2014).

Sanford, W.E., Pope, J.P., 2013. Quantifying groundwater's role in delaying improvements to Chesapeake Bay water quality. Environmental Science and Technology 47 (23), 13330–13338.

Smith, E.T., 2007. Early successional habitat. U.S. Department of Agriculture, Natural Resource Conservation Service, Washington, DC, and Wildlife Habitat Council, Silver Spring, MD. Fish and Wildlife Habitat Management Leaflet, Number 41. 16p.

The Sampson Group, Inc., 2000. Sustainable Forest Management Plan for the Chesapeake Forest Project. Maryland Department of Natural Resources, Salisbury, MD. 186 p.

Chapter 18

Durango State University Forest Las Bayas, Durango, Mexico

J. Ciro Hernández-Díaz,[1] Javier L. Bretado-Velázquez,[2] J. Javier Corral-Rivas,[1]
Héctor M. Loera-Gallegos[2] and Eusebio Montiel-Antuna[2]

[1]*Instituto de Silvicultura e Industria de la Madera, Universidad Juárez del Estado de Durango, Durango, Mexico,* [2]*Universidad Juárez del Estado de Durango, Durango, Mexico*

ABBREVIATIONS

FMP   Forest Management Program
UJED   Universidad Juárez del Estado de Durango

MANAGEMENT SETTING AND BACKGROUND

Las Bayas is a timber estate belonging to the Universidad Juárez del Estado de Durango (UJED) since 1988 when the land was donated to UJED by previous owners. Now UJED, represented by the rector, is the permit holder of the harvesting rights and is also the legal and technical entity responsible for proper conservation and management of this forest resource. The property is located in the municipality of Pueblo Nuevo, 100 miles (162 kilometers) south of the city of Durango (Figure 18.1). It lies between the parallels 23°22′15″ and 23°29′50″ north latitude and the meridians 104°48′45″ and 104°53′00″ west longitude. Following the criteria of the National Water Plan 2001–2006 (Comisión Nacional del Agua, 2001), this property is part of seven microbasins of the Hydrological Region No. 11, called Rio Presidio-Rio San Pedro. This property has been subject to timber harvesting activities for several decades, almost without interruption. It is known that in 1979 the Forest Management Unit No. 9, "La Flor," developed a forest plan to justify the authorization of timber harvesting until 1990. In 1991, the same unit, with a different name (UCODEFO No. 9), developed a new forest management plan, which was in force until 2004 (UCODEFO No. 9, 1991a,b). In 1995, UJED became technically responsible for the property for the first time through the Faculty of Forestry (Universidad Juárez del Estado de Durango, 1996), which in 2006 formulated this proposal, named Forest Management Program (FMP), which will remain in force until 2016 (Universidad Juárez del Estado de Durango, 2006).

The total area of the land in Las Bayas is 11,666 acres (ac) (4,721 hectares (ha)), yet only 5,992 ac (2,425 ha) are considered in the current FMP, of which just 3,024 ac (1,224 ha) are suitable for timber production. In this FMP, priority is given to research, teaching, and training functions in the areas of forestry and environmental sciences and also to implement research projects through intra- and inter-institutional agreements. These objectives strengthen the activities of academic groups within the Faculty of Forestry and the Institute of Forestry and Wood Industry, which are both academic UJED units. Research activities in Las Bayas include the establishment of a greenhouse, an inventory of biomass, the maintenance of permanent research sites, and the development of growth and yield models, among others. A proposal for the establishment of a management unit for the Conservation of Wildlife at the site was also developed. In addition, reforestation as well as sanitation felling and pruning have been implemented as complementary treatments to the implementation of the existing FMP.

The current FMP was developed with the Forest Planning System SPF-SUG-2001 (MARS SOFTWARE SA de CV, Técnica Informática Aplicada, 1991), which is licensed for use by the Faculty of Forestry. A design of stratified simple random sampling without replacement was used, based on the division of the forest in stands and substands generated for the property. Each stand is defined by permanent physical characteristics (similar slope and aspect), while the substands

Forest Plans of North America. http://dx.doi.org/10.1016/B978-0-12-799936-4.00018-7

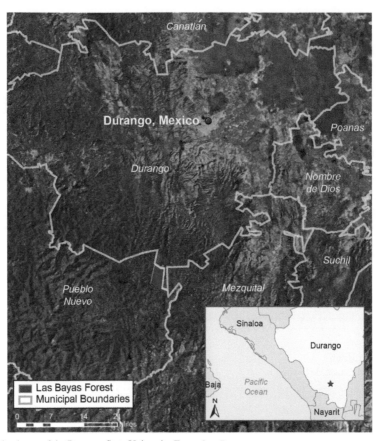

FIGURE 18.1   A map of the land area of the Durango State University Forest Las Bayas.

are stand divisions defined by similar vegetation characteristics and could be as small as an acre (about half a hectare) and as large as the whole stand.

For the forest inventory, circular sites of 1,000 m² randomly distributed in each stand were used. A sample intensity of 3.8% on average was applied, verifying that there were at least three plots for each minimum management unit (substand). If the allowable sampling error goal (less than 10%) was not achieved, the necessary additional sites were applied.

Even more important than the commercial aspect of timber harvesting activities is the implementation of measures to promote, protect, and conserve other associated resources. There is a short-term goal of certifying the forest management of this property and its chain of custody. This certification depends on the analysis and opinion of international organizations such as the Forest Stewardship Council (FSC).

PLANNING ENVIRONMENT AND METHODOLOGY

In forest management programs designed for long-term commercial use of the resource, it is important to ensure that the rate of resource use does not exceed the rate of recovery in the medium- and long-term. In the FMP of Las Bayas, the basic objectives being pursued are to ensure the conservation, protection, recovery, sustained use and development of forest resources, under management regimes that are silviculturally, economically, socially, and environmentally compatible and sustainable. Given this, alternatives for each management unit (substand) were proposed that are based on their productive capacity and silvicultural conditions in order to ensure current and future development without detriment to the ecosystem. Further, the protection of forest areas from pests, diseases, fire, and grazing is promoted, and necessary restoration is taken into account before the occurrence of unforeseen adverse weather phenomena. The conservation of areas of low productivity is also sought, and the recovery of areas with high rates of soil and forest degradation is encouraged to incorporate these areas into the timber production program.

In addition to protecting the flora, fauna and, scenic beauty, one of the main goals of this management program is to prevent soil erosion. Therefore, it is proposed to prevent the cutting of trees in water erosion-prone areas as well as in those substands containing water channels and springs. Another main set of constraints in the plan involves the selection of a management method and associated silvicultural treatments. The conditions of the forest to be handled, in combination with the management objectives for each area, are the main aspects to be considered in formulating the respective prescriptions. In this property, the initial proposal was to apply regular (even-aged) forest management, with the main silvicultural regeneration treatment being the use of seed trees, better known as "single parent trees." For this reason, the landscape classification was developed taking into account not only the definition of stands (based on permanent physical characteristics of the land), but also substands (based on the current characteristics of the vegetation). However, due to physical, ecological, and structural conditions, it was decided to apply an irregular forest (uneven-aged) management method, using single-tree selection silvicultural treatments. One advantage of applying single-tree selection treatments in substands is that the cutting intensities are assigned more accurately at the substand level than what would be possible at the stand level. Among the main constraints that influenced decisions to apply single-tree selection treatments are features such as:

1. Substands having a slope greater than 50% (maximum cut intensity is 25% of the volume).
2. Substands having a medium level of soil degradation, in which timber harvesting is feasible, but with conservative cut levels (maximum cut intensity is 25%).
3. Substands with overmature dominant trees (80 years or more), where cutting will be directed toward these trees, leaving standing younger individuals (maximum cut intensity is 40%).
4. Substands having pests or diseases that compromise the quality and survival of trees, and where sanitation cuts are necessary (maximum cut intensity is 60%). In heavily infested substands, the cutting rate can be raised and may not guarantee the recovery of the forest during the cutting cycle, in exchange for achieving a sanitation goal.
5. Substands having a low stock of standing timber (less than 40m^3rta per ha) or very low site-index values, where there is a risk to restocking the forest and soil deterioration (maximum cut intensity is 25%). In substands of low density and poor site quality, the cutting rate is sometimes lower than the average, because of the possibility of slow recovery.
6. Substands that are valuable for water purposes in the upper part of basins (their cut intensity is lower in order to protect and preserve the current hydrological regime).

Three main groups are involved in the management of the forest, including the technical directorate of the property, the holder of the forest harvesting rights, and other institutions. The directorate develops strategies for implementing the forest management program, trains staff, trains and informs the property owners, applies silvicultural treatments, and monitors actions to ensure compliance with laws and regulations. The holders of the harvesting rights select and hire professionals for the technical directorate, provide financial resources, and are responsible for adhering to the laws, regulations, and guidance suggested in the FMP. Other institutions supervise and ensure that the FMP is being implemented correctly in the field and employ some restoration measures in degraded areas. They also act to enforce laws, regulations, and other norms, coordinate with other agencies on policies and operational programs, and monitor and evaluate the plan as it is being implemented.

Guiding Laws, Regulations, and Policies

The FMP is coordinated and supervised by institutions responsible for ensuring that the laws and official rules are observed. The federal government institutions that directly attend the forestry sector are the Ministry of Environment and Natural Resources (SEMARNAT), Federal Attorney for Environmental Protection (PROFEPA), and the National Forestry Commission (CONAFOR). This latter federal unit has the main function of providing incentives to licensees for promoting good forestry practices. Within the state government, the Ministry of Natural Resources and Environment (SRNyMA) provides support to forestry programs and oversees and monitors the activities undertaken by federal agencies in the area of their competence. The main laws that must be observed in forestry are primarily the Constitution of the United Mexican States (Carbonell, 2008; Poder Ejecutivo de la Nación, 2008) and the Agrarian Law (Anes, 1995), which is a sectorial law derived from the Constitution. Furthermore, with respect to utilization and other correlated activities, the General Law of Ecological Equilibrium and Environmental Protection, which is also statutory of the Constitution (Presidencia de la República, 1988), General Law of Sustainable Forest Development (Presidencia de la República, 2003) and General Law of Wildlife (Presidencia de la República, 2000a) and their respective regulations are applicable. In particular, the Regulation of the General Law of Ecological Equilibrium and Environmental Protection in Respect of Environmental Impact (Presidencia de la República, 2000b) and the Regulation of the General Law of Sustainable Forest Development (Presidencia de la República, 2005) are important. In addition to laws and regulations, there are several official norms governing the activities in forest areas. The main Mexican Official Norms are:

- NOM-059-ECOL-1994: "determine the species and subspecies of terrestrial flora and wild fauna endangered and aquatic threatened, rare and specially protected species and establish specifications for their protection" (Instituto Nacional de Ecología, 1994a).
- NOM-062-ECOL-1994: establish "the specifications to mitigate adverse effects on biodiversity that are caused by the change of land use of forest land to agriculture" (Instituto Nacional de Ecología, 1994b).
- NOM-EM-015-SEMARNAP/SAGARPA-1997: "technical specifications for methods of use of fire in forest land and agricultural land use" (SEMARNAT, 2009).

Within the same regulatory framework, the Agreement of Technical Ecological Norm NTE-CRN-001/92, for the protection of forest soils (Secretaría de Desarrollo Urbano y Ecología, 1992) also applies.

The Las Bayas forest is in the process of developing the necessary information to seek and achieve certification of forest management through the evaluator company Rainforest Alliance, following the principles of the Forest Stewardship Council (FSC). Through this process a large number of theses, research studies, student practices, and agreements with other universities have helped support various elements of management and social impacts. However, the observation of the Rainforest Alliance is that forest documentation needs to describe the impacts of management in terms of jobs generated and increased incomes from forestry, highlighting how many people have benefited from management activities by getting involved in the planning and implementation of forestry operations.

Data Development and Analysis

To develop the management plan, an inventory was conducted and the following objectives were sought: (a) to provide information on forest structure and forest growth (at the substand level) and the general characteristics of each site, and (b) to provide reliable estimates of volume and growth of commercial forests, in order to plan the potential production of roundwood that helps supply the local sawmill industry.

Before going to the field, the number of sites required for each basic management unit (substand) and its location was defined. This was performed using the following steps:

- Determine the area of the stands and substands studied
- Determine the number of sample sites that should be located within the stand
- Randomly select the location of the sampling sites, as well as a surplus for the adjustment if required in the field
- Locate the sites in the plans (maps) and aerial photographs of the area
- Plan the route to get to the correct places of each of the sampling sites

The distribution of the sites within each stand was selected using a stratified random sampling scheme. The graphical distribution of sample sites was carried out by applying a software program (Técnica Informática Aplicada, 1991). For operational purposes, an allocation table of sites within each stand and substand was developed, which was distributed to the field-work brigades. This table also facilitated the capture and processing of information. The weighted average of sampling intensity used throughout the study area was 3.8%, and included a total of 630 sample sites of about 0.25 ac (1,000 m²) in 154 substands. An important consideration was that all substands should have at least three sample sites to validate their statistical analysis; therefore, in substands with less than 25 ac (10 ha) the sampling intensity was increased considerably. Furthermore, if the substand area was greater than 25 ac (10 ha) the sampling intensity may decrease and be less than the overall average. Ultimately, the weighted sampling error was 6.6%.

During the inventory, basic attributes of trees were collected. With this data, an inference process was conducted to estimate several attributes at the site level, such as number of trees by species group, diameter and height classes, and the forest type structure for each substand. The current annual increment in volume was estimated using the volumetric average growth rate for individual trees. Results were obtained for each substand by species group (pine, oak, other conifers, hardwoods, and dead trees) and diameter category within each group (<10 in (25 cm), 10–14 in (25–35 cm), etc.) and were presented as weighted average values for each substand. Because the entire property is the integral entity with defined objectives along with a specific production and ecological framework, property-level estimates of forest resources were determined using the accumulated substand average value estimates.

The field inventory data were based on the stands and substands delineated using photogrammetric methods. This allowed fast and accurate location of each site in the field. Two-man teams (brigades) were then formed. Each brigade was given the appropriate equipment as well as forms and instructions for registering information at the location of each site. To facilitate data analysis, field data was captured in a database that later was validated and debugged, correcting errors in the data. An information processing program known as the Forest Planning System (SPF-SUG-2001), developed by MARS SOFTWARE S. A. DE C. V (Técnica Informática Aplicada, 1991), was used to manage the data collected.

In order to determine the forest area that could be allocated to timber production, land was classified into 26 categories, which were defined in terms of the variability of forest physical characteristics, vegetation, and current use of the soil detected in each substand. With this procedure, the classification process divided land into categories such as:

- *Areas of low productivity.* These are areas with low site-index values, less than 28 feet (ft) (8.5 meters (m)) using a base age of 60 years. These areas present climatic or soil limitations on the growth of pine forests, favoring conditions of low density, low wood stocking, and reduced commercial value.
- *Areas with low density or low stocking levels.* These are areas that, in spite of having medium site-index values, are segregated from the forest harvesting program due to their low density, as reflected in crown coverage values lower than 10% and a crown competition factor below 75%. These areas represent a high risk of disturbance to the soil. In this category are land areas considered to also have pine timber stock under 60 m³rta per ha.
- *Areas of forest restoration.* These areas are currently forested or suitable for forestry, with good site-index values, but which have deteriorated. They include those areas having inadequate pine regeneration or that need to be rehabilitated through reforestation or restoration measures in order to regain and maintain all or part of their biodiversity, soil, and hydrological dynamics.

The mensuration information of each tree allows formulating a proposed prescription for each site. With this information, biometric databases are analyzed with the support of a silvicultural growth simulator that allowed the planning team to project conditions into the future. This distance-independent model is critical to the analysis and selection of management regimes (Técnica Informática Aplicada, 1991) and has four main modules: (a) a regimes generator, (b) an increase predictor, (c) a removal prescriber, and (d) a removal valuator. The regimes generator uses logging information and current dasometric and silvicultural information to simulate management methods and develop a set of proposed basic management schemes. The increase predictor module estimates dominant heights, diameter and height increments, and mortality. The removal prescriber module simulates harvest volume as a function of the diameter distribution according to the treatment assigned and to the silvicultural structure. The removal valuator module has a submodule for product distribution based on commercial characteristics, and another submodule for economic valuation.

To determine the cutting age for each even-aged substand, the silvicultural simulator determines when the mean annual increment is maximized. In substands managed with the uneven-aged selection system under this FMP, the entry cycle was determined by evaluating the average age at which trees reach a minimum commercial diameter (about 8 in, or 20 cm). Areas to be harvested were assigned a cutting intensity consistent with productivity conditions and current density and age of each substand, so as to ensure recovery of the residual forest. Timber harvesting recommendations for the first period (years 1 to 10) of the plan, for each stand and substand, including prescriptions and cutting intensities, are then defined.

Silvicultural Strategies

In areas of the Las Bayas property where commercial timber is harvested, foreseeing the need to sufficiently restock harvested sites is very important. Restocking is basically conducted through natural regeneration processes, but in some substands the silvicultural prescription includes the need for artificial reforestation. Reforestation is accomplished with native tree species, since they have better development potential given the natural conditions of the area. For artificial reforestation, the field location of each seedling is prepared manually (holes 12–16 in (30–40 cm) deep). The distribution of the seedlings is not necessarily regular, but depends on the local terrain features, ensuring that the distance between seedlings is approximately 5 to 10 ft (1.5 to 3.0 m). Reforested areas are also fenced to prevent livestock crossing and grazing. Following reforestation, the number of freezing days and the number of frosts is registered because seedlings may die as a result of this natural phenomenon. The development of seedlings is also monitored for vigor, height, diameter, and other factors. In case of excessive mortality, seedling replacement is performed in the following year to ensure greater survival of reforestation. Natural regeneration is less expensive and is often advantageous to artificial reforestation, because seedlings that are born naturally on the site are better adapted to site conditions such as climate, altitude, latitude, and soil type. The reestablishment of forests through natural regeneration is the most common strategy for the replacement of forests. Regeneration is stimulated by performing soil treatments, such as soil plowing (where possible), so that greater amounts of water and nutrients are captured.

Throughout the forest, firebreaks are also built. Rakes are used to clear litter and branches in strips of 6 to 12 ft (2 to 3 m) wide on the ground. In support of natural regeneration, and to reduce the risk of forest fires, other complementary treatments are applied. These include waste control, where branches and tips of trees left as waste of harvesting are chopped and rearranged perpendicular to the slope. This helps to retain runoff and promotes higher rainwater infiltration. Also, controlled burning is used to complement other activities on about one-tenth of the annual cutting area.

OUTCOMES OF THE PLAN

A land classification, developed through photogrammetric processes and supported by field data, is presented in Table 18.1. As a result of this classification, only 3,024 ac (1,224 ha) are considered available for commercial logging. The timber production areas of the property are generally covered with pine-oak forests, on which most of the commercial pine trees have ages of 51–55 years (28.1%), 46–50 years (27.8%), and 56–60 years (21.6%). Only 6.9% of the forest is aged less than 45 years and 13.6% is older than 60 years (Figure 18.2). Site index (base age 60 years) in the Las Bayas forest ranged from 33–49 ft (10–15 m) for 7.7% of the property, 49–66 ft (15–20 m) for 60.2% of the property, and 66–82 ft (20–25 m) for 24.9% of the property (Bretado and Díaz, 1994). A crown cover of 25–50% is found in 48.1% of the harvestable area, and 38.8% of the property has less than 25% crown cover.

According to the inventory, stocks of pine standing timber (which is the most valuable commercial timber) are below 50 m³rta per ha in 41.5% of the suitable areas for harvesting and are between 50 and 100 m³rta per ha in another 41.9% of these areas. The total volume of allowable cut for the proposed cycle (10 years) is 7,942.7 m³rta of pine, 8,552.6 m³rta of oak, 812.0 m³rta of other conifers, 411.9 m³rta of other hardwoods, and 525.1 m³rta of dead pine. This means that a total of 18,244.3 m³rta will be cut in 10 years (1.5 m³rta per ha per year) out of which only 43.5% is live pine. Over 77.5% of the trees are between 46 and 60 years of age (Figure 18.2), which denotes a marked deficiency of young trees, risking the possibility of achieving long-term sustainable harvest.

In the Las Bayas forest (Figure 18.3), no pests and diseases were detected. However, much of the necessary information to carry out control measures has been collected should harmful agents present themselves and threaten the forest resource.

TABLE 18.1 Classification of the Las Bayas Forest Based on its Use or Condition

Land Feature	Area (ac)	Area (ha)	Percent of Total Area
Nonforested areas			
Agriculture	18.85	7.63	0.16
Grassland	68.42	27.69	0.59
Rock outcrops or glades	256.37	103.75	2.20
Human settlements	14.16	5.73	0.12
Sub-total	357.80	144.80	3.07
Unusable forested areas			
Areas in recovery (low volume)	3,050.87	1,234.67	26.15
Areas with high soil deterioration	884.22	357.84	7.58
Recessed areas[a]	1,380.62	558.73	11.83
Sub-total	5,315.71	2,151.24	45.56
Harvestable forest areas (included in the FMP)			
Special use areas for ecological monitoring	1,234.76	499.70	10.58
Areas for sanitation and protective treatments	557.06	225.44	4.77
Areas in reserve of utilization[b]	1,177.56	476.55	10.09
Commercial timber production areas	3,023.89	1,223.75	25.92
Sub-total	5,993.26	2,425.44	51.37
Total area	11,666.78	4,721.48	100.00

FMP: Forest management plan.

[a] Areas designated for hydrological services, biodiversity, etc.
[b] Areas affected by fires, storms, etc.

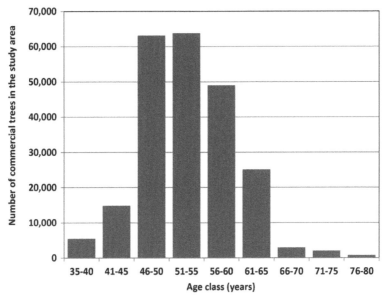

FIGURE 18.2   Number of pine trees per age class in the study area (3,024 ac).

FIGURE 18.3   Panoramic view of the Durango State University Forest Las Bayas.

DISCUSSION AND CONCLUSIONS

Learnings and Insights

The current condition of the forest is the result of having applied intensive logging during the two previous cutting cycles before the formulation of this FMP. The current forest is composed of young- and medium-age pine trees with good health conditions and over-mature, large oak trees that require harvesting or removal in order to open spaces that facilitate re-stocking with pine species. Further, inventory data show that the age structure of the pine trees is far from approaching a normal forest. On the one hand, there is a great shortage of young age classes, and also there may be many other categories of middle-age classes with a surplus or a deficit. These conditions must be improved through proper management in the medium- and long-term periods.

Sustainability Issues

To restock the forest, it is necessary to ensure the establishment of seedlings in the field, either from natural regeneration or from reforestation. For this purpose, it is important to make complementary treatments to both the soil and the vegetation. From an economic point of view, in the Las Bayas area, pine has a better market value, therefore, it is necessary to focus efforts to repopulate the area with valuable species and to reduce the currently high percentage of low-value hardwoods in the forest composition. From an environmental point of view, there is the need to continue strengthening management programs that encourage investment in the forest in order to protect it, promote it, and grow it. It is desirable to develop and implement better forest practices programs and improve forest management in general, and priority should be given to genetic improvement. In the current FMP, when harvesting occurs, managers leave in place a sufficient number of dead trees and hardwoods in order to maintain appropriate conditions for nesting requirements of wildlife. From a socioeconomic point of view, the FMP promotes the development, commercialization, and industrialization of small-diameter pines and hardwood tree species even if they are less profitable than pines. It is also necessary to harmonize the different productive activities (e.g., agriculture, livestock production, and forestry) according to topographic conditions, land uses, and soil fertility.

Plan Development Challenges

Although it has frequently been mentioned that sustainability is a fundamental objective of managing the forest environment of Mexico, economics is often regarded as separate from the technical aspects of forestry activities, especially when these activities are wholly or partially funded by government programs. To determine whether forestry is profitable and competitive, it is necessary to conduct economic and financial analyses of each stage of a management regime or activity, not only in relation to the production process (felling, yarding, transportation, etc.) but also in relation to the supporting activities of production. These include issues related to the preparation and conduct or performance of the management program, the genetic improvement program, the reforestation program, programs aimed at protection against pests, diseases, and fires, the road management program, and other complementary activities.

It is also necessary to implement training programs for the holders of the property, especially for the operating personnel who perform different stages of forest harvesting, in order to improve their level of awareness and preparedness and to mitigate the negative impacts of human activities on the forests and its associated resources.

These are general challenges for forest management in Mexico, but since Las Bayas is a forest managed by a university it is viewed as a model for others to follow. Management should create good precedents and positive experiences that may be used by other landowners and technicians responsible for other properties.

Plan Implementation Challenges

To optimally implement this management program, it is necessary to develop and implement public awareness programs regarding the importance of forest conservation and development. Further, it is necessary to update and train technical and operational staff at all levels for the implementation and monitoring of the FMP. The managers need to perform more detailed studies about costs and benefits generated from all forestry activities, considering the actual value of the local timber market. In addition, they need to install temporary and permanent monitoring plots to assess the ecological and environmental impacts on natural resources caused by forestry activities, mainly in areas allocated for regeneration felling and other intensive treatments. To take advantage of both the positive and negative experiences, and thus to improve forest management, it is necessary to consider the impacts not only for the current cutting cycle of the current FMP program, but also for two or more consecutive cutting cycles. This monitoring program needs to be designed to track and assess not only what happens at the whole forest level, but also at the stand level.

Other Interesting Issues Related to the Plan

Within the study area, there is an extensive network of roads covering almost all of the cutting areas. This results in lower costs for implementing the management program. However, it is necessary to analyze in detail the road network in relation to methods and equipment available for yarding and to choose the most efficient combination. This ensures that the products are obtained with the greatest economic competitiveness, while also maintaining the least possible number of roads since roads are often where the largest environmental impacts originate. Finally, the development of a database that allows the maintenance and storage of experiences, actions, and results derived from the implementation of the FMP over time is important. With support of information from research sites, and by permanently keeping the same delimitation of the forest stands in consecutive cutting cycles, silvicultural, ecological, and economic rules can be created to manage forests according to their growth historical dynamics.

REFERENCES

Anes, G., 1995. La ley agraria. Alianza.

Bretado, V., Díaz, V., 1994. Evaluación preliminar de la calidad de sitio y la densidad en el bosque "Las Bayas" de la UJED. Tercera Reunión Regional de Investigación en la UJED, Durango, Dgo.

Carbonell, M., 2008. Constitución Política de los Estados Unidos Mexicanos. Editorial Porrúa.

Comisión Nacional del Agua, 2001. Programa Nacional Hidráulico 2001-2006. Comisión Nacional del Agua, México, 128 p.

Instituto Nacional de Ecología, 1994a. Norma Oficial Mexicana NOM-059-Ecol-1994, que determinan las especies y subespecies de flora y fauna silvestres terrestres y acuáticas en peligro de extinción amenazadas, raras y las sujetas a protección especial, y que establece especificaciones para su protección. Diario Oficial de la Federación, 6 de marzo, México DF.

Instituto Nacional de Ecología, 1994b. NORMA Oficial Mexicana NOM-062-ECOL-1994, que establece las especificaciones para mitigar los efectos adversos sobre la biodiversidad que se ocasionen por el cambio de uso del suelo de terrenos forestales a agropecuarios. Diario Oficial de la Federación, 13 de mayo, México DF.

Poder Ejecutivo de la Nación, 2008. Constitución Política de los Estados Unidos Mexicanos. Constitución publicada en el Diario Oficial de la Federación el 5 de febrero de 1917. Última reforma publicada DOF 7 de febrero, 2014, México DF.

Presidencia de la República, 1988. Ley General del Equilibrio Ecológico y la Protección al Ambiente. México. Diario Oficial de la Federación, México DF.

Presidencia de la República, 2000a. Ley general de vida silvestre, México. Diario Oficial de la Federación, México DF.

Presidencia de la República, 2000b. Reglamento de la Ley General del Equilibrio Ecológico y la Protección al Ambiente en Materia de Evaluación del Impacto Ambiental. Diario Oficial de la Federación, 30 de Mayo. México DF.

Presidencia de la República, 2003. Ley General de Desarrollo Forestal Sustentable. México. Diario Oficial de la Federación. Febrero, México DF.

Presidencia de la República, 2005. Reglamento de la Ley General de Desarrollo Forestal Sustentable. Nuevo Reglamento. Diario Oficial de la Federación, México DF.

Secretaría de Desarrollo Urbano y Ecología, 1992. Acuerdo de la norma técnica ecológica NTE-CRN-001/92 para la protección de los suelos forestales. Diario Oficial de la Federación Jueves 9 de abril, México DF.

SEMARNAT, 2009. Norma Oficial Mexicana NOM-EM-015-SEMARNAP/SAGARPA-1997, que establece las especificaciones técnicas de métodos de uso del fuego en los terrenos forestales y en los terrenos de uso agropecuario. Diario Oficial de la Federación. Viernes 16 de enero de, México DF.

Técnica Informática Aplicada, 1991. Metodología para la elaboración del plan de manejo integral de recursos forestales. Mimeografiado, 8 p.

UCODEFO No. 9, 1991a. Planteamientos del Manejo Integral Forestal en la Unidad de Conservación y Desarrollo Forestal No. 9 "La Flor". Mecanografiado inédito, Durango, Dgo.

UCODEFO No. 9, 1991b. PMF para las Áreas de Corta 1991–2004 para el Predio Particular "Las Bayas"—UJED. Mecanografiado inédito, Durango, Dgo.

Universidad Juárez del Estado de Durango, 1996. Propuesta Técnica para la Modificación de la 6ª. Área de Corta del P.P. "Las Bayas"—UJED. Escuela de Ciencias Forestales—Institute of Forestry and Wood Industry (ISIMA), Mecanografiado inédito. Durango, Dgo, 40 p.

Universidad Juárez del Estado de Durango, 2006. Programa de Manejo Forestal para el P. P Las Bayas. de la UJED Municipio de Pueblo Nuevo, Dgo, 103 p.

Chapter 19

Garcia River Forest, California, United States of America

Scott Kelly[1] and Kevin Boston[2]

[1]*North Coast Forest Conservation Program, The Conservation Fund, Caspar, California, USA,* [2]*Department of Forest Engineering, Resources and Management, Oregon State University, Corvallis, Oregon, USA*

ABBREVIATIONS

ERN   Ecological Reserve Network
IRMP   Integrated Resource Management Plan
SOD   Sudden Oak Death
THP   Timber Harvest Plan
TMDL   Total Maximum Daily Load
TNC   The Nature Conservancy

MANAGEMENT SETTING AND BACKGROUND

The Conservation Fund (the Fund) acquired the 23,780 acre (ac) (9,624 hectare (ha)) Garcia River Forest in 2004. The forest (Figure 19.1) is located in the redwood (*Sequoia sempervirens*) region of the coastal mountain range of southwestern Mendocino County, California. The forest encompasses approximately one-third of the entire 72,000 ac (29,138 ha) Garcia River watershed, which it shares with a major industrial timberland owner, Mendocino Redwood Company. Like many forested areas in California, approximately 10% of the watershed is held in smaller ranches and subdivisions, and approximately 15% of the watershed area is in agricultural uses.

The Garcia River Forest (Figure 19.2) is composed of redwood and Douglas-fir (*Pseudotsuga menziesii*) with minor amounts of grand fir (*Abies grandis*), sugar pine (*Pinus lambertiana*), and western hemlock (*Tsuga heterophylla*). The major hardwood species present are tanoak (*Notholithocarpus densiflorus*) and Pacific madrone (*Arbutus menziesii*). Tanoak and Pacific madrone are naturally occurring hardwood species in the redwood region and are generally considered to be early- to mid-successional tree species. The forest has been managed for commercial timber production since the early 1950s and has had multiple landowners who left a legacy of timber depletion and resulting habitat loss related to coniferous forests. Harvesting practices in the riparian zones resulted in increased stream sedimentation and loss of habitat for coho salmon (*Oncorhynchus kisutch*) and steelhead trout (*Oncorhynchus mykiss*). Much of the logging damage occurred with the introduction of the crawler tractor, widely used for timber harvesting after World War II. A 1966 report on stream damage surveys (Fisk et al., 1966) describes the impact from uncontrolled logging on steep hillsides laced with skid trails and logging roads. There had been little effort to keep heavy equipment out of the stream channels, often resulting in moving streams out of their natural channel. Due to past harvesting practices, the hardwood component, especially tanoak, is unnaturally high. Tanoak sometimes comprised up to 60% of the basal area in some stands at the time of acquisition.

When the Fund became a forest landowner in the Garcia River Basin, it adopted the inherent responsibilities of landownership, including the development of an Integrated Resource Management Plan (IRMP), an erosion control plan, and independent forest stewardship certification. This transaction was different from most others completed by the Fund as its traditional model is to conserve important lands by purchasing a piece of property with the intent to convey it to a local partner, such as a land trust, federal, state, or local agency, for the purpose of long-term management. Often, the land to be

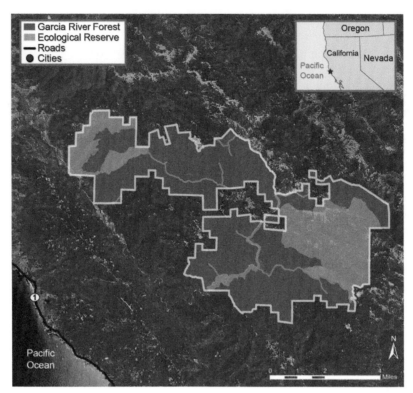

FIGURE 19.1 Location map of the Garcia River Forest.

FIGURE 19.2 Redwood trees within the Garcia River Forest.

protected is adjacent to a federal forest or state park, and the land transfer makes good management sense. In these cases, the Fund may act as an intermediary to secure funding that may not otherwise be available to these governmental organizations. In other cases, the Fund can make a quick purchase, thus securing the property while the partner raises the necessary funds or locates government funding. The Garcia River Forest did not fit the Fund's usual business model, as the forest is not adjacent to any federal or state lands. The state of California and the U.S. Forest Service were not interested in the property due to its remoteness, and long-term management funds were not available. It was obvious the Fund needed a new approach; one that took into account the unrelenting annual costs of maintenance and management and the need for restoration of habitat for coho salmon and steelhead trout. In addition, the ability to maintain the forest as a working commercial timber property and contribute to the local economy was of primary importance to the Fund.

The question to be answered was: Could a large, understocked tract of coastal California forest be returned to sustainable timber production and ecological vitality through patient management by a nonprofit organization, in partnership with private and public agencies and community stakeholders? Because of past harvesting practices, the forest was incapable of producing much income at the time of acquisition, but would it be sufficient to sustain a not-for-profit landowner? Furthermore, the Fund could raise additional funds from a variety of philanthropic sources to assist in the management of the forest and perform the actions necessary to rebuild the commercial inventories, upgrade road systems, and restore the streams to their natural condition.

To test this idea, the Fund partnered with the State Coastal Conservancy, the Wildlife Conservation Board, and The Nature Conservancy (TNC) to acquire the 23,780 ac (9,624 ha) Garcia River Forest in 2004 for a purchase price of $18 million U.S. dollars (USD). At the time of this purchase, it established the largest nonprofit-owned working forest in the western United States. The goal was to demonstrate that non-profit ownership and management of a large, formerly industrial timberland could be transformed into a viable working forest with a conservation strategy while supporting the local economy. This would help offset the current trend of forested tracts being converted into vineyards, subdivisions, or other non-forest uses.

To protect the public interests, a conservation easement on the forest was required by the funders due to the unprecedented nature of the project and the controversial nature of forestry in redwood the region, caused by the harvest of old-growth forests (Chase, 1995). The funders appropriately asked that an Integrated Resource Management Plan (IRMP) be prepared that would describe the goals and the specific actions to be deployed to achieve them. Working with the local community, public agencies and nonprofit partners, the IRMP for the forest was developed. The IRMP establishes the following broad goals:

- Restore and protect a productive and relatively natural coastal California forest ecosystem.
- Protect and restore fish and wildlife habitat associated with this ecosystem, in particular oak woodlands, native grasslands, which host many endemic plant communities, redwood and Douglas-fir forests, and spawning habitat for coho salmon and steelhead trout.
- Protect significant water resources, springs, and water quality.
- Maintain the capacity of the forest for productive forest management, including the long-term sustainable harvest of high-quality forest products, thus contributing to the economic vitality of the state and region.
- Maintain the use of the forest for outdoor recreation by providing limited public access.

The conservation targets for the Garcia River Forest relate to redwood and Douglas-fir forests, anadromous fish-bearing streams, oak woodlands and grasslands, non-riverine freshwater wetlands, and northern spotted owl (*Strix occidentalis caurina*) habitat. Protection and enhancement of these targets is expected to further provide for the conservation of 80 additional "nested" conservation targets for the forest that would not be practical to assess individually.

In addition to the IRMP, the funding sources wanted assurance of the quality of management of the forests. They required that the forest be certified as a "well-managed forest" through an independent third-party forest management certification program. The Fund subsequently pursued and received certification through both the Forest Stewardship Council and the Sustainable Forestry Initiative programs.

The Garcia River Forest is primarily composed of second- and third-growth redwood and Douglas-fir, ranging in age between 30 and 55 years old. The forest can be characterized as an even-aged stand resulting from past silvicultural practices; primarily clear-cutting in the 1950s followed by shelterwood removal harvests in the 1980s. The western portion of the forest is subject to coastal influence and experiences regular summertime fog during the morning hours, generally cool days and cool nights. The climate is conducive to the development of forests dominated by redwood. The middle-eastern portion of the forest is much hotter and drier, and is dominated by Douglas-fir. On the very east end, the conifer forest gives way to mixed conifer and hardwoods with some natural grassland intermingled. Precipitation within the forest generally falls as rain although occasional snow may fall at the higher elevations. Rainfall amounts average approximately 60 inches (in) (1,524 millimeters (mm)) of rain per year, falling primarily between the months of November and April.

The north coast of California is geologically young. The landscape has been shaped by the collision of the Gorda and North American tectonic plates, resulting in steep terrain and deep drainages. The lower Garcia River follows the San Andreas Fault for nearly 10 miles before entering the ocean. The upper watershed areas are deeply incised by tributaries. The result of the geologic forces is a steep and weak parent material easily susceptible to surface erosion from large rainfall events, large debris slides, and rotational failures commonly occur that also shape the landscape. High rainfall and the steep gradient of the streams give them a high capacity to generate and transport sediment from old logging roads and skid trails. The watershed east of the San Andreas Fault, including all of the Garcia River Forest, is entirely composed of the Franciscan Complex. The parent rock in these formations is often weakly consolidated or sheared, leading to a high erosion risk. The exception on the forest is the Inman Creek sub watershed, which is comprised of a more erosive metamorphic geology that has resulted in a higher clay component in the soils. These formations do not readily support conifer forests and are dominated by grasslands and oak woodlands.

The forest is home to various native wildlife species, the most notable of which are coho salmon, steelhead trout, and the northern spotted owl, which are listed under the California Endangered Species Act and the Federal Endangered Species Act as threatened. Other important species include the California black bear (*Ursus americanus*) and mountain lion (*Puma concolor*). There are several nonnative species of concern that have the potential to harm the restoration efforts, including feral pigs (*Sus scrofa domestica*), along with a host of non-native plants. The most notable invasive weeds are French broom (*Cytisus monspessulanus*) and Jubata grass (*Cortaderia jubata*).

PLANNING ENVIRONMENT AND METHODOLOGY

The IRMP had to be completed prior to any timber harvesting; therefore, the development of the plan was the first priority for management of the forest. The plan was jointly developed by the Fund in collaboration with TNC staff and many other local partners. Various consulting foresters and local environmental groups provided key inputs into the planning process. Numerous public meetings were held to seek out public opinion and approval of forest management policies. The goal was to establish guidelines that everyone agreed to. The controversies in the redwood region at the time required this reestablishment of the social license. The plan emphasized the use of uneven-aged silvicultural systems and harvesting less than growth from the forest each year. It was felt that maintaining adequate forest cover in all age classes would generally provide a level of protection of the ecosystem values. Stands composed of larger trees will provide good habitat for large mammals such as bear, mountain lion, and the endangered northern spotted owl. Younger age classes, including some very early seral stages dominated by hardwood species, would provide vital foraging habitat for dusky-footed woodrats (*Neotoma fuscipesa*), favored prey for the northern spotted owls. Additionally, these hardwood stands are the summer home for migratory song birds and provide forage for bear and deer (Johnson and Franklin, 2013).

As part of the purchase agreement, TNC contributed $4 million USD to purchase a conservation easement on the forest. The agreement with TNC was that one-third of the forest was to be placed in an Ecological Reserve Network (ERN) in which the development and maintenance of late-seral stage forest was the primary management goal, with no active management activities to be performed in these areas unless they can help the forest recover more quickly. The IRMP allows for limited timber harvests, primarily thinning from below within the ERN, to further the goal of late-seral stage development. The remaining two-thirds of the forest are managed as a working forest in which the maintenance of important ecological attributes and habitat is required, but the primary emphasis is not on the exclusive development of late-seral stage forest.

Guiding Laws, Regulations, and Policies

In addition to the self-imposed constraints contained in the conservation easement and outlined in the IRMP, the Fund is obligated by law to conform to the various environmental regulations imposed by the federal, state, and local governments. California has a reputation for an extensive set of state regulations and the reputation is well deserved. Harvesting in California is primarily controlled by two separate acts, the Z'Berg-Nejedly Forest Practices Act of 1973 (California Public Resources Code-Division 4, Chapter 8) and the California Environmental Quality Act (Pub.Res.C. 21,000 et seq.). Thus, the timber harvest plan (THP) satisfies the California Environmental Quality Act requirement for an Environmental Impact Report. The California Forest Practices Act has promulgated many additional rules that govern timber harvest plan development. These include minimum rotation ages for even-aged management of between 50 years for high-site lands and 80 years for low-site lands. One feature of the act is that it places limits on much of the practice of forestry, including the preparation of THP by a Registered Professional Forester.

In addition, the Forest Practice Rules require timber harvest plans be submitted to the state for review by a multi-disciplinary review team consisting of a member of the California Department of Forestry and Fire Protection (Forest Practice Division), California Department of Fish and Wildlife, Division of Mines and Geology, and the Regional Water Quality Control Board; some counties also have THP review authority. Thus, forest plan and operations must be in compliance with both the state and federal wildlife acts such as the California Endangered Species Act (Fish & G.C. 2050 et seq.) and the Federal Endangered Species Act of 1973 (16 U.S.C. 1531–1544, 87 Stat. 884). California has some of the most restrictive water quality acts with both the Federal Clean Water Act, Clean Water Act of 1972 (33 U.S.C. §1251 et seq.) and California's Porter-Cologne Water Quality Control Act (California Water Code, Division 7).

One unique aspect of the Garcia River watershed is that it was one of the first areas to challenge the authority of the United States Environmental Protection Agency (EPA) to regulate water quality using total maximum daily loads (TMDL), in cases that were brought forth through the Pronsolino Cases (Pronsolino v. Nastri, 291 F.3d 1123 (9th Cir. 2002); Pronsolino v. Marcus 91 (F. Supp. 2d 1337 (2000))). The rulings upheld the EPAs ability to develop TMDLs for non-point source pollutants. Section 303(d) of the federal Clean Water Act and 40 CFR §130.7 require states to identify water bodies that do not meet water quality standards and are not supporting their beneficial uses. These waters are placed on the Section 303(d) List of Water Quality Limited Segments (List), also known as the 303(d) List of Impaired Water Bodies. The List identifies the pollutant or stressor causing impairment and establishes a schedule for developing a control plan to address the impairment. Placement on this List generally triggers development of a pollution control plan (guided by TMDLs) for each water body and associated pollutant or stressor on the List. The Garcia River was listed as sediment and temperature impaired for salmonids, specifically coho salmon, chinook salmon (*Oncorhynchus tshawytscha*), and steelhead trout. The EPA partnered with the North Coast Regional Water Quality Control Board to develop the Garcia River Action Plan for sediment reduction. The action plan shall be implemented by each landowner within the Garcia River Drainage to meet the requirements of the TMDL.

The *Action Plan* for the Garcia River TMDL (California Department of Environmental Protection Agency, 2011) is an amendment to the Basin Plan developed by the EPA and the North Coast Regional Water Quality Control Board, which includes the TMDL, an implementation plan, and a monitoring plan for the Garcia River. The Action Plan was approved by the State Office of Administrative Law on January 3, 2002, and by the EPA on March 7, 2002. The Action Plan has been in effect since State Office of Administrative Law approval on January 3, 2002.

The Garcia River Action Plan contains waste discharge prohibitions that apply within the Garcia River watershed. The prohibitions are as follows:

1. "The controllable discharge of soil, silt, bark, slash, sawdust, or other organic and earthen material from any logging, construction, gravel mining, agricultural, grazing, or other activity of whatever nature into waters of the State within the Garcia River watershed is prohibited."
2. "The controllable discharge of soil, silt, bark, slash, sawdust, or other organic and earthen material from a logging, construction, gravel mining, agricultural, grazing, or other activity of whatever nature to a location where such material could pass into waters of the state within the Garcia River watershed is prohibited."

The Action Plan provides a list of numeric targets that landowners in the Garcia Basin are obligated to meet through one of three options provided in the plan. One is to comply with an approved Erosion Control Plan and Site Specific Management Plan. The Erosion Control Plan requires the creation of an inventory of controllable sediment delivery sites on a property, an assessment of unstable areas, and a plan for monitoring the effectiveness of the sediment reduction effort. The Garcia River Forest's Erosion Control Plan's sediment reduction schedule is set such that on average 20% of the estimated volume of sediment must be corrected every four years, which results in a 20-year period to correct all of the identified anthropogenic sites.

The planning team for the IRMP consisted of members of TNC science staff and the Fund's forestry staff charged with incorporating all of the relevant state and federal statutes and other management goals into a comprehensive document. The primary issues were forest management policies, silviculture, and compliance with the Garcia River TMDL. A conservation planning approach developed by TNC was used. This is known as the Conservation Action Planning process, or the "5-S" planning approach (The Nature Conservancy, 2003). This process is designed to assist in the identification and development of conservation targets, the strategies to achieve those targets, take actions, and measure success, and to form the basis for adaptive management. The process begins with the selection of the general conservation objectives or targets that form the central underpinning of the management planning process.

Key ecological attributes (components that most clearly define or characterize a conservation target, limit its distribution, or determine its variation over space or time) were identified for each target. Because key ecological attributes are often difficult to measure directly, indicators that can be reasonably and efficiently measured or manipulated were also identified to support implementation.

Once the conservation targets were determined, TNCs Conservation Action Planning process was used to identify the environmental factors that must be maintained to ensure the long-term viability of the targets, including forest structure and forest composition; the second was the interactions between abiotic and biotic processes. Management of the forest both in and out of the ERN will seek to enhance and maintain ecological processes critical to achieving the conservation targets. Monitoring of these indicators and factors for the achievement of these conservation targets will be used to evaluate and adjust management actions on the forest using the principles of adaptive management. The targets can be adjusted as new data is collected on how the environment and management the activities react.

OUTCOMES OF THE PLAN

For the Garcia River Forest, five main objectives were incorporated into the management plan:

(1) Protection and enhancement of anadromous fish-bearing streams through the development and protection of the riparian area.

The Fund adopted riparian protection measures in excess of the California Forest Practice Rules to provide better stream protection than would have normally occurred. To improve near-stream shading and increase potential for large wood recruitment into streams a 25 foot (ft) (7.6 m) no-harvest buffer was adopted on all fish-bearing and perennial streams. Further, a 200 ft (61 m) stream buffer was adopted for fish-bearing streams in which silviculture had to promote the development of a late-seral stage forest similar to the Ecological Reserve. The resulting 400 ft (122 m) buffers eventually became part of the Ecological Reserve Network.

In addition to the increased riparian protection, the Fund initiated an aggressive program to place large wood in the stream channel to attempt to restore channel complexity. Additionally, large wood can help reduce sediment movement, develop deep scour spools, and provide direct cover for rearing fish. Large wood projects are generally conducted through a cooperative program administered by the California Department of Fish and Wildlife known as the Fisheries Restoration Grant Program in which partial grant funding is provided to willing landowners to place logs in streams. Logs can be placed with equipment or by direct felling into the active stream channel. To date, we have treated approximately 11 stream miles (mi) (17.7 kilometers (km)), which includes all of the high-priority coho salmon streams, and added approximately 500 logs to improve fish habitat.

The reduction of road-born sediment is also an important consideration for fisheries restoration and is required by the Garcia River TMDL. Therefore, the Fund developed a comprehensive road improvement and abandonment plan and subsequently embarked on an aggressive sediment reduction program. The primary method used for sediment reduction is to upgrade road watercourse crossings by replacing undersized culverts and adding sufficient rock armor to prevent crossing fill failures. Roads are also outsloped or treated with rolling dips to reduce accumulated water runoff. Near streams, roads and landings are abandoned wherever possible, and future harvests will use cable skyline logging methods to access those areas. To date, more than 88 mi (142 km) of forest roads have been upgraded or decommissioned, and 632 erosion sites have been treated to reduce erosion with an estimated reduction in sediment production of 118,000 cubic yards (yd^3) (90,204 cubic meters (m^3)).

TNC has initiated a stream habitat monitoring program using the Environmental Monitoring and Assessment Program. Using this program, stream reaches are randomly selected and a statistically valid assessment of in-stream conditions can be made. Physical stream metrics such as pool depth and occurrence, water temperature, large wood, sediment size, aquatic species, and benthic invertebrates are also recorded. Through the Environmental Monitoring and Assessment Program monitoring, forest managers will eventually be able to determine with statistical validity whether the management prescriptions are working to improve water quality and if some further adjustment to management is necessary:

(2) The return to historic density of redwood and Douglas-fir in the forest.

The IRMP contains several guidelines that are designed to achieve this goal, primarily a commitment to harvesting less than growth. The forest is committed to harvesting approximately one-third of growth for the first 20 years of ownership; thereafter, the harvest rates will gradually increase and by year 80 harvest should equal growth. The timber marking is designed to maintain or restore the original species balance, so harvest of redwood and Douglas-fir is in proportion to their occurrence in each stand. In stands where the species balance is off due to previous harvest favoring redwoods, Douglas-fir or pine are selectively removed and redwood is favored as retention trees.

As noted earlier, many of the stands are hardwood-dominated primarily by tanoak. In addition to the ecological imbalance, conifer growth and regeneration is impaired by the preponderance of tanoak. To control hardwoods and promote conifer growth, two methods are employed. The first one relies on herbicide treatments: up to 100 ac (40 ha) of tanoak-dominated stands are treated per year with Imazapyr, using the stem injection method or "hack and squirt." This is performed primarily in conjunction with THPs but Imazapyr has been used in young oak-dominated stands where the conifers are too small to harvest. Herbicides are used to release existing conifers under the hardwood canopy, however, Imazapyr has also been used for site preparation by injecting oak-dominated stands followed by conifer underplanting of the treated oak stands. The second methods involves hardwood felling: this practice is becoming increasingly important as other ways to control tanoak and to restore the forest species balance are being explored. Therefore, direct felling of oaks is practiced within harvest units to release the existing conifers. Conifers targeted for release by direct oak felling are well-established and 10–20 ft (3–6 m) in height ensures tanoak sprouts do not overtake the target trees and that the release effort will be successful.

An inventory of the forest was conducted in 2006 using traditional air photo interpretation methods to identify unique timber strata as part of a stratified random sample. The goal was to determine the baseline timber stocking. The forest was re-sampled in 2010 as part of a carbon certification effort. The 2010 inventory was a simple stratified sample, however. Rather than using air photos, LiDAR imagery was used to stratify the forest and to develop a new and more accurate stratification system and inventory (Golinkoff et al., 2011). Future inventory efforts will confirm or refute that the management strategy resulted in increased conifer stocking.

(3) Protect and restore the oak woodland and grassland.

The major oak species are black oak (*Quercus kelloggii*), shreves oak (*Quercus parvula*, var. *shrevei*), and coast live oak (*Quercus agrifolia*). All true oaks, madrone and tanoaks 20 in (51 cm) or greater in diameter at breast height are retained throughout the forest for wildlife habitat. The primary danger to oak woodlands in California is the occurrence of Sudden Oak Death (SOD), caused by a pathogen (*Pytophthora remorum*) that targets most true oaks and tanoak. SOD has proven difficult to control and only responds to very intensive eradication efforts that are not cost effective on a landscape basis. The pathogen has an alternate host, California bay (*Umbellularia californica*), which is unaffected by the pathogen and thus it can persist for years on bay trees. The pathogen is airborn during the spring and can travel from tree to tree, but is most pervasive when crowns are in contact with each other. SOD is present on the Garcia River Forest. Observations of the pathogen suggest that it seems to spread slowly in a managed forest. Perhaps this is because oaks are actively removed and the management regimes do not promote complete crown closure. To date, the Garcia River Forest has no definitive plan to control SOD.

Native grasslands existing on the eastern portion of the forest are suffering from Douglas-fir encroachment and have been overrun in some cases with non-native grasses. Plans to cut back the encroaching Douglas-fir are being made. The restoration of native grasses may be possible through the use of prescribed fire. Grass seeds are spread through animal and bird droppings so maintaining native grasses will be an ongoing operation. Yellow star thistle (*Centaurea solstitialis*) is a non-native, invasive weed common in California, and control of yellow star thistle in the meadows is an important consideration. Herbicide trials to control yellow star thistle are planned, however, the effects of the herbicide on the native grasses will have to be evaluated prior to developing a propertywide eradication policy.

(4) Protect and preserve non-riverine wetlands.

Non-riverine wetlands include the alluvial floodplains, permanent seeps or springs, and perennial watercourses. Protection of the flood plains and perennial watercourses are generally captured in the anadromous fish-bearing stream protection measures that were adopted by the California Department of Forestry and Fire Protection in 2011 to provide additional protection to the fisheries resource. Permanent springs and seeps are given a 50 ft (15 m) buffer in which no equipment operations are allowed and where at least 50% of the existing overstory canopy is retained. The obvious exceptions are springs that are related to road cut banks. Because these features are static on the landscape, the initial protection applied is sufficient to meet the target.

(5) Promote the recovery of the northern spotted owl.

Due to the past harvest history, the northern spotted owl habitat on the Garcia River Forest is low or poor quality compared to the rest of the redwood region. To evaluate the northern spotted owl across the property, a survey was conducted of the entire forest using the latest U.S. Fish and Wildlife Service survey protocol. It was determined that there were 10 occupied activity centers on the forest. An activity center is defined as an area consistently occupied by an owl or owl pair. The U.S. Fish and Wildlife Service requires a minimum of 500 ac (202 ha) of habitat be maintained within a 0.7 mi (1.1 km) radius of each activity center; of that, at least 200 ac (81 ha) shall be nesting and roosting habitat as defined by the U.S. Fish and Wildlife Service. Ultimately, northern spotted owl habitat is tracked through the forest inventory. Future inventory efforts will confirm or refute whether the management strategy employed results in increased conifer stocking and improved northern spotted owl habitat.

DISCUSSION AND CONCLUSIONS

The IRMP was published in August 2006 and was driven by clearly defined goals or targets set by the landowner and the regulatory environment in which the managers work. The goals are subject to change, depending on the outcome of forest monitoring efforts and changes in the regulatory environment. In addition to the silvicultural and resource goals enumerated above, the IRMP includes provisions for public access and adaptive management. The plan is strictly a guidance document for the management of forest resources and is not a sustained yield plan.

The long-term sustained yield plan for the Garcia River Forest indicates that there will be an increase in forest stocking over the 100-year planning horizon (Table 19.1). The Forest and Stand Evaluation Environment (California Growth and Yield Modeling Cooperative, 2011) growth model was used to model the growth and harvest of the forest while considering the forest management constraints mentioned above. The pre-harvest standing inventory increases for the entire planning period; however, harvest levels decrease after about 80 years because an Ecological Reserve Area (approximately 6250 ac (2529 ha)) was reserved from the harvest schedule. The projected harvest level is about 55% or less of the growth of the forest over the next 100 years (Figure 19.3).

Sustainability Issues

Since the development of the initial IRMP, the Garcia River Forest has been certified by both the Sustainable Forestry Initiative and Forestry Stewardship Council programs as a well-managed forest. Additionally, one of the positive outcomes of increasing conifer stocks on the property is that the forest qualified as a forest carbon offset project through the Climate Action Reserve. Through the implementation of the IRMP, which suggests harvesting less timber than is grown on the forest, and the fact that a conservation easement was established on the forest, the Garcia River Forest was one of the first large timberland owners in the state to be certified by the Climate Action Reserve. The Garcia River Forest is also now one of the largest producers and sellers of registered forest carbon offsets in the state. The sale of carbon offsets will allow for funds to be used for additional restoration work in the future.

TABLE 19.1 Growth, Standing Volume, and Harvest Levels for the Garcia River Forest

Time period (years)	Pre-harvest standing volume (MBF)	Projected harvest volume (MBF)	Projected growth (MBF)
2014–2018	252,291	11,304	48,695
2019–2023	289,682	13,209	59,073
2024–2028	335,546	15,225	69,643
2029–2033	389,964	19,140	76,733
2034–2038	447,556	19,628	69,522
2039–2043	497,450	22,991	69,199
2044–2048	543,659	26,512	69,562
2049–2053	586,710	28,790	69,528
2054–2058	627,447	32,587	69,258
2059–2063	664,118	34,227	68,840
2064–2068	698,730	36,794	68,132
2069–2073	730,068	30,508	67,950
2074–2078	767,511	36,988	67,209
2079–2083	797,732	35,394	66,526
2084–2088	828,864	31,843	66,099
2089–2093	863,121	26,051	65,897
2094–2098	902,967	10,910	66,809
2099–2103	958,866	7,981	67,885
2104–2108	1,018,770	11,933	68,615
2109–2113	1,075,452	11,810	69,260

MBF: Thousand board feet.

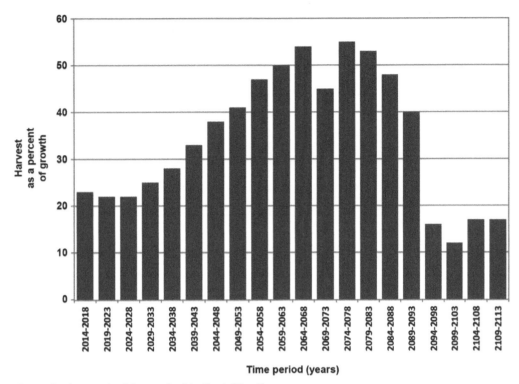

FIGURE 19.3 Harvest levels as a ratio of the growth of the Garcia River Forest.

ADDITIONAL READING AND RESOURCES

This chapter represents a synthesis of the management plan for the Garcia River Forest, managed by the Conservation Fund. To view the planning documents described in this chapter, please visit this Internet site, which was available on June 12, 2014:

http://www.conservationfund.org/our-conservation-strategy/focus-areas/forestry/north-coast-conservation-initiative/

If in the future the link to the site appears broken, search the Internet using the title of the plan and the keywords provided.

REFERENCES

California Environmental Protection Agency, 2011. Garcia River TMDL. California Environmental Protection Agency, North Coast Regional Water Quality Control Board, Santa Rosa, CA. http://www.waterboards.ca.gov/northcoast/water_issues/programs/tmdls/garcia_river/ (Accessed June 12, 2014).

California Growth and Yield Modeling Cooperative, 2011. FORest and Stand Evaluation Environment (FORSEE) beta version computer program. http://www.cagym.com (Accessed June 12, 2014).

Chase, A., 1995. In a dark wood: The fight over forests and the rising tyranny of ecology. Houghton Mifflin Co., Boston, MA. 535 p.

Fisk, L., Gerstung, E., Thomas, J., 1966. Stream damage surveys. California Department of Fish and Game, Sacramento, CA. Inland Fisheries Administrative Report No. 66-10.

Golinkoff, J., Hanus, M., Carah, J., 2011. The use of airborne laser scanning to develop a pixel-based stratification for a verified carbon offset project. Carbon Balance and Management 6, 9.

Johnson, K.N., Franklin, J.F., 2013. Recommendations for future implementation of ecological forestry projects on BLM western Oregon forests. Department of Forest Ecosystems and Society, College of Forestry, Oregon State University, Corvallis, OR. Report completed under OR/WA BLM Assistance Agreement L10AC20355. 45 p.

The Nature Conservancy, 2003. The Five-S framework for site conservation: A practitioner's handbook for site conservation planning and measuring conservation success. The Nature Conservancy, Arlington, VA.

U.S. Environmental Protection Agency, 2002. Research strategy, environmental monitoring and assessment program. Office of Research and Development National Health and Environmental Effects Research Laboratory, Research Triangle Park, NC.

Chapter 20

Indigenous Community of Nuevo San Juan Parangaricutiro, Michoacán, Mexico

Alejandro Velázquez,[1] Gerardo Bocco,[1] Alejandro Torres,[2] Adolfo Chavez Lopez[3] and Felipe Aguilar Gómez[3]

[1]*Centro de Investigaciones en Geografía Ambiental, Universidad Nacional Autónoma de México, Morelia, Michoacán, Mexico,* [2]*Comisión Nacional de Áreas Nacionales Protegidas, Andador Vicente Guerrero, Jalpan de Sierra, Querétaro de Arteaga, Mexico,* [3]*Comunidad Indígena de Nuevo San Juan Parangaricutiro, Municipio de Nuevo San Juan, Nuevo San Juan Parangaricutiro, Michoacán, Mexico*

ABBREVIATIONS

ICNSJP	Indigenous Community of Nuevo San Juan Parangaricutiro
SAS	Statistics Analysis Software
SEMARANT	Ministry of Environment and Natural Resources (Secretaria del Medio Ambiente y Resursos Naturales)
UNAM	National Autonomous University of Mexico

MANAGEMENT SETTING AND BACKGROUND

The Indigenous Community of Nuevo San Juan Parangaricutiro (ICNSJP) is a Purepecha community devoted to the management of its temperate forest resources. The area (Figure 20.1) is located approximately 15 kilometers (km) (9.3 miles (mi)) west of Uruapan in Michoacán, Mexico. The climate of the area is temperate, with a summer rainy season. Soils and landforms are of recent volcanic origin. The altitudes range from 1,800 meters (m) (5,906 feet (ft)) to 3,000 m (9,843 ft) above sea level. The area of the ICNSJP is about 190 km^2 (46,959 acres (ac)), of which about 100 km^2 (24,710 ac) are under forest cover and devoted to wood extraction. About 35 km^2 (8,649 ac) are dedicated to traditional agriculture, 5 km^2 (1,236 ac) are dedicated to intensive orchard production (mainly avocado and peach), and 50 km^2 (12,355 ac) are covered by the lava and ashes produced by the eruption of the Paricutin Volcano that occurred between 1943 and 1952 (Figure 20.2).

The ICNSJP was first established at Pantzingo, the most relevant marketplace of the Purepecha Empire during the fifteenth century, immediately before the conquest by Spain. In 1535, Fray Juan de San Miguel was entrusted by the Spanish Crown to outline the urban settlement of the original location of the community (Moheno, 1985). Since 1715, and until the mid-nineteenth century, the ICNSJP retained its land tenure titles granted by the Spanish Crown.

The Catholic Church has always played a crucial role in the governance of the community. By the end of the nineteenth century, the Church was instrumental in granting access to land to a group of families so that few members became landlords in spite of the collective land tenure system established on paper. This was common practice in rural Mexico and was at the root of a very unfair system just before the Mexican agrarian revolution (1910–1920). Despite the land reform in the 1930s that followed the revolutionary process, landlords at the ICNSJP still retained their privileges, although some plots were subdivided and inherited or sold to other tenants as private lands. Contradictions and turmoil between the federal government and Church-sponsored interests triggered the so-called *Guerra de los Cristeros* (the "Christian War"), which affected the study area. This movement was confronted by the federal government during the 1930s and 1940s. Lázaro Cárdenas, an emblematic president of Mexico, eagerly promoted landowner dispossession and empowerment of rural people (Velázquez, 2013). The Cristeros movement, however, was interrupted by a volcanic eruption. The former urban settlement of San Juan Parangaricutiro was abandoned following the eruption of the Paricutin Volcano in 1943. In 1946, community members resettled in three different locations; one of those was the Valley of the Rabbits, which became the current location of the ICNSJP. The new settlement was consecrated to *El Señor de los Milagros* (the "Lord of Miracles"),

Forest Plans of North America. http://dx.doi.org/10.1016/B978-0-12-799936-4.00020-5

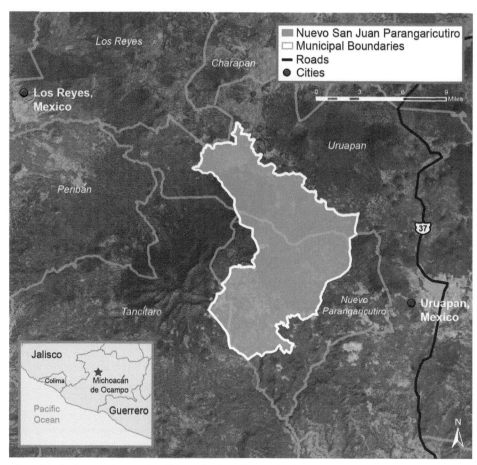

FIGURE 20.1 Location of the Indigenous Community of Nuevo San Juan Parangaricutim.

FIGURE 20.2 Old lava flow partially covering the church in the former San Juan's settlement. This catastrophic event changed social and cultural conditions as well as traditional communal governance structures making, as a result, sustainable forest management possible.

whose image survived the eruption of Paricutin Volcano in 1943. During 1949, a restructured ICNSJP began a claim for its collective land tenure rights. These were granted in 1991; however, several plots of land remain disputed.

The Purepecha Empire is well known for its resistance against the Spanish colonization and for its outstanding richness in forest resources. These resources can be regarded as providers of timber and non-timber forest products as well as forest ecosystem services. Until the end of the nineteenth century, local stakeholders carried out a low-impact exploitation of these resources. Thus, no severe depletion of forest had occurred. At the beginning of the twentieth century, landlords contracted a private enterprise (Michoacana Transportation Company) to conduct systematic forest exploitation using community members as labor. We carried out in-depth interviews with older community members who recalled these events and especially recalled what their parents and grandparents recalled to them when they were young. We are grateful to these honorable members of the ICNSJP, and their oral history, which provides the insight into the deforestation in the region that followed. In addition, workers were poorly treated, infestations of defoliators such as caterpillars occurred, and a wide application of DDT was used to limit the damage. As a result, efforts were made to expel the Michoacana Transportation Company from the area. Gradually, social and natural systems recovered. Interviews revealed that by the end of the 1970s, the social situation improved, landlords lost power because of the Paricutin eruption, and the company left the region. The forests recovered on hilly terrain originally affected by Paricutin volcanic ash, a fact also indicated by Rees (1970), who researched the social response to the volcanic catastrophe.

Between 1975 and 1979, the ICNSJP joined a federation of forest communities of the Meseta Purepecha and earned some experience in this type of organization. In 1979 following the request by a small group of visionary community members, the community was granted the right to develop its first forest management plan. A forest management plan ("Plan de Manejo Forestal") is the official term used by the Mexican government to authorize forest exploitation on the basis of a thorough technical forest survey. According to the Mexican Constitution, both forest and water are properties of the Nation. In the beginning, the ICNSJP was not able to carry out the technical forest survey and relied on private consultants. In 1981, a hybrid model was established, and a private consultant was responsible for performing the forest survey while the head of the community board dealt with the political and agrarian issues. Both worked together and reported to the Board and to the General Assembly of the community on a monthly basis. The operation of a Board and of a General Assembly of community members is reinforced by Mexican agrarian legislation for every rural community.

In 1983, the ICNSJP established its first sawmill, partially subsidized by external investors. The aim was to add value to timber forest products and to create jobs for community members. But above all, the sawmill was the seed to create a community-based enterprise devoted to managing the community forest and to avoid the need for selling the timber without further profits. During the first years, the profits were split equally among community members. Very soon, consensus was reached to reinvest in order to create capacity building and purchase equipment to ensure the processing of forest products. Thanks to the acquisition of technical skills following investments, in 1988 the community was granted the right to carry out its own forest technical services, a prerequisite reinforced by federal law to authorize an entity to manage the forest on the basis of a technical management plan. Thus, the enterprise was built on a strong legal and technical basis. Timber extraction increased gradually, and in 1993 the ICNSJP reached its maximum yearly harvest of about $100,000\,m^3$ (3,5 million ft^3) (Alvarez-Icaza, 1993). The enterprise was able to create nearly 800 permanent jobs, an additional 250 seasonal jobs, and an annual income of more than half a million U.S. dollars (USD). Timber products created, such as wooden floors and furniture, reached national and international markets. A land cover change analysis (Velázquez et al., 2001) indicated that management was sound enough to maintain the same forest cover area during a 25-year time span (1975–2000). By the mid-1990s, the Forest Stewardship Council granted the ICNSJP international certification as a successful community forest enterprise.

PLANNING ENVIRONMENT AND METHODOLOGY

In 1994, the effects of the North American Free Trade Agreement on the forest markets were felt in Mexico. The enterprise subsequently decreased in value by about 20%. Mexican forest enterprises were not prepared for competition from abroad (Velázquez et al., 2001). However, the ICNSJP reacted positively and the enterprise, under the leadership of a very capable manager, developed a plan to explore alternative economic activities, other than those related to timber forest production. In this context, the ICNSJP established a contact with the National Autonomous University of Mexico (UNAM) and asked for support to transform the traditional forest management plan requested by law to an integrated community-based resource management plan much like a land-use planning program. The objectives of this plan were three-fold:

1. Automate forest management through the use of geographic information systems (GIS) and statistical forest data handling, and ensure that forest would not be badly managed.
2. Look for non-timber forest products and access markets for other services provided by the forest resource.
3. Perhaps most importantly, strengthen technical capacities and leadership within the enterprise and the community, to ensure that the plan would not rely on external technical personnel.

A participatory landscape-based forest management plan was co-developed by enterprise technical staff and UNAM researchers (Bocco et al., 2000), authors of this chapter. The plan was submitted to and approved by the Board, The *Council of Caracterizados* ("wise community members"), and the General Assembly. The Council (common in well-organized communities) is a traditional governance committee tasked with advising the Board and the enterprise. It is formed by honorable and wise members of the community, usually, but not only, elderly people. To become a member of the Council, current members decide who to invite. All significant decisions to be made at the General Assembly are previously discussed at the Council. It is a very dynamic body, and by law the Assembly gathers frequently and engages in decision making on the first Sunday of every month. Financial resources were made available by the enterprise, UNAM (yearly research grants through five years), and a one-year grant from the U.S. Fish and Wildlife Service. A mixed technical team comprised by researchers and community members was created to jointly discuss and perform every task needed to achieve plan objectives. The team reported weekly to the head of the enterprise, bi-monthly to the Board and Council, and every six months to the General Assembly. Implementation, also led by the authors, took place from 1995 until 2000; the first two years required full-time involvement, while the remaining three required a half-time commitment by the joint team.

The entire approach was based on a simple idea: the well-established notion of land evaluation or land capability assessment (Rosete et al., 2003), in this case, was to be jointly operated by the technical team through a hands-on training procedure. The approach assumes that a terrain analysis coupled with a land cover and soils evaluation (jointly conforming to a land unit type) are able to provide the key indicators to predict the best use of land based on the highest potential of given land-use types. In other words, a land suitability survey was developed, which contrasted the availability of land units against the needs by land-use types. The survey also provided a practical ground to gather the local knowledge of community members, not only on forestry but also on agriculture and cattle-grazing. The land-use types evaluated included:

- Rain-fed traditional agricultural areas
- Perennial grasslands
- Avocado and peach orchards
- Forests (according to quality)

The entire procedure, encompassing field data creation and gathering, edition, and analyses, was implemented in a GIS environment, a task that also involved hands-on training (Bocco et al., 2001). For the purpose of this chapter, we will focus on the new forest management plan, a model that is still used by the ICNSJP.

Plan Development Process

Based upon conventional photo interpretation of a 1997 panchromatic black and white aerial photo survey at 1:25,000 approximate scale, 1,271 relatively homogeneous forest stands (polygons) for timber management purposes were delineated on the basis of (dis)similarities in forest cover, terrain, and tree density. Forest stands were digitized and orthorectified with the aid of GIS. All forest stands were assessed to measure potential wood volume using site index as the main variable. The site index helps evaluate growth and yield and is based upon average height of the dominant and co-dominant trees at 50 years. A systematic sampling scheme was conducted on 4,662 sampling circular units of approximately 36 m (118.1 ft) in diameter in order to cover 1,000 m² (10,750 ft²) of land. In every sampling unit, 30 variables were measured including elevation, slope aspect, slope length, and slope steepness as well as forest-stand parameters such as inventory of tree species, diameter of each individual tree at breast height (1.30 m, 4.27 ft), height, and basal area (see Velázquez et al., 2003 for details). Emphasis was placed on commercial tree species:

- Smooth-bark Mexican pine (*Pinus pseudostrobus*)
- Montezuma pine (*Pinus montezumae*)
- Sacred fir, or pinabete (*Abies religiosa*)
- Oaks (*Quercus* spp.)
- Mexican alder (*Alnus firmifolia*)

Harvestable volume, based on a multiple regression model, was developed for each commercial tree species. The Schumacher growth algorithm was selected as the most robust algorithm for stand-height prediction with the aid of Statistics Analysis Software (SAS) (Cody and Smith, 1987).

Each forest stand was classified according to its site index and the outcome of the Schumacher model. Ten clusters were chosen, and all forest stands regrouped on the basis of their yearly harvestable wood volume. The 10 clusters were represented spatially using the GIS. The geodatabase served to determine forest stands to be harvested on a yearly basis along a 10-year period. This information was also used to fulfill the requirements demanded by the Ministry of Environment and Natural Resources (Secretaria del Medio Ambiente y Resursos Naturales or SEMARNAT) to grant the permit to conduct the 10-year forest management plan.

The authors participated in the forest management plan from 1994 through 2003. Because this authorization should be requested every 10 years, the described database and its geographic expression were kept as a baseline to update the information. The assessments to update the characteristics of the forest stands to be harvested from 2005 until present were substantially less time consuming and obviously less expensive. All the data gathering, data analysis, and data handling in the GIS were conducted by community members; supervision by university staff was given on demand.

In addition, permanent monitoring plots to estimate annual yields were established. On this basis, the ongoing 10-year forest management plan (2005–2014) was granted by SEMARNAT and represents one of the outstanding successful forest management experiences in Mexico (Bray et al., 2004).

OUTCOMES OF THE PLAN

Twelve community members were fully trained in at least one of the following procedures: geospatial data handling, forest assessment, soil surveying, biodiversity monitoring, and ecotourism. A larger group was enrolled in other activities to build capacity, leadership, and to strengthen awareness on the relevance of the integrated management of community natural resources other than timber forest production.

Concerning the implementation of geospatial data handling, two community technicians were fully trained in the use of a GIS package still operational at the enterprise. A large relational and spatial database was compiled and encompassed data from more than 11,000 tree individuals, from 4,662 sampling units, and 1,271 forest stands. Other layers such as soils, landforms, land cover, and vegetation were also integrated into the GIS database.

With regard to forest inventory, three community technicians with a background in forestry were trained in statistical data analysis. In addition, other technicians received training in sustainable forest management. To illustrate this further, a permanent forest study was designed to assess the net effect of harvesting. In this experiment, 15 forest stands were selected in areas representative of the environmental conditions of the community lands. Subsets of three plots randomly selected were chosen to apply five intensities of harvesting: 100%, 80%, 60%, 40%, and 20% wood extraction. The wood volume per intensity was established, and the experiment is repeated on a yearly basis to determine the net impact on yield. The outputs are yet to be completed since the experiment was meant to last 25 years. Preliminary results have allowed the managers to adjust harvesting intensities so that forest management is sustainable by extracting the maximum amount of wood without surpassing the annual growth. The ICNSJP became very well known and recognized at regional and global levels since only a few examples of semi-automated forest harvest schemes existed worldwide.

Biodiversity surveys and management activities were conducted with the aid of two technicians, who were trained in inventorying and monitoring of birds and mammal populations—namely, those listed as endangered. Ongoing research activities are using these inventories as a baseline for monitoring harvesting performance and ecological impact on the functional ecosystem integrity. Environmental sustainability awareness among community members also was a major target of the cooperation. Interviews with community hunters suggested that the density of the white-tailed deer (*Odocoileous virginianus*) population had decreased with the intensity of wood harvesting. Besides its ecological importance as an indicator of forest ecosystem health, deer is important culturally as an offering during Catholic feasts having a strong Purepecha tradition. To cope with this, a white-tailed deer breeding program was launched. After the first two years of effort, the population increased from 6 in 1995 to 60 deer in 1997. Originally, hunters became stewards of the deer population and led a large ecotourism program featuring a visit to the nursery as a key attraction. In addition, visits to the Paricutin Volcano and lava flow and hiking paths attract the attention of thousands tourists who can find full lodging in wooden cabins. Besides raising awareness on forest conservation, ecotourism has allowed the creation of seven full-time and three part-time jobs.

The ICNSJP extracts about 65,000 m³ (2.3 million ft³) of wood and 1,100 tons of resin annually. About 10% of the wood extracted is used to produce furniture, floor boards, and other derived products. The remaining 90% is used to produce boards, wooden boxes, and other products. Resin is chemically processed to produce turpentine, pitch, and modified resins for the chemical industry. On the whole, 850 permanent jobs exist and total yearly revenue of $11.5 million USD is obtained.

Currently, the ICNSJP and the enterprise are undergoing a large management reform. One single manager is no longer responsible for all productive actions in the community. Rather, 10 different independent enterprises were established, but their managers report monthly to the General Assembly. The main target is job creation and improving family income. Innovations among these enterprises include improving avocado and peach orchards, and semi-enclosed livestock production, and ecotourism enterprises (Figure 20.3).

FIGURE 20.3 The Pantzingo ecotourism center is a new enterprise serving annually thousands of visitors to Nuevo San Juan's Community who want to experience its holistic forest management.

DISCUSSION AND CONCLUSIONS

Polycentricity (Ostrom, 2003) best describes the theoretical framework underlying this case study. Polycentricity is a social system that has many decision centers, each with its own limited and autonomous prerogatives, yet each operating under an overarching set of rules (Aligica and Tarko, 2012). According to Bray (2013), there are specific principles that explain the successful operation of most Mexican forest community organizations, including the ICNSJP. These principles basically deal with two issues: clear land boundaries and rights to access resources and clear rules to conduct resource management and conflict resolution. Community enterprises such as the ones created in San Juan are also embedded in this framework. In San Juan, three major governance structures interact. The General Assembly, which approves major decisions concerning community life and enterprises management; the Council, which provides advice to the Assembly on strategic community and economic issues; and the Board, which is the agrarian authority that also decides on key positions in the enterprises (Aligica and Tarko, 2012). In other words, the old agrarian society has merged with a modern productive structure, and rules derived from tradition are used to govern community life and economic activity. Cultural and institutional identities are maintained. All community members have the right to apply for enterprise jobs or to become managers of the enterprises based on their technical and leadership capabilities. Resilience, in its social (Adger, 2000) and ecological (Gunderson and Holling, 2002) versions, has often lacked empirical, spatially explicit examples (Folke et al., 2010). This study case may well serve as a resilient forest enterprise, since the main features of this promising concept apply as well.

Learnings and Insights

The recognition granted and ratified by the Forest Stewardship Council to the ICNSJP as a successful collective enterprise and its operational framework relies upon two major issues. First, robust natural resource knowledge and management governed by means of consensus, target-oriented issues transversally and openly discussed. Second, reinvestment of a substantial proportion of the enterprise's income, which can be redistributed through job creation and capacity governments, provides access to credit and to rural or forest production projects. In addition, the ICNSJP has always counted on the support of academic institutions providing advice and technical support when needed.

Sustainability Issues

To conclude, the forest management plan in ICNSJP plays a key role as an instrument of sustainable governance; and vice versa, forest management would be unsustainable without the sound governance prevailing at ICNSJP. Bearing in mind the critical situation of the Mexican countryside, and that of Michoacán in particular, the ICNSJP has shown strong evidence of social resilience as described by Adger (2000) and Folke et al. (2010), where sound governance and forest management operate simultaneously (Durán et al., 2011).

Plan Development Challenges

It is irrefutable that ICNSJP stands out as a model community forest organization, yet organizational performance may not be stable over the long run. To ensure improvement, and to facilitate future plan development processes, two major issues are foreseen as critical. The first concerns the lack of an institutional initiative for a higher education strategy within the ICNSJP. A critical analysis of the ICNSJP's organization, conducted by members of the same community, has yet to be accomplished. The second concerns the lack of a monitoring framework to assess organizational execution and effectiveness.

Clear goals should be defined for the medium and long term, especially to gain insight and experience and to improve communication of this forest management experience, where the community forest is transforming into a consortium of enterprises.

Plan Implementation Challenges

There are key challenges that threaten the long-term functioning of the ICNSJP. Education and lack of commitment among younger generations are perhaps the most striking ones. Evidence over the last 15 years showed a limited number of community members engaged in higher education programs; furthermore, a large percentage of the successful ones reaching university do not get involved in the community decision-making processes. A thorough analysis of both is outside the scope of this chapter; however, we think that a major effort should be given to understanding both, particularly by the Board and the General Assembly. In addition, evidence of ecological impoverishment has been observed. It is clear that a number of commercial tree species are favored, and more than 90% of forested plots experience understory clearing. The actual impact of this is yet to be ascertained.

ACKNOWLEDGEMENTS

Financial support came from UNAM (DGAPA-IN 202214-3). María Guadalupe Lira assisted in formatting the final version of the chapter. Antonio Navarrete helped prepare Figure 20.2.

REFERENCES

Adger, N., 2000. Social and ecological resilience: Are they related? Progress in Human Geography 24, 347–364.

Aligica, P.D., Tarko, V., 2012. Polycentricity: From Polanyi to Ostrom, and beyond. Governance 25 (2), 237–262.

Alvarez-Icaza, P., 1993. Forestry as a social enterprise. Cultivation Surveys 17 (1), 45–47.

Bocco, G., Rosete, F., Bettinger, P., Velázquez, A., 2001. Developing a GIS program in rural Mexico. Community participation equal success. Journal of Forestry 99 (6), 14–19.

Bocco, G., Velázquez, A., Torres, A., 2000. Comunidades indígenas y manejo de recursos naturales. Un caso de investigación participativa en México. Interciencia 25 (2), 9–19.

Bray, D.B., 2013. When the state supplies the commons: Origins, changes, and design of Mexico's common property regime. Journal of Latin American Geography 12 (1), 33–55.

Bray, D.B., Merino, L., Barry, D., 2004. Los bosques comunitarios de México: Manejo sustentable de paisajes forestales. Instituto Nacional de Ecología-SEMARNAT, Mexico City, DF. 444 p.

Cody, R.P., Smith, J.K., 1987. Applied Statistics and the SAS Programming Language, second ed. SAS Institute Inc, NC. 280 p.

Durán, E., Bray, D.B., Velázquez, A., Larrazábal, A., 2011. Multi-scale forest governance, deforestation, and violence in two regions of Guerrero, Mexico. World Development 39, 611–619.

Folke, C., Carpenter, S.R., Walker, B., Scheffer, M., Chapin, T., Rockström, J., 2010. Resilience thinking: Integrating resilience, adaptability and transformability. Ecology and Society 15 (4), 20.

Gunderson, L.H., Holling, C.S., 2002. Panarchy: Understanding Transformations in Human and Natural Systems. Island Press, Washington, DC. 507 p.

Moheno, C., 1985. Las Historias y Los Hombres de San Juan. Colegio de Michoacán, Mexico. 187 p.

Ostrom, E., 2003. How types of goods and property rights jointly affect collective action. Journal of Theoretical Politics 15 (3), 239–270.

Rees, J.D., 1970. Paricutin revisited: A review of Man's attempts to adapt to ecological changes resulting from volcanic catastrophe. Geoforum 4, 7–25.

Rosete, F.A., Sánchez E, J.F., Bocco, G., 2003. El sistema automatizado de evaluactión de tierras. In: Velázquez, A., Bocco, G., Torres, A. (Eds.), Las enseñanzas de San Juan: Investigación participativa para el manejo integral de recursos naturales. Instituto Nacional de Ecología-SEMARNAT, Mexico City, DF. pp. 437–472.

Velázquez, A., 2013. Review on revolutionary parks. Conservation, social justice, and Mexico's national parks, 1910–1940. Studies in Social Justice 7, 169–171.

Velázquez, A., Bocco, G., Torres, A., 2001. Turning scientific approaches into practical conservation actions: The case of Comunidad Indígena de Nuevo San Juan Parangaricutiro, México. Environmental Management 5, 216–231.

Velázquez, A., Fregoso, A., Bocco, G., Cortés, G., 2003. The use of a landscape approach in Mexican forest indigenous communities to strengthening long-term forest management. Interciencia 28 (11), 632–638.

Chapter 21

Jackson Demonstration State Forest, California, United States of America

Helge Eng

California Department of Forestry and Fire Protection, Sacramento, California, USA

ABBREVIATIONS

Board	California Board of Forestry and Fire Protection
Department	California Department of Forestry and Fire Protection
EIR	Environmental Impact Report
JAG	Jackson Advisory Group
JDSF	Jackson Demonstration State Forest
SCA	Special Concern Area

MANAGEMENT SETTING AND BACKGROUND

The California Department of Forestry and Fire Protection (the Department) manages approximately 71,000 acres (ac) (28,733 hectares (ha)) of Demonstration State Forests on behalf of the people of California. The Jackson Demonstration State Forest (JDSF), a 48,652 ac (19,689 ha) second-growth redwood (*Sequoia sempervirens*) forest located in Mendocino County, between Fort Bragg and Willits, is the largest forest in the Department's Demonstration State Forests system.

The JDSF was purchased from the Caspar Lumber Company in 1947 after nearly 90 years of management for timber production. The establishment of the JDSF was predicated upon declining volumes of old-growth timber and the fact that large areas of potentially productive timberland in California were not producing a satisfactory growth of young timber. When the state acquired Caspar Lumber Company's holdings, most of the coastal watersheds, such as Caspar and Hare Creek, had regenerated to even-aged stands of 15–60-year-old second-growth forests, though post-logging fires had burned through many of the regenerated stands. Caspar Lumber Company started partial cutting on the east end of the forest in the 1930s. After acquiring the forest, the state continued partial cutting on the east end during the 1950s and 1960s. This first round of partial harvest was an individual marked tree cut that removed about 70% of the coniferous volume. As a result, most of the large old-growth trees were removed. This initial cut was followed by a diameter-limit harvest that removed most remaining coniferous trees greater than 22 inches (in) (56 centimeters (cm)) in diameter. This harvest pattern on the east end of the forest resulted in an irregular uneven-aged stand structure, characterized by a relative abundance of hardwoods, pole timber, small young second-growth conifers, and individual scattered residual old-growth conifers. This kind of irregular stand structure is typical of current stands on the eastern portion of the forest, and distinguishes it from the western portion. Although the western portion of the forest was subject to partial cutting of the second-growth stands, it has retained a more uniform stand structure due to the early history of large-scale clear-cutting within the coastal watersheds.

For the purposes of growing trees, the JDSF is primarily composed of site class II and III lands. Redwood and Douglas-fir (*Pseudotsuga menziesii*) trees dominate the forests. Other conifers include grand fir (*Abies grandis*), western hemlock (*Tsuga heterophylla*), and Bishop pine (*Pinus muricata*). Hardwoods comprise substantial secondary components in this type and are represented principally by tanoak (*Lithocarpus densiflorus*) and madrone (*Arbutus menziesii*). The majority of trees are less than 120 years old. Approximately 459 ac (186 ha) of old-growth redwood groves remain on the JDSF and are protected from harvesting. Old-growth residual trees, which were left standing when the forest was first harvested and

Forest Plans of North America. http://dx.doi.org/10.1016/B978-0-12-799936-4.00021-7

during subsequent harvests, can be found as isolated individuals or in small aggregations. Redwood becomes less dominant as one moves inland, as Douglas-fir and hardwood increase in composition. Some of the inland areas would be classified as a Douglas-fir forest series by Sawyer and Keeler-Wolf (1995) and Holland (1986). Tanoak and madrone dominate young Douglas-fir and redwood stands in some areas and exist within most conifer stands at the mid- and lower-canopy levels. Hardwoods are more prevalent toward the central and eastern portions of the forest. The western portion of the forest can contain relatively pure stands of red alder (*Alnus rubra*). Alder, bigleaf maple (*Acer macrophyllum*), and willow (*Salix* spp.) are generally restricted to riparian areas. Additional hardwoods found on the JDSF include California bay (*Umbellularia californica*), chinquapin (*Chrysolepis chrysophylla*), and canyon live oak (*Quercus chrysolepis*).

The Mendocino pygmy forest is a unique ecological community that occurs only in coastal Mendocino County. The California Natural Diversity Database (California Department of Fish and Wildlife, 2007) recognizes it as a community that is "rare and worthy of consideration." The Pygmy Cypress community covers approximately 613 ac (248 ha) of the JDSF near the western end of the forest. The California Department and Parks and Recreation cooperate to manage some of this area.

The JDSF contains 142 continuous forest inventory plots that have been measured at 5-year intervals since 1959. Forest growth is in the range of 900–1,300 board feet per ac (5.2–7.4 cubic meters (m^3) per ha) per year in most areas. Harvest levels average 15–25 million board feet (MMBF) (35,400–59,000 m^3) per year. The standing inventory of the JDSF averages approximately 43,000 board feet per ac (251 m^3 per ha). The average density of snags on the JDSF is 1.9 per ac (4.7 per ha). Slightly more than half of the snags (57%) are from conifers. However, the most common species of snag is tanoak (23%), followed by young growth Douglas-fir and Bishop pine (20% each), madrone (15%), and young-growth redwood (8%). The diameter at breast height (DBH) of the snags averages 17.6 in (44.7 cm) with a maximum of 44 in (111.8 cm) and does not differ appreciably between conifers and hardwoods.

The Pacific Ocean is a moderating influence on the climate of the region. The JDSF has a Mediterranean climate, characterized by a pattern of low-intensity rainfall in the winter and cool, dry summers. Fog is a dominant climatic feature, generally occurring frequently during the summer months and less frequently during the rest of the year. About 90% of the precipitation in this area falls between October and April. Mean annual precipitation is 39 in (991 millimeters (mm)) at Fort Bragg (California Department of Water Resources, 2013), but ranges from 39 to 55 in (991–1,397 mm) across the forest. The rainfall, runoff, and stream discharges in this region are all considerably lower than the wetter redwood forest areas in Humboldt and Del Norte counties to the north.

In general, the landscape is characterized by moderate to high relief. Elevations range from less than 100 feet (ft) (30.5 meters (m)) within stream valleys along the western edge of the JDSF, to a maximum of 2,092 ft (638 m) in the southeast corner. The area drains directly into the Pacific Ocean. The local stream pattern is reminiscent of a "trellis," where short tributary streams flow into larger streams at roughly right angles. The stream pattern is controlled in part by structural patterns in the bedrock. As is true throughout Coast Range, the predominant structural pattern of streams trends northwesterly. Thus, many of the principal watercourses in the area are oriented from northwest to southeast.

Debris slide slopes, followed by rockslides, are the features covering the greatest amount of area. Mass wasting on the JDSF is dominated by shallow debris slides associated with roads and landings and slides in inner gorges and steep colluvial filled hollows. Surface erosion for the JDSF planning watersheds has been estimated from field surveys, results from the Caspar Creek watershed study, and erosion hazard ratings. The eastern planning watersheds have the highest percentage of land in the high or extreme categories. Sediment delivery to stream channels has been estimated to come from heavily used gravel-surfaced roads within 200 ft (61 m) of streams. Overall, average sediment delivery from surface erosion associated with the JDSF riparian roads is 50% of the total estimated from all sources. The legacy effects of old streamside roads were found to be substantial and continue to be a focus of the JDSFs road management efforts.

The most significant impact to stream channels located within the JDSF has been the widespread removal of large woody debris from low-gradient stream channels from the 1950s to the early 1990s. This large woody debris removal has reduced pool frequency and depths and overall habitat complexity, which has in turn reduced the quality of over-summering and over-wintering habitat for anadromous fishes. Where wood has been removed, stored sediments have been flushed, resulting in channel lowering and entrenchment disconnecting channels from floodplains and reducing backwater habitats thought to be important refuges for fish during strong winter storms. Additionally, older logging practices that occurred until the mid-1970s resulted in large inputs of sediment into stream channels. Some channels have shown slight recovery from aggradation but, overall, most continue to show evidence of high sediment input, increased entrenchment, and reduced large woody debris levels. Restoring large woody debris to streams continues to be a management and research focus at the JDSF.

The JDSF and the surrounding forested area provide habitat for a number of listed and sensitive fish and wildlife species, including the northern spotted owl (*Strix occidentalis caurina*), coho salmon (*Oncorhychus kisutch*), and steelhead trout (*Oncorhychus mykiss*). In addition, the JDSF has the potential to provide habitat for several listed or sensitive species that are not currently known to occur on the forest. These species include the marbled murrelet (*Brachyramphus marmoratus*), Pacific fisher (*Martes pennanti*), and Humboldt marten (*Martes americana humboldtensis*). As such, the JDSF, in conjunction with other parcels of public land in central Mendocino County, represents a valuable resource of potential re-occupancy and sustainability for at-risk wildlife species. In 2011, the JDSF was designated as critical habitat for the marbled murrelet.

Historically, coho salmon and steelhead trout occurred in all of the planning watersheds within the forest. On the JDSF, there are about 90 miles (mi) (145 kilometers (km)) of streams with fish habitat, and within the planning watersheds draining the JDSF there are about 192 mi (309 km) of this habitat. Steelhead trout are found in all 15 planning watersheds reviewed, and coho salmon are found in 12 of the 15 planning watersheds. Coho generally use stream channels with less than 4% gradient and were found in 92 mi (148 km) of the Class I watercourses found in the 15 planning watersheds (i.e., about 48% of the total Class I stream system present).

The JDSF receives an estimated 61,000 recreational visitors per year. There are more than 60 individual campsites, many miles of riding and hiking trails, and more than 200 mi (322 km) of forest road utilized by the public. Other common recreational activities conducted on the forest include picnicking, hunting, swimming, wildlife viewing, and target shooting. The forest also is a local source of firewood and other minor forest products such as mushrooms and greenery for both personal and commercial use.

PLANNING ENVIRONMENT AND METHODOLOGY

The eight Demonstration State Forests in California are self-funded. The Board of Forestry and Fire Protection (the Board) specifies that the purposes of all the state forests are research and demonstration, timber production, and recreation. Given the overarching mandate of research and demonstration, timber production is the primary land use on the JDSF, while recreation is recognized as a secondary but compatible land use. Funding to support all activities on the forests, including salaries and overhead expenses, comes from revenues generated on the forests, primarily timber sales, but also recreation fees and fees from the sale of minor forest products. Revenues from all the Demonstration State Forests are pooled in a joint fund and distributed to individual forests annually. The JDSF, by virtue of being the largest forest in the Demonstration State Forests system, is the major contributor of revenue to the fund.

Management plans are developed for each Demonstration State Forest through a public process and are subject to mandates of the California Environmental Quality Act. Similar to the federal National Environmental Policy Act (42 U.S.C. § 4321 et seq.), the California Environmental Quality Act requires that environmental values be integrated into the decision-making processes by considering the environmental impacts of proposed actions and reasonable alternatives to those actions. This requirement is met by preparing an environmental impact report (EIR) or other similar documents. The approval process for an EIR allows for public input. All forest management activities on the Forest must also comply with the California Forest Practices Act and regulations promulgated by the Board.

Additionally, management plans are reviewed every 5 years by the Board, and policy requires the Demonstration State Forests to operate under a current management plan. In an open public process, the Department presents to the Board a thorough review of each existing plan at least every 5 years. After each review, the Board may direct the Department either to continue management under the existing plan, to prepare amendments to the plan, or to prepare a new plan for public review and Board approval. During the management plan review and update process, Board policy permits the Department to continue to manage state forests under existing management plans with appropriate consideration for changes in law or regulation until amendments or new plans are approved by the Board.

Management Objectives

The Department formulated several guidelines and objectives for the management plan. These objectives were derived from a number of sources. Existing laws and regulations, Board policy, science, professional experience, advisory groups, and public input all played a role in formulating the guidelines and objectives. The JDSF will be managed as a working forest with regular timber harvesting operations consistent with environmental constraints related to public trust resources such as watershed, wildlife, fisheries, aesthetics, and recreational enjoyment. The annual allowable harvest level will be determined by a spatially explicit long-term harvest schedule.

The JDSF will maintain as wide a range of forest structures as possible to maintain diversity and maximize options for future research and management. This will be achieved by creating and maintaining a dynamic matrix of habitats and seral stages that moves across the landscape over time. A late-seral stand can be harvested and reset to a younger stand and be replaced by a candidate late-seral stand somewhere else in the forest, reflecting the paradigm that disturbance is the rule rather than the exception in most ecosystems (Botkin, 2006). The JDSF will be made available to educational institutions and other agencies for conducting research and demonstration projects. Demonstration areas will be developed. These will incorporate a wide range of forest management approaches within a compact, easily accessible area. Finally, the management plan specifies that a Forest Learning Center will be established at the JDSF to support and facilitate forest management research and learning activities.

Forest management will focus on increasing the amount of older forest structure and late-seral forest available for terrestrial wildlife, including areas adjacent to aquatic habitats. Maintaining corridors of contiguous habitat will improve habitat connectivity and reduce forest fragmentation. Activities will use a range of management techniques to compare natural and accelerated forest restoration approaches in areas designated for development of late-seral forest characteristics. Designated old-growth reserves will be managed for maintenance of late-seral habitat values.

Riparian areas will be managed for late-seral habitats, while allowing for flexibility to conduct research on riparian protection zones. Recovery habitat for listed species will be created or naturally developed. Forestry practices that maintain stability of hill slopes will be used. A control program will be developed to limit sources of mass wasting and surface erosion. Finally, a comprehensive road management plan will be developed to reduce sediment production, including upgrading roads remaining in the permanent transportation network and properly abandoning high-risk riparian roads where possible.

Constraints and Opportunities for the Planning Area

Special concern areas (SCA) at the JDSF are areas where management is restricted in order to protect certain resources. They include unique habitats, habitat for species of concern, riparian areas, recreational areas, areas near residences and parks, research areas, water supplies, and sensitive slopes. Figure 21.1 illustrates several of the SCAs located on the JDSF.

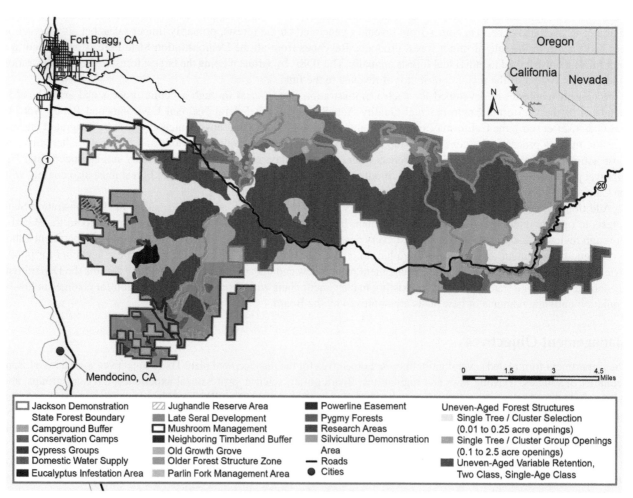

FIGURE 21.1 Forest management areas and special concern at the Jackson Demonstration State Forest.

Many SCAs physically overlap. Examples include the power line rights-of-way crossing through the watercourse and lake protection zones or the uneven-aged management areas; the overlap of pygmy forest and the Jughandle Reserve; and road and trail corridors located within the Woodlands Special Treatment Area. The areas noted below are those that are assigned to each SCA independently; thus, the total is more than the total forest area affected by SCAs. The most restrictive limitations will be applied during implementation of the management plan. The research and demonstration mandate coupled with public trust resource protection has resulted in the following SCAs on the Forest:

1. An *Older Forest Structure Zone* (6,803 ac, 2,753 ha) that consists of areas designated for management to connect specific old-growth groves, late-seral development areas, watercourse protection zones, and upland forest to form a contiguous area of habitat with structural characteristics of older forest, such as large trees, snags, down logs, and a high degree of vertical and horizontal diversity. Where timber harvest is proposed adjacent to the Old Forest Structure Zone, a buffer will be applied. No even-aged silvicultural systems may be used within 300 ft, (91 m) and only single tree selection may be used within the first 100 ft (30 m) adjacent to these areas.

2. *Cypress groups* (253 ac, 102 ha), or stands dominated by pygmy cypress, that occur on sites with generally unproductive soils (i.e., sites that are considered non-timberland), but not considered to be true pygmy forest. These areas will not be harvested. Conifer stands containing cypress that occur on more productive sites may be subject to harvesting and are not included in this SCA.

3. The *Pygmy forest* (613 ac, 248 ha) is a unique type of dwarf vegetation occurring on a unique soil profile that includes an underlying hardpan and highly acidic soils that offers an inhospitable environment for species and greatly stunts growth. Pygmy forest can be found on old marine terraces dominated by pygmy cypress and other specially adapted species. These areas will not be harvested.

4. The *Jughandle Reserve* (247 ac, 100 ha) is an administrative area designated to protect a tract of pygmy forest within the JDSF and to manage recreational access to these lands in a manner compatible with human use in the adjacent Jughandle State Reserve. This SCA lies almost entirely within the pygmy forest SCA, and there will be no harvesting within this SCA.

5. A *eucalyptus infestation area* (300 ac, 121 ha) is an area in the Caspar Creek planning watershed that includes eucalyptus species mixed with the native species. This is an area of special management concern because of the need to control eucalyptus to allow regeneration of conifers in this stand and to prevent the spread of this exotic species on the Forest. The JDSF intends to convert this area to native conifer species.

6. *Inner gorges* are steep slopes adjacent to streams that are prone to mass wasting and have a high potential for sediment delivery to stream channels. These areas are subject to silvicultural limitations, such as no harvest or limited single tree selection, depending on the results of a site review during timber harvest plan preparation.

7. *Northern spotted owl nest areas* are buffers around known nest site locations that will be managed to minimize disturbance to these sites and enhance their value as nesting habitat for the northern spotted owl.

8. *Osprey nest areas* are buffers around known nest site locations that will be managed to minimize disturbance to these sites and enhance their value as nesting habitat for osprey.

9. *Watercourse and lake protection zones* (7,440 ac, 3,011 ha) are areas designated for special management to protect aquatic and riparian resources, maintain terrestrial habitat connectivity for wildlife, and promote development of late-successional forest stand conditions. Silviculture is limited to no harvest or uneven-aged regimes designed to promote development of late-successional forest stand conditions.

10. *Woodlands Special Treatment Areas* (2,511 ac, 1,016 ha) are a special management area adjacent to the Mendocino Woodlands. Silvicultural activities, with limited exceptions, are focused on promoting late-successional forest conditions, maintaining aesthetic qualities, and limiting impacts on the operation of Mendocino Woodlands.

11. *Domestic water supply areas* (195 ac, 79 ha) are designated areas for domestic water supply in the JDSF that are sensitive to disturbance. Only a limited range of silviculture is allowed in these areas.

12. *Buffers adjacent to non-timberland neighbors* (875 ac, 354 ha) are areas along the boundary of the JDSF that are adjacent to non-industrial timberland owners where a buffer zone is designated to minimize impacts on neighbors. Only a limited range of silviculture is allowed in these areas.

13. A *power line right-of-way* (89 ac, 36 ha) that bisects the forest, is generally parallel to Highway 20, and is operated by the Pacific Gas and Electric power company. The right-of-way is not available for timber production.

14. *State Park Special Treatment Areas* (415 ac, 168 ha) are areas adjoining state parks where the application of silvicultural systems must take the values of the parks into consideration.

15. *Reserved old-growth groves* (459 ac, 186 ha) includes the existing mapped old-growth grove reserves. These areas will not be harvested.

16. *Late-seral development areas* (2,762 ac, 1,118 ha) will be managed to develop late-seral habitat conditions potentially suitable for the marbled murrelet. These areas will be managed to promote development of late-seral stand conditions to help buffer the adjacent old-growth groves and to enhance the value of these areas for wildlife species that are associated with late-seral forests. A buffer will be applied if timber harvest is conducted near late-seral development areas. No even-aged silvicultural systems may be used within 300 ft (91 m), and only single tree selection may be used within the first 100 ft (30 m) adjacent to these areas.

17. *Campground buffers* (133 ac, 54 ha) are areas immediately adjacent to campgrounds that are managed for public safety and esthetic enjoyment. Even-aged silviculture is not allowed within the campground buffers.

18. *Conservation camps* (43 ac, 17 ha) are areas occupied by the Parlin Fork and Chamberlain Creek Conservation Camps. These areas will not be managed for timber production.

19. *Road and trail corridors* (4,790 ac, 1,938 ha) are buffer areas along trails and roads intended to maintain aesthetic qualities valued by the public. Only a limited range of silviculture is allowed in these areas.

20. *Parlin Fork Management Area* (279 ac, 113 ha) is an area adjacent to the Parlin Fork Conservation Camp that is used as a demonstration area for small woodlot management.

21. *Research areas* (1,680 ac, 680 ha) are areas set aside for various existing research studies.

22. *Areas with a high relative landslide potential* are areas identified from California Geological Survey geology and geomorphology maps as having a high relative landslide potential using the best available data and assessment methodologies. These areas will be reviewed on the ground following established guidelines. They are potentially subject to limitations on road construction, yarding methods, and silviculture and may need to be evaluated by a certified engineering geologist.

23. *Mushroom Corners Management Area* (330 ac, 134 ha) is an area particularly important to the mycological research community, in part, due to its ease of access and presence and abundance of a diverse number of species.

Plan Development Methods

The planning team for developing the management plan was drawn from both the JDSF and Department specialists and included specialists in forestry, silviculture, measurements, growth and yield, harvest scheduling, forest economics, wildlife biology, hydrology, geology, and cultural resources.

The analysis used to develop the management plan was driven by simultaneous consideration of the multiple goals and objectives identified for the JDSF. SCAs that contain unique resource values were first identified and mapped, and management regimes were tailored to the resource values of each. An example of this analysis is provided by the case of the marbled murrelet. Although the marbled murrelet has not been observed at the JDSF, sightings have occurred on nearby state parks land; therefore, activities in the southwestern portion of the forest were restricted to silvicultural methods intended to accelerate the development of late-seral forest conditions, which can be suitable habitat for this species.

A major focus during management plan development was identifying the desired future forest conditions in terms of forest structure, seral stages, and wildlife habitat values. The uncertainty surrounding future management situations and research priorities, and limitations of the science, in particular on wildlife values, led to general consensus on creating and maintaining as wide a range of forest structure conditions as possible, so as not to foreclose on future management and research options.

Wildlife in the Northern California Ecological Sub-Region, which includes the JDSF, is relatively diverse, although few species are endemic (occurring nowhere else) to the region. With the exception of heavily studied species such as the northern spotted owl, there is only limited information on the role of forest composition and forest patch, or stand, juxtaposition on population dynamics. Maintaining a diverse forested mosaic that helps support the many species in the region is beneficial for both forest management and research. Habitat protection and restoration of relatively rare habitat types is also an important element of forest management. Not all large trees have the same value to wildlife; nor does age alone determine a tree's value to wildlife. Many slow-growing redwood trees retained after logging activity in the 1880s and early 1900s are no longer shaded by neighbors and now grow vigorously, appearing to be younger second-growth trees. When the core of these trees is examined, the actual age would date the tree back to pre-European settlement. These trees lack the structural elements, large limbs, and cavities that other old trees possess, and they have wildlife value equivalent to second-growth trees. Because of the favorable climate and soils of the region and history of harvest, some young trees are now larger than trees that are indeed old. In general, as time progresses, tree size will vary more with the resources (sunlight, nutrients, water) available for growth rather than the number of growth years. Because of redwood's potential to reach great size and age, assessing age only by size is likely to be inaccurate.

With the SCAs identified, a plan was formulated to maintain or enhance ecological functions in all areas, to create diverse forest types, to produce high levels of sustainable timber growth, and to create the diverse range of forest structures from early- to late-successional stages (Figure 21.2). The plan required the development of a high-quality research and demonstration program. The forest was divided into management areas where the boundaries approximated watershed boundaries. The range of silvicultural methods permitted within each SCA were then identified. Those watershed areas not covered by SCAs were designated to receive a larger range of potential management regimes. Some watershed areas will be selectively harvested, while others will incorporate a component of even-aged management dispersed in time and space to maintain a variety of forested habitats. Still, other watershed areas may be left unmanaged for short or long periods of time to act as controls for experiments.

A long-term harvest schedule with a planning interval of 100 years was developed to estimate the sustainable annual harvest level. Decision variables in the harvest schedule included vegetation, site class, land types (including the SCAs), silvicultural methods, and time of harvest entries. Following the completion of the long-term plan, a short-term harvest plan was formulated for the next 5 years with the objectives of identifying feasible harvest areas and volumes. This short-term harvest plan was spatially explicit and accounted for timber sales already in the planning stages, the road management plan, the complete set of constraints noted above, and other logistical constraints.

Guiding Laws, Regulations, and Policies

This management plan meets the requirements of the California forest practice rules (California Department of Forestry and Fire Protection, 2013). These rules are promulgated by the Board under the 1973 Z'berg-Nejedly Forest Practice Act. A companion document to the management plan, the JDSF "Option A" plan, the sustained yield plan requirements in the Forest Practices Act, meets the regulatory requirements for estimating sustainable harvest levels on the JDSF under Section 913.11(a) of the forest practice rules. Timber harvest plans, written by a California Registered Professional Forester, are required for individual timber sales developed under this management plan. The Department has the statutory authority to approve all timber harvest plans. The JDSF timber harvest plans are approved by the Department's forest practice program by Department employees who have had no involvement in writing them.

In addition to protection measures in the Forest Practice Rules, the California Porter-Cologne Water Quality Control Act and the federal Clean Water Act control management activities to protect water quality. The California Endangered Species Act and the Federal Endangered Species Act (16 U.S.C. 1531–1544) control management activities to protect wildlife and habitat.

The Forest's management direction derives from state statutes and from policies set forth by the California Board of Forestry and Fire Protection. The Public Resources Code specifies the overarching direction for the JDSF of research into economical forest management, timber production, and recreation. Board policy describes the JDSF as *commercial timber-land areas managed by professional foresters who conduct programs in timber management, recreation, demonstration, and investigation in conformance with detailed management plans.*

More specifically, Board policy states that the primary purpose of the JDSF is to conduct innovative demonstrations, experiments, and education in forest management; that timber production will be the primary land use; and that recreation is recognized as a secondary but compatible land use on the JDSF. Further noteworthy Board policy directions include: research and demonstration projects shall include silviculture, mensuration, logging methods, economics, hydrology, protection, and recreation; and research and demonstration projects shall be directed to the needs of the general public, small forest landowners, timber operators, and the timber industry.

The Department is in the process of obtaining Forest Stewardship Council certification for the JDSF. Many of the mills in the area are certified by the Forest Stewardship Council, thus there is an incentive for the JDSF to obtain this certification. Additionally, many of the certification requirements are already met by the California Forest Practice Rules and the California Environmental Quality Act public comment requirements.

OUTCOMES OF THE PLAN

Planned harvest actions are designed to achieve desired forest structural conditions in addition to achieving growth and revenue goals. Under this management plan, standing timber volumes per unit area will continue to build over time (Figure 21.3), while providing a significant contribution to the local economy through the harvest and processing of timber. The average annual harvest levels during the next decade are estimated to be slightly more than 25 MMBF per year. This harvest level represents approximately half the total annual growth increment, or about 1% of the total inventory on an annual basis.

FIGURE 21.2 Berry Gulch, on the Jackson Demonstration State Forest.

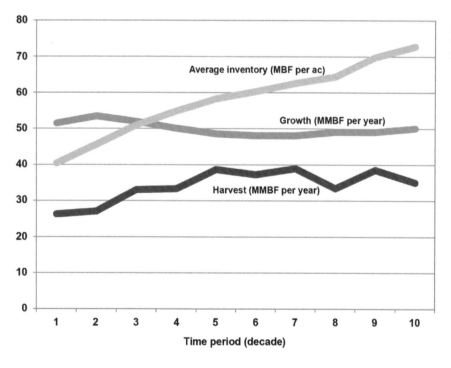

FIGURE 21.3 Inventory, growth, and harvest over time for the coniferous resource at the Jackson Demonstration State Forest.

The Forest Learning Center located at the JDSF was completed in 2009, with partial funding from the U.S. Department of Agriculture Forest Service. It is a 2,000 ft^2 (186 m^2) facility strategically located in the middle of the Forest with easy access close to a public highway. It provides a meeting location, and it is a convenient facility for researchers to stay in while conducting research at the JDSF.

The JDSF management plan establishes desired future conditions or targets for management. The central goal is to create and maintain a diverse range of forest structure conditions, a set of forest structures that to the maximum extent possible preserves options for future research and management. The management plan provides long-term goals for the establishment of particular forest structures over time that include the following:

- Late-seral or old-growth forests: 15–25%
- Older forest structure: 10–20%
- Forests with mature or large trees: 5–15%
- Mixed ages and sizes of forests: 30–40%
- Regeneration and forests with pole-sized younger trees: 10–20%
- No specific structure assigned: 0–10%

The structure goals allow a 5% deviation on either side of these figures, recognizing the long time it will take to reach these targets and the inevitable changes in management situation and natural disturbances along the way. The desired future conditions represent a dynamic matrix of seral stages designed to move across the landscape over time. Any particular area that is not permanently protected from management can be harvested and a different area can take its place to represent a given seral stage or forest type.

Given the current low level of older forest in the redwood region, a significant portion of the structural goals are oriented toward accelerating the development of older forest structures. The management plan specifies sustained yield of high-quality sawtimber, emulating natural processes, and a broad diversity of forest structures and habitats. Silvicultural systems are just one of the management tools that will be used to help achieve these desired future conditions. The management plan emphasizes that restoration of natural ecosystems is a high priority. A range of watershed management measures are prescribed to reduce sediment inputs to streams and improve large woody debris inputs. The management plan includes an aggressive road management plan.

This management plan preserves all existing old-growth groves, augmenting most of them to provide large, contiguous areas of older forest habitat. It leverages existing late-seral and old-growth areas by connecting these areas in a contiguous corridor of forest with older forest structural characteristics that extends from the west to the east and the north to the south. Individual large old-growth trees and smaller residual old-growth trees with unique habitat attributes are protected. The management plan sets goals for increased retention of structural habitat elements such as snags, downed logs, and large green trees.

The allocation of silvicultural methods to be used to achieve the desired forest structure conditions (Table 21.1) suggests that even-aged management will be used as necessary to achieve the forests with mature and large trees, as well as the regeneration and pole-sized forests. Uneven-aged forest management practices will be used to facilitate the development of the other forest structure classes. Even-aged management also may be used to address forest health and problematic regeneration conditions, as well as for research and demonstration purposes.

TABLE 21.1 Planned Distribution of Silvicultural Methods on the Jackson Demonstration State Forest

Method	Land Area (ac)	Land Area (ha)	Proportion of Forest (%)
No harvest[a]	1,350	546	3
Late seral development and older forest structure prescriptions	15,801	6,395	33
Uneven-aged management[b]	8,933	3,615	18
Uneven-aged management[c]	7,325	2,964	15
Uneven-aged or even-aged management[d]	12,788	5,175	26
Unclassified[e]	2,455	994	5
Total	48,652	19,689	100

[a] *Old-growth groves, pygmy forest, cypress groups, conservation camps.*

[b] *Single tree or cluster (of trees) selection.*

[c] *Group selection or single tree/cluster selection.*

[d] *Single tree/cluster selection, group selection, variable retention, two-aged management, or one-aged management.*

[e] *Research areas (variable silvicultural treatments) and power line rights-of-way.*

In response to public concerns, limits were placed on the use of even-aged management. The total area receiving any form of even-aged silvicultural treatments will not exceed 2,700 ac (1,093 ha) per decade, or 5.5% of the JDSF area. Clear-cutting will be conducted only where necessary for the purposes of research, demonstration, addressing forest health, or addressing problematic conditions for regeneration.

Herbicide use in forestry is often the subject of public controversy in California. The management plan describes in detail the measures required to minimize the use of herbicides. There are four management situations where herbicides will be used at the JDSF:

1. In road maintenance treatment of native vegetation.
2. In reforestation treatments that target native shrubs.
3. In the control of hardwoods to adjust the conifer/hardwood stocking ratio.
4. In the control of invasive weed species as part of an Integrated Weed Management program.

A total ban on herbicide use would compromise the research and demonstration value of the JDSF and could result in adverse environmental consequences, such as expansion of the area, on and off of the forest, occupied by invasive species. Herbicides and other vegetation control methods may be used without restrictions in research and demonstration projects. In an operational context, herbicides will be used only when no other feasible control methods are available.

DISCUSSION AND CONCLUSIONS

California's Demonstration State Forests, and notably the JDSF, meet an important need to advance research and demonstration into sustainable forestry practices. This is particularly critical in helping address the pressures associated with a rapidly growing population, increasing demands on forest lands for recreation, forest products, environmental protection, and conversion of forest land to residential and other uses. Given the often controversial role of timber production in California, the JDSF fills an important role in helping create solutions to difficult and controversial environmental issues related to forest management.

The management plan has helped the Department to coordinate daily activities, to plan research activities, and to focus public discussions and involvement. This chapter could only describe the main elements of the JDSF management plan. More information about the JDSF, including the complete management plan and accompanying EIR, can be found using the Internet links provided at the end of the chapter.

Sustainability Issues

The management plan specifies that the JDSF will continue to move toward a late-seral forest condition. There is some concern about growing the forest into a condition where the resulting logs are too large for local sawmills to process. There are currently two sawmills within a reasonable hauling distance from the JDSF that are capable of processing logs greater than 48 in (122 cm) in diameter. An annual harvest level that is approximately half of annual growth on the forest may also result in challenges for the creation and distribution of forest structures envisioned. The management plan will be updated over time to achieve a trajectory toward the forest conditions specified above in the Outcomes section.

Plan Development Challenges

Developing the management plan took 6 years, exceeding the normal 5-year Board review interval. This led to logistical challenges as the regulatory and legal environment, and stakeholder representatives and interest, changed during the development of the plan. Developing the EIR took more time than developing the management plan itself, reflecting a climate of ongoing litigation during the development of the management plan and EIR.

The management plan was developed to reflect the best science pertaining to JDSF management. The science pertaining to many of the resources at the JDSF, in particular watersheds and threatened species, evolved during the development of the management plan, requiring re-analysis in some instances. In the case of the JDSF, characterized by rapid changes in public preferences, science, laws and regulations, there is clearly a premium on minimizing the amount of time spent on developing or updating a management plan.

Plan Implementation Challenges

Because of the many resource values associated with coastal redwood forests, the JDSF and its management is the subject of considerable public interest. In 2001, the Dharma Cloud Foundation filed a lawsuit against two timber sales at the JDSF, alleging that the Department's 1983 management plan for the JDSF was out-of-date and therefore invalid. This effectively halted all timber harvest operations on the JDSF along with its potential revenue. In November 2002, a new management plan and associated EIR for the JDSF was approved by the Board, and timber harvesting resumed. The Dharma Cloud Foundation along with Forests Forever then filed a new lawsuit against the Department in 2003, this time alleging that the EIR prepared for the management plan was inadequate, and sought an injunction on forest management activities. The judge ruled that the EIR should have been approved by the Board not the Director of the Department, the "regional setting" cumulative impacts were not sufficiently addressed, and the environmental setting was not adequately addressed. This lawsuit halted two active timber sales until the issues raised in the lawsuit were addressed and, again, effectively shut down all timber sales on the JDSF. The Department decided to rewrite the entire EIR, which was completed in November 2005. The EIR went out to public review for 45 days and public comments were addressed. A revised management plan was completed in 2007, addressing the judge's directives. In January of 2008, the Board approved this management plan, and both the management plan and EIR were then sent back to the judge to clear the court case via a settlement agreement. Timber harvests finally resumed on the JDSF in the summer of 2009.

As part of approving the management plan, the Board authorized the formation of the Jackson Advisory Group (JAG) and charged it with reviewing the management plan. The 12-member JAG consisted of a wide range of stakeholders, from the principal of the Dharma Cloud Foundation, the original litigant against the Department, to timber companies and loggers. The JAG presented its report, *A Vision for the Future*, to the Board in 2011 (Jackson Demonstration State Forest Advisory Group, 2011). The Board adopted the JAG landscape allocations and associated goals as well as the general direction for research and demonstration and adopted a research governance process for the JDSF. As a part of the research governance process, the Board charged the Department with developing a research plan for the JDSF. The Department is presently completing a research plan and incorporating the Board's findings into the management plan. The Board stated that the next planned full review of the management plan will occur in 2016 unless there are further changes that result from the research governance process. The set of documents consisting of the 2008 management plan, the JAG recommendations, and the research plan effectively constitute the management plan direction for the JDSF until the next scheduled Board review of the management plan in 2016.

ADDITIONAL READING AND RESOURCES

This chapter represents a synthesis of the 2008 forest management plan for the Jackson Demonstration State Forest in California. To view the plan itself, please visit this Internet site, which was available on July 16, 2014:

http://calfire.ca.gov/resource:mgt/resource:mgt_stateforests_jackson_mgtplan.php

The EIR for the Jackson Demonstration State Forest can be found by visiting this Internet site, which was also available on July 16, 2014:

http://calfire.ca.gov/resource:mgt/resource:mgt_stateforests_jackson_deir.php

If in the future the links to these sites appear broken, search the Internet using the title and the keywords provided.

REFERENCES

Botkin, D.B., 2006. Studying life from a planetary perspective: the character of nature. Proceedings of the Society of American Foresters 86th National Convention. Society of American Foresters, Bethesda, MD.

California Department of Fish and Wildlife, 2007. CNDDB Maps & Data. California Department of Fish and Wildlife, Sacramento, CA. http://www.dfg.ca.gov/biogeodata/cnddb/mapsanddata.asp (Accessed May 30, 2014).

California Department of Forestry and Fire Protection, 2013. California Forest Practice Rules. California Department of Forestry and Fire Protection Resource Management, Forest Practice Program, Sacramento, CA. 340 p.

California Department of Water Resources, 2013. California Data Exchange Center—Precipitation. California Department of Water Resources, California Data Exchange Center, Sacramento, CA. http://cdec.water.ca.gov/snow_rain.html (Accessed June 8, 2014).

Holland, R.F., 1986. Preliminary Descriptions of the Terrestrial Natural Communities of California. California Department of Fish and Game, Sacramento, CA. 156 p.

Jackson Demonstration State Forest Advisory Group, 2011. A vision for the future, The report of the Jackson Demonstration State Forest Advisory Group. California Board of Forestry and Fire Protection, California Department of Forestry and Fire Protection, Sacramento, CA. 127 p, http://www.calfire.ca.gov/resource:mgt/downloads/Final_JAG_Report_January2011.pdf (Accessed June 3, 2014).

Sawyer, J.O., Keeler-Wolf, T., 1995. A Manual of California Vegetation. California Native Plant Society, Sacramento, CA.

Chapter 22

Mission Municipal Forest, British Columbia, Canada

Don Reimer

D.R. Systems Inc., Nanaimo, British Columbia, Canada

ABBREVIATIONS

FSP	Forest Stewardship Plan
MoFLNR	Ministry of Forests, Lands and Natural Resource Operations
TFL	Tree Farm License
THLB	Timber-Harvesting Land Base

MANAGEMENT SETTING AND BACKGROUND

The Mission Municipal Forest is the oldest community forest in British Columbia. During the 1930s, many parcels of privately owned forest and rural lands reverted to the District of Mission (Mission) due to nonpayment of taxes. By 1945, there were about 1,200 hectares (ha) (2,965 acres (ac)) of this municipally owned land. In 1946, Mission, in response to citizen interest in having more say in management of local forest lands, proposed the establishment of a community forest composed of a blend of municipal lands (1,200 ha) and local, provincially owned forest land (7,800 ha (19,274 ac)). At the time, this proposal was rejected by the Province of British Columbia. In 1948, Mission established the Mission Municipal Forest Reserve, which was comprised of 1,076 ha (2,659 ac) of municipally owned land. Over the next 6 years, Mission continued to work with the Province on a way to establish a local community forest. In 1954, an agreement in principle was reached between the Province and the District of Mission, and in 1958 the Mission Municipal Forest was established as Tree Farm License (TFL) 26. In British Columbia, TFLs are a form of long-term, renewable tenure based on a combination of private forest land (or in this case municipal ownership) and provincially owned forest land. TFL licenses require the development and implementation of an approved long-term forest management plan that must address all resource objectives on a long-term sustainable basis. Every 5 years, a new updated plan, as well as documented performance relative to the objectives and requirements of the last plan, must be submitted for review to the Provincial Ministry of Forests, Lands and Natural Resource Operations (MoFLNR). The new plans must reflect the latest provincial and federal environmental and forest management regulations and must address any shortfalls that were not achieved during the last 5-year plan. Upon approval of the proposed forest management plan, the TFL license is renewed for another 25 years.

The Mission Municipal Forest is located in the northern part of the District of Mission, a municipality of about 36,000 people in the north-central Fraser Valley. The current total area of the municipal forest (Figure 22.1) is 10,538 ha (26,039 ac) of which 88% are provincially owned lands and 12% are municipally owned lands. The standing volume for the municipal forest, based on the forest inventory to January 1, 2000, was estimated at 2,580,391 cubic meters (m^3) (over 91 million cubic feet (ft^3)). The terrain (Figure 22.2) is variable with most of the area between 100 and 700 m (328 to 2,297 ft) above sea level. However, a section in the northwest corner reaches up to the highest point in Mission, known as Mt. Crickmer, at 1,356 m (4,449 ft) above sea level. Not all of the land within the Municipal Forest is forested; a small amount is non-forest land. Almost 7% is deemed nonproductive, and all land classes deemed not suitable for commercial forestry total 10.1%. In addition, a significant portion of the municipal land base is designated for land uses other than traditional commercial forestry. These classifications include lands set aside for habitat, biodiversity, and

Forest Plans of North America. http://dx.doi.org/10.1016/B978-0-12-799936-4.00022-9

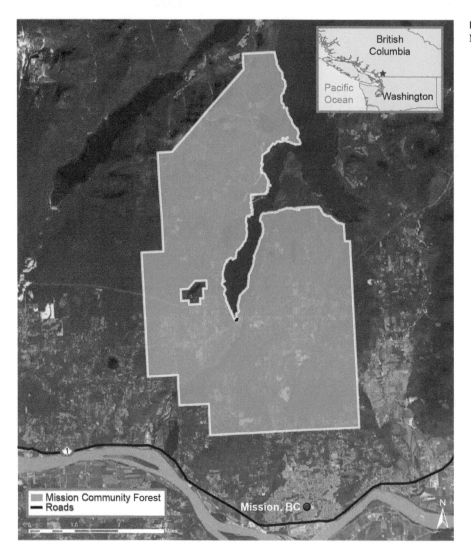

FIGURE 22.1 Location map of the Mission Municipal Forest.

FIGURE 22.2 Western and northern portions of the Mission Municipal Forest.

other ecological requirements as well as the area occupied by roads and trails. These include environmentally sensitive areas and riparian areas. The timber-harvesting land base (THLB) represents a little more than 69% of the forest. The THLB is composed of areas of the municipal forest that are suitable for commercial forest management, yet must also be managed carefully, meeting all applicable environmental, social, and cultural requirements. The breakdown of areas within the forest includes the following:

- Timber-harvest land base (THLB): 69.1%
- Nonproductive, noncommercial, inoperable, nonmerchantable: 10.1%
- Environmentally sensitive areas: 7.8%
- Deciduous leading: 6.9%
- Wildlife tree patches: 1.8%
- Riparian areas: 1.6%
- Roads and trails: 1.3%
- Specific geographic areas: 1.3%

Western hemlock (*Tsuga heterophylla*) and Pacific silver (balsam) fir (*Abies amabilis*) dominate the tree species on about 60% of the THLB, including 5% in high elevation western hemlock and subalpine fir (*Abies lasiocarpa*) areas, most of which is on medium and good productivity sites. Douglas-fir (*Pseudotsuga menziesii*) and western red cedar (*Thuja plicata*) cover 24% and 13% of the THLB, respectively and are also located on medium and good productivity sites. After harvest, 60% of the hemlock and balsam fir-dominated stands are expected to be converted to Douglas-fir and western red cedar. The remaining stands will be regenerated to the existing leading species mixes. The current age-class composition (Figure 22.3) of the total forested area and the THLB shows that the majority of forests are 80 years of age and younger. Forested areas within the municipal land base but not part of the THLB may contribute to landscape ecological and biodiversity targets but are not eligible for harvesting. Approximately 15% of the THLB is less than 20 years old and about 68% of the THLB is between 21 and 80 years old. Approximately 17% of the THLB is in stands older than 80 years. Only 1.45 % of the THLB is older than 250 years.

PLANNING ENVIRONMENT AND METHODOLOGY

There are two levels of planning required for nongovernmental organizations managing public forest lands in British Columbia. Since the Mission Municipal Forest includes provincially owned public land and municipally owned land, and is managed by the District of Mission under a TFL agreement with the Province, the Mission Municipal Forest is required to have an approved long-term (20-year or longer) forest management plan and an approved short-term (5-year) forest

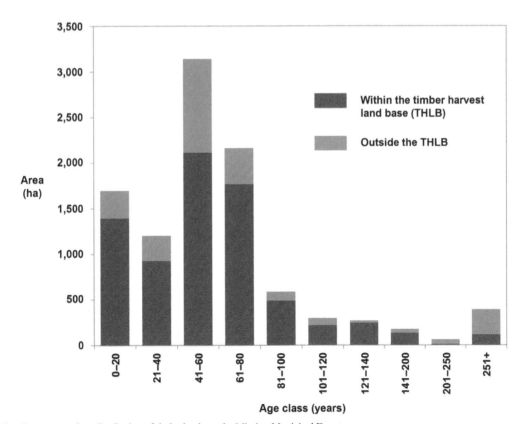

FIGURE 22.3   Current age-class distribution of timberlands on the Mission Municipal Forest.

stewardship plan. Currently, the Mission Municipal Forest is operating on Management Plan #9 (District of Mission, 2010), which was formally approved by the Provincial Ministry of Forests in 2010. The plan report is formally titled "Mission Tree Farm License 26, Mission Municipal Forest, Management Plan 9." Management Plan 9 utilizes the work done in developing the last 20-year plan for the Mission Municipal Forest. That work is described in the report titled "District of Mission, Mission Municipal Forest, Timber Supply Analysis and Twenty-year Plan, Tree Farm License #26," which was approved in 2001. Management Plan #9 was constructed within the management strategies developed and approved in the 20-year plan, taking into account developments and activities undertaken in managing the municipal forest since 2001 and any applicable changes in regulations or legislation.

As mentioned, in addition to an approved long-term Forest Management Plan, licensees are also required to develop a 5-year Forest Stewardship Plan (FSP). FSPs contain detailed statements on how the municipal forest will be operated within the framework established in the long-term 20-year plan. All FSP planning, budgeting, and scheduling of harvesting and silvicultural activities are developed, monitored, and reported on an annual basis to the MoFLNR using Phoenix^Pro, a software package used for tactical and daily management of forest lands developed by D.R. Systems Inc. (2008). All operational land management activities are implemented with reference to provincial and municipal objectives on biodiversity, old-growth management, management of riparian zones, silvicultural practices, management of wildlife habitat, management of community watersheds, visual quality objectives, management of cultural heritage sites, management of invasive plants, and management of recreation resources. FSPs are subject to public review and are also formally approved by the MoFLNR. Currently, the Mission Municipal Forest is operating on an FSP that was formally approved in 2007. The FSP was updated and amended in 2011 and approved in 2012 for an additional 5 years to 2017.

The forest is almost 100% second-growth timber and is actively being managed for a wide variety of objectives. One of the long-term objectives of the municipal forest management plan is to increase the percentage of the total land base in old forests over 200 years of age to meet the provincial-recommended percentage of forested land in old forests. Currently, this objective is set at 15%. Under present forest management strategies this objective will be achieved over the next 120 years as older stands are reserved from harvest, allowing them to grow into the old age classes.

The Forest Simulation and Optimization System was used as the timber supply model (Forest Ecosystem Solutions Ltd., 2006). The long-term harvest levels, which can be sustained by the current forest management strategies and silvicultural practices, will continue to increase over the next 165 years as the current, naturally regenerated second-growth forest is harvested and replaced by managed stands. Currently, the harvest level is set at 45,000 m³ (1.59 million ft³) per year, but is expected to rise to about 50,000 m³ (1.76 million ft³) in the long-term.

The objectives for the analyzed area were to develop and implement a balanced, sustainable approach to management of the municipal forest that meets all provincial regulatory and environmental objectives while providing local employment, local sources for forest products, and revenue to the municipality. In addition, the management plan would be developed to ensure that environmental benefits, including water, air quality, fish and wildlife habitat, biodiversity, and natural beauty would be accommodated, as would cultural and spiritual values for society, religions, and ethnic groups, and societal values that include recreation, education, employment, and the local and regional use of forest products.

There are a number of constraints affecting forest management and land use within the municipal forest. The municipal forest contains a number of smaller lakes and fish-bearing streams, some of which are the source water supply for domestic consumption. It borders, in part, on Stave and Hayward Lakes, both of which are relatively large popular recreation areas. A number of the smaller lakes are also very popular recreation sites. All of the riparian areas require buffer zones and special management considerations. Approximately 13% of the Mission Municipal Forest land base is in riparian and special management zones. The Mission Municipal Forest is also unique in that a large number of private residential and rural properties exist within the municipal forest boundary. As a result, the municipal forest shares a common boundary with many different private landowners. The municipal forest staff work very closely with all landowners to develop suitable forest management practices and land and road uses that are compatible with its neighbors.

The municipal forest is one of the closest public forest areas to Vancouver and the lower mainland population centers. As a result, the municipal forest has a very high number of general recreation users. In recent years, the forest has been hosting more than 1.4 million visitor-days per year. This requires special attention to access, forest protection, and scheduling of forest management activities relative to silviculture and harvest planning in order to minimize conflicts between forest management objectives and general recreation interests of the visitors. Finally, Provincial requirements for biodiversity, wildlife habitat, and old-growth management areas must be accommodated in all silvicultural and harvest planning and activities. At present, approximately 5% of the Mission Municipal Forest is in old-growth reserves. The long-term objective is to raise this to 15% by growing second-growth forests into these older age classes over time.

Guiding Laws, Regulations, and Policies

The specific requirements for TFL forest management plans and forest stewardship plans are contained in Provincial legislation. The legislation prescribes content requirements, including planning and reporting requirements, submission timing as well as recommended plan development, and public consultation processes to be used for developing both the long-term forest management plans and the 5-year forest stewardship plans. All plans must follow the legislated requirements and procedures in order to qualify for review and approval. For the Mission Municipal Forest, Management Plan #9 includes the following content:

- A general description of TFL 26 including size, ownership between the municipality and the province, description of the location, and a map of the TFL
- A brief history of TFL 26 from inception to current status
- The title and description of each of the publicly available planning documents used to guide the forest management operations in the TFL
- A timber supply analysis to establish a long-term sustainable harvest level and associated forest management strategy in support of the long-term harvest level
- The supporting documents for the timber supply analysis
- A description of the steps used to conduct the public review of Forest Management Plan #9
- A summary of the public comments received from the public review
- Appendices that include details referenced in the plan

The Mission Municipal Forest is licensed and managed under a British Columbia Provincial TFL tenure as that was the only tenure deemed suitable for managing a community forest when the Mission Municipal Forest was established (1958). The current TFL tenure requirements of the licensee, as well as those of the MoFLNR, are part of the British Columbia Forest Act of 1978. The activities and responsibilities associated with managing a TFL are described in the British Columbia Forest and Range Practices Act of 2002, and the requirements for planning and reporting are contained in the Tree Farm License Management Plan Regulation (B.C. Reg. 280/2009).

In 1998, the Province introduced a new tenure and legislation for community forests. The new tenures are called Community Forest Agreements. Licenses are issued for a minimum of 25 years and are renewable on a 10-year cycle. As of April 2010, 45 new community forests have been gradually established under this new legislation. Older community forests, such as the Mission Municipal Forest, now have a choice as to whether they wish to stay with the TFL tenure system or switch to the newer Community Forest Agreement tenure. The Mission Municipal Forest is reviewing the pros and cons of moving to this alternative tenure system.

The Mission Municipal Forest is certified under ISO14001, and the District of Mission Forestry Department is certified by the BC Forestry Safety Council as a certified SAFE organization. The municipal forest has not yet sought sustainable forest management certification as such certification is not required by either provincial or local regulation. However, the intent is to apply and obtain such certification when markets or environmental issues arise that require such certification.

Organizational policies include environmental policies that:

- Maintain the protection of riparian areas and sensitive sites
- Use variable retention harvesting techniques
- Maintain or enhance biodiversity through harvesting and silvicultural practices
- Maintain or enhance wildlife habitats and habitat capabilities
- Prohibit use of herbicides
- Prohibit broadcast slash burning (air quality standards)

Cultural/social organizational policies include policies that:

- Maintain a steady level of forest management activities to provide a steady level of jobs and revenues for the community
- Protect culturally sensitive sites
- Maintain or expand recreation opportunities as funds, land base capabilities, and public use allow
- Maintain communications and working relationships with adjacent landowners, residents, local First Nations, and stakeholders for all forest management activities, including development of management plans

Finally, the Mission Municipal Forest has several financial policies:

- Be a self-funding department with surplus revenue available for the community
- Maintain a financial reserve (to a recommended minimum level of $1,000,000 Canadian dollars (CAD)) to continue all essential forestry operations, and the associated employment, in poor economic times without burdening the taxpayer
- Maximize revenues over a 5-year cut-control period versus trying to maximize revenues on a yearly basis, and, in doing so, recognize the effects of business cycles on log prices and enable the municipal forest to take advantage of market opportunities

OUTCOMES OF THE PLAN

The proposed forest management plan was reviewed in detail by the MoFLNR and by the office of the Chief Forester of British Columbia. Based on those reviews, the Deputy Chief Forester makes the decision on approval of the proposed forest management plan and sets the maximum harvest level (Annual Allowable Cut) to be used for the 10-year term of the plan. In addition, the Deputy Chief Forester also can make recommendations on actions that the Mission Municipal Forest should undertake during the new plan period to address issues that the Deputy Chief Forester deems important to the long-term success of the community forest.

The following decision and recommendations were made by the Deputy Chief Forester with reference to the proposed Forest Management Plan #9 and Tree Farm License 26 (British Columbia Ministry of Forests and Range, 2010):

1. The long-term harvest level remained the same (45,000 m³ (1.59 million ft³) per year) as was determined for the last management plan, reflecting the ongoing success of the overall forest land management program the Mission Municipal Forest staff have planned and implemented and the fact that estimated forest growth on the municipal forest exceeds 55,000 m³ (1.94 million ft³) per year.
2. The above decision was based on the recognition that the forest management program and associated inventory and silvicultural data have been consistent with few uncertainties or changes in assumptions since the last major analysis and plan was developed.
3. The Deputy Chief Forester recommended the municipal forest undertake the following tasks over the next 10 years to help reduce any risks and uncertainties with respect to future estimates of timber supply and the attendant supporting forest management program and practices:
   a. Update the inventory for the TFL, including information on environmentally sensitive areas, and information regarding the use of improved seed and associated genetic gain adjustments.
   b. Work with the Provincial Ministry's Forest Analysis and Inventory Branch staff to determine an appropriate and feasible methodology to incorporate improved site productivity information.
   c. Review information regarding cultural heritage resources, archaeological sites, and wildlife habitat resources to ensure their protection and management is reflected in the timber supply analysis.
   d. Work with the Provincial Integrated Land Management Bureau to spatially locate and establish old-growth management areas.
   e. Review and update maps of established wildlife tree patches.

In a previous analysis (2001), the Deputy Chief Forester identified several factors that could result in an increased sustainable harvest level. The factors identified included the harvest of deciduous forest stands, improved estimates of site productivity, improved harvest utilization standards, improved estimates of landscape-level biodiversity requirements, and more flexible management of domestic water supply watersheds. The municipal forest staff confirmed that these factors are relevant, and they are planning or implementing work on most of the recommendations. They have developed harvest plans for limited harvesting in deciduous stands as markets allow; they are planning on implementing a study of site productivity along with updating their forest inventory within the next 2 to 3 years; and harvest utilization standards have been changed to enable utilization of smaller previously unmerchantable volumes for engineered wood products and biomass. There are no immediate plans for harvesting in riparian zones or in the one watershed that provides domestic water supplies.

DISCUSSION AND CONCLUSIONS

Learnings and Insights

Managing the municipal forest from a long-term sustainability viewpoint has been of significant benefit to the local community as well to the regional economy. Establishment and maintenance of a Reserve Fund is a critically important component of the Mission Municipal Forest long-term management strategy. The reserve fund has enabled the municipal forest to maintain a relatively uniform level of forest management and, therefore, jobs and log flows during times of economic recession without drawing any funds from municipal coffers. Consistent management opportunities have in turn enabled longer term investment strategies in forest management and infrastructure. The following list contains examples of the major funding provided to the local community from municipal forest surplus revenues (all CAD) in recent years:

- $685,000 toward the community library and archives
- $132,000 toward Firehall #3 and fire truck
- $170,000 toward ice rink conversion
- $1,200,000 for municipal budget stabilization fund
- $66,000 yearly for local and regional arts and culture grants
- Contribution of $250,000 (approximately $50,000 per year over the last 5 years) to the operating funds of other municipal departments

The following quote describes the public appreciation for the municipal forest:

"The Mission Municipal Forest helps support the educational, recreational, safety, social, cultural and economic heart of the community ... an excellent example of community forestry."

The establishment and maintenance of a Reserve Fund has been a major factor in the long-term success of the municipal forest, including its recognition by the community of its value. During times of economic recession, having a Reserve Fund to use for maintaining the forest management and silvicultural programs of the municipal forest without any financial assistance from the District of Mission has been critical to those programs. The Reserve Fund has enabled the municipal forest to provide steady jobs for local residents as well as to maintain a reasonable level of log supplies to regional sawmills and processing plants without compromising the long-term management plan for the municipal forest and without drawing any monies from municipal coffers.

Managing harvest levels over a 5-year cut-control period has given the municipal forest some added flexibility in taking advantage of business cycles by harvesting more volume when log prices are higher in economic boom times and lowering harvest volumes in times of economic recession. A significant percentage of the area and volume in the municipal forest is western hemlock—a species that traditionally has lower value—especially in times of poor markets. The ability to adjust harvests, both in total volume harvested as well as across species, and to take advantage of good economic times has been very useful in maximizing revenues from hemlock log sales.

One of the major outreach successes of the municipal forest has been its educational program. This program focuses a significant amount of effort on educational field tours for grade school and high school students and their teachers in the District of Mission. In addition, the municipal forest hires summer students to assist in operational forest management activities as well as in promotion and delivery of the school-based education program and recreational opportunities in the municipal forest. As a result, over the last 30-plus years, an outstanding level of awareness and appreciation of the values of a working community forest to the residents of Mission has been achieved.

Sustainability Issues

There are no known sustainability or biodiversity issues on the municipal forest. The municipal forest has been managed for more than 55 years (since 1958) using a long-term sustainability viewpoint.

Plan Development Challenges

There are a few challenges that affect long-term planning in British Columbia. These factors also could potentially affect the planning and strategies for managing the Mission Municipal Forest. First, there is some uncertainty with respect to the land base covered in the planning process due to First Nations land claims. Aboriginal groups have not settled their land claims and ownership issues with the Province of British Columbia. As a result, First Nations retain the right to dispute developments and land-use activities on Provincially owned lands. Since 88% of the Mission Municipal Forest is Provincially owned, there is some degree of risk with respect to tenure rights and the opportunity to continue to manage those lands as part of the municipal forest. There is a provincial land claims settlement process in place. However, progress in settling land claims has been very slow, and resolution of the issue is considered to be a number of years into the future. In the meantime, the management of the Mission Municipal Forest continues as required by legal and regulatory statutes related to the management of a TFL, which is the tenure type of the Mission Municipal Forest.

In addition, over the last 55 years, the municipal forest has had reasonably consistent support from the municipality's elected officials. However, as the population continues to grow, it is clear that elected officials are becoming less well-informed about the benefits of the municipal forest. As a result, the Mission Municipal Forest management team has begun work toward creating an independent enterprise company that is owned by the municipality but has an independent board of directors. While the municipality's elected officials will have representation on the Board of Directors, they will not have control, so that the municipal forest staff can act in the best interest of the municipality and the resource base over the long-term, in light of the political environment of short-term elected local municipal officials.

Provincial politics are also a concern. The province is responsible for handling consultation with First Nations as well as negotiations with First Nations land claims. While consultations with respect to the Mission Municipal Forest on forest management plans have been relatively smooth, the settlement of land claims is moving very slowly. This continues an air of uncertainty over long-term planning. In addition, during the recent economic recession, provincial resource priorities shifted from timber to energy. As a result, the majority of provincial attention is focused on energy projects. This has subtle secondary effects on factors such as technical support from the MoFLNR on forest research, forest inventory, and silviculture.

Plan Implementation Challenges

One interesting issue related to plan implementation is the large number of private ownerships within the municipal forest boundaries. There is a wide range of ownership types from small housing developments to larger single-owner plots and parcels owned by companies and organizations for uses from gravel quarries to rod and gun clubs. Owners have a wide range of concerns with respect to forest management and road access. The municipal forest staff spends a great deal of time communicating with the various landowners with respect to all forest management activities, including road construction, road use during harvesting, and silviculture activities. The evidence of success in this regard is a very high level of cooperation and respect from private landowners as evidenced by the lack of negative comments on management and forest stewardship plans.

Increasing and more costly regulation versus that of competing timber markets is another plan implementation challenge. British Columbia has extensive environmental and regulatory requirements for managing forest lands, both public and private. British Columbia restricts higher percentages of productive forest land from commercial forest harvesting activities than many competing jurisdictions. In addition, forest tenure holders must pay all of their own costs of road and bridge construction and maintenance as well as all silviculture costs related to planting and managing harvested lands until such time as the harvested lands are considered successfully restocked and the new stands of trees are free to grow with respect to minimal competing vegetation. As a result, forest land management costs are higher in British Columbia than most other places in North America and the world.

Finally, the growing urbanization of Mission and the lower mainland (Vancouver, British Columbia and surrounding cities) causes some concern with respect to implementation of the forest management plan. Vancouver and the lower British Columbia mainland are one of the most desirable places to live in the world. As a result, population in the lower mainland area is growing quite rapidly. British Columbia advertises the province as having a clean environment with well-managed cities and world-class forest lands as part of super-natural British Columbia. The Mission Municipal Forest is the first readily accessible public land on the eastern side of the lower mainland. As a result, the municipal forest has a rapidly growing number of visitors from the large urban and city environments in the Vancouver area. In addition, the costs and requirements related to accommodating the increasing numbers of visitors is increasing the costs of managing the municipal forest, both with respect to providing additional recreation opportunities as well as in clean-up and maintenance of the accessible forest areas. Many of these visitors have little or no appreciation of what is entailed in managing and maintaining a working forest that also has high recreation and aesthetic values. Visitors appreciate the recreation and aesthetic opportunities, but they do not recognize that the land base must also generate the necessary revenues to pay for management and maintenance of the entire landscape. As the percentage of visitors and potential residents from the bigger urban and city environments increases, the political ability to maintain a balanced, triple-bottom-line approach to managing the municipal forest (where the municipal forest is being managed to sustainably meet all environmental, social, cultural, and economic objectives) will become more difficult.

ADDITIONAL READING AND RESOURCES

This chapter represents a synthesis of the 2010 Management Plan for the Mission Municipal Forest in British Columbia. To view the plan materials, please visit this Internet site, which was available on June 30, 2014:

http://www.mission.ca/wp-content/uploads/MP-9-for-TFL-26-October-18-2010-Approved-Version.pdf

If in the future the link to the site appears broken, search the Internet using the title of the plan and the keywords provided.

REFERENCES

British Columbia Ministry of Forests and Range, 2010. Tree Farm License 26 held by the Corporation of the District of Mission, rationale for allowable annual cut (AAC) determination. British Columbia Ministry of Forests and Range, Victoria, BC. http://www.for.gov.bc.ca/hts/tfl/tfl26/tsr3/26tf10ra. pdf (Accessed June 30, 2014).

District of Mission, 2010. Mission Tree Farm License No. 26. (Mission Municipal Forest). Management Plan 9. District of Mission, Mission, BC. http:// www.mission.ca/wp-content/uploads/MP-9-for-TFL-26-October-18-2010-Approved-Version.pdf (Accessed June 30, 2014).

D.R. Systems Inc, 2008. Phoenix Professional, Forest Activity and Obligation Tracking System. D.R. Systems Inc, Nanaimo, BC. http://www.drsystemsinc. com/desktop_solutions.php (Accessed June 30, 2014).

Forest Ecosystem Solutions Ltd, 2006. Forest Simulation and Optimization System (FSOS). Forest Ecosystem Solutions Ltd, North Vancouver, BC. http://65.61.221.40/technology_fsos.html (Accessed June 30, 2014).

Chapter 23

San Pedro El Alto Community Forest, Oaxaca, Mexico

Héctor M. de los Santos-Posadas,[1] J. René Valdez-Lazalde[1] and Juan Manuel Torres-Rojo[2]

[1]Colegio de Postgraduados, Postgrado en Ciencias Forestales, Montecillo, Estado de México, Mexico, [2]Centro de Investigación y Docencia Económicas, Col. Lomas de Santa Fe, México DF, Mexico

ABBREVIATIONS

COFF	Community-owned Forest Firm
FCE	Forest Community Enterprise
SICODESI	System for Conservation and Forest Development
SPF	San Pedro Forest

MANAGEMENT SETTING AND BACKGROUND

Mexico has close to 65 million hectares (ha) (over 160 million acres (ac)) of forests (Food and Agriculture Organization of the United Nations, 2010). Temperate forests cover 51% of this area and are dominated by pine and pine-oak mixes. Pine forests are composed of numerous species, as Mexico is a major center of biodiversity for pine species (Styles, 1993). Equally complex are the pine-oak forests, since there are more than 150 species of oaks in the country. The San Pedro Forest (SPF) is well known for integrating part of this forest diversity.

Near 61% of the forests in the country are under communal ownership, acquired either by ancient rights recognized and granted by the Mexican Government to indigenous communities (*comunidades*) or as part of the land grant concessions given to groups of people after the Mexican Revolution (*ejidos*). San Pedro El Alto is one indigenous community out of 2,340 forest ownerships of this type in the country. There are close to 17,800 communal forests (owned by *ejidos* or indigenous communities) in Mexico, with an average size close to 2,200 ha (5,436 ac). Temperate community forests have an average area close to 2,100 ha (5,189 ac) with a forest cover of about 63%. Nearly 43% of these communities own forest areas smaller than 1,000 ha (2,471 ac), making forest management and timber production activities inefficient compared with other parts of the world. These conditions make timber extraction, in most cases, a temporal or seasonal activity with low returns (Torres-Rojo and Magaña-Torres, 2006). Within this context, the SPF is an indigenous forest community that can be considered above the 90th percentile given its total size of 30,047 ha (74,246 ac) and forest cover of 28,868 ha (71,333 ac).

Community forestry accounts for almost 84% of the timber production in Mexico. However, only 4.4 million ha (about 10.87 million ac), or 6.5% of the total forest cover in the country have timber management as an objective (Robles-Berlanga, 2012). These lands have an average productivity estimated at 3.8 m^3 per ha (54.3 ft^3 per ac) per year and a forest stock close to 60 m^3 per ha (857 ft^3 per ac) for coniferous forests. Within this context, the forest area at San Pedro El Alto is relatively more productive compared to the national average, and is composed mostly of mature timber. The forest stock averages 290 m^3 per ha (4,145 ft^3 per ac), almost five times the national average for community forests. Hence, in terms of forest resources availability, San Pedro El Alto is not a typical community forest. It not only has a considerable extent of forest but it also is a highly profitable forest.

The SPF is a cold, temperate forest located in southwestern Mexico, in the state of Oaxaca, in the municipality of Zimatlán de Álvarez (Figure 23.1). The geographical coordinates are approximately 16°35'13" to 16°50'18" north in latitude, and 97°00'56" to 97°12'22" west in longitude. It forms part of the Sierra Madre del Sur forest (Southern Highlands

Forest Plans of North America. http://dx.doi.org/10.1016/B978-0-12-799936-4.00023-0

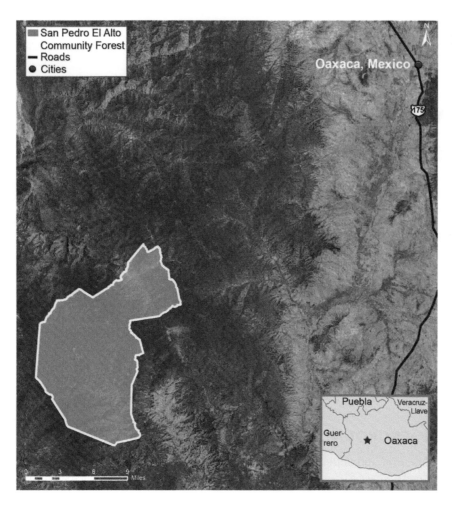

San Pedro El Alto Community Forest
— Roads
● Cities

Oaxaca, Mexico

Puebla | Veracruz-Llave
Guer-rero | ★ Oaxaca

FIGURE 23.1 Location of the San Pedro Forest in the Mexican state of Oaxaca.

of Oaxaca, or Sierra Sur de Oaxaca) and has an area of 30,047 ha (74,246 ac). Three generic types of vegetation are recognized in the SPF: pine forests (19,567 ha or 48,350 ac), pine-oak forests (1,169 ha or 2,889 ac), and deciduous oak forests (8,991 ha or 22,216 ac). The pine forest is composed mostly of mixed pine stands that include:

- Smooth bark Mexican pine (*Pinus pseudostrobus*)
- Patula pine (*Pinus patula*)
- Montezuma pine (*Pinus montezumae*)
- Mexican white pine (*Pinus ayacahuite*)
- Douglas pine (*Pinus douglasiana*)
- Mexican yellow pine (*Pinus oocarpa*)
- Oaxaca pine (*Pinus oaxacana*)
- Michoacan pine (*Pinus devoniana*)
- Pringle's pine (*Pinus pringleii*)
- Ocote pine (*Pinus teocote*)
- Chihuahua pine (*Pinus leiophylla*)

Also present is an endemic species, Oaxaca fir (*Abies oaxacana*). The pine-oak forests consist of a mixture of the pine species mentioned above and several species of oaks (*Quercus* spp.), some of which are not fully identified. The preferred regeneration method used is seed-tree silviculture with occasional planting to overcome the lack of naturally established seedlings 5 years after a regeneration harvest in most stands.

Nearly the entire property (28,868 ha, 71,333 ac) is wooded with pine and oak species (Figure 23.2) with an age distribution of mostly mature (50–80 years), surplus (older than the desired rotation age) forest (Figure 23.3). The forest areas with no timber production (3,578 ha, 8,442 ac) have been allocated a reserved or conservation status during the current

FIGURE 23.2   A pine-oak forest in the San Pedro Forest.

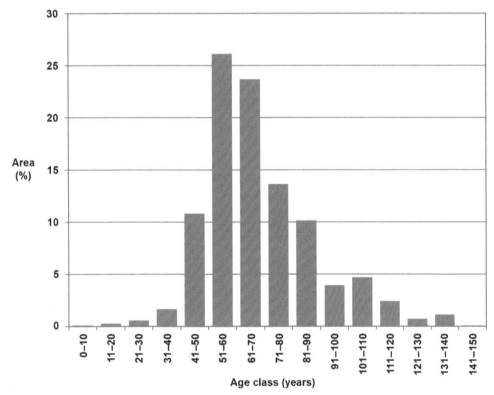

FIGURE 23.3   Age-class distribution of the San Pedro Forest.

10-year planning period. The rest of the land has agricultural or urban use (Table 23.1). Most of the commercial harvesting concentrates on pine species: 80% of the harvest volume for lumber production, 10% for pulpwood or chip-n-saw products, and the remaining 10% is considered noncommercial tree tops. For the oak species, these proportions are 60% for lumber production, 30% for pulpwood, and 10% for noncommercial tree tops.

Recent history of timber harvest in the SPF is similar to what has occurred in most of the forests of Mexico. Until the early 1950s timber and non-timber forest resources were primarily for domestic uses such as housing, firewood, and charcoal (Secretaría del Medio Ambiente, Recursos Naturales y Pesca, 2000). In 1954, the SPF was granted in concession for 28 years by the Government of Mexico to produce timber for a private enterprise called Forestry Company of Oaxaca S. de R.L. During the concession period the "comuneros" (community owners) received an income, called "derecho de monte" (stumpage). In addition, the community owners were employed by the enterprise to perform the forestry field work. Decisions concerning forestry and other aspects of forest management, including the opening of primary and secondary roads, were defined by the concessionaire firm. At the end of the concession in 1982, the "comuneros" of the SPF fought before the federal authorities for the right to perform the planning and management of the SPF, constituting a legal entity called *Specialized Economic Unit for Using the San Pedro El Alto Communal Forest*.

TABLE 23.1 Land Classification/Productive Activity in the San Pedro Forest

Land Classification	Area (ac)	Area (ha)	Area (%)
Production	57,082.13	23,100.82	76.88
Protection	8,842.05	3,578.33	11.91
Roads	512.26	207.31	0.69
Agriculture	1,633.36	661.01	2.20
Induced grassland	496.15	200.79	0.67
Plantations	316.36	128.03	0.43
Restoration	9.41	3.81	0.01
Urban	219.52	88.84	0.29
Area under dispute	5,085.10	2,057.91	6.85
Other uses	49.79	20.15	0.07
Total	74,246.13	30,047.00	100.00

Currently, the San Pedro El Alto community is a highly evolved one, if compared to average communities in Mexico. The level of organization and silvicultural practices applied were sufficient to obtain a certification of forest management practices from the Forest Stewardship Certification (FSC). The driving force responsible for this leadership role is the community-owned forest firm (COFF) that has catalyzed self-organization, or the development of social capital and capacity building among community members. The COFF is responsible for supervising all of the activities along the timber production process. In addition, the COFF fosters the participation of community members in many parts of the forest production process such as protection, conservation, and restoration activities, connecting people to the environment where local and basic ecosystem management principles are easier to understand and apply. This level of involvement has made members more knowledgeable about the ecosystems and the sustainable management options for their lands. Additionally, such level of organization has allowed the community to build a proper governance structure, where principles of equity, transparency, and accountability govern the community and the performance of the community leaders. San Pedro El Alto is a good example of organization at community level. The community has organized itself not only for timber management operations and vertical integration of the lumber production chain, but also for the development of many other economic activities, from mining and agriculture to the involvement of women in other economic activities, such as sewing and handcrafting. Recently, a bottled-water plant has initiated operations to allow the community to produce and market its own brand.

PLANNING ENVIRONMENT AND METHODOLOGY

The SPF is managed under a common property-based governance system, which means that 380 "comuneros" (heads of family in a community of 1,890 people) are the co-owners of the forest, and they plan and decide all aspects regarding the use or exploitation of the natural resources existing on their property. The existence of the community has been recorded since the Colonial era (1600 A.D.), when it was formed by local indigenous inhabitants identified as Zapotecos. During the Colonial era, they were granted property rights to the land, and in 1954 the Mexican government, through a presidential resolution, granted them legal possession of their land.

It was not until 1990, when the community with support of Mexicans and Finnish foresters, developed the first integrated forest management plan for the SPF with a time horizon of 5 years. In 1995, the community developed a second 10-year forest management plan, which was subsequently approved by the federal government. Currently, a third forest management plan is in operation; it was approved for a 10-year period up to 2016, and includes logging and silvicultural activities scheduled on 5,862 ha (14,485 ac), generating an annual harvest of 50,000 m³ (1.77 million ft³) of roundwood. The three management plans implemented since 1995 have focused basically on timber harvesting, but are aiming to integrate the ecological and social context of the community.

The three management plans that were developed for the SPF in the last two decades, including the one being carried out currently, followed the methodology outlined in the *Method for Silvicultural Development* (MDS). The method is based on even-aged management principles. The rotation age is fixed (50–60 years), and two or three thinnings occur every 10 years before the final harvest. The regulation criterion is area-volume check, and diverse social and environmental constraints are introduced to define the allowable cut. MDS was developed to expand the use of even-aged methods and intensive silviculture in Mexico, particularly in highly productive communal (community/*ejido*) forest lands that were not harvested previously or were composed by surplus forests.

Mexican legal framework allows any silvicultural system and management planning method as long as it guarantees sustainability of forest resources. Despite this openness to use any criterion to define harvest flows, regulation principles, or silvicultural systems, most of the forest plans developed in the country follow one of three methods: MDS, the *Mexican Method of Forest Regulation* (MMOB), and *Mixed Strategies*. The MMOB system was developed in the 1940s, and it was originally a single-objective (timber production) method. It has undergone many changes since its development, but is still in use today in approximately 45% of the forest communities under management in Mexico. The MMOB was developed to recognize the complexity and diversity of Mexican forests and the lack of data needed to manage them (such as estimates of forest growth rates). Hence, it follows a very conservative approach using a *continuous forest cover* or irregular type of management strategy in which the total removal of the forest cover is not allowed.

The MDS system was developed in the early 1980s (Rosales-Salazar et al., 1982) as a single-objective (timber production) management method. It follows a Rotation Forestry (RF) strategy or regular system of management in which the extraction of most trees in the forest stand is allowed when they reach maturity at the established rotation age. MDS is used in 34% of the community forests under management in Mexico, and mainly in those areas with high productivity where more homogeneous, single genus (pines mainly) stands are found (Torres-Rojo and Magaña-Torres, 2006). The MDS system follows an area regulation approach and has defined type, timing, and intensity for thinnings during the rotation. By the end of the 1990s, MDS incorporated many of the principles of multiple-use management and was incorporated into a decision support system named *SICODESI*, for System for Conservation and Forest Development. In general, the allowable cut for the SPF is estimated by using the principles of either the MMOB or the MDS according to the characteristics and management objectives of each stand. Forest planning is required at tactical levels, and in most cases forest management plans have short-term planning horizons as defined by the law (10 years maximum), with very conservative allowable cut estimates. Strategies to increase harvest flows by accelerating the harvest of overmature stands or by optimizing harvest volumes are not usually well received by federal authorities unless *a good technical reason* supports those strategies.

Despite these nationwide and institutionally accepted timber management approaches, highly variable management schemes are being applied in the real life, not only because of ecological diversity but also because of the type and extent of forest ownership, institutional arrangements within the ownership to manage the forest, and the way harvesting and silvicultural operations are conducted.

Guiding Laws, Regulations, and Policies

The 27th Article of the Mexican Constitution gives the Mexican Nation the right to organize land property in three categories: private, communal, and *ejido*. Private ownership of forests is limited, however, to a maximum of 800 ha (1,977 ac), although private corporations are allowed to own up to 20 times the maximum size of the private ownership. This makes communal lands and *ejidos* the only legal, large forest-track ownerships in Mexico. The main regulatory law that establishes guidelines for timber and non-timber management in natural forests in Mexico is the General Law for Sustainable Forestry Development (Ley General de Desarrollo Forestal Sustentable). This law was first passed by the Congress in 1997 and represents the seventh forest law since 1926. It was last modified in 2003 to include new environmental guidelines, and as this chapter is being written, it is being modified again as part of the 2012–2018 Federal Government Development Plan.

In 2006, to address significant differences between government agencies regulating the General Law for Sustainable Forestry Development content and the foresters charged with developing the management plans, the Ministry of Environment and Natural Resources (Secretaría del Medio Ambiente y Recursos Naturales (SEMARNAT)), which has the responsibility to regulate and promote forestry in the country, issued a legal protocol (Norm) called the Official Mexican Norm NOM-152-SEMARNAT-2006. This Norm outlines detailed requirements for forest management plans submitted for approval by the Federal Government (SEMARNAT). The management plan for the San Pedro El Alto Community Forest was thus developed (Dirección Técnica Forestal "San Pedro el Alto," 2006).

It is fair to say that Mexico's forest lands are heavily regulated when owners follow the applicable laws and regulations in the management of their resources, but such approach also has some advantages. Having an approved forest management plan allows the owner to harvest and transport timber to the open market or industry of preference, while performing silviculture and the recommended environmental protection practices (such as wildlife conservation, soil conservation, watershed protection). Property owners with an approved forest management plan can also apply for federal or state funding to implement and improve silvicultural practices (such as reforestation, thinning, pruning, wildfire prevention, or pest control) and to receive compensation for providing some environmental services to society, mostly watershed protection (called "servicios hidrológicos"), even though some other services (wildlife conservation) are also eligible for compensation.

Community Organization

Community forests in Mexico are different from community forests around the world because the community members own and have real dominion and control over (have property rights for) the land they usufruct (have the right to use and enjoy), as stated by the Agrarian Law. If a forest community wishes to legally harvest timber, they usually form a Forest Community Enterprise (FCE), which is a private business owned by community members (or even other investors outside of the community) with rights over the land grant. These social firms resemble community forestry operations in other parts of the world because the decision-making process involves all the community members with property rights to the land, and decision making is performed over the whole land grant regardless of its current use. For the case of timber management, decisions such as harvesting rules, silvicultural systems, conservation criteria and areas, logging systems, and the use of resources are prepared according to technical principles and are presented by an accredited forester to the Community's General Assembly (equivalent to a Community Counsel, which in addition is the highest authority in the community) for discussion and approval. At this stage, it is possible to find a broad diversity in community's capabilities to get involved in the forest planning process, and such involvement depends on the community's ability to self-organize, its knowledge of forestry principles, and its environmental constraints. Usually, during the decision-making process, a forest management plan is modified as management objectives involve the improvement of the well-being of the community members, jobs, conservation of special places (e.g., religious or traditional sites), conservation or promotion of specific species (including noncommercial species), and the mixture and promotion of different economic activities (e.g., orchards, livestock, agriculture) with forestry. Hence, a FCE tends to have more constraints on the timber harvest levels, and usual timber management goals (such as achieving forest regulation, efficiency, maximization of profit, or a sustained yield of timber) might be less important than other community goals (Antinori and Bray, 2005). This level of involvement of the community members is highly variable among forest communities, with extremes ranging from the unique involvement of the community representative and COFF staff, up to the direct involvement of the whole community, not only in the process of discussion and approval of the forest management plan, but also in the discussion of all the production activities related to land use within the ownership.

The San Pedro El Alto forest community has a vertically integrated FCE. Their members get involved in all stages of the planning process and operational activities. Its indigenous status allows all community members to partake in the benefits derived from the use of their resources, reducing equity and moral hazard problems. The community has clear regulations for the use of resources in the whole ownership, ranging from the use of certain parts of the forest to collect non-timber forest products, to the timing and constraints for grazing on communal lands and the monitoring of livestock, and the actions necessary to minimize the impact of other land uses on forestry. The community has also introduced strict rules to involve community members in practices such as production of seedlings, tree plantings, thinnings, and road maintenance activities.

Decision Support System

The first version of a decision support system designed to assist the development of a forest management plan under MDS was released in 1986 under the name of SICODESI. The system was developed from a joint effort between the federal Mexican government and the Finnish Agency for Technological Development Aid. SICODESI facilitates the calculations required to incorporate multiple-use legal and technical requirements into forest management plans through the use of simulation techniques for valuing alternative scenarios. The distinct characteristic of this system is that it uses forest inventory data from all over Mexico and incorporates a set of forest growth and yield functions for specific regions and species. This facilitates the use of the system to a wide range of forest types and conditions across the country, providing reasonable results.

At the operational level, SICODESI follows a two-stage forest inventory. The first stage uses a two-plot sampling cluster with circular $300\,\text{m}^2$ (0.074 ac) plots (forming a $600\,\text{m}^2$ (0.148 ac) cluster) to sample most of the commercial and noncommercial stands. This stage is called the *Strategic Inventory,* and its results determine the total inventory and the allowable harvest for the whole forest. The original SICODESI system specified that at least 10% of the clusters should be used to monitor forest development and silviculture, but re-measurements of this sub-sample were never obtained.

In the second stage, an *Operational Inventory* is implemented to gather detailed data for harvest scheduling planning. Taken into account is the information of the previous forest plan, the information of the harvested areas in the current plan, and the data provided by the strategic inventory. Based on this information, the COFF suggests a harvest schedule and other silvicultural activities to be applied in the coming 10-year planning period. Then, the Community Counsel analyzes the alternatives and defines the active forest management area for the given 10-year cycle. Once the management area is defined, a relascope sampling system is performed for all management units on the approved area using a factor of $2\,m^2$ per ha ($8.7\,ft^2$ per ac). The average management unit covers 15 ha (37 ac) and there are more than 1,900 management units in the commercial forest alone.

During this process, the Community Counsel has the final say for the 10-year planning cycle. The Counsel decision is primarily oriented to preserve stocking, since the general perception is that the outcome of most silvicultural practices applied to date is still uncertain. In the last two cutting cycles, only 50% of the estimated allowable harvest has been approved by the Counsel. The general belief is that well-stocked to overstocked stands somehow guarantee the sustainability of forest harvest and provide market flexibility, since large diameter trees are becoming less abundant. This approach, however, constrains thinnings to be light in young and mature stands, which may require more intense harvests later to guarantee adequate composition and spacing.

Forest management planning for the SPF is one of the cases in Mexico that is being carried out under the SICODESI decision support system framework. As timber production is the main objective of the SPF, SICODESI was used to define:

1. The design and implementation of two forest inventories. One defined as strategic and applied across the whole forest tract, and the other called operational, that includes only the harvesting area defined by the community. Both have timber production as the main objectives.
2. The projection of timber volumes for six 5-year periods (30 years into the future) based on regression equations.
3. The management alternatives for each stand, based on their productive capacity and forest condition.
4. A harvest schedule consistent with the socioeconomic needs of the community and its workforce and operating capacity.
5. A program of silvicultural activities for the conservation, protection, and promotion of associated resources, including conservation areas, river buffers, and wildlife, among others.

As with many other automated support systems, SICODESI has some limitations. Because the software platform and underlying equations have had very few updates over the years, SICODESI has become somewhat obsolete. In several parts of Mexico, it is being replaced by privately owned forest management programs that are licensed to foresters and that also comply with the NOM-152 regulations.

OUTCOMES OF THE PLAN

The main economic activity of the owners of the SPF is forestry, which represents about 80% of the entire productive activity of the community. To organize these activities, the community has created a technical forestry staff, which is led by a community-native certified forester appointed chief technical director. One of the director's main tasks is to design a proposal for a *Forest Management Program* that must comply with federal and state legislation on the matter. The program must be submitted to the community assembly for approval. Later, the forest management program is submitted for approval before the federal authority.

As with most of the forest management plans implemented in Mexico, the one in force for the SPF defines as its purpose *to ensure the conservation, protection, recovery, development, and promotion of the forest resources of the community, under silvicultural management that is economic, social, and ecologically sustainable (Dirección Técnica Forestal "San Pedro el Alto," 2006. 2 p).*

This is a purpose that has been verified by national and international agencies such as FSC, which awarded a certificate for good management to the community.

Profits obtained from timber harvesting allow for job creation within the community and the hiring of personnel from other communities. A portion of the profits is distributed equitably among the comuneros, another part is invested in civic and religious infrastructure and activities of interest to the community and for the development of community-based businesses. This enables the community to diversify its productive activities and facilitate the employment of women. To date, the community has invested in a bottled-water plant, a fleet of buses to serve the community and its neighbors, and a sawmill.

DISCUSSION AND CONCLUSIONS

The planning process for the SPF will face many challenges in the future. As community leaders are increasingly aware of the changes happening in their surroundings, they are more prone to invest in their forest capital. Following a highly conscious, but very conservative approach, only half of the estimated allowable harvest, which is based on a very precise inventory estimates, is approved. This "stock saving" is perceived as an investment, since keeping a well-maintained forest stock may overcome the ups and downs of the timber market and guarantees a sustained income as well (perhaps this is the main force behind the final harvest schedule). On the other hand, such a conservative management approach can be considered as inefficient in financial terms and risky in terms of forest health and other likely ecological problems. The main technical challenge is to find a balanced forest management proposal that community owners can understand and be willing to accept with calculated income failure risk.

Technological changes as well as rising production costs, pose both a threat and an opportunity for the SPF and its COFF. A more efficient logging system, paired with more intensive silvicultural practices, have already been put to work and may be the answer in a local market constantly pressured by timber imports that are commonly less expensive than locally produced timber. Still, the Mexican market has enough room to accommodate local producers since legally produced timber roughly accounts for only 20% of the total timber consumption estimated for the country (25 million m^3, or 883 million ft^3).

Under the current silvicultural scenario, the compact area determined for every cutting cycle may be replaced by a more landscape-dispersed, yet spatially constrained, approach that will keep harvesting costs low and allow intensive silviculture in selected stands with associated shorter rotation periods. Other areas would then be managed under a less intensive approach and include large conservation areas where little or no harvesting is performed over long periods of time to maintain composition or forest coverage, biodiversity, and environmental benefits.

Learnings and Insights

The constraints on harvest flows and areas to be harvested are strongly supported by the community and are a reflection of a strong governance structure. Nevertheless, this may represent a toll to the community's total income, since the approach is not based on an economic criterion or a long-term strategy. However, this may be the best approach, given the 10-year horizon restriction in the Mexican forest law.

Sustainability Issues

Mexican legal framework allows any silvicultural system and management planning method to be employed as long as the method guarantees the sustainability of forest resources. The SPF is composed of well-stocked to overstocked stands, and these seem to guarantee the sustainability of forest harvests and provide market flexibility for the community. The size of the property and the particular way in which it is managed make it very important to develop a long-term silviculture monitoring system (using research plots and continuous inventory plots) by the community in order to obtain the necessary information to base future silvicultural decisions, such as a reduction of the rotation age or the establishment of multiple rotations, according to site quality. However, the success of a community forestry program depends on the decisions made by the community members. Wrong decisions regarding harvest flows, the definition of harvest tracts, or the relaxation of conservation or protection constraints may jeopardize the sustainability of the forest land. Maintaining good decisions requires an investment in capacity-building in areas such as environmental issues, managerial tools, marketing, and social capital building.

Plan Development Challenges

The SPF is located in one of the most productive forest areas in Mexico, yet its management has proven to be very conservative and oriented toward timber stock maintenance. The community has not yet determined an appropriate stocking level or a minimum inventory strategy that may boost the income flow in the future and increase the flexibility of the management plan. Highly productive stands can reach yields over 800 m^3 per ha (11,435 ft^3 per ac) at age 70, and comprise 30% of the total commercial forest area. However, the main idea of the COFF is to reduce rotation ages whenever possible and intensify silviculture when and where possible. Harvesting and silvicultural operations of any kind have to be first approved by the federal government, which currently foster a shift to more intensive silviculture in the country since approval has to be sought every 10 years or less. In any case, the development of a long-term strategy to liquidate surplus forest stocks may be necessary not only for increasing forest returns but also to reduce the risks related to maintaining large areas of mature forests. In addition, the new sawmill acquired by the FCE will require a higher timber volume to operate efficiently. One important challenge of plan development is to link not only upgraded logging and hauling technology to forest tracts, but also to link those harvest volumes to the future developments of the sawmill and other wood products that the FCE might decide to produce. In the very long run, some areas currently under conservation uses might require some management to prevent the attack of pests or to reduce the risk of wildfires, while some others could be incorporated into the timber harvest schedule.

Plan Implementation Challenges

Enforcing new environmental diversity regulations may pose an operational challenge for both community and law enforcement agencies. The highly conservative approach of the SPF, with high stocking volumes and continuously maintained crown coverage, may be beneficial for some wildlife species seeking interior forest habitats. If one of the species occurring in these areas is listed as an endangered species in the NOM-059, then there will be changes to forest management that restrict logging operations. This has not yet happened in the SPF due to the relatively low-intensity silviculture, but in time it may pose a threat to the economic sustainability of the timber operation.

No other challenges for implementing a new forest management plan are visualized as technical personnel are experienced and the road network extensively covers the forest. However, a more intensive approach to silviculture, as well as the increasing demand for logs for local sawmills, may require a more coordinated logging effort.

Other Interesting Aspects of the Plan

The economically diverse structure that is now occurring in the local community as a result of the development of complementary and supplementary economic activities around forestry may give way to a loss of governance in the community, and might reduce the local support to maintain the current goals for management of the community forest. Poorly based forest management strategies and the low social capital of communities around the SPF could threaten the forest through illegal logging or increase the risk of forest fires.

The high profile of the community members in areas related to forest conservation issues (at national and international levels) is one of the strengths of the SPF, but this has yet to be communicated under an adequate public relations strategy. Given the high proportion of timberland under conservation, the SPF could qualify for, and obtain support from, many conservation funds (public and private) linked to the mitigation of climate change or biodiversity conservation. However, the high profile of the SPF has made the community forest a target for takeover attempts, as recently a land-rights lawsuit has been filed that favors a neighboring community and threatens a potential loss of about 2,000 ha (4,942 ac). These kinds of lawsuits are common in Mexico and are directed mainly at rich and well-organized communities, since they generally settle cases by giving up low-value land to maintain their timber operations. However, it may be just the beginning of other lawsuits from other communities that may eventually weaken the governance structure in the San Pedro El Alto community.

Despite all of these possible threats and challenges, the SPF is an example of good management of a community forest that can be achieved through a governance structure where rapid liquidation of forest stock surpluses is not necessarily a goal for a community that has a long history in the area.

REFERENCES

Antinori, C., Bray, D.B., 2005. Community forest enterprises as entrepreneurial firms: Economic and institutional perspectives from Mexico. World Development 33, 1529–1543.

Dirección Técnica Forestal "San Pedro el Alto", 2006. Programa de manejo forestal maderable persistente periodo 2006–07 al 2016 (Nivel avanzado) para el aprovechamiento de los recursos forestales maderables en el predio de San Pedro el Alto, Zimatlán, Oaxaca. Mier y Terán No. 603-altos, Centro, Oaxaca, Mexico. 253 p.

Food and Agriculture Organization of the United Nations, 2010. Global Forest Resources Assessment 2010. Food and Agriculture Organization of the United Nations, Rome, Italy, FAO Forestry Paper 163. 340 p.

Robles Berlanga, H.M., 2012. Ejidos y comunidades en México: Problemas y perspectivas. In: de México, Ciudad, Reyes, J.A., D'Acosta (Eds.), Memorias del Taller "Propiedad Social y Servicios Ambientales". Proyecto de Cooperación RAN-IICA-CCMSS-CONAFOR, México, DF, 8 de noviembre de 2011.

Rosales-Salazar, P., Olayo-Martínez, M.A., Morales, J.A., Alvarez, R., Martínez, I., Castro, S., 1982. El Método de Desarrollo Silvícola: una Alternativa en la Producción Forestal y Ordenación de Bosques. Unpublished Thesis, Universidad Autónoma Chapingo, Departamento de Bosques, Chapingo, México.

Secretaría del Medio Ambiente, Recursos Naturales y Pesca (SEMARNAP), 2000. Conservación y manejo comunitario de los recursos forestales en Oaxaca. SEMARNAP Delegación Oaxaca-PROCYMAF, Oaxaca, Oax. Mexico. 212 p.

Styles, B.T., 1993. Genus Pinus: A Mexican Purview. In: Ramamoorthy, T.P., Bye, R., Lot, A., Fa, J. (Eds.), Biological Diversity of Mexico: Origins and Distribution. Oxford University Press, New York, pp. 397–420.

Torres-Rojo, J.M., Magaña-Torres, O.S., 2006. Management of Mexican community forests with timber production objectives. Allgemeine Forst und Jagdzeitung 177, 63–71.

Chapter 24

Sand Hills State Forest, South Carolina, United States of America

Scott Danskin[1] and Brian Davis[2]

[1]*South Carolina Forestry Commission, Wedgefield, South Carolina, USA*, [2]*Sand Hills State Forest, South Carolina Forestry Commission, Patrick, South Carolina, USA*

ABBREVIATIONS

RCW   Red-cockaded Woodpecker
SCFC   South Carolina Forestry Commission
SHSF   Sand Hills State Forest

MANAGEMENT SETTING AND BACKGROUND

Prior to acquisition by the U.S. Government in the early- to mid-1930s, the land that is now the Sand Hills State Forest (SHSF) was primarily abandoned agricultural land. The low fertility and high infiltration rates of the characteristic deep, sandy soils precluded the long-term establishment of agricultural fields, and many areas were allowed to revert back to natural vegetation after only a few seasons in row crops. The SHSF originated from the Federal Resettlement Administration of the 1930s. This government program sought to purchase worn-out farmland and relocate struggling families to more productive farms. Many families were willing to sell their impoverished farms, which left this eroded, fire-swept, cut-over, and abused land to be managed by Federal agencies. Approximately 92,000 acres (ac) (37,232 hectares (ha)) of this land, located predominantly in Chesterfield County, SC, were evenly divided into the Carolina Sandhills National Wildlife Refuge (U.S. Fish and Wildlife Service (USFWS)), and the Sand Hills State Forest (South Carolina Forestry Commission (SCFC)), and were to be managed under a long-term lease arrangement with the U.S. Department of the Interior.

Under the original contractual agreement that began in 1939, the SCFC managed the 46,000 ac (18,616 ha) state forest autonomously, and also managed all timber practices on the adjacent Wildlife Refuge as well. Through the early 1940s, the Civilian Conservation Corps contributed manpower and expertise to the developing property. In 1991, the SCFC was granted fee simple title (legal possession) to the state forest.

From the outset, the SCFC agreed to operate the property as "a demonstration conservation area, embodying the principles and objectives of multiple-use management." Part of the long-range goal was also to provide local jobs and stimulate local industry through forest production. The local community also benefited from a SCFC operations tax, wherein 25% of all proceeds generated from forest products must be given to the local county school district in lieu of taxes. The SHSF has been totally self-supporting since 1967, and this forest products tax is still in place, and has provided millions of dollars in revenue to the local counties.

Throughout the years, traditional forest products (sawtimber, poles, and pulpwood) have been regularly harvested from the forest with predominantly softwood lumber being generated. Due to the presence of natural longleaf pine (*Pinus palustris*) stands on the property, other products, including pine tar, turpentine, fence posts, and pine straw have played important economic roles over varying time periods of SCFC management. Currently, pine straw management is still a major source of operational revenue. Associated enhancement programs that support the long-term development of stands for pine straw production have also been successful at producing stand conditions favorable for the red-cockaded woodpecker (RCW) (*Picoides borealis*) habitat.

Forest Plans of North America. http://dx.doi.org/10.1016/B978-0-12-799936-4.00024-2

As indicated in the land transfer agreement, the SHSF must manage the land for the benefit of endangered plants and animals that occur, and must consult with the USFWS regarding the impacts of its management practices upon endangered and threatened species as if the SCFC were a federal agency with RCW habitat management in the forefront, as directed by the RCW Recovery Plan (U.S. Fish and Wildlife Service, 2003). Long-term management of the site had resulted in the discovery of most, if not all of these sensitive sites, species, and communities, and thus their protection has been an integral ongoing part of the management process.

The SHSF (Figure 24.1) has a total of about 40,000 ac (16,188 ha) of forestland that is managed primarily for pine, with longleaf pine the primary species (Table 24.1). Loblolly pine (*Pinus taeda*) is also managed, but accounts for less than 10% of the SHSF forests. Longleaf pine is the predominant species of choice when planting new ground on SHSF, but, in some cases, loblolly pine may be planted due to site conditions. Previously, slash pine (*Pinus elliottii*) was widely planted across the SHSF, but these forests are currently being replaced, through stand conversion processes, to longleaf pine forests. Improvements to RCW habitat are the primary driver of this change in approach, and slash pine is also considered an off-site species that is susceptible to insect, disease, and storm damage. The conversion of slash pine to longleaf pine forests on the SHSF is nearly complete, with only remnant patches left to be removed. All of the slash pine plantations will be clear-cut once they reach maturity, and replanted with longleaf, which is native to the area and grows best on deep, sandy soils. Overall, 55% of the forestland is made up of pine plantations while the remaining 45% consists of mostly natural pine stands (Figure 24.2). A small percentage of this is also in mixed-pine and hardwood stands or natural bottomland hardwood.

The ability to successfully harvest pine straw without negatively impacting RCW populations is well established, provided that certain pine basal area constraints and rotation age sufficient for long-term species viability are maintained (Roise et al., 1991). In fact, since the inception of the pine straw enhancement program in 1998, the quality of habitat for the RCW has improved greatly on the SHSF. There are about 21,000 ac (8,499 ha) of longleaf pine forests in which the understory has been controlled, and where greater than 90% of the scrub oaks (primarily turkey oak (*Quercus laevis*) and bluejack oak (*Quercus incana*)) have been eradicated. These stands are maintained in this *park-like appearance* primarily through the use of prescribed fire. Following this reduction in hardwood understory and coupled with a regular fire regime, the SHSF has observed the return of native wiregrass (*Aristida stricta*) and flowers that were once abundant in the longleaf

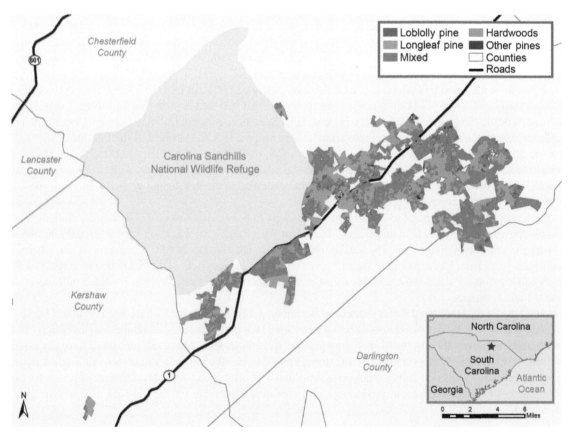

FIGURE 24.1 Forest cover types and general location of Sand Hills State Forest.

TABLE 24.1 Sand Hills State Forest Cover Type, Area, Average Site Index, and Establishment Method

Cover Type	Area (ac)	Average Site Index[a]	Establishment Method	
			Natural Regeneration (ac)	Planted (ac)
Oak-hickory	8	65	8	
Bottomland hardwood	2,556	100	2,556	
Upland hardwood	19	87	19	
Total hardwood	2,583		2,583	
Pine-hardwood	10	70	10	
Total mixed	10		10	
Loblolly pine	1,679	78	87	1,592
Longleaf pine	18,357	69	790	17,567
Slash pine	104	82		104
Sand pine[b]	11	70	11	
Mixed pine	17,304	71	17,304	
Mixed pine bottom	4,665	90	4,665	
Total pine	42,120		22,857	19,263
Total	44,713		25,450	19,263

[a]*50-year base age for hardwoods, 25-year base age for conifers.*
[b]*Pinus clausa.*

FIGURE 24.2 An aerial view of planted pine, natural pine, bottomland hardwood, and wetlands located on the Sand Hills State Forest (Google Earth).

pine ecosystem. Also, the control of the understory through the pine straw enhancement program is important in reducing the risk of a catastrophic wildfire that could negatively impact RCW habitat, along with associated losses in timber values and other areas.

RCW populations on the SHSF have been estimated several times. Early estimates in the 1980s put the population at approximately 70 groups. However, these early estimates were based on unreliable surveys. Since 1991, the South Carolina Forestry Commission and the South Carolina Department of Natural Resources have cooperated on RCW management on the forest and more reliable population figures have been gathered.

In 1992, a thorough survey of the SHSF documented 36 active groups. Much of the habitat was in poor condition and RCWs had a small pool of potentially suitable cavity trees to choose from. In 1992, an active banding and artificial cavity installation program was initiated in an effort to expand the RCW population on the SHSF. As a result of this active management program, the population grew steadily to 61 groups, as verified during the 2005 survey year. There have been nearly 300 artificial cavities installed thus far under the program.

Though the number of active groups at SHSF has increased steadily, reproductive output has remained relatively stable. This may be due in part to the habitat mosaic at the SHSF. Extensive off-site tree species conversion and privately owned in-holdings have resulted in a highly fragmented landscape. Such an interrupted landscape may limit foraging efficiency and make it difficult for dispersing individuals to locate breeding vacancies (Ferral, 1998). As longleaf pine stands become reestablished, the problems associated with fragmentation should abate. In addition to problems with fragmentation, habitat quality may have been affected by the absence of fire in certain areas of the forest. Although prescribed fire has been used to decrease the hardwood understory in pine stands, fire suppression activities have resulted in hardwood midstory encroachment in many stands. However, currently most of the hardwood problems have been addressed on the SHSF.

PLANNING ENVIRONMENT AND METHODOLOGY

As previously described, SHFS forestland is a natural upland sand ridge habitat suitable for longleaf pine and associated species (Table 24.1). The forest, however, is almost equally divided in naturally occurring and planted longleaf pine, a result of many years of stand conversion from slash pine. Additionally, the natural stands are more commonly mixed pine stands, though they often contain large, mature longleaf pine desirable for RCW cavity construction. These two stand types are also bimodal in age distribution, with the planted stands much younger, and thus potentially unavailable for cavity construction for many years. In mixed stands, thinning practices have consistently targeted other pine species, accelerating the development of pure, longleaf pine in open stand condition. In younger planted stands, managers are often challenged in conducting thinnings too early as those stands conditions are most desirable for pine straw production. The development of this management plan and associated harvest model are both tools used by managers to assess these conditions and make decisions beyond the current time scale which, if correctly implemented, should ensure adequate habitat across the forest while this bimodal age distribution is converted to a steady-state mosaic of desirable age and condition classes.

Previously, to meet the SHSF objectives and to be in compliance with the RCW recovery program, management plans had been developed in a way that identified the acquirable goals and the processes necessary to meet them. A long-range management plan, tailored to RCW recovery, but also inclusive of other known or potential endangered species, was completed in 2006. While this plan thoroughly addressed forest management activities and their associated ability to promote RCW population growth and habitat improvement, opportunities for optimizing harvest scheduling without violating RCW constraints led to the additional implementation of a harvest scheduling model. The requirements for RCW habitat are laid out in the RCW recovery plan (U.S. Fish and Wildlife Service, 2003), and impose restrictions to harvest activity based on maintaining specific basal area requirements within RCW cluster partitions. Due to the complexity of these requirements, the SHSF elicited an independent contractor to develop a spatially explicit harvest schedule model. Model development was exclusive to the State Forest System and its specific habitat requirements and spatial constraints, but was inclusive of two other forests with regular harvest activity, and constraints varied across forests as needed. Harvest recommendations generated from this model were reviewed internally and by the USFWS before approval for any harvest activity was given. The harvest scheduling model was initiated in 2008 and has since been updated to reflect the eventual changes and deviations from earlier model proposals. The next model iteration is underway, and final model outputs are expected in 2015 or 2016. The SCFCs relatively new approach to harvest scheduling, along with associated improvements to other portions of the 2006 management plan that could potentially dovetail with requirements of both the SFI and ATFS initiatives, motivated the SHSF to update the forest management plan beginning in the fall of 2012.

Guiding Laws, Regulations, and Policies

The SHSF must be managed for RCW recovery as outlined in the RCW recovery standard, but this is conditional also on federal mandate as set forth in Section 109 (f) of Pub. L. 101-593. This provision requires that the SHSF, which is unique to the SCFC State Forest Division, must be managed to meet certain federal requirements that other state forest lands do not meet. State requirements to manage and protect for rare, threatened, and endangered species are also in place as well as similar rules on issues that may be impacted by forest activities (wetlands, road development, etc.). South Carolina Best Management Practices (BMPs) are currently voluntary; however, as a representative of the State Forest system, practices have continually been implemented to provide a working example of their success. Internally, a long-range plan has served as an umbrella document to individual forest plans, and policy within the plan has provided valuable guidelines for viewsheds, harvest activity, harvest levels, recreational use, and how these and other activities are integrated on the forest.

The SHSF has a goal of updating the existing management plan to continue to be compliant with the Sustainable Forestry Initiative (SFI) and American Tree Farm System (ATFS) certification programs, and it became certified under the two standards (American Tree Farm System, 2014; Sustainable Forestry Initiative Inc., 2014) in 2013. To meet SFI standard requirements for long-term planned conditions of forest stands, the harvest scheduling model provided a tactical solution at the stand level that would continuously maintain a certain level of acceptable RCW habitat. However, other issues that were required by the SFI standard were either not clearly defined in the previous plan or were not incorporated, as they were addressed by other divisions within the agency. For instance, a methodology was in place for compliance evaluation of forestry BMPs, but this process was under the purview of the South Carolina Forestry BMP program. In such cases, language was developed to more clearly define these relationships, and often improvements were made to the methodology, either along lines of communication or in the data development processes.

OUTCOMES OF THE PLAN

Owing to the development of a harvest schedule model, the timing of harvest activity and harvest levels at the individual stand level did not need to be addressed within the management plan, as the long-term stand conditions were parameters in the model's development. Instead, the SHSF views the harvest schedule model as an addition to the plan, which allows for updating that portion of the operational directive on a time scale independent of other planning needs or requirements. With RCW recovery, the overall main objective, the timing of harvest activities was instead driven by the management of stands within recommended basal area conditions as set forth by the USFWS (U.S. Fish and Wildlife Service, 2003), while maximizing the net present value (NPV) and ensuring that annual harvest incomes are consistent across all harvest years. Previous modeling had included potential revenue from other activities (i.e., pine straw raking), but these factors were not considered in the current model, which kept the focus solely on wood production. The immediate result of model implementation was a recommendation to increase thinning activity, as much of the younger stands had basal areas beyond those considered good quality foraging habitat for the RCW. Also, due to the SHSF having a large amount of land with older longleaf pine forests, some harvesting of these stands was recommended earlier than the maximum allowed rotation age, so that future forest conditions could become a mosaic of stand age classes and desirable conditions, preventing a possible falldown situation where large amounts of timber become overmature at the same time.

As mentioned previously, all planned harvest activity must be determined to be in compliance with the RCW recovery standard, and must be explicitly approved by the USFWS. These requirements adopt a spatial allocation of basal area and standing timber volumes by diameter class, and are further detailed to which level of recovery a given property has been assigned. The SHSF is considered a secondary recovery site, which requires specific basal area levels that have significant impact of how much timber harvesting can be conducting. This spatial allocation is applied over a 0.5 mile (0.8 kilometer) radius around a minimum convex polygon of RCW cluster trees. The minimum convex polygon (Figure 24.3) must be at least 10 ac (4 ha), and this results in a partition which, unimpeded, covers approximately 500 ac (202 ha). Cluster areas often occur within distances of each other that cause impediment of that partition, and a process of equally dividing areas of overlap results in increasingly smaller partitions. Another impact is the general trend of new cluster establishment, known as budding, to occur within a distance of the home cluster that would result in establishment of a new partition, and impediment of all partitions nearby. Another factor relevant to managers of public lands is that partitions that exceed the public land boundaries are still required to meet the RCW recovery standard goals. Where clusters exist near property boundaries, the impact of this restriction can be severe.

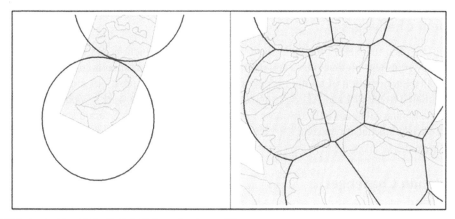

FIGURE 24.3   Impeded versus unimpeded red-cockaded woodpecker partitions.

DISCUSSION AND CONCLUSIONS

The SHSF was one of five state forests that transitioned to SFI and ATFS standard compliance. As anticipated, many challenges arose during this process (Moore et al., 2012), and most were specific to the particular State Forest in question. At the SHSF, it was found that the earlier long-range plans, with specific and obtainable objectives, made the transition to a new plan more easily accomplished than with other state forests. However, several issues arose that needed to be address in the revised plan, as well as one particular issue that will require more long-term planning and involvement with the USFWS.

Learnings and Insights

In a manner consistent with the experiences of many other resource managers that have adopted the SFI standard, the SHSF found that documentation both across state forests and within individual forests was either inconsistent, or not robust enough in both data capture rate and detail (Moore et al., 2012). Documentation of resources and conditions in many cases did exist, but was not maintained in an easily verified location or format. Standardization of a new set of organizational forms and documents was conducted through interagency exchange, and this has been made available through internally (to the agency) shared datalinks, which has helped to ensure consistency in record keeping. In some areas, where documentation could be used to establish actions and their associated compliance with the RCW recovery standard, BMPs, and other similar concerns, the improved documentation was worth the effort involved in adopting the SFI standards.

Another noticeable improvement the SHSF experienced was the streamlining and enhancement of processes for the ensured protection of threatened and endangered species that have the potential to occur on the forest, but have not previously been discovered. Given the long-term management of the forest for wildlife habitat development, the SHSF benefited from continual involvement and support of biological researchers, as well as internal expertise of the land base. However, the possibility of unknown threatened or endangered populations is considered high priority in operational planning. Through habitat restoration, driven primarily by a prescribed fire regimen, it is anticipated that some wildlife species may, over time, be reestablished. These factors and others led to a more complete and updated database of all species of concern for the SHSF, as well as an improvement in the involvement of personnel through identifiable training goals that emerged from the revised documentation process. In addition, the SHSF developed a methodology for spatially determining which planned harvest areas could incur a significant risk before the start of the annual harvest cruising, so personnel would have clear knowledge of what areas may require additional assessments before harvest activity is conducted.

Sustainability Issues

Through revision of the management plan and through compliance with the SFI and ATFS standards, The SHSF became certified under the standards in the fall of 2013. As a working forest under the SCFC organization, the SHSF is a role model for other forests, especially where the challenge of harvesting timber with RCW habitat concerns are foremost. Through compliance with the standards, the SHSF provides an example of how these can assist in specifically addressing resource use concerns. Certification illustrates the commitment of both the SCFC and State Forest system to the principles of sustainability and responsible land stewardship. Further, it provides the privilege of marketing timber sales under the SFI and ATFS banner. This has improved the state's ability to address issues and concerns as they arise, and more consistently as an agency, not just as an individual state forest.

Plan Development Challenges

During the planning process, the SHSF was able to identify the major concerns associated with long-term planning. A personnel training program was implemented, and the development of a role-specific training matrix has become more defined, providing both documentation of training events and guidance as new positions are filled.

The SHSF is committed to the SFI and ATFS programs, and will continue to work to update its management plan over time, either in response to internal needs or in response to planned changes in certification standards. The overall planning process has been beneficial and educational, and has better prepared the SCFC to meet a wide variety of existing and emerging environmental concerns.

Plan Implementation Challenges

When RCW partitions are unimpeded, the ability to stay within the RCW recovery standard goals is relatively easy to accomplish. A larger land base provides more area, and thus volume, of desirable species to occur, as well as an increased probability that preferred variation in diameter classes will occur. As impediment diminishes this land base, partitions may approach, and even exceed physical limits for the partition to meet recovery standard goals. These reduced-area partitions are allowed to have much less harvest activity conducted and, in some cases, no harvest conducted at all. Additionally, they may impact future harvest modeling attempts due to the effects of spatial restrictions (Murray, 1999).

In some instances, which are only now emerging as a significant management issue for long-range planning, the SHSF is faced with two equally undesirable options: conduct harvest activity in violation of the recovery standard stand volume level requirements, or not conduct harvest activity at all and be in violation of the recovery standard habitat requirements. Currently, the SHSF accepts the latter option as restrictions on harvesting by the USFWS are more explicit, and their potential occurrence is very transparent due to the harvest approval request process. However, it should be noted that the effects of simply reserving the land are not acceptable because stands are dynamic and will change over time into conditions that may not be favorable (Roise et al., 1990). Experienced impact of this effect has been limited, with only two partitions of great impediment. However, the more successful the SHSF is at managing for RCW habitat and population growth, the less able the forest will be to harvest timber products. As established previously, the SHSF is independently funded, primarily from sales of these products. Research on similar habitat limiting conditions has been shown to have the potential to greatly reduce the net present value of forest plans (Bettinger et al., 2003). It is expected that the conflicts in the RCW recovery standard goals (habitat management and volume or basal areas goals) will also be experienced by other land managers concerned with RCW habitat. However, the additional financial impact has increased uncertainty in the SHSFs ability to meet long-term management goals.

REFERENCES

American Tree Farm System, 2014. About American Tree Farm System. American Forest Foundation, Washington, DC. https://www.treefarmsystem.org/about-tree-farm-system (Accessed April 17, 2014).

Bettinger, P., Johnson, D.L., Johnson, K.N., 2003. Spatial forest plan development with ecological and economic goals. Ecological Modelling 169, 215–236.

Ferral, D.P., 1998. Habitat quality and the performance of red-cockaded woodpecker groups in the South Carolina sand hills. Master of Science thesis, Clemson University, Clemson, SC.

Moore, S.E., Cubbage, F., Eicheldinger, C., 2012. Impacts of Forest Stewardship Council (FSC) and Sustainable Forestry Initiative (SFI) forest certification in North America. Journal of Forestry 110, 79–88.

Murray, A.T., 1999. Spatial restrictions in harvest scheduling. Forest Science 45, 45–52.

Roise, J., Chung, J., Lancia, R., Lennartz, M., 1990. Red-cockaded woodpecker habitat and timber management: productions possibilities. Southern Journal of Applied Forestry 14, 6–12.

Roise, J.P., Chung, J., Lancia, R., 1991. Red-cockaded woodpecker habitat management and longleaf pine straw production: An economic analysis. Southern Journal of Applied Forestry 15, 88–92.

Sustainable Forestry Initiative, Inc., 2014. SFI Standard. Sustainable Forestry Initiative, Inc, Washington, DC. http://www.sfiprogram.org/sfi-standard/ (Accessed May 16, 2014).

U.S. Fish and Wildlife Service, 2003. Recovery Plan for the Red-Cockaded Woodpecker (*Picoides borealis*) Second Revision. U.S. Fish and Wildlife Service, Southeast Region, Atlanta, GA. 296 p.

Chapter 25

Sessoms Timber Trust, Georgia, United States of America

Yenie Tran,[1] Dotty S. Porter,[2] Norris Mattox,[2] Bob Izlar[1] and Jacek P. Siry[1]

[1]Warnell School of Forestry and Natural Resources, University of Georgia, Athens, Georgia, USA [2]Sessoms Timber Trust, Homerville, Georgia, USA

ABBREVIATION

FLPA Forestland Protection Act

MANAGEMENT SETTING AND BACKGROUND

The Sessoms Timber Trust is a family-held timber company, consisting of approximately 50,000 acres (ac) (20,235 hectares (ha)) of pine timberland, located in the lower coastal plain of southeast Georgia (Figure 25.1). The economy of southeast Georgia is historically and currently particularly dependent on forest markets, with Savannah as a major port allowing access to international markets. Historically, this area was the leading producer of turpentine, a product derived from resin primarily from pine trees. By the mid-1960s, turpentine production decreased significantly due to increased labor production costs and its production as a by-product of the pulp and paper making process; and landowners in the region started to shift more into pulpwood production. Important timber species in the region include slash pine (*Pinus elliotti*), loblolly pine (*Pinus taeda*), and longleaf pine (*Pinus palustris*) (Tran et al., 2014).

The history of the Sessoms Timber Trust (hereafter "the Trust") spans 135 years, starting with Alexander Sessoms (1834–1910) who started amassing the land holdings in 1879. It was his son, Alexander Kelly Sessoms (1881–1944), who had the vision that has led to five generations of sustainable forestry on the family forestland. Much of the impetus began when Alexander Kelly Sessoms (A.K. Sessoms) moved to northern Clinch County, Georgia upon his father's death. After fulfilling his responsibility of dividing up the estate among his siblings, he was determined to reacquire the holdings from his siblings. Thereafter, with entrepreneurial spirit, he began the production of turpentine and timber harvesting. In the process, he established the community of Cogdell, Georgia, which became a thriving settlement, complete with a lumber mill, turpentine still, a railroad depot, a cattle farm, housing for employees, a company store, a school, and numerous churches (Izlar, 2007a).

Savannah, Georgia has historically provided plentiful market opportunities for timber products, as it still does today, but transportation of the products to Savannah proved difficult during A.K. Sessoms' time. In response to the challenge, Sessoms formed a company in 1911 to construct the Waycross & Western Railroad, a railroad line running from Waycross about 27 miles (mi) (about 43 kilometers (km)) from Cogdell to Milltown (known today as Lakeland, Georgia), and this railroad provided a means of transport for Sessoms timber products to Savannah.

In 1913, A.K. Sessoms married Edna Sirmans, the daughter of F.B. Sirmans, a Georgia state senator and a large landowner in the southeastern part of the state. This relationship with the Sirmans family further helped establish the rail company. Through the years, A.K. Sessoms became one of Georgia's largest private landowners, as he continued to acquire land throughout his lifetime, predominately in Clinch County, but also in Atkinson, Lanier, and Ware counties in Georgia.

Forest Plans of North America. http://dx.doi.org/10.1016/B978-0-12-799936-4.00025-4

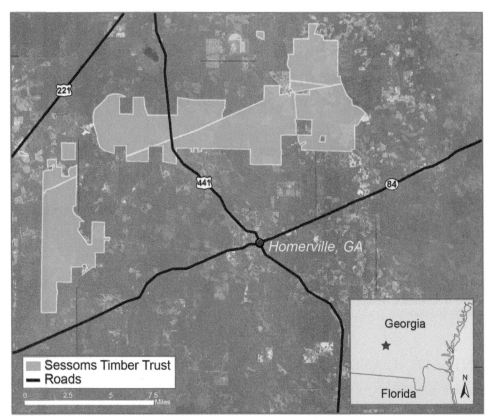

FIGURE 25.1 Map of the land area of the Sessoms Timber Trust.

Innovation was integral to the success of the Sessoms companies. Naval stores, which originally applied to ships' stores in general, came to mean the by-products from pine trees like pitch, tar, and turpentine (Herndon, 1968). This industry became an important export product of Georgia. However, inefficient distillation and nonsustainable extraction methods, such as box cutting, caused overproduction and practically guaranteed the destruction of pine forestland, and hence the naval stores industry (Reed, 1982). The box cutting method involved chopping a cup-like cavity at the base of a pine tree to catch crude turpentine, which flowed from scarified cuts on the trunk above. This method of collecting the turpentine weakened trees, leaving them vulnerable to disease, wind, and fire. Charles Herty, an American chemist, professor, and researcher, pondered over the problem to develop a more practical system to collect turpentine. After a visit to the Sessoms family, Herty invented a cup-and-gutter system, which was a more sustainable and efficient method of collecting naval stores from pine trees (Reed, 1982). A.K. Sessoms' uncle became an early Florida adopter of and proponent of the Herty System in 1902. A.K. Sessoms' educational background in engineering contributed to his development of a steam-powered turbine distilling plant that further revolutionized the naval stores and turpentine industry. By the 1920s, Sessoms' Timber Products Company in Cogdell (Izlar, 2007a) was producing lumber, spirits of turpentine, resin, and crossties. It sold dead trees and stumps to Hercules Powder Company. Sessoms' Timber Products Company had professional, university educated foresters, an absolute requirement that all crops of boxes be worked with the Herty System, and a modern steam turpentine distillation plant. The company had two motorized fire patrols inspecting its lands at all times, and the company maintained fire breaks and fire corridors through the forest. It also leased land for cattle grazing and had 80 mi (129 km) of fences to maintain (Izlar, 2007a).

Soon after the visit, Herty wrote a report on the management of Sessoms' Timber Products Company, which was published in *Manufacturer's Record*. The report suggested that Sessoms' Timber Products Company was a model for the South's new industry. Herty and Sessoms ardently believed that by suppressing fire and encouraging natural regeneration of slash pine, a forest landowner could maintain a perpetual income stream. Sessoms, and those like him, needed an outlet for small-diameter timber generated through slash pine thinning operations, and Herty felt the answer was the development of a white paper and newsprint industry. Sessoms provided the 10- to 15-year-old slash pine pulping samples, and Herty shopped them around the country to several private, government, and industry labs. The lab trials confirmed what Herty thought: young southern yellow pines had very low resin content and no heartwood. This was the beginning of the Deep South pulp and paper industry (Izlar, 2007a).

Prior to his death in 1944, A.K. Sessoms and his wife Edna Sirmans Sessoms worked to ensure the longevity of sustainable forest management on their forest lands. Two trusts were established. The first was the A.K. Sessoms Trust, which was established in 1937 and comprised approximately 38,000 ac (15,378 ha) of pine timberland. The second trust was the E.S. Sessoms Trust, which was established in 1939 by Edna Sessoms and comprised approximately 11,000 ac (4,452 ha) of land she inherited from her father. These two trusts today make up Sessoms Timber Trust. A tenet put forth in both trust agreements is that the trust shall remain intact for 21 years after the death of the last surviving child of the donor. This occurred in December 2008, which set into motion the countdown to the dissolution of the Trust in 2029. In addition to a manager of operations, three trustees currently oversee the management of the trust, including one family member, one corporate member, and one professional forester as a nonrelated member.

The principal timber species on the managed area of the Trust, which comprises 67% of the total land area, are slash pine (75% of the managed area) and loblolly pine (25% of the managed area) (Figure 25.2). The remaining 33% of the total area is composed of unmanaged mixed pine, cypress (*Taxodium* spp.), and hardwoods. Age classes among the managed areas are fairly even among the 0- to 5-year through 21- to 25-year classes (Table 25.1). Half of the volume of merchantable wood is concentrated in slash pine forests (Figure 25.3). The soil types for the Trust's properties primarily consist of the poorly drained and moderately permeable soils of Mascotte fine sand (61%), Surrency muck fine sand (26%), Olustee loamy fine sand (7%), Mascotte mucky fine sand (5%), and Leefield loamy sand (1%). The poorly drained nature of these soils presents a challenge to the transport of wood during very wet and rainy weather. The elevation is generally flat throughout the property, ranging from 190 to 220 feet (58 meters (m) to 67 m) above sea level. The climate is warm and humid most of the year, due in part to its southern latitude and its proximity to the Atlantic Ocean and the Gulf of Mexico. The average rainfall is about 52 inches (in) (1,321 millimeters) annually. Thunderstorms and associated lightning strikes commonly occur from mid-May through mid-September. Most wildfires occur during this period. Water levels also play a very important factor in determining wildfire risk.

Widespread pine tip moth (*Rhyacionia frustrana*) infestation was observed during the early 1990s in about 8,000 ac (3,238 ha) of planted loblolly stands that were less than 10 years old, chiefly in stands that were planted in sites more suited for slash pine, which is generally resistant to *R. frustrana* attack (Asaro et al., 2004). This was remedied by switching to

FIGURE 25.2   Pine stand on the Sessoms Timber Trust property.

TABLE 25.1 Summary Statistics for the Principle Forest Types and Age Classes on Managed Areas of the Sessoms Timber Trust Properties

Age Class (Years)	Slash Pine (%)	Loblolly Pine (%)	Total (%)
0–5	24.3	–	18.2
6–10	26.0	–	19.4
11–15	10.9	18.9	13.0
16–20	9.0	47.8	18.8
21–25	5.8	26.9	11.1
26–30	5.4	5.6	5.4
31–35	–	0.8	0.2
36+	18.6	–	13.9
Total	100.0	100.0	100.0

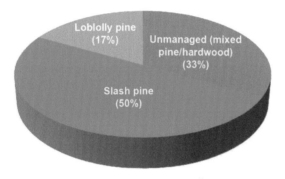

FIGURE 25.3 Allocation of total volume within three principle forest types within the Sessoms Timber Trust property.

slash pine stock. Southern pine beetle (*Dendroctonus frontalis* Zimmermann) has been limited to isolated incidents among lightning damaged trees. White-tailed deer (*Odocoileus virginianus*) is the predominant game species on the property, with about a dozen hunting clubs that pay annual hunting fees and help monitor and manage the associated wildlife.

PLANNING ENVIRONMENT AND METHODOLOGY

The current forest management plan was developed for the period from 2011 through 2030 with the use of a forestry consultant. This simplified the planning process due to the foresters' experience with modern technology, making resources such as digitalized maps more readily available. The plan is detailed and addresses management at the individual stand level and contains harvest years and projected volumes to be harvested. However, it is still flexible enough to adjust for uncertainty in markets, weather, fire, and other unplanned events. For instance, a fire can set the management plan back 5–10 years, depending on how much timber is damaged or destroyed. The Trust document has the stated goal, "Under centralized control with application of proper forestry practices and proper and timely disposition of timber and timber products, dependable and continuous profit will be made," which is restated in the management plan as the overarching goal of the plan (A.K.S. Trust, 1937; E.S.S. Trust, 1939).

The markets for timber have changed in recent years along with the method of selling timber employed by the Trust. In 2010, a decision was made by the trustees to eliminate the use of dealers to sell their wood, which in turn meant changing the method of timber sales from sealed bids to negotiated sales. The Trust now manages the sale and delivery of wood directly to mills, rather than using wood dealers. The Trust had to first establish and build relationships with wood buying facilities. In the long run, this has resulted in a more consistent cash flow to manage trust assets. The Trust thus works directly with a logging crew, which cuts and delivers wood exclusively for the Trust. The logging company contracted by the Trust was recognized as the Georgia Forestry Association Logger of the Year in 2013. The Trust has since become its own wood dealer through direct marketing of logs to local mills. This change has been well received in the market because the Trust has been able to reliably deliver wood every week. Further, the Trust has recognized time savings, more consistent cash flows, improvements in long-term returns, and increases in its ability to control operations.

According to the National Interagency Fire Center, the southern United States experienced a large number of fires over the 5-year period from 2009 to 2013, with an average of more than 32,000 per year (National Interagency Fire Center, 2013). Forest fires in the South tend to be surface fires where the mid- and understory are consumed. The overstory canopy generally does not burn unless fires occur during extreme wind or drought conditions. This is principally true in the southeastern region of Georgia where the Trust's ownership lies, but not in the Okefenokee Swamp, which is susceptible to surface and subsurface fires, due to the high concentration of peat deposits (Izlar, 1984). Severe fires in this area tend to be associated with severe drought (Izlar, 1984; Yin, 1993). Forest fires have therefore been a major threat for the Trust. The most fire-prone area on the property is in Arabia Bay, which is a 5,000 ac (2,023 ha) Carolina Bay. Carolina Bays are elliptical wetland depressions with sand-rimmed bays that are formed in poorly drained sediments and generally filled with peat (Caldwell et al., 2007). The most recent fires experienced at Arabia Bay were in 2000, 2004, 2007, with the worst one in 2011. The deep layer of peat causes a fire to smolder for an extended period of time. Ordinarily, wet winter weather conditions would saturate peat and suppress the spread of fire, but the 2010–2011 winter season was unusually dry. Further, the area is inaccessible to firefighting equipment, thus the fire spread and burned a total of 11,000 ac (4,452 ha), of which about 3,275 ac (1,325 ha) were Trust lands.

Many sandy and loamy soils in the southeastern United States have compacted subsoil layers, commonly called hardpan, which occur at an approximate depth range of 12–24 in (30–61 centimeters). Hardpan is a condition in which soil grains become cemented together by bonding agents such as iron oxide and calcium carbonate, forming a hard, impervious mass. A hardpan can naturally occur or be the result of mechanical compaction of the soil. It does not impede logging operations, but in wet conditions log skidders can damage the top layer of the soil, which can lower the site index and be aesthetically unattractive. The Trust requires its loggers to use mats or to lay a layer of wood on the surface of the soil where logging skidders are used, which reduces the soil damage. Site preparation is mechanized, and the Trust is working on 3-year contract terms for site preparation operations, rather than the 1-year contract terms that were used in the past. This change provides stability and financial security for the Trust as well as for the contractor.

There are about 200 mi (322 km) of company roads that run through the Trust property. Maintenance on these roads is performed by the Trust. Tractor-trailers are the standard logging truck used today throughout North America. These rigs are typically allowed to haul 80,000 pounds (36.3 metric tons) gross vehicle weight (GVW) in most states including Georgia. Trucking costs are affected by haul distance, payload, profit margins, insurance, capital costs, and the idle time spent unloading at the mill and loading in the woods. Longer haul distances can be more easily justified if mill or woods turnaround times are short and/or payloads are high. However, with the increase in fuel prices, more efficiency has been needed in hauling wood or limiting the haul distance.

Guiding Laws, Regulations, and Policies

Georgia is the ninth most populous state and is also one of the fastest growing states in the United States (Mackun and Wilson, 2010). In Georgia, like many other states, ad valorem taxes are levied on all tangible property, including forestland and harvested timber. The tax is based on the value of the land and also serves as one of the most important sources of revenue for financing local governments and supporting county school systems. Property taxes levied have increased 73% over a 9-year period from 2000 to 2009, which is roughly an 8% increase per year (Izlar et al., 2011). Population growth is the primary contributing factor for the increase in property taxes levied. Georgia's population has increased 20% over the same 9-year period, which has led to increased demand for county-based services by residential property owners (Izlar et al., 2011).

The revenue generated by the Trust's hunting leases at one time was enough to cover the annual property taxes; however, this is no longer possible today with the dramatic increase in property taxes levied over the past decade. In 2009, a Georgia constitutional amendment known as the Forestland Protection Act (FLPA) became law. FLPA allows any forest landowner with 200 contiguous ac (81 ha) to enroll in the program (Izlar et al., 2011), and expands the eligibility for a Conservation Use Valuation Assessment (CUVA) to be applied to land by removing the maximum area of 2,000 ac (809 ha). It also institutes a 15-year covenant requirement. FLPA has been helpful in lowering taxes for many landowners. The Trust has carefully scrutinized its decision concerning tracts to include in the program to avoid breaking the covenants created by FLPA. Due to available technology, identifying the different soil types and productivity classes has led to a fairer tax evaluation, although it is unlikely that the Trust will recognize tax savings. Arabia Bay generally contains nonproductive land, and land owned there continues to be a tax burden on the Trust.

The Sessoms Timber Trust forestland has been independently third-party certified under the American Tree Farm System (ATFS) for more than 50 years. ATFS is a member of the Programme for the Endorsement of Forest Certification (PEFC), which is an international, nongovernmental organization dedicated to promoting independent third-party certification. ATFS is the oldest and largest forest certification program in the United States. All land certified through ATFS is required to conform to the American Forest Foundation's Standards of Sustainability for Forest Certification, have a written forest management plan, and be assessed by an ATFS inspecting forester to ensure proper forest management that includes the conservation of soil, water, and wildlife. The primary benefit to certification is that the Trust can supply the markets that require certified wood. For instance Georgia Biomass, a pellet mill in Waycross, Georgia, requires wood to be certified. In addition, the Trust faithfully follows Georgia's Best Management Practices for forestry to ensure sustainable and sound conservation practices (Georgia Forestry Commission, 1999). This has led to the need for continual innovation, flexibility in the management plan, and more efficient operations to respond to changing market conditions.

OUTCOMES OF THE PLAN

The Trust has always operated under a forest management plan, but the plans have grown in complexity through the years. The latest iteration includes descriptions of stand types, land areas, and species. This information has helped the Trust manage the resource more precisely. The goal of the Trust's operations is to stay within the harvest area suggested by the plan, which is the total number of acres the Trust can harvest annually. This number was derived from the management plan based on a 20-year rotation to ensure sustainability. The goal of staying within the allotted harvest area has been accomplished thus far.

The most common product delivered to markets in the U.S. South today is in tree-length form. Pine stems are often processed further into logs or shortwood pulpwood bolts before being transported from the woods. The main timber products derived from the Trust's land today are pulpwood (50%), chip-n-saw (40%), and sawtimber (10%). Thinning practices were removed from the management plan about 5 years ago because the rotation age had been shortened from 28 years to about 20 years to focus more on pulpwood markets rather than on sawtimber markets. Having a detailed management plan made it easier to adapt to this change.

After the forest fires of 2011, the Georgia Forestry Commission sponsored a town meeting in Homerville, Georgia to discuss the aftereffects. It was there that Sessoms Timber Trust representatives were introduced to easement possibilities for Arabia Bay. In 2012, the trustees of the Trust and surrounding landowners signed an agreement with the U.S. Department of Agriculture's Natural Resources Conservation Agency (NRCS), committing Arabia Bay to the Wetland Reserve Enhancement Program, a program of the NRCS Wetland Reserve Program (WRP). This program allow states (including political subdivisions or state agencies), nongovernmental organizations, and Indian Tribes to partner with U.S. Department of Agriculture in the selection and funding of contracts, as long as the selected contracts meet the purposes of the WRP (7 U.S.C. §2206, 2008). The WRP provides technical and financial assistance to eligible landowners to restore, enhance, and protect wetlands through permanent easements, 30-year perpetual easements, or restoration cost-share agreements. Landowners participating in the WRP continue to control access, have use of nondeveloped recreational activities such as hunting and fishing, and maintain the right to lease the recreational uses of their land for financial gain, provided the use does not conflict or affect any of the uses prohibited by the easement deed. The Trust, along with adjacent landowners, had to agree to the contract terms to make the easement work. The final contract for a perpetual easement was signed in late 2013. The NRCS will work with the landowners to restore Arabia Bay back to its natural state. Ditch plugs will be installed where the land was canalled years ago, which will control the water in the wetland, thereby keeping the peat saturated and greatly reducing future fire risk. A buffer around Arabia Bay will be installed, and about 500 ac (202 ha) of longleaf pine will be planted around the rim on the southern and eastern sides. In addition, Arabia Bay provides habitat for several threatened and endangered species, such as the gopher tortoise (*Gopherus polyphemus)*, the flatwood salamander (*Ambystoma cingulatum*), and the wood stork (*Mycteria americana*).

The Trust, like other large landowners, has investigated the opportunities to diversify its products. Other landowners have diversified into the production of blueberries, making Georgia one of the top producers of blueberries in the nation. Instead of using its land for intensively managed agricultural products, the Trust has looked into diversifying its business into pine straw operations as a way to address increasing property taxes. About 500 ac (202.3 ha) has been set aside for this effort, but the pine straw market has since declined. The Trust will continue to explore other market opportunities.

DISCUSSION AND CONCLUSIONS

Markets are the biggest consideration to the implementation of the management plan. The markets for pine sawtimber and chip-n-saw products have declined in recent years. However, pulpwood has increased due to the demand in Europe for wood pellets. The Georgia Biomass pellet plant, located in close proximity to Trust lands, has helped with the market for pine pulpwood. Although two separate accounting books must be kept for the two individual trusts that make up Sessoms Timber Trust, the land from the two trusts are managed as one under a single management plan. Technology has made performing some administrative tasks much easier, particularly in financial record keeping, inventory, and mapping.

Sustainability Issues

The Trust continues to replant pine trees in areas where clear-cut harvests occurred, even though these trees will never be harvested by the Trust, due to the impending dissolution date (2029). The Trust has maintained a rich history of entrepreneurship throughout several generations, and the current generation of family members continues to seek innovative ways to sustainably manage the forests for future generations of the family and the local, regional, and global economies that depend on these resources.

Plan Development Challenges

With respect to the management of the Sessoms Trust lands, the same family members are associated with two different legally defined trusts. A challenge has been to manage their expectations, since goals and objectives can vary among family members. The current management plan and its associated transparency have promoted an openness of communication among the family. The end date of the Trust (2029) has prompted the family to discuss future planning, including the benefits associated with the current resources and enhancing other business opportunities.

Plan Implementation Challenges

Following the management plan during extreme weather conditions is a challenge, as it is for many landowners in the southeastern United States. Weather conditions can hinder the ability to provide a consistent wood supply to markets, and market conditions are dynamic, both locally and abroad. After the U.S. housing slump that began in 2006, sawtimber prices dropped, but the price of pulpwood began to increase due to biomass demand spurred by European interests. Rising fuel costs have also increased the challenge of transporting wood supply to the markets. Further, wildfire concerns continue to increase as Georgia's population grows and urbanization increases. Air quality has become a concern, and prescribed fire has been targeted as one of the many sources of harmful emissions (Georgia Forestry Commission, 2013; Izlar, 2000). Firefighting abilities have diminished over the years, which have not been helped by budgetary pressures on the Georgia Forestry Commission.

The top three things that would help to overcome present challenges to plan implementation are higher stumpage prices (or better markets), lower site preparation costs, and lower property taxes. There is a large quantity of timber on the Trust's land that could be used as chip-n-saw products, but the price is currently $10–$15 (U.S. dollars) per ton less than it was 15 years ago. Reforestation is among the costliest expenses. Because of the wet soil conditions on the property, double-bedding is used rather than single-bedding, which increases expenses, but also increases seedling survival by 20%.

REFERENCES

A.K.S. Trust, 1937. Trust Indenture. August 18, 1937, A.K.S. Trust, Clinch County, GA.

Asaro, C., Sullivan, B.T., Dalusky, M.J., Breisford, C.W., 2004. Volatiles associated with preferred and nonpreferred hosts of the Nantucket pine tip moth, Rhyacionia frustrana. Journal of Chemical Ecology 30, 977–990.

Caldwell, P., Vepraskas, M.J., Gregory, J.D., 2007. Physical properties of natural organic soils in Carolina Bays of the southeastern United States. Soil Science Society of America Journal 71, 1051–1057.

E.S.S. Trust, 1939. Trust Indenture. December 27, 1939, E.S.S. Trust, Clinch County, GA.

Georgia Forestry Commission, 1999. Georgia's Best Management Practices for Forestry. Georgia Forestry Commission, Macon, GA. 68 p.

Georgia Forestry Commission, 2013. Georgia's Sustainable Forests: A Resource for All Generations. Georgia Forestry Commission, Macon, GA. 35 p. http://gatrees.org/resources/publications/2013Sustainability%20Report%20Final%20LowRes.pdf (Accessed April 22, 2014).

Herndon, G.M., 1968. Naval stores in colonial Georgia. Georgia Historical Society 52 (4), 426–433.

Izlar, R.L., 1984. Some comments on fire and climate in the Okefenokee swamp-marsh complex. In: Cohen, A.D., Casagrande, D.J., Andrejko, M.J., Best, G.R. (Eds.), The Okefenokee Swamp: Its Natural History, Geology and Geochemistry. Wetlands Surveys, Los Alamos, NM. pp. 70–85.

Izlar, B., 2000. Honey, I think I shrunk the drip torch!. In: Moser, W.K., Moser, C.F. (Eds.), Tall Timbers Fire Ecology Conference Proceedings. Fire and Forest Ecology: Innovative Silviculture and Vegetation Management, vol. 21. Tall Timbers Research Station, Tallahassee, FL. pp. 10–11.

Izlar, B., 2007a. Enlightened entrepreneurs: Alex Sessoms, Charles Herty and how they reversed 100 years of indifference to save Georgia's forests. In: Sessoms Timber Trust 70th Anniversary Keynote Address. 70th Anniversary of Sessoms Timber Trust, Sessoms Timber Trust, Cogdell, GA. pp. 11–13.

Izlar, B., Dangerfield, C., Izard, J., Coe, C., 2011. Property Tax Incentives for the Georgia Landowner. University of Georgia, Center for Forest Business, Athens, GA. Center for Forest Business Research Note No. 3.

Mackun, P., Wilson, S., 2010. Population distribution and change: 2000 to 2010. 2010 Census Briefs. U.S. Department of Commerce, Census Bureau, Washington, DC. C2010BR-01. http://www.census.gov/prod/cen2010/briefs/c2010br-01.pdf (Accessed April 3, 2014).

National Interagency Fire Center, 2013. Wildland Fire Summary and Statistics Annual Report. National Interagency Fire Center, Boise, ID.

Reed, G., 1982. Saving the naval stores industry: Charles Holmes Herty's cup-and-gutter experiments 1900–1905, Journal of Forest History 26 (4), 168–175.

Tran, Y., Siry, J.P., Harris, T.G., Izlar, B., 2014. How forest landowners connect us to the world. Georgia Forestry Today 10 (1), 2–24.

Yin, Z.Y., 1993. Fire regime of the Okefenokee Swamp and its relation to hydrological and climatic conditions. International Journal of Wildland Fire 3 (4), 229–240.

Chapter 26

South Willapa Bay Conservation Area, Washington, United States of America

Liane Davis,[1] Tom Kollasch,[1] Kyle M. Smith[1] and Kevin Boston[2]

[1]The Nature Conservancy, Astoria, Oregon, USA, [2]Department of Forest Engineering, Resources and Management, Oregon State University, Corvallis, Oregon, USA

ABBREVIATIONS

Conservancy   The Nature Conservancy
Refuge   Willapa National Wildlife Refuge
SWBCA   South Willapa Bay Conservation Area

MANAGEMENT SETTING AND BACKGROUND

In July of 2003, The Nature Conservancy (the Conservancy) and U.S. Fish and Wildlife Service signed a Memorandum of Understanding for the purpose of "collaborating to accomplish forest management goals and objectives" on properties managed by both parties in Pacific County, Washington (U.S. Fish and Wildlife Service, 2011). This began a partnership to restore young, managed forestlands to old-growth conditions at a landscape scale across the Conservancy's 6,931 acre (ac) (2,805 hectare (ha)) Ellsworth Creek Preserve and the 7,230 ac (2,926 ha) terrestrial portion of the neighboring Willapa National Wildlife Refuge (the Refuge). Collectively, this area (Figure 26.1) represents the roughly 14,161 ac (5,731 ha) South Willapa Bay Conservation Area (SWBCA). This low-elevation area is entirely within the Sitka spruce (*Picea sitchensis*) forest zone (Franklin and Dyrness, 1973). Historically, the area was dominated by unmanaged temperate old-growth forests. Extensive forest harvest has profoundly changed ecological conditions within the landscape. Currently, the landscape is dominated by young (10–60 years old) managed forest stands; these stands were commonly densely planted, even-aged conifers of a single species (Figure 26.2). Because of the rarity and biological significance of old-growth forest ecosystems in the region, the Conservancy and Refuge are working together to test and implement methods to accelerate forest development processes to restore a forested landscape that once again supports ecological functions representative of those of the historical old-growth forest landscape. The South Willapa Bay Forest Landscape Restoration Plan was prepared to provide specific goals and management guidance over a 20-year period to promote this restoration effort.

The Refuge was established in 1937 as a refuge and breeding ground for migratory birds and other wildlife in and around Willapa Bay (Executive Order 7541, dated Jan. 22, 1937). The Refuge currently manages approximately 15,000 ac (6,070 ha) of land that includes coastal dunes and beaches, intertidal mudflats, saltwater and freshwater marshes, grasslands, and forestlands. As mentioned, the terrestrial forested portion of the Refuge is approximately 7,230 ac (2,926 ha) including 362 ac (147 ha) designated as a Research Natural Area.

The Nature Conservancy is an international nonprofit 501(c)(3) conservation organization whose mission is to conserve the lands and waters, on which all life depends. The Washington chapter of the Conservancy was established in 1979 and began acquiring properties that formed the Ellsworth Creek Preserve in 1998. Currently, the Ellsworth Creek Preserve encompasses almost the entire 5,000 ac (2,024 ha) Ellsworth Creek watershed and includes upland forest and freshwater stream systems and estuarine habitats. Overall, both ownerships are dominated by structurally simple managed forests younger than 60 years of age since harvest (Figure 26.3).

Forest Plans of North America. http://dx.doi.org/10.1016/B978-0-12-799936-4.00026-6

225

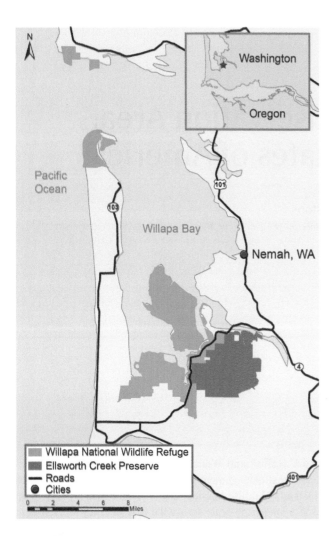

FIGURE 26.1   Map of the South Willapa Bay Conservation Area.

FIGURE 26.2   An even-aged coniferous forest within the South Willapa Bay Conservation Area, containing remnant forest structure.

Forests within the SWBCA are mainly composed of Douglas-fir (*Pseudotsuga menziesii*), western hemlock (*Tsuga heterophylla*), western red cedar (*Thuja plicata*), Sitka spruce, and red alder (*Alnus rubra*). Older stands of western hemlock, Sitka spruce, and western red cedar can contain more than 90 thousand board feet (MBF) per ac (525 m³ per ha). In younger age classes of managed conifer stands, the density of trees can exceed 1,900 per ac (4,695 per ha), while research indicates that many old-growth stands likely initiated and developed at densities as low as 30–50 trees per ac 74–124 trees per ha (Poage and Tappeiner, 2002). In terms of structural components that are important contributors to wildlife habitat, old-growth coniferous forests often contain more than 60 snags per ac (148 per ha) that have an average diameter at breast

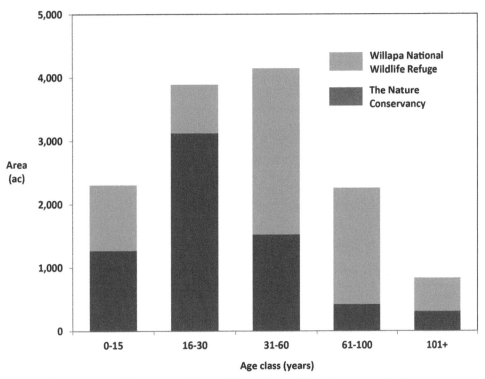

FIGURE 26.3   Forest age-class distribution of the inventoried forests within the South Willapa Bay Conservation Area.

height (DBH) of 12 inches (in.) (29.7 centimeters (cm)), and perhaps 200–300 down logs per ac (494–741 per ha). Such structures are often absent in young, managed stands.

PLANNING ENVIRONMENT AND METHODOLOGY

Given the highly simplified forest conditions that existed at the time of purchase, it was expected to take long periods of time for the SWBCA to naturally redevelop complex forest canopy and understory structures, high levels of standing and downed wood, dynamic and complex stream habitats, diverse species communities, and resilience to natural disturbances that are typical of unmanaged late-successional forest landscapes in the Pacific Northwest (Franklin and Spies, 1991; Naiman et al., 2000). The intent of management within the SWBCA is to test and implement techniques to accelerate development of ecological processes that create and eventually maintain dynamic and resilient naturally functioning forest and stream systems. The Conservancy and Refuge propose to do this by actively abating threats to the landscape and sources of habitat degradation. The major identified threats include simplified forest habitat and the hydrologic impacts of an extensive forest road systems, degraded stream habitats, increased sediment loads, and invasive species that are often tied to intensive timber harvest. Restoration and management practices will be based upon the best science available and the level of active management will be varied across the landscape. Monitoring and refinement of management practices will occur as a key component of the restoration process.

Left untreated, young-managed forest landscapes may independently develop ecological conditions that are comparable to unmanaged or late-successional forest landscapes found within the same physiographic province, but it is assumed that this will occur over very long time periods (>200 years). This landscape restoration project aims to test whether active management of these landscapes might accelerate the development of these conditions. It is recognized that existing unmanaged forest landscapes developed under unique environmental conditions and that those histories cannot be replicated (Spies et al., 2002). Thus, metrics from the remnant forests will only be used as an initial template for comparison, not as an ultimate target to reach and maintain throughout the landscape.

The Nature Conservancy's goal is more than just to restore the landscape to favor the development of late-successional forests found in the physiographic province. The Conservancy aims to develop methods to accelerate these characteristics in young stands through a series of planned experiments whose knowledge may be used by other organizations that have similar goals to restore and recreate resilient landscapes in the Pacific Northwest.

There are two forest management goals for the Refuge: (1) to preserve and protect unique ecosystems associated with Willapa Bay and (2) to manage for the conservation and recovery of threatened and endangered animals in their natural ecosystems. These goals are intended to be accomplished through a set of specific objectives for their forest management program. The primary objective is to restore ecological function to Refuge forests by creating a natural distribution of stand structure, composition, and successional stages while promoting old-growth and late-successional characteristics to benefit forest dependent wildlife—especially the marbled murrelet (*Brachyramphus marmoratus*). This will be achieved by decommissioning unnecessary forest roads to reduce or eliminate stream impacts and reduce the fragmentation of forest habitat. Also, the Refuge has adopted forest management practices designed to change thickets of western hemlock over a period of time to resemble old-growth in its structural complexity and to reduce fuel loads and fire risk. As a federal agency with responsibility for a variety of species, the Refuge aims to protect and, where appropriate, restore associated stream habitat to prevent further declines of anadromous fish stocks and enhance native amphibian populations and other stream-dependent wildlife species. Their final objective is to reduce the risk from insects and disease where endemics are likely, referring to remaining patches of endemic forest habitat.

Constraints for the Analyzed Area

After analyzing the ecological conditions and management goals for the property, the following management constraints were identified during plan development. One set of constraints is imposed by limits on staff resources that can be devoted to management of the area. The second set of constraints is limited by the financial resources that can be spent on restoration by the Conservancy. While producing revenue is not the primary management objective, the ability of the Conservancy and Refuge to fund and implement road decommissioning, road maintenance, and forest and stream restoration activities will be significantly affected by budget and staffing levels. The source of these funds will be from available grants and the amount of revenue that can be produced from forest restoration activities. And while the decision of when and how to treat stands will be driven primarily by ecological criteria, the costs and potential revenues from alternative treatments must be factored in when considering what treatments are economically feasible. Financial constraints on restoration activities arise most significantly because of the existing road system. At the time of purchase, the existing road access for silvicultural operations ranged from good to poor throughout the landscape, and many segments in the road systems were in need of significant maintenance and upgrades at considerable expense. Roads presenting a high risk for erosion and landslide will need to be decommissioned, and removal of these roads will limit access to some portions of the landscape. Also, the rugged and steep terrain constrains the harvest systems that can be used, requiring the use of more expensive cable yarding systems as opposed to the lower cost, ground-based systems.

The Nature Conservancy has devised a long-term experimental research project, designating eight sub-basins within a portion of its ownership as active, passive, or control areas, known as the Ellsworth Creek Adaptive Management Area. Active basins receive road upgrades and thinning, while passive and control basins receive no silvicultural intervention. The Conservancy will remove and decommission all roads within the passive basins and upgrade those within the control basins. While this will allow for a comparison between the treatment areas over time, it imposes significant financial constraints where potential revenue from forest thinning is foregone and increased costs are incurred where roads must be removed.

Both the Conservancy and Refuge are valued places for public recreational and educational purposes. Until recently, these areas were largely in industrial timber production and are still valued by many in the community as a place for forest-sector jobs and revenue. Maintaining public access and minimizing negative economic impacts on the local community are also important considerations for the Conservancy and Refuge in management decisions.

Guiding Laws, Regulations, and Policies

The Conservancy, as a private forestland owner, must comply with Washington State Forest Practice Act and water quality laws. This requires the Conservancy to apply for permits under Forest Practice Act regulations for forest management actions that may affect the resources of the state. The Refuge, as a federal agency, is not required to obtain state permits for similar work (the Refuge nonetheless strives to conduct work at or above these standards). The Refuge is required, through the National Environmental Policy Act, to conduct a review of significant management actions. The National Environmental Policy Act also requires the Conservancy to consult with federal natural resource management agencies prior to using federal funds to conduct management actions or where federal funds have been used to acquire portions of the SWBCA.

The Conservancy and Refuge must follow all applicable laws and regulations pertaining to active management, which could impact sensitive, threatened, or endangered species listed under the federal Endangered Species Act and Washington State protected species regulations (WAC 232-12-297). A particular emphasis is placed on avoiding disturbance to listed species. All forest management activities will be aimed at increasing suitable habitat over time for listed species.

The principal operational constraints on forest management activities pertaining to listed species occur in relation to marbled murrelets and, to a lesser degree, northern spotted owls (*Strix occidentalis caurina*). Both species have specific protection measures codified within the Forest Practice Act regulations. These regulations are largely intended to control the level of impact from forest practices, such as clear-cut harvesting, where listed species may be present. Because forest restoration is the primary goal within the SWBCA, alternative practices may be appropriate. The Conservancy consults with the appropriate state and federal regulators prior to implementing alternative practices.

The Nature Conservancy has been a member of the Forest Stewardship Council (FSC) since 2001 and holds a certified forest manager credential for over 250,000 ac (101,174 ha) nationwide. The Conservancy requires itself to obtain FSC certification on its lands before conducting extensive forest management activities. The plan directs the Conservancy to obtain FSC certification at its Ellsworth Creek Preserve as a primary goal; thus, all management actions undertaken in the SWBCA must comply with FSC-US Forest Management Standards using the Pacific Coastal Standards (Forest Stewardship Council U.S., 2010).

Plan Development Methods

The context for management planning was developed through the evaluation of the physiographic setting (climate, geology, and soils), the conservation significance, the site history, current condition of the surrounding landscape, and a landscape recourse assessment. Three primary assessments informed the essential elements of the plan. The first was a forest vegetation structure assessment, essentially a modified forest inventory cruise that gathered standard and nonstandard attributes of forest composition, cover type, and health. The second assessment was a complete survey of stream habitat that included a computation of the biotic integrity index that represents the health of large wood volumes, channel complexity, and channel substrate suitability. The final assessment surveyed the condition forest roads cataloging culverts and fish passage barriers, drainage conditions, and erosion or mass-wasting risk as well as surfacing, ditch line, and cut-slope variables.

Forest inventory data collected by the Conservancy from 2002 to 2004 was used to assign stands in the SWBCA into forest age and stand type classifications (Table 26.1). General stand statistics, including structural metrics for each inventoried stand were then generated. A Modified Old-growth Index was developed and used, based on four structural variables associated with old-growth forests: tree size, number of snags, volume of down woody debris, and tree size diversity. The resulting index value was based on the following criteria:

TABLE 26.1 Distribution of Stand Types Inventoried in 2002–2004

Stand Type	Age (Years)	The Nature Conservancy Area (ac)	Willapa National Wildlife Refuge Area (ac)	Total Area (ac)
WH-SS-RC-1	0–15	870	840	1,710
WH-SS-RC-2	16–30	1,692	127	1,819
WH-SS-RC-3	31–60	1,512	2,497	4,009
WH-SS-RC-4	61–100	420	1,627	2,047
WH-SS-RC-5	101+	292	534	826
Douglas-fir-1	0–15	448	115	563
Douglas-fir-2	16–30	1,428	454	1,882
Alder-1	0–15	73	77	150
Alder-2	16–30	2	221	223
Alder-3	31–60	118	136	254
Alder-4	61–100	0	220	220
Non-forest	—	76	382	458
Total		6,931	7,230	14,161

WH, western hemlock; SS, Sitka spruce; RC, western red cedar.

1. Number of large trees per unit area greater than 39.4 in (100 cm) in DBH.
2. Number of large snags per unit area greater than 19.7 in (50 cm) DBH and greater than 49.2 feet (15 meters) tall.
3. Volume of down woody debris per unit area.
4. Tree size diversity or number of trees within four different diameter classes.

The old-growth index values also proved to be fairly well correlated with age class, and the scores were relatively low for most of the inventoried stands as would be expected given the SWBCA's management history. Examples of Modified Old-growth Index values are around 15–32 for young (0–15-year-old) and medium-aged (16–30-year-old) Douglas-fir forests, 25–32 for red alder forests, and 48–62 for older (61–100 years and 100+ years) western hemlock, Sitka spruce, and western red cedar forests.

The desired future condition for the SWBCA landscape was discussed and agreed upon by partners through a facilitated workshop that identified the following elements to be created: a forested ecosystem resistant and resilient to environmental perturbation at multiple scales; a landscape reflective of high spatial and temporal heterogeneity; maintenance of functional landscape linkages where natural movement of materials, species, and ecological processes can occur across forests, stream, and estuarine systems; and provision of habitat for late-successional species and species of concern.

The silvicultural system for active forest restoration was defined, encompassing four broad silvicultural treatment categories: (1) drop and leave, (2) biomass removal, (3) understory management, and (4) decadence acceleration. The drop and leave system implies that trees will be left on the ground following a thinning operation. In younger forests, the operation is considered a variable density thinning. In older forests, it involves individual tree selection. The biomass removal system implies that most of the trees will be removed from the site and that some may be sold for revenue generation purposes. These operations include variable density thinnings, individual tree selection systems, and group selection harvests. The understory management system is intended to accelerate the development of diverse understory structure and composition while promoting overstory tree species diversity. This system involves planting, shrub control, browse protection, variable density thinning, and invasive species control treatments to the forests. The decadence acceleration system is intended to increase the number of snags and facilitate large down wood recruitment processes. Treatments in this system include snag creation practices, coarse wood creation activities, and fungal and mistletoe inoculation of trees.

Areas of high value to landscape processes, areas of high biodiversity, or other unique or sensitive areas were first identified and the appropriate type of management determined. The experimental treatment designations for each sub-basin within the Ellsworth Creek adaptive management area were then incorporated. Next, the remaining portions of the landscape were analyzed to determine what categories of treatments were appropriate, given stand conditions and landscape scale considerations. A prioritization framework to provide guidelines for future planning and scheduling efforts was therefore developed. Since prioritization is rarely a linear, formulaic process, it must involve consideration and balancing of many overlapping, conflicting, or interconnected factors that operate on multiple scales. The framework for prioritizing stands was based on five factors:

1. *The contribution to major landscape goals.* Components included expanding and connecting blocks of late-seral forests, restoring and maintaining stream function, contributing to marbled murrelet habitat, and restoring landscape windthrow dynamics.
2. *The coordination of treatments with road system needs.* Components included grouping stands for treatment within a road system, putting roads "to bed" when activities were complete, and treating stands where road removal was a high priority.
3. *The need to address stand-level structural condition.* Components included addressing stand density issues, addressing windthrow susceptibility, assessing potential responses to thinnings, scheduling to meet thinning windows, and addressing potential tree species losses.
4. *The need to acknowledge economic factors.* Components involved understanding markets, harvesting costs, and revenue potential and weighing these against silvicultural trade-offs.
5. *The need to address management concerns.* Components involved addressing experimental design needs, work-flow needs, and revenue expectations.

Stands were categorized into management designations based on the above analysis. In most cases, designations were clearly prescribed by the management objectives, legal requirements, or practical considerations, such as road access or social factors. These include Ecological Reserves like the existing blocks of old-growth forests larger than 5 ac (2 ha) and Research Natural Areas. The only management intervention that may take place is fire suppression and removal of invasive non-native species. The Reserve category establishes the control and passive basins for ongoing research on the impact of the restoration activities. The passive basins within the Ellsworth Creek Adaptive Management Area will never be thinned and all roads will be removed. Similarly, the control basins will not be thinned for at least the next 10–20 years, but the roads will remain to have an experimental control to evaluate against the effects of road removal and restoration silviculture in other parts of the landscape. They also provide for landscape heterogeneity by ensuring that a portion of young stands remains in an untreated condition. Another management designation category is Limited Management Areas, which are stands where biomass removal treatments are rarely appropriate given regulatory requirements or management constraints. Other silvicultural treatments, such as drop and leave, decadence acceleration, or understory management treatments may be more appropriate to achieve ecological objectives. Finally, Unreserved Management Areas represent the remaining portion of the landscape where all categories of restoration silviculture may be fully applied. The types of treatment used in specific areas are driven by a decision model that considers the location of stands, whether they are in "road removal" areas, the potential to meet desired future stand conditions, stand age, the appropriateness of silvicultural treatments, and impacts on ecological objectives.

Growth models, economic analysis tools, and wind models were used to evaluate the ecological and economic trade-offs between management activities under alternative treatment scenarios. Stands were then scheduled for treatment based on the treatment categories, management designations, the generalized prioritization framework, economic analyses, Forest Vegetation Simulator (Pacific Coast variant) (Dixon, 2013) results, and other factors outlined in the plan. Five time periods were used for scheduling: an annual basis for the first 3 years (2007, 2008, and 2009), the subsequent 7-year period (2010–2016), and concluding with a final 10-year period (2017–2026).

OUTCOMES OF THE PLAN

There were three main outcomes of the plan. First, the plan facilitated open communication and dialog between the Conservancy and the Refuge, helped to establish roles, responsibilities, and a common understanding of goals, objectives, and actions needed to achieve the common results. This has fostered effective and efficient cross-ownership collaboration on forest and road management. Second, it allowed the Conservancy to obtain FSC certification for the Ellsworth Creek Preserve. Third, it allowed for significant progress toward implementation of active forest restoration treatments and road upgrades and maintenance along with road decommissioning. These activities included the following:

- 20.2 miles (mi) (32.5 kilometers (km)) of road upgrades
- 15.7 mi (24.3 km) of decommissioned or obliterated roads
- 3,831 ac (1,550 ha) of drop and leave silvicultural practices
- 731 ac (296 ha) of biomass removal thinning practices

The plan enabled implementation and has elevated the exposure of the project as a real-world example of restoration management. Other entities and organizations interested in similar restoration management regularly request the plan as well as field tours of the SWBCA.

However, the potential for revenue generation is limited by the inherent young age of many of the forest stands on this forest. Many of the stands will not be coming "online" for commercial thinning for one to two decades. This is a problem that cannot be overcome by anything but time and growth. If active restoration is the goal of forest management and it is expected that operations will "pay for themselves," then attention must be paid to age-class distributions at the time of purchase. The goal is to have thinning prescriptions that can provide significant revenue to support restoration operations over time, but the margin is tight and is often not amenable to unforeseen expenses. Revenue from other sources (e.g., carbon credits) may help to provide some cushion in the future but is not currently a viable revenue source.

The current regulatory environment is often not well suited to encourage restoration; it is designed to restrict damage and disturbance and to be risk-averse. In this sense, it can also be limiting to restoration activities which, while well intentioned, are often inherently experimental. The plan, including the experimental design, enabled partners to negotiate flexibility, in an otherwise restrictive regulatory environment, for creative restoration treatments implemented within potential murrelet habitat or near occupied habitat. Obtaining such flexibility would have likely been greatly hindered if an official plan did not exist.

DISCUSSION AND CONCLUSIONS

Learnings and Insights

Planning was instrumental in clarifying roles and responsibilities as well as in making certain everyone was clear on what management activities would occur and why, including the impact of constraints and considerations that would be used for scheduling or implementing activities. It clarified the decision-making tools and framework for restoration. In hindsight, the planning process suffered from a lack of expertise in operational forestry, as TNC did not hire a forester until late in plan development. As such, many of the operational assumptions that went into harvest scheduling and road maintenance were incomplete or inaccurate. The original schedule was discarded within a year as realities about the condition of stands and roads were better understood and inefficiencies in the original schedule came to light. A new schedule was developed once the interactions of operational constraints (e.g., road conditions) and forest stand conditions were better understood.

The project has resulted in increased capacity, as Refuge crews became highly skilled in road decommissioning processes and equipment operation, while TNC staff became highly skilled in watershed science, forest planning, and forestry operations. Merging these strengths across ownerships has translated into increased capacity and efficiencies for both partners.

Sustainability Issues

The inclusion of an experimental research project to evaluate active and passive restoration pathways was an important facet of the Conservancy's Ellsworth Creek project, but balancing restoration, operational, financial, and research objectives presents a continuous challenge. Most significantly, it presents challenges to financial sustainability by limiting the area available for commercial thinning and increasing the funding needed for road decommissioning. Maintaining institutional support and commitment for long-term projects can be challenging within any organization or agency. Human nature tends to have shorter attention spans than a project intended to restore a late-seral forest ecosystem, so there is a need to produce frequent examples of project benefits and results.

The presence of a threatened or endangered wildlife species can pose similar challenges for restoration activities as it does for other types of forest management. A major goal of the project is to restore habitat conditions critical to some listed species, but restoration activities intended to achieve those goals are somewhat constrained by the possibility of inadvertently disturbing or affecting potential habitat. A set of conservation measures has been agreed upon by the Conservancy and the U.S. Fish and Wildlife Service to protect these species while enabling restoration activities, but this required a lengthy negotiation process that other managers may not be well suited to or able to engage in successfully.

Plan Development Challenges

The planning process and implementation schedule fostered a dialog around issues and challenges associated with restoration management. One important reason for the project's existence has been exporting the restoration insights learned here. However, quantifying the impact of the project beyond its boundaries in a way that can be meaningfully communicated to people has proven difficult. In addition, the initial scheduling of forest restoration activities was made based on limited data of stand ages, conditions, and initial road assessments. Once these factors were better understood, forest restoration activities had to be rescheduled and prioritized. Thus, developing and maintaining an inventory is critical to plan implementation.

Plan Implementation Challenges

Inclusion of the experimental design in the planning process permitted engagement of a diverse set of academic and agency interests that may have otherwise been overlooked in the project. It enabled the collection of an extensive amount of baseline pretreatment data that is rarely available and provides a great opportunity to actually practice adaptive management. It has also helped to establish trust and credibility both with people hesitant to harvest trees as a restoration tool as well as regulators concerned with resource and species protection. However, it did present some operational and financial constraints and challenges to implementation and sustainability as discussed previously.

In addition to these issues, access to the funds necessary to initiate implementation activities has been challenging. Upfront costs necessary for road maintenance and upgrades prior to logging were needed before revenue could be generated from thinning operations to offset restoration costs. The road system was in much worse condition than initially anticipated, which required a significant amount of funding to reroute, decommission, and upgrade to a new and improved road network. Many forest restoration activities were delayed until roads were upgraded to a suitable condition to support logging equipment and log hauling trucks. We recommended that road assessment and budget be part of data collected during the plan development process. Further, the experimental design, by designating unthinned control and passive sub-basins, placed some additional financial constraints on the project by reducing the total area that could have potentially been thinned for the generation of revenue. This constraint is unlikely to be encountered by other landowners not engaged in an adaptive management experiment.

As was learned, timber markets have a huge and immediate impact on the financial feasibility of implementation of the plan. Project implementation was delayed due to the recent recession and has been able to function over the past couple years largely due to good markets. There is such a tight profit margin with thinning that managers must be prepared to act when market conditions are favorable. There is a limited economic return from sale of FSC certified logs—the local market is not well developed and is not yet providing a strong premium for certification.

Finally, the administrative and data management infrastructure needed for forest management is complex (e.g., contracting and legal mechanisms, internal project approval processes and policies, log accounting mechanisms, forest inventory data management). Such infrastructure, particularly for conservation organizations, is often not well suited to handle the specific forest management infrastructure needs. A delay in project implementation may occur if significant capacity changes or additions are needed within an organization.

REFERENCES

Dixon, G.E., 2013. Essential FVS: A User's Guide to the Forest Vegetation Simulator. U.S. Department of Agriculture, Forest Service, Forest Management Service Center, Fort Collins, CO. 226 p.

Forest Stewardship Council U.S., 2010. FSC-US Forest Management Standard (v1.0). Forest Stewardship Council U.S, Minneapolis, MN. https://us.fsc.org/download.fsc-us-forest-management-standard-v1-0.95.pdf (Accessed April 27, 2014).

Franklin, J.F., Dyrness, C.T., 1973. Natural vegetation of Oregon and Washington. U.S. Department of Agriculture, Forest Service, Pacific Northwest Forest and Range Experiment Station, Portland, OR. General Technical Report PNW-8. 427 p.

Franklin, J.F., Spies, T.A., 1991. Composition, function and structure of old-growth Douglas-fir forests. U.S. Department of Agriculture, Forest Service, Pacific Northwest Research Station, Portland, OR. General Technical Report PNW-GTR 285.

Naiman, R.J., Bilby, R.E., Bisson, P.A., 2000. Riparian ecology and management in the Pacific coastal rain forest. BioScience 50, 996–1011.

Poage, N.J., Tappeiner II, J.C., 2002. Long-term patterns of diameter and basal area growth of old-growth Douglas-fir trees in western Oregon. Canadian Journal of Forest Research 32, 1232–1243.

Spies, T.A., Reeves, G.H., Burnett, K.M., McComb, W.C., Johnson, K.N., Grant, G., Ohmann, J.L., Garman, S.L., Bettinger, P., 2002. Assessing the ecological consequences of forest policies in a multi-ownership province in Oregon. In: Liu, J., Taylor, W.W. (Eds.), Integrating Landscape Ecology into Natural Resource Management. Cambridge University Press, Cambridge, UK. pp. 179–207.

U.S. Fish and Wildlife Service, 2011. Willapa National Wildlife Refuge, Final Comprehensive Conservation Plan and Environmental Impact Statement. U.S. Department of the Interior, Fish and Wildlife Service, Willapa National Wildlife Refuge, Ilwaco, WA.

Chapter 27

Umbagog National Wildlife Refuge, New Hampshire, United States of America

Thomas LaPointe, Paul Casey, Ian Drew and Sean Flint

Northern Forest Land Management Research and Demonstration Program, Umbagog National Wildlife Refuge, Errol, New Hampshire, USA

ABBREVIATIONS

CCP   Comprehensive Conservation Plan
FMP   Forest Management Plan
HMP   Habitat Management Plan
NWR   National Wildlife Refuge

MANAGEMENT SETTING AND BACKGROUND

Established in 1992, Umbagog National Wildlife Refuge (Refuge) straddles the border of New Hampshire and Maine in the counties of Coos (New Hampshire) and Oxford (Maine), approximately 30 miles (mi) (about 48 kilometers (km)) south of the United States and Canada border (Figure 27.1). Umbagog Lake, the centerpiece of the Refuge, is an impounded lake at the headwaters of the Androscoggin River. The Emergency Wetland Protection Act of 1979 was one of three acts that allowed for the original acquisition of land in the Umbagog Lake area by the U.S. Fish and Wildlife Service in 1992. As a result, most of the initial purchases were in wetland areas, including emergent and forested wetlands, and contained relatively little upland forest. In recent years, the Refuge has acquired additional forested uplands surrounding these wetlands. Current holdings include 22,500 acres (ac) (70%, 9,106 hectares (ha)) of forested uplands, 4,000 ac (12%, 1,619 ha) of forested wetlands, and 5,500 ac (17%, 2,226 ha) of freshwater wetlands.

The Refuge is located within an industrial forest landscape where much of the land is managed by private landowners. Conserved lands are becoming more common, but are still a minor component of the patchwork of ownership. Much of the forested land purchased by the Refuge within the last decade (~10,000 ac (4,047 ha)) had the majority of commercial timber removed prior to acquisition. Earlier acquisitions contain a greater abundance of commercial timber creating opportunities to use commercial forest management to achieve the biological goals of the Refuge. The economy of the area is dependent on forestry and seasonal recreation.

The Refuge is within a transition zone between boreal and deciduous forests (Figure 27.2) and contains a variety of forest communities associated with both types of forest. Black spruce (*Picea mariana*) woodland bogs, northern white cedar (*Thuja occidentalis*) swamps, floodplain forests, and spruce-fir-tamarack (*Larix laricina*) wetlands typically occur on hydric soils with a peat component. Lowland spruce-fir communities dominated by red spruce (*Picea rubens*) and balsam fir (*Abies balsamea*) typically occur on poorly drained, shallow soils common on gentle slopes and flats. Northern hardwood communities occur on dryer nutrient rich soils and include species such as American beech (*Fagus grandifolia*), paper birch (*Betula papyrifera*), yellow birch (*Betula alleghaniensis*), red maple (*Acer rubrum*), sugar maple (*Acer saccharum*), quaking aspen (*Populus tremuloides*), bigtooth aspen (*Populus grandidentata*), and mountain ash (*Sorbus americana*). Northern hardwood-conifer communities typically occur along gradients between deciduous and coniferous communities and are dominated by species from both. Montane spruce-fir communities occur on high elevation and rocky talus soils with characteristic stunted red spruce and balsam fir trees.

Forest Plans of North America. http://dx.doi.org/10.1016/B978-0-12-799936-4.00027-8

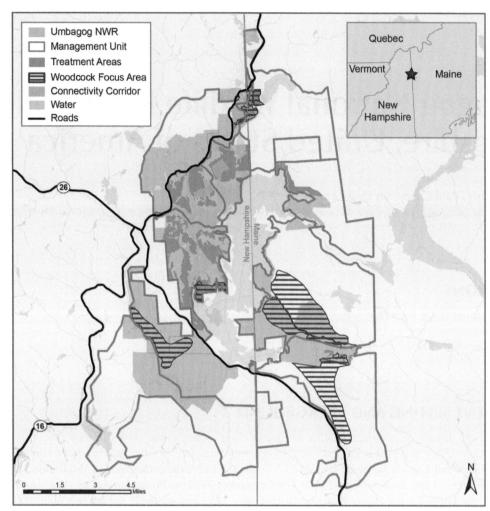

FIGURE 27.1 A map of the Umbagog National Wildlife Refuge.

FIGURE 27.2 A northern forest landscape in New Hampshire and Vermont.

The current forest composition and structure is the result of a history of timber harvesting, natural events, and introduction of diseases. Eastern white pine (*Pinus strobus*) and spruce-fir forests were harvested extensively following European settlement, and over the last century spruce-fir was harvested intensively for paper production reducing the extent and maturity of the forest. A spruce budworm epidemic during the 1970s caused a massive and sudden damage in mature spruce-fir stands. The result was a change in the species dominance, with balsam fir now more prevalent in areas that historically supported greater proportions of red spruce. Eastern white pine, a timber species desired throughout history, remains a component of the forest, but once abundant shoreline stands have been reduced to scattered groups and individual trees. Historically, forests were managed using predominately even-aged techniques, and, as a result, age variation now occurs between stands rather than within stands. Forests with late-seral stage characteristics are absent from the landscape except for a few marginal conifer stands that contain large-diameter trees, snags, and rotten logs on the forest floor. Beech trees, important to wildlife for habitat and food, are under-represented in the Refuge's northern hardwood forests and are declining in northern New England at a rapid rate due in part to beech bark disease.

Wildlife conservation is a top priority at the Refuge. The Refuge's 32,000 ac (about 12,950 ha) (Figure 27.1) provide regionally significant breeding and migratory habitat for waterfowl and land birds. The Refuge contains a mosaic of spruce-fir, northern hardwood, and northern hardwood-conifer forest communities that provide habitat for a large number of important species including the blackburnian warbler (*Dendroica fusca*), Canada warbler (*Cardellina canadensis*), black-throated green warbler (*Setophaga virens*), and American woodcock (*Scolopax minor*).

PLANNING ENVIRONMENT OR METHODOLOGY

The mission of the National Wildlife Refuge System is to "…administer a national network of lands and waters for the conservation, management, and, where appropriate, restoration of fish, wildlife, and plant resources and their habitats within the United States for the benefit of present and future generations of Americans" (U.S. Fish and Wildlife Service, 1997 and 2009, pp. 1–7). The 1997 National Wildlife Refuge System Improvement Act (Improvement Act) initiated a renewed vision for the future of the refuge system where:

- Wildlife comes first
- Refuges are anchors for biodiversity and ecosystem-level conservation
- Lands and waters of the System are biologically healthy
- Refuges are national and international leaders in habitat management and wildlife conservation

The Improvement Act required that all National Wildlife Refuges complete a Comprehensive Conservation Plan (CCP) (U.S. Fish and Wildlife Service, 2009). These plans must comply with the standards of the National Environmental Policy Act (NEPA) of 1969, which requires all federal agencies to examine the environmental impacts of their actions by incorporating environmental information and public participation. The 2009 Umbagog NWR CCP is the culmination of a 7-year planning process that guides the management of all refuge programs over the next 15 years. The CCP articulates the Refuge's vision and conservation role within the landscape. It defines biological, recreational, educational, and partnership goals for the Refuge, and outlines objectives and strategies for implementation. The CCP is an overarching plan often requiring additional *step-down* plans that describe implementation in greater detail.

The 2010 Umbagog Habitat Management Plan (HMP) addresses (steps down) the biological goals of the CCP. It provides specific guidance on managing habitats critical for the resources of concern at Umbagog NWR. In addition to the CCP and HMP, the Umbagog NWR developed a Forest Management Plan (FMP). The FMP provides specific information on where, when, and how to achieve HMP Objective 3.1, which states "…sustain well-distributed, high quality breeding and foraging habitat for species of conservation concern…". Also, where consistent with management for those refuge focal species, "protect critical deer wintering areas and provide connectivity of habitat types for wide-ranging mammals." Achieving this objective will in turn lead to achieving CCP Goal 3, which states "Manage upland forest habitats, consistent with site capability, to benefit Federal trust species and other species of conservation concern." The FMP prescribes specific forest management for a series of treatment areas that will be managed over the next 15 years. Prior to a management action an implementation plan is written for each treatment area and includes site-specific information and stand-level prescriptions.

Management Units and Management Zones

The Refuge is divided into 10 management units (Figure 27.1). Management unit boundaries were defined primarily by habitat conditions, yet the definition further included the practical and logistical aspects of managing habitats. The purpose was to create units that could be managed ecologically, be recognized by Refuge staff, be analyzed and assessed individually, and made sense logistically. Topographic and geographic features such as roads and waterways, and occasionally political lines such as town and state boundaries, were used to define management unit boundaries. Management units included Refuge lands and lands within the CCP expansion boundary (U.S. Fish and Wildlife Service, 2009).

Management zones govern the type of activities that can be applied to the land. Management zones provide protection for a variety of forest and non-forest resources identified in the Umbagog HMP. Refuge lands were divided into General Management (low resource sensitivity), Special Management (moderate resource sensitivity), and Restricted Management (high resource sensitivity and inoperable) zones. In the General Management zone, forest management decisions are afforded the greatest flexibility to diversify forest age class and structure to benefit focal species. Within the Special Management zone, forest management must consider seasonal closures of operations, the maintenance of closed canopy conditions, and the retention of coarse woody debris and/or snags. Within the Restricted Management zone, no heavy equipment is generally allowed, and no harvest may occur. However, individual trees may be felled, girdled, or otherwise treated for the benefit of wildlife. This zone includes forested wetlands and areas containing rare plants. This zone also includes areas otherwise considered unique, exemplary, or inoperable with heavy equipment (e.g., excessively steep (>30%) slopes, or hydric soils).

Desired Forest Conditions

A variety of important habitats and forest characteristics were identified in the HMP and used to develop *desired forest conditions*. Desired forest conditions inform the type of forest management thought to best create conditions considered lacking. In the HMP, four migratory bird species were selected to represent a variety of avian species and taxa that utilize similar habitats. These species, referred to as *focal species*, are blackburnian warbler, Canada warbler, black-throated green warbler, and American woodcock. Providing suitable habitat for focal species also provides habitat for a variety of associated wildlife and plant species. Also used to develop desired forest conditions are other important habitats and forest characteristics identified in the HMP, such as mixedwood forests with a high proportion of conifers, complex forest structure, deer wintering areas, and within stand features (e.g., coarse woody debris, raptor and colonial bird nest trees).

In the northern forest, the blackburnian warbler is associated with mature coniferous forests and in mixed coniferous and deciduous forests. Males sing from the tops of the tallest conifers, preferably over 60 feet (ft) (18.3 meters (m)). The warbler nests and gleans insects in the upper canopy of conifers, especially spruce and eastern hemlock (*Tsuga canadensis*) (DeGraaf and Yamasaki, 2001). It is considered a forest interior species and is susceptible to forest fragmentation and short-rotation timber harvesting (50 years or less) (Hagen et al., 1996; Morse, 2004). Blackburnian warblers are found in higher densities in upland mixed forest with a high conifer component, in contrast to wet, bottomland spruce-fir forest. Removal of large conifers decreases local populations of this species. A similar species, tied more closely to mixed forests, the black-throated green warbler occurs at highest densities in closed canopy middle-to-mature-aged forests with a significant conifer component. This foliage-gleaning warbler generally forages high in the canopy, but at a lower height than blackburnian warblers (Morse, 1967). Spruce (particularly red spruce) and paper birch are favored foraging substrates. Although it will nest in deciduous trees, preferred nest sites are in dense conifer foliage on a limb or tree fork, at a height of about 20 ft (6 m) (DeGraaf and Yamasaki, 2001; Foss, 1994). Large spruce trees are favored male singing perches (Morse, 1993). Black-throated green warblers appear to need large forest patches in a forested landscape (Norton, 1999). Askins and Philbrick (1987) found that they disappeared from a 250 ac (101 ha) forest tract that became isolated from other forested habitat, and Morse (1993) found that densities declined in heavily thinned forest. Structurally heterogeneous forests that include small gaps provide improved foraging opportunities for this warbler (Smith and Dallman, 1996). Consequently, quality breeding and foraging habitat for blackburnian and black-throated green warblers must have large contiguous upland forests greater than 250 ac (101 ha) and with an age greater than 50 years old, a high conifer component with over 70% canopy cover, and more than 60 ft (18.3 m) coniferous super canopy trees with small gap openings.

American woodcock habitat requirements are much different than those of blackburnian and black-throated green warblers. American woodcock require several different habitat conditions in close proximity to one another both on uplands and wetlands. These include small areas clear of trees and debris for courtship, large openings for night roosting, young, second-growth hardwoods (15–30 years) for nesting and brood-rearing, and young shrubby wetland foraging areas (Keppie and Whiting, 1994; Sepik et al., 1981). *Courtship areas* are diverse and include natural openings, clear-cuts, roads, pastures, lawns, cultivated fields, or reverting farmlands. The use of an opening by a male depends on the quality of the surrounding brood and nesting cover. Females choose nest sites near good brood cover and tend to nest within about 492 ft (150 m) of a singing ground. Females typically nest in young, open, second-growth forests. Good brood cover includes high stem densities of alder (*Alnus* spp.) or aspen dominant forests on moist soils (U.S. Fish and Wildlife Service, 1996). Woodcock prefer stands dominated by deciduous trees with loamy soils that are more likely to harbor earthworms. The provision of quality breeding and foraging habitat for American woodcock requires a mosaic of second-growth aspen dominant forests less than 30 years old near open fields, clearings, and alder-dominant wetlands.

Conifer forests and mixedwood forests with a high proportion of conifers, especially red spruce, were considered the most important ecological contribution the Refuge could make to the Upper Androscoggin River watershed, the Northern Forest, and the Refuge System. Also important is a complex forest structure. All uneven-aged forest management treatments attempt to create forests that include attributes similar to those that developed through natural processes such as greater than 70% canopy closure; a range of tree ages, including some more than 120 years old; shade-tolerant plant species; large-diameter trees; cavity trees; snag trees; super canopy trees; small openings of varying size; a variety of understory layers; and coarse woody debris on the forest floor.

The HMP also identified white-tailed deer (*Odocoileus virginianus*) as an important consideration because they are at the northern edge of their range and are limited by harsh winter conditions. Several areas on and adjacent to the Refuge are considered critical *deer wintering areas*. Deer wintering areas have two important habitat components: (1) conifer forests (excluding tamarack and black spruce) greater than 35 ft (10.7 m) tall, greater than 100 ft^2 per ac (23 m^2 per ha) of basal area, and at least 70% crown closure (Flatebo et al., 1999; Reay et al., 1990) and (2) woody regeneration near or within the conifer forests to provide browse. Deer wintering areas often represent connectivity corridors on the landscape and provide important links between patches of available habitat. These corridors are important at both coarse and fine scales, as connectivity provides safe movement particularly for rare species that may suffer from inbreeding or loss of genetic variation if movements between isolated populations are restricted (Flatebo et al., 1999).

Finally, habitat for colonial birds such as the great blue heron (*Ardea herodias*) and raptors, including the bald eagle (*Haliaeetus leucocephalus*) and osprey (*Pandion haliaetus*), was identified in the HMP. These species utilize mature dominant trees or snags for nesting, perching, and hunting. Trees taller than the surrounding canopy (super canopy trees) within close proximity (about 250 ft, or 76 m) to ponds, lakes, and large rivers are critical components of these species habitat. Super canopy trees within 1 mi (1.61 km) of waters that provide a consistent food source are important for continued protection of these species.

OUTCOMES OF THE PLAN

Forest management activities will occur over the next 15 years at 15 identified treatment areas that average 315 ac (127 ha) in size. Two of the treatment areas are within woodcock focus areas and utilize area control to provide desired forest conditions (early-successional habitat) in locations with site characteristics necessary for woodcock to successfully breed and rear young. The remaining 13 treatment areas contain habitat characteristics important to blackburnian and black-throated green warblers, which are generally associated with interior forests and complex forest structures. In these treatment areas, uneven-aged management is prescribed.

Plan implementation used aerial photography to identify forest stands that could be managed for the desired forest conditions using commercial forest management. Tree height and canopy density were used as a proxy for timber merchantability. Forest stands with trees greater than 30 ft (9.1 m) in height and greater than 60% canopy closure were considered potentially merchantable. Vegetation data was available for approximately 16,400 ac (6,637 ha) at the time the FMP was created. Within General and Special Management zones 4,773 ac (1,932 ha) (Table 27.1) met the criteria for merchantability. These lands were stratified into *treatment areas* assuming a 15-year timeframe for conducting harvests. The annual harvest during the next 15 years was estimated to be approximately 953 cords (3,454 m^3) of various forest products. This was considered a conservative estimate based on the best available information and forest growth and yield modeling and represented the preponderance of uneven-aged forest management. Harvests using area regulation were expected to contribute more than what was accounted for in the model estimate. Recent timber sales have exceeded estimates for the initial harvest, but we expect timber sale yields to more closely track model estimates over time.

Each treatment area is comprised of one or more broad forest types. Forest types, often referred to as *stands*, are evaluated at the time the treatment area is assessed. To facilitate implementation, the FMP includes prescription guidelines. Guidelines include traditional silvicultural descriptions and important considerations for providing desired forest conditions. The prescription guidelines are refined for each treatment area after stand level inventories are completed to reflect attributes unique to each stand. For example, the prescription for uneven-aged management in hardwood and mixedwood forests includes single-tree and group selection harvesting to transition even-aged forests to multi-aged and multistructure forests with a minimum of three age classes and a diameter distribution approaching the slope of $q = 1.3$. The q-metric (diminution quotient) is the ratio between the number of trees in successively smaller diameter classes and is a measure of the slope of the diameter distributions approximated by the reverse J-shaped curve (Leak et al., 1987). This guidance produces a stand (Figure 27.3) that has an approximate basal area distribution of 53 ft^2 per ac (12.3 m^2 per ha) in 5–11 inch in (12.7–27.9 centimeter (cm)) diameter trees, 22 ft^2 per ac (5.1 m^2 per ha) in 12–15 in (30.5–38.1 cm) diameter trees, and 26 ft^2 per ac (6.0 m^2 per ha) in 16+ in. (40.6+ cm) diameter trees. The prescription distributes 0.2–0.5 ac (0.08–0.2 ha) group

TABLE 27.1 Resource Statistics of Forested Habitats Within the General, Special, and Restricted Management Zones of the Umbagog National Wildlife Refuge (Acres)

Height Class (ft)	Crown Class (%)	Hardwood		Mixedwood		Softwood	
		Gen & Spc[a]	Restricted	Gen & Spc[a]	Restricted	Gen & Spc[a]	Restricted
0–10	81–100	27	15	36	41	–	–
	61–80	51	17	64	59	–	–
	31–60	–	–	33	4	–	–
	0–30	51	2	–	–	34	15
10–30	81–100	257	64	126	99	–	–
	61–80	80	11	42	20	23	87
	31–60	33	3	4	21	2	143
	0–30	–	–	–	1	2	7
30–50	81–100	1,428	240	973	290	101	140
	61–80	366	152	1,122	825	425	371
	31–60	178	27	394	306	105	262
	0–30	49	23	149	68	36	18
50+	81–100	8	–	–	–	–	–
	61–80	347	147	2	70	–	7
	31–60	369	371	71	53	33	–
	0–30	38	6	–	–	7	–

■: land that meets the criteria for timber merchantability.
[a]General and Special Management Zones.

selection harvests throughout the stand, and, when warranted, patch clear-cuts approximately 1–2 ac (0.4–0.8 ha) in size and spaced widely apart (>1000 ft (305 m)). The group and patch openings in aggregate should not exceed 10–15% of the stand, and single-tree selection between groups is applied where appropriate. Fifteen-year harvest intervals were suggested, with residual basal area goals of 100 ft^2 per ac (23.0 m^2 per ha) for mixedwood forests, and 70 ft^2 per ac (16.1 m^2 per ha) for hardwood forests.

The uneven-aged prescription emphasizes the importance of promoting a predominately closed canopy (>70% closure) stand with a variety of age classes. Further emphasized are:

- The retention and promotion of coniferous tree species, particularly red spruce
- The retention of beech trees, especially those that exhibit potential resistance to beech bark disease
- The retention and promotion of super-canopy trees, especially large pines, but also eastern hemlock and red spruce
- The retention and promotion of trees with large horizontal branching
- The retention of uncommon tree species

Approximately 13% of the Refuge is designated as a Woodcock Focus Area. The Refuge uses even-aged management in these areas to provide suitable breeding habitat for American woodcock and other species of conservation concern with similar habitat needs. The even-aged prescription guideline for American woodcock is to manage where site conditions are best suited and include all height classes in area regulation. For area regulation, 40-year rotations are used to create and maintain four forest age classes distributed in approximately 5 ac (2 ha) patch sizes. Ten-year harvest intervals are scheduled for each patch, and patches are distributed in a fashion that creates a mosaic of ages throughout the area being regulated. Initial harvests are focused on areas with aspen that are mature and at risk of declining in population or vigor. American woodcock is a priority within the focus areas, but because of the limitations of spatial analysis lands within them may have

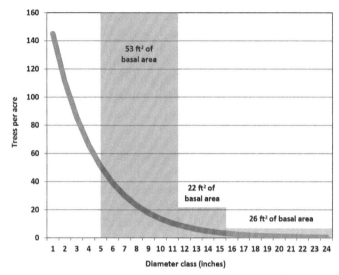

FIGURE 27.3 An uneven-aged forest diameter distribution using a q-factor of 1.3 and containing 26 ft^2 per ac (6.0 m^2 per ha) of basal area in 16+ in (40.6+ cm) tree diameter classes.

marginal opportunities to provide ideal woodcock habitat. The extent of land best suited to management for woodcock is determined prior to implementation when area regulation is being designed. Even-aged management that provides habitat for other species of conservation concern, such as the Canada warbler, is applied elsewhere within the Woodcock Focus Areas.

The FMP includes guidelines for within-stand features (e.g., snag trees, vernal pools, and downed wood), roads, and landings. Approximately 7 ft^2 per ac (1.6 m^2 per ha) of basal area (approximately six trees per ac (14.8 trees per ha)) is considered as reserve trees with the expectation that they will contribute to snag, cavity, and coarse woody debris objectives. Ideal snag and cavity recruitment trees are those with diameters in excess of 12 in (30.5 cm) and one in excess of 18 in (45.7 cm). Preferably, at least three trees per ac (7.4 trees per ha) would meet these criteria. For coarse woody debris, the guideline is to retain all debris and root wads found resting on the forest floor and to leave topwood, branchwood, and other cull wood (especially hollow logs from harvested trees) on the forest floor. For vernal pools, seeps, and streams the guideline maintains >70% canopy closure in *outer* management buffers that include

- 1,000 ft (305 m) from the edge of vernal pools
- 150 ft (46 m) from the stream bank of first- and second-order streams
- 400 ft (122 m) from the stream bank of third-order streams
- 1,000 ft (305 m) from the stream bank of fourth-order streams
- 100 ft (30.5 m) from the edge of seeps
- 1,000 ft (305 m) from the edge of non-forested wetlands or ponds
- 1,320 ft (402 m) from raptor/great blue heron nests sites

Guidelines also include no-cut *inner* buffers such as

- 100 ft (30.5 m) from the edge of vernal pools
- 50 ft (15.2 m) from the stream bank of first- and second-order streams
- 100 ft (30.5 m) from the stream bank of third- and fourth-order streams
- 50 ft (15.2 m) from the edge of seeps
- 100 ft (30.5 m) from non-forested wetlands or ponds less than 10 ac (4 ha)
- 300 ft (91 m) from the edge of non-forested wetlands or ponds greater than 10 ac (4 ha)
- 600 ft (183 m) from raptor/great blue heron nest sites

DISCUSSION AND CONCLUSIONS

The USFWS plays a vital role in the conservation of the nation's natural resources and focuses on fish, wildlife, and plants, while it includes all aspects of ecosystems upon which life depends (e.g., clean air and water). The Umbagog National

Wildlife Refuge is an example of how a National Wildlife Refuge analyzes, assesses, plans, and implements management actions for forest ecosystems within the context of the region and landscape to contribute to conservation of resources of concern. The Umbagog CCP planning process involved people affected by the implementation of the plan and included specialists and experts that helped find creative ways to resolve conflicts and craft viable solutions. The FMP benefited from the outlines in the CCP and HMP, making it more strategic in its design and use of forest management with greater assurance the Refuge will contribute an appropriate amount of desired forest conditions within the landscape.

Learnings and Insights

Much was learned about developing and implementing management plans that addressed the spatial and temporal complexity of forested ecosystems. Important to consider was the practicality of the strategies being proposed at broad scales. This often resulted in developing all or portions of step-down plans when developing the CCP. Also important was the use of spatial analysis. Spatial analysis helped reduce uncertainty for many decisions and was useful for interpreting landscapes and prioritizing; although, it was not used exclusively because of potential for inaccuracies. It was important to be clear and descript but to retain flexibility within the plans to adapt to local-scale information identified during field evaluation. A key idea that informed planning and drives implementation is site capability. Site capability speaks to the plant community most suited to thrive in the various combinations of abiotic features that compose a site. The Refuge intends to provide suitable habitat where such habitat is able to grow, and in doing so expects habitats to be more resilient to environmental stress.

Sustainability Issues

The Refuge is federally owned public land allowing ownership longevity and a unique opportunity within the landscape to use and study forest management that sustains wildlife habitat, biological diversity, and ecological integrity. To this endeavor, science is used for planning and implementation and adaptive management is used to ensure long-term biological goals are met. The Umbagog forest management plan was developed to provide sustainability of priority habitats and forest characteristics, and through the use of forest management, provide a sustainable supply of forest products. Forests are dynamic and stochastic systems being managed throughout the landscape for a variety of landowner goals. Climate change and significant stand-replacing natural events will impact the forest conditions the Refuge intends to sustain, and changes in the forest products industry and surrounding land use may change the biological goals of the Refuge, but through research and monitoring, future forest management at the Refuge will adapt accordingly.

Plan Development Challenges

The Umbagog Refuge was one of the first predominately forested refuges in the U.S. Fish and Wildlife Service's northeast region to complete a CCP, HMP, and FMP. Developing the series of Refuge plans was challenged by the amount of time needed to assess the landscape, review all pertinent scientific literature, meet USFWS planning and NEPA requirements, and design a strategy that balances the conservation and recreational interests of the Refuge, its partners, and the public. During development, few examples of successful plans were available to use as a guide or template, which caused additional time committed to creating formats and testing processes. A geographic information system was an essential tool for making informed decisions; however, the limited availability of applicable spatial data at times caused delay. Landscape data was not as readily available or as consistent as current datasets, creating a need to develop spatial data before conducting analysis. This was further complicated by portions of the Refuge being bisected by the New Hampshire and Maine state boundary, which created inconsistencies in data availability and interpretation. Development of the FMP was hampered by the purchase of additional lands where data was unavailable or inconsistent. This proved a minor barrier to the process, and has become less challenging now that the Refuge has acquired the majority of land within its authorized expansion boundary and has been developing comprehensive datasets. Future reviews and revisions are expected to be much less time consuming.

Plan Implementation Challenges

The application of uneven-aged management, designed to improve wildlife habitat rather than produce a sustainable flow of forest products, has proved challenging. The uneven-aged prescriptions are complex due to the difficulty of translating language describing wildlife habitat and biological diversity (e.g., complex forest structure) into forest mensuration and silviculture. The discussed FMP is an attempt to bridge this divide to allow technicians applying the management prescriptions a foundation for making decisions. Conservation success starts with successfully translating forest management prescriptions into marked, sold, and cut sales. This process can be jeopardized if natural resource professionals with adequate knowledge and experience are not involved. At the Refuge, training was needed for staff involved with implementing forest management despite their diverse environmental backgrounds. It is critical that staff with forestry knowledge and experience be involved in the planning and implementation processes, but it is challenging for people with forestry backgrounds to apply forest management that values the intricacies of wildlife habitat higher than the growth and yield, and economic potential of trees. It was just as challenging for people without forestry backgrounds to apply forest management that incorporate forest ecology, tree vigor, timber value, and logistical aspects of a timber harvest (e.g., harvesting equipment and infrastructure), all of which are essential. Implementation has also required additional project specific planning, preparation, administration, inventory and monitoring surveys, and possible research. In an era with continued budget reductions and loss of staff, resource availability has been challenging; although, it was not a barrier.

The timber offered from the Refuge has been sold through a bid process. This requires additional processing and risks of failure if not sold. The bid process is influenced by a number of variables beyond the control of the Refuge including market conditions, weather, and operator availability. The Refuge is in a region that has a variety of markets and operators and operating periods skewed to very short periods of time with frozen ground conditions. Harvesting that involves high operator costs and low productivity is difficult to sell, especially if constrained to frozen conditions, or by other provisions. This stresses the importance of economic considerations during sale layout and when applying the prescriptions.

Other Interesting Issues Related to the Plan

Traditional forest metrics associated with stand heterogeneity are being used as a proxy for the complex structure required by our resources of concern. The values used in the prescriptions represent an effort to describe the desired wildlife habitat and stand characteristics, and to translate these characteristics into traditional silvicultural ideas. For example, the diminution quotient (q) common to forest management is used only as a proxy for forest heterogeneity, an attribute important to a host of wildlife species. The authors of the forest management plan recognize forest ecosystems are complex and dynamic systems, and that expected wildlife and plant responses to management actions can vary. However, to facilitate the practical application of forest management, it was important to translate biological terminology into language commonly used by foresters, and to create management goals that were based on common forest mensuration and silviculture terms. Forest management at the Umbagog NWR differs from traditional forestry and silviculture that are based on timber production. It is more closely aligned with concepts of ecological forestry. Very little of the refuge forests provide the quantity or quality of desired habitats, so initial efforts are perhaps better described as restoration or rehabilitation rather than a particular silvicultural method.

REFERENCES

Askins, R.A., Philbrick, M.J., 1987. Effect of changes in regional forest abundance on the decline and recovery of a forest bird community. Wilson Bulletin 99, 7–21.

DeGraaf, R.M., Yamasaki, M., 2001. New England Wildlife: Habitat, Natural History, and Distribution. University Press of New England, Hanover, NH.

Flatebo, G., Foss, C.R., Pelletier, S.K., Elliott, C.A., 1999. Biodiversity in the Forests of Maine: Guidelines for Land Management. University of Maine, Orono, ME. Cooperative Extension, Bulletin #7147.

Foss, C.R., 1994. Atlas of Breeding Birds in New Hampshire. Audubon Society of New Hampshire, Chalford Publication Corporation, Dover, NH, 414 p.

Hagen, J.M., Vander Haegen, W.M., McKinley, P.S., 1996. The early development of forest fragmentation effects. Conservation Biology 10, 188–202.

Keppie, D.M., Whiting Jr., R.M., 1994. American Woodcock (*Scolopax minor*). In: Poole, A., Gill, F. (Eds.), The Birds of North America, 100. The Academy of Natural Sciences and The American Ornithologists' Union, Philadelphia, PA and Washington, DC.

Leak, W.B., Solomon, D.S., DeBald, P.S., 1987. Silvicultural Guide for Northern Hardwood Types in the Northeast (Revised). U.S. Department of Agriculture, Forest Service, Northeastern Forest Experiment Station, Broomhall, PA, Research Paper NE-603. 36 p.

Morse, D.H., 1967. The contexts of songs in black-throated green and blackburnian warblers. Wilson Bulletin 79, 64–74.

Morse, D.H., 1993. Black-throated green warbler (*Dendroica virens*). In: Poole, A., Gill, F. (Eds.), The Birds of North America, 55. The Academy of Natural Sciences and The American Ornithologists' Union, Philadelphia, PA and Washington, DC.

Morse, D.H., 2004. Blackburnian warbler (*Dendroica fusca*). In: Poole, A. (Ed.), The Birds of North America Online. Cornell Laboratory of Ornithology, Ithaca, NY. http://bna.birds.cornell.edu/BNA/account/Blackburnian_Warbler/ (Accessed April 4, 2014).

Norton, M.R., 1999. Status of the black-throated green warbler (*Dendroica virens*) in Alberta. Alberta Environment, Fisheries and Wildlife Management Division, and Alberta Conservation Association, Edmonton, AB. Wildlife Status Report No. 23. 24 p.

Reay, R.S., Blodgett, D.W., Burns, B.S., Weber, S.J., Frey, T., 1990. Management Guide for Deer Wintering Areas in Vermont. Vermont Department of Fish and Wildlife, Waterbury, VT.

Sepik, G.F., Owen Jr., R.B., Coulter, M.W., 1981. A Landowner's Guide to Woodcock Management in the Northeast. University of Maine Life Sciences and Agriculture Experiment Station, Orono, ME. Miscellaneous Report 253.

Smith, R., Dallman, M., 1996. Forest gap use by breeding black-throated green warblers. Wilson Bulletin 108, 588–591.

U.S. Fish and Wildlife Service, 1996. American Woodcock Management Plan: Eastern Region. U.S. Fish and Wildlife Service, Hadley, MA.

U.S. Fish and Wildlife Service, 1997. National Wildlife Refuge Improvement Act 1997. Public Law 105-57-Oct.9, 1997.

U.S. Fish and Wildlife Service, 2009. Umbagog National Wildlife Refuge Comprehensive Conservation Plan. Umbagog National Wildlife Refuge and U.S. Fish and Wildlife Service, Coos County, NH and Hadley, MA.

Chapter 28

Weaverville Community Forest, California, United States of America

Alex Cousins,[1] Colleen O'Sullivan[1] and Patrick Frost[2]

[1]Trinity County Resource Conservation District, Weaverville, California, USA, [2]Shasta College-Trinity Campus, Weaverville, California, USA

ABBREVIATIONS

PRC   California Public Resources Code
TCRCD   Trinity County Resource Conservation District
WCF   Weaverville Community Forest

MANAGEMENT SETTING AND BACKGROUND

Trinity County is a rural county in northern California. Roughly 70% of its 2.2 million acres (ac) (0.89 million hectares (ha)) are federally managed, with the majority of that managed by the U.S. Forest Service (USFS) (1.48 million ac, 598,140 ha). Another 10% (220,000 ac, or about 89,033 ha) are industrial timberlands. Most of the industrial timberlands are owned by one entity, Sierra Pacific Industries. The remainder includes a mix of nonindustrial forests, agricultural lands, and residential areas. Trinity County is sparsely populated, with approximately 14,000 people (according to the 2010 U.S. Census). Trinity County does not have any incorporated towns or cities. Weaverville, with a population of approximately 3,500, is the county's largest community and the County Seat. It was founded in 1848 during the gold-mining era of California. Much of the downtown area is a historic district, and tourism is important to the town's economy. The town is situated in a small valley or basin surrounded by a mix of public (USFS and Bureau of Land Management (BLM)) and private forestlands that are similar in ownership allocations as the county at large. In the Weaverville Basin (Figure 28.1), there are 17,656 ac (7,145 ha) of federally managed public land (USFS and BLM), 6,289 ac (2,545 ha) of private, industrial timberlands, and 7,853 ac (3,178 ha) of other land uses that include nonindustrial forestlands, rural and urban residential areas, local government infrastructure, and others. This chapter focuses on the partnership that was developed between the Trinity County Resource Conservation District (TCRCD), the BLM, and the USFS for the management of 13,000 ac (5,261 ha) of forestland in Trinity County (Trinity County Resource Conservation District, 2011).

Trinity County and the Weaverville Basin are located in the Klamath Bioregion, or Klamath Province. It is generally characterized as a mixed coniferous forest with species composition influenced by elevation, slope, and aspect. The elevation ranges from about 2,000 feet (ft) (610 m) to 3,000 ft (914 m), and as a basin the slopes face in all directions. The region has a Mediterranean climate, with a distinct winter rainy season and hot, dry summers. Wildfire is a natural component of the region, with vegetative assemblage communities that are adapted to fire. However, fire has been excluded from much of the landscape since the early decades of the 1900s, and only recently has there been an effort to return fire to the landscape as a management tool.

Riparian forests are associated with the streams, which are tributaries to Weaver Creek, itself a tributary of the Trinity River. Riparian species include willows (*Salix* spp.), white alder (*Alnus rhombifolia*), bigleaf maple (*Acer macrophyllum*), with an overstory that includes Douglas-fir (*Pseudotsuga menziesii*), and ponderosa pine (*Pinus ponderosa*). Himalayan blackberry (*Rubus procerus*), a nonnative, has invaded and overgrown much of the riparian understory. There is an oak woodland component at lower elevations. The dominant species of oak is Oregon white oak (*Quercus*

Forest Plans of North America. http://dx.doi.org/10.1016/B978-0-12-799936-4.00028-X

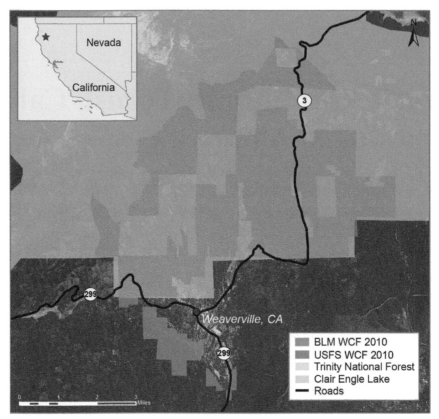

FIGURE 28.1 Map of the Weaverville Basin and the Weaverville Community Forest area.

garryana). On hotter, drier slopes, the oak woodland includes gray pine (*Pinus sabiniana*) and various "California lilac" (*Ceonothus* spp.) and manzanita (*Arctostaphylos* spp.). Elsewhere, ponderosa pine and Douglas-fir are intermixed in the oak woodland. Above the oak woodlands, the forests are a mix of Douglas-fir, ponderosa pine, and incense cedar (*Calocedrus decurrens*).

The Weaverville Basin has been altered greatly by humans. The streams and hillsides were intensively and extensively mined for gold. Tailings and ditches can be found throughout the basin. Forests were harvested to the point of denuding the slopes during the mining era (1850–1930). Therefore, there is almost no primary forest remaining, except in very isolated and inaccessible pockets. The forests typically found in the Weaverville Basin are considered second- or even third-growth. The species composition and the density of the vegetation have both been influenced by the exclusion of fire, making the forests less resilient to wildfire and more apt to burn more intensely than before fire exclusion. The mixed conifer stands are quite dense, generally with a closed canopy, and this has also affected the general health of the forests as trees must compete for limited resources, and many have poor height-to-crown ratios. The exclusion of fire also means that the understory is not very open, but contains dense thickets of conifer seedlings.

PLANNING ENVIRONMENT AND METHODOLOGY

In 1999, citizens in the community of Weaverville raised concerns over a proposed land exchange. The BLM advertised its intent to trade 1,000 ac (405 ha) of land in the Weaverville Basin to a timber company. These lands had been identified by the BLM for disposal in its Resource Management Plan, because they were isolated from other BLM lands and therefore difficult to manage. Of particular concern to the community were the potential visual effects that intensive logging would have on Weaverville's viewshed and by extension, the effects that would have on the community's quality of life and tourism. Members of the community began to meet informally to discuss alternatives to the land trade. Before long this loose-knit group of neighbors became the *Weaverville 1000*, with a goal of finding a mechanism to prevent the trade and keep these lands in public ownership. They petitioned the County Board of Supervisors to intervene. The Board submitted a request to the BLM to put the trade on hold while the community explored its options. The BLM agreed to a 2-year hold on the land trade, but offered the community only one path to pursue—acquisition of the lands from the BLM.

Given this direction from the BLM, the Weaverville 1000 pursued the acquisition option, which included three elements: (1) a public outreach campaign to build support locally and regionally; (2) an assessment of sustainable management as a mechanism for financing the acquisition; and (3) a search for a public entity with the authority to own land, and one that also could be trusted to manage the resources with a sensitivity to the community's concerns, which involved protecting the viewshed through low-impact, sustainable forest management methods.

The Weaverville 1000 included key community leaders who assumed the role of spokespersons. They kept the issue in front of the public, with regular updates to the Board of Supervisors, articles and editorials in the local newspaper, speeches at local service clubs, and attendance at the BLM's quarterly meetings of its regional Resource Advisory Committee. They used petition drives to show the broad support within the community, and they expanded the list of benefits of public ownership from simply protecting the viewshed to:

- Enhancing recreational opportunities (there was a network of 50 miles (mi) (80 kilometers (km)) of recreational trails on the public lands surrounding the town)
- Providing logs to the local mill through sustainable harvesting (the Trinity River Mill in Weaverville was the last mill operating in a county that once had more than 40 mills)
- Reducing the risk to the community from catastrophic wildfire through sustainable forest management

This last issue rose in importance to the community in August 2001 when the "Oregon Fire," a 2,000 ac (809 ha) wildfire, burned on the ridge toward the western edge of town. Driven by high winds, it destroyed several homes in Weaverville and caused the evacuation of neighborhoods, the local hospital, and the county jail, and essentially threatened the entire town.

The Weaverville 1000 was sensitive to the political context of forest management in northern California. The region had been torn by "timber wars" for more than 20 years, most recently due to the controversy surrounding the listing of the northern spotted owl (*Strix occidentalis caurina*) as an endangered species and the subsequent development of the Northwest Forest Plan. The plan was viewed as a detailed and complicated document that dictated management of federal forest lands in the range of the northern spotted owl and centered around strategies that emphasized protection and restoration of late-seral (old growth) forest types in the Pacific Northwest, including Trinity County in northern California.

A local nonprofit group, the Trinity Resource Conservation and Development Council (now the Northwest California Resource Conservation and Development Council), secured a small grant from the U.S. Department of Agriculture, Natural Resources Conservation Service (NRCS) to obtain the services of local consulting foresters and to conduct an inventory of the forest. The effort was undertaken to develop an estimate of the sustained yield that could be expected through active management at levels supported by the community on the 1,000 ac (405 ha) and to determine the viability of using sustainable harvests as the means for financing the acquisition. This inventory and assessment identified the likely constraints on management, including:

1. The community perceptions and the level of acceptance of commercial harvesting, especially where the forest abutted neighborhoods or was associated with recreational trails.
2. The regulatory programs that were established to help with the recovery of northern spotted owl populations.
3. The streams and the associated habitat for species of salmon (*Oncorhynchus* spp.) that were listed under the Endangered Species Act (ESA), or that were at risk of being listed.
4. The portions of land that were identified as having significant cultural resources, primarily from the gold-mining era, and therefore required protection.

Following the transfer of the land from federal ownership to nonfederal ownership the need to adhere to the requirements of the National Environmental Policy Act (NEPA) was eliminated as there would unlikely be any significant federal action that would occur under normal activities. Perhaps if a Habitat Conservation Plan was prepared, a NEPA analysis would be required. But normal forest management actions in California must still follow the multiple state-mandated environmental regulations, such as the California Environmental Quality Act and the California Forest Practices Act. The Forest Practices Act is functionally equivalent to the California Environmental Quality Act for projects involving forest management, especially where forest products are sold. It was determined that these constraints likely would regulate active forest management to varying degrees.

There are few organizations or local government entities located in Trinity County, such as the county government and local special districts. The Board of Supervisors for the county had no interest in acquiring the land, because it already was financially stretched in its effort to provide essential services (public health and safety, law enforcement, transportation, land use planning and permitting, etc.). Special districts are established to provide specific services to specific geographic areas (e.g., Weaverville Sanitary District providing sewage treatment, Weaverville Fire District providing fire protection, etc.). They are overseen by boards of directors, who are elected or appointed from within their boundaries. These special district boards are required to follow state laws governing the operations of all local governments. These regulations are designed to provide transparency and open, public participation in local government. The TCRCD is a special district and was established in 1956 under Division 9 of the California Public Resources Code (PRC) to assist in the delivery of soil and water conservation practices to landowners throughout the county. From 1956 until the mid-1980s, the TCRCD primarily acted as a bridge between federal agencies and local landowners to provide technical assistance to improve conservation practices on private land. From the mid-1980s until the present time, the TCRCD used its Division 9 authority to expand its role in natural resources management to include working with federal land and resource management agencies in what could be called *government-to-government* cooperative agreements.

By the year 2000, the TCRCD had developed a skilled, professional, in-house workforce specializing in natural resources planning and management and had parleyed this skill set to establish a number of long-term cooperative agreements with the BLM, Bureau of Reclamation (both in the U.S. Department of the Interior), USFS, and NRCS (both in the U.S. Department of Agriculture) under which the TCRCD essentially acted as a local arm of the federal agencies to implement on-the-ground management on their behalf. The TCRCD often leveraged federal funds to secure state and local funds that would augment annual work plans. The work plans included watershed improvement and restoration, forestry practices primarily directed toward reducing forest fuel loads, and environmental education activities. Division 9 of the PRC authorizes conservation districts to own easements and land if the ownership is consistent and furthers the mission of the district. The Weaverville 1000 approached the TCRCD Board of Directors about assuming the ownership role of the land if successful acquisition could be accomplished. The TCRCD Board agreed to participate, if the community could acquire the property, and if there seemed to be public support for the concept.

Stewardship Contracting

The Weaverville 1000 continued to work on refining the concept of the land acquisition, but the finances looked tenuous at times. It was estimated that the acquisition, either through a direct purchase or through purchase of lands desirable to the BLM and then affecting a land trade, could cost about $2 million USD (U.S. dollars), and it did not look like the community's *light touch on the land* approach to sustainable harvests on the 1,000 ac (405 ha) could support the acquisition costs. Undaunted, the Weaverville 1000 and TCRCD continued to promote the concept with the local press and organized a community meeting to build a broad, consensus-driven vision and strategy for a community forest. They investigated other models, such as the Arcata Community Forest in adjacent Humboldt County, California.

At the same time (circa 2002–2004), stewardship contracting and agreements were being applied by the federal land management agencies, and they were gaining some popularity. A Congressionally mandated Stewardship Contracting Pilot Program evolved into a regular contracting tool for the USFS and BLM (Section 323, P.L. 108-07). Under these stewardship authorities, federal agencies develop contracts or agreements that blend the sale of harvestable material (e.g., saw logs from public lands) with service work (e.g., erosion control work, invasive weed removal, forest fuel reduction, watershed restoration, etc.) on public lands, and the agencies pay for these projects with receipts received from the harvested material. In using this approach, the receipts paid to the federal government do not go to the U.S. Treasury, but stay within the management unit, giving an added incentive to agency staff to use these authorities. An essential component of the stewardship authorities is that the primary purpose of each project must focus on resource stewardship that maintains or improves the natural resources, and not timber production geared toward maximizing yield from the public land. Timber harvesting is therefore a secondary benefit in the context of stewardship contracting.

The TCRCD invited the community to a day-long meeting in May 2004 with an outside facilitator paid for by NRCS. The meeting was attended by more than 50 individuals representing a wide range of interests and views (Figure 28.2). Three important outcomes of this meeting included:

1. A clearly articulated consensus vision for the Weaverville Community Forest (WCF). In fact, this was the first time the term *Community Forest* was used. The vision centered around management of the lands to protect the viewshed of Weaverville, promotion of a more fire-resilient forest to help protect Weaverville from catastrophic wildfire, facilitation of sustainable timber harvests for merchantable timber delivered to the local mill, maintenance and improvement of the recreational trail system, and use of these resources as an outdoor classroom.
2. A decision that the TCRCD should be the manager of the community forest.
3. An alignment of the community vision for the land with the guidelines for implementing the stewardship authorities, which led the BLM to offer its stewardship authorities as an alternative to the community. Under this scenario, the BLM retained ownership of the land and its resources, but the land management decisions and some of the management costs would fall within the purview of the TCRCD, with regular input from the community.

FIGURE 28.2   A community meeting designed to build a broad, consensus-driven vision and strategy for the community forest.

The community meeting ended with a consensus decision for the TCRCD to negotiate a multiyear, stewardship agreement with BLM that would embrace the community vision for the management of the WCF. The BLM and the TCRCD entered into a 10-year stewardship agreement for the WCF in September 2005. The agreement process involved the local BLM staff, the BLM state forester, California Director of the BLM, and Washington Office of the BLM. While the BLM had previously entered into some stewardship contracts to implement specific projects on other BLM lands, this agreement was unique in that it was between the federal government and a local government, and not a contract with a timber purchaser. The agreement did not identify any specific projects to be implemented or project schedule, but set up a process by which the TCRCD would seek community input to guide the timing, location, and design of projects within the WCF, including timber harvests and the other stewardship work that would be performed over time. A broad goal was to implement harvests on approximately 200 ac (81 ha) every other year and use the money paid to BLM for the merchantable timber (retained receipts) to fund the other types of projects (e.g., invasive weed management, trail maintenance, watershed restoration, forest fuels reduction, etc.) during alternating years.

The community visioning meeting in May 2004 concluded with the selection of three individuals to negotiate with BLM: the TCRCD Manager, a TCRCD Board member, who had been an early opponent of the land trade, and a community member associated with the recreational trails system. The initial meetings with the BLM and negotiation team focused on defining the area to be included in the stewardship agreement and, therefore, the WCF. The community's negotiating points for selecting BLM parcels for inclusion were:

- Include the parcel that was originally the subject of the land trade to protect the community's viewshed
- Include parcels adjacent to neighborhoods to enhance protection from wildfire
- Protect the recreational trail system (existing and proposed trails)
- Select parcels that needed restoration due to the 2001 wildfire

The volume of merchantable timber on the parcels was considered by the community but was secondary to the above objectives.

The BLM had its own objectives, most particularly:

- Include parcels that were isolated from other BLM lands making them difficult to manage
- Include parcels adjacent to neighborhoods
- Exclude parcels adjacent to industrial timberlands to reserve them for future trade considerations

The result was the mutual agreement after only one meeting to include several parcels, including the original "trade" parcel, totaling approximately 1,000 ac (405 ha). This set the stage for the next step—the development of a justification for a stewardship agreement between BLM and TCRCD. The BLM had a specific format for this process. The Stewardship Project Review Checklist identified seven stewardship objectives contained in the stewardship law. The BLM State Forester and TCRCD Manager reviewed these objectives against the community vision and found support for the following objectives:

1. *Maintain roads and trails to improve water quality.* The TCRCD had an existing program reducing erosion from roads into streams.

2. *Protect and improve soil productivity and habitat for wildlife and fisheries.* Active forest management and fuel load reduction would generally benefit the forest and stream ecosystems and would specifically address endangered species (the northern spotted owl and coho salmon (*Oncorhynchus kisutch*)).
3. *Remove certain vegetation to promote healthy forest stands and to reduce fire hazards.*

Four other stewardship objectives were reviewed, but these were considered needing evaluation in the future.

1. *Use prescribed fire to promote healthy forest stands and to reduce fire risk.* The current conditions of the forests did not make them candidates for prescribed fire until certain vegetation was removed.
2. *Direct watershed restoration* (in-stream work). This was not considered a high priority in the community vision.
3. *Restore wildlife and fisheries habitat.* This was seen as ancillary to the above objectives.
4. *Control noxious (invasive) weeds.* There were only minor infestations and this was seen as a low priority that would be addressed later.

Two other criteria were assessed, and it was determined that they supported the stewardship agreement.

- The overall vision of the WCF would result in the restoration and maintenance of ecological processes
- The strong sense of collaboration between the community and the BLM was supported by a local organization (TCRCD) that had the rural community's needs at the center of the organization's mission

The Stewardship Checklist was completed by January 2005 and submitted to the BLM State Director and then the BLM Washington Office for review and approval, which took until summer of 2005. An initial concern on the part of BLM in the national office centered on the proposed 10-year duration, the maximum allowed under the law. The State Forester prevailed, explaining that forest management is a long-term proposition and that the success of the stewardship agreement hinged on timber prices, which can fluctuate greatly from year to year.

The final step was to develop the actual stewardship agreement, which was executed in September 2005. The agreement was a standard inter-agency assistance agreement. TCRCD agreed to assign a project manager to oversee implementation of a project management plan that was to be developed between the TCRCD and the BLM in concert with the community. That plan would look broadly to the entire 10-year timeframe, but would include specific projects for the first 2 years. TCRCD would lead design and layout of the first forest health timber harvest plan on 200 ac (81 ha) adjacent to a neighborhood. The BLM would provide seed money to develop the harvest plan and complete the NEPA. The TCRCD would implement the harvest and initiate the other nonharvest projects agreed upon with community, also seeking outside funds.

In developing this agreement, the TCRCD established procedures for keeping the public involved and informed. Further, the funds collected by the BLM from timber harvests are to be maintained by the BLM in a WCF Stewardship Account to be drawn upon through an annual work plan. The TCRCD actively seeks additional funds from nonfederal sources to augment the stewardship account to complete service projects. The TCRCD and the BLM also work together with the community to develop a long-range strategic plan for the community forest. The stewardship agreement allows the BLM to negotiate timber sales directly with the TCRCD. The TCRCD then implements the harvests. The agreement also authorizes the BLM to use money collected from the timber sales to pay the TCRCD to design and implement the stewardship projects like invasive weed control and trail maintenance. Finally, the BLM retains the responsibility for completing NEPA requirements, but the TCRCD provides technical assistance for the design and layout of projects (including timber harvests) and the acquisition of permits, when needed.

The WCF has been managed with the guidelines established in the Stewardship Agreement, and organizational policies are detailed in the operational policies of the TCRCD. As a special district in California, the TCRCD has well-developed financial policies including those for contracting, purchasing, fiscal management, and oversight. The TCRCD has open public meetings every month and has added the community forest as a standing report from staff. The operations of the community forest are generally overseen by the manager of the TCRCD with guidance provided by the Community Forest Steering Committee. The Steering Committee is an inclusive rather than exclusive group, and members of the public can attend, participate, and have an equal voice. That said, a group of about eight individuals have been the core members of the Steering Committee. A member of the TCRCD Board of Directors and the TCRCD Manager have led the group, which includes representatives of local environmental groups, the timber industry, the county office of education, the Board of Supervisors, a registered professional forester, and members of the federal agency staff. Decisions are made by consensus, including the development of the strategic plan. Project development and design is a product of collaborative Steering Committee deliberations and annual community meetings. The annual community meetings are open to the general public, and each includes an overview of the status of projects by the TCRCD and then wide-ranging, round-table discussions during which community members identify resource concerns or issues, project ideas, or their concerns about future projects. Community field trips and educational tours are scheduled throughout the year to look at completed, ongoing, and future projects.

OUTCOMES OF THE PLAN

As of the development of this chapter, the strategic plan (Trinity County Resource Conservation District and Bureau of Land Management, 2006; Trinity County Resource Conservation District, Bureau of Land Management, and U.S. Forest Service, 2010) and the Stewardship Agreement with the BLM are currently in their ninth year. The most significant outcome has been the expansion of the WCF onto lands managed by the USFS. The success of the use of stewardship authorities by the BLM led the Steering Committee to pursue a similar agreement between the TCRCD and the USFS. A 10-year stewardship agreement for 12,000 ac (4,856 ha) was executed by the TCRCD and the Shasta-Trinity National Forest in December 2008. The result is that the WCF has expanded from the original 1,000 ac (405 ha) of BLM land to 13,000 ac (5,261 ha) of federal lands, or about 90% of the federal land within the Weaverville Basin.

There are different ways to measure the outcomes of the WCF. One way is to look at measurable accomplishments and outputs (areas treated, money earned and spent, volume of logs sent to the mill). Four timber harvests have been completed since 2005. Three of these were located on BLM land and one on USFS land, resulting in 2,700 MBF (6,372 m³) of wood sold to the local mill. These harvests resulted in treating 650 ac (260 ha) of the forest to make them more fire resilient. About $150,000 USD were deposited in the BLM stewardship account, and the TCRCD secured six dollars for every stewardship dollar obtained, for a total of nearly $1,000,000 USD in leveraged funds to complete projects in every category of service work identified in the WCF Strategic Plan. As an example of leveraged funds, in 2009 the WCF was able to utilize around $343,000 USD (Figure 28.3) for various projects within the community forest area (Trinity County Resource Conservation District, 2010). The success of the WCF for accomplishing work through the two stewardship agreements has made the WCF a priority area for the BLM and the USFS to allocate additional funds to implement WCF projects, such as trail construction and fuels reduction work, above and beyond what was completed through the stewardship work.

Another set of outcomes are less measurable, but just as real. They include the changes in public opinion or attitude, the sense of local ownership in the community forest, and the desire to see on-the-ground projects. The sense of ownership and results-oriented philosophy by the community was a change in the way of thinking by community members. Trinity County is in the heart of northern spotted owl territory. In 2005, when the stewardship agreement was signed by the BLM and the TCRCD, there had not been a successful timber harvest of living trees in the Weaverville Basin since the adoption of the Northwest Forest Plan. Most attempts by the federal land managers to complete NEPA requirements and implement a timber harvest anywhere in Trinity County had been met with litigation. The WCF model of engaging the public, making room for different perspectives and honoring minority opinions through a consensus process, has resulted in completion of all of the projects put forward by the community and the Steering Committee.

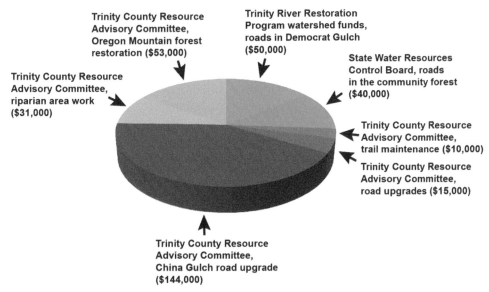

FIGURE 28.3   An example of funds leveraged for projects on the Weaverville Community Forest in 2009.

DISCUSSION AND CONCLUSIONS

The WCF received national recognition in 2009, being selected for a *Partners in Conservation* award from Ken Salazar, Secretary of the U.S. Department of the Interior. Locally, real estate agents now advertise property as being *adjacent to the Community Forest* in order to emphasize a value-added amenity. The local swim team sells Christmas wreaths made from boughs collected in the Community Forest. The WCF was even featured on travel show that is broadcast state-wide on public television, and annual bus tours visit Weaverville to view the WCF. Regional ultra-marathons, mountain bike races, and equestrian competitions highlight the recreational attributes of the Community Forest. All of these under-score exactly what the Weaverville 1000 originally set out to protect and highlight for the community. Finally, a Trinity County Collaborative was established in 2013 by the Board of Supervisors and was endorsed by the Secretary of the U.S. Department of Agriculture, Tom Vilsack. The goal of the Trinity County Collaborative is to build consensus around the management of the federal lands within Trinity County (BLM and USFS). The model that they began to organize around was the WCF.

Sustainability Issues

The underpinnings of the WCF are in the Stewardship Authorities, which are embedded in federal law. However, laws can expire, and they can be changed for the better or the worse. The early successes are linked to individuals in the community, within the TCRCD, and within the BLM and the USFS. Unfortunately, there is no succession plan in the federal agencies, and as key individuals move on or retire there is concern about who will continue to *carry the torch* for the community forest concept. The long-term sustainability of the community forest is therefore in the hands of future generations. Building relationships between youth and young adults in the community and the Steering Committee and helping develop their appreciation for the community forest and its resources is critical. When things are going smoothly, the community becomes distracted with more pressing issues and becomes complacent. Public outreach continues to be as important to a mature and successful program as it was when it was struggling to emerge.

The federally managed forestlands around Weaverville were not being managed at all until the WCF was established. Today, there is active management. The four stewardship objectives that were deferred initially are now being met. The timber harvests are improving wildlife habitat and reducing fuel loads. Prescribed fire is being used. Noxious weeds are being treated and stream habitat projects implemented. Most importantly, there is broad public support for these projects and their continuation into the future. The WCF Steering Committee is negotiating an extension of the Stewardship Agreement with BLM, which expires in 2015, for an additional 10 years and an expansion to include all of the BLM parcels that were excluded in 2005. That extension and expansion will ensure that the community vision of *light on the land* harvests every two years will continue. The first 200 ac (81 ha) that was harvested will be reinventoried and likely reentered for a selective harvest that will improve forest health and further protect the adjacent neighborhood.

The success of the WCF also means that the federal agencies are more comfortable using stewardship contracting else-where. The TCRCD and the BLM entered into a 10-year stewardship agreement for approximately 17,000 ac (6,880 ha) in the eastern edge of Trinity County, and the first forest health harvest since the public obtained ownership in 1992 is being planned. It also has meant that the WCF is being used as a model for community-based collaborations elsewhere in California and on larger areas. Similar projects, such as the Burney Hat Creek Basins, the Amador-Calaveras Consensus Group, the Dinkey Landscape Restoration Project, and the Lakeview Stewardship Landscape all have similar foundations to the WCF.

Plan Development Challenges

Much has been learned in the development of the Stewardship Agreement for the community forest. First and foremost, it takes a great deal of time and dedicated individuals to effect change, especially when working with large bureaucracies. This is not meant to be a negative reflection on the BLM or the USFS, but more of a congratulatory statement for the original community members, the Weaverville 1000, who volunteered their time and energy to the effort. This community engagement is commendable since it took 5 years to accomplish the agreement with the BLM and the TCRCD.

Plan Implementation Challenges

Different interpretations of the same law can be confounding. For example, the Stewardship Authorities are administered by the BLM and the USFS differently. The agreements with the TCRCD are quite different. The BLM allows the TCRCD to implement service work under task orders, while the USFS does not. The BLM allows the TCRCD to negotiate timber sales directly, but the USFS requires harvests to be sold through a bidding process. Similarly, NEPA requirements are interpreted differently by different personnel (line officers) in the different agencies. Line officers have different comfort levels for the level of documentation and analysis required for the same task, and this can affect the amount of time and money spent in completing NEPA requirements and the cost of implementing specific projects.

The vagaries of the economy, and therefore demand for processed lumber, have been the greatest challenge. The stewardship agreement model hinges on log prices. The recent recession greatly depressed the price of saw logs, making it difficult for the TCRCD to recover its costs implementing harvest plans on BLM land, and making it difficult for the USFS to attract bidders to its harvest projects. Another challenge has been the loss of workforce capacity in Trinity County. There simply are not the skilled workers or contractors locally that there were 20 years ago when the forest industry was more robust.

Learning and Insights

The WCF planning process demonstrates a method and a mind-set that can break down barriers between disparate opinions. In this regard, small successes build the framework for larger successes, and education of community members increases not only awareness but also increases acceptance for active forest management. The Weaverville Community also learned that the definition of *community* must be broadened beyond the town limits when dealing with a national resource, like forestlands managed by federal agencies. The community forest is literally situated in the backyards of the residents of Weaverville, but individuals throughout California and throughout the country may have an interest in *our forest*, and they have the legal right to participate in the decision making through the NEPA process.

REFERENCES

Trinity County Resource Conservation District. Weaverville Community Forest. 2011. Trinity County RCD, Weaverville, CA. http://www.tcrcd.net/wcf/index.htm (Accessed April 10, 2014).

Trinity County Resource Conservation District, 2010. Weaverville Community Forest Stewardship Project. FY 2009 annual report Trinity County Resource Conservation District, Weaverville, CA, 7 p. http://www.tcrcd.net/wcf/pdf/WCF_2009_Annual_Report.pdf (Accessed April 10, 2014).

Trinity County Resource Conservation District and Bureau of Land Management, 2006. Weaverville Community Forest Strategic Plan (2006–2009). Trinity County Resource Conservation District and Bureau of Land Management, Weaverville, CA; Redding, CA. 15 p.

Trinity County Resource Conservation District, Bureau of Land Management, and U.S. Forest Service, 2010. Weaverville Community Forest Strategic Plan (2010–2015). Trinity County Resource Conservation District, Bureau of Land Management and U.S. Forest Service, Shasta-Trinity National Forest, Weaverville, CA; Redding, CA. 17 p.

Chapter 29

Yale School Forests, New England, United States of America

Mark S. Ashton, Marlyse C. Duguid, Alex L. Barrett and Kristofer Covey

Yale School of Forestry and Environmental Studies, New Haven, Connecticut, USA

MANAGEMENT SETTING AND BACKGROUND

Starting in the early 1900s, The Yale School of Forestry and Environmental Studies began to acquire forestlands throughout New England, mostly through donations from alumni. Throughout the first several decades of ownership, these lands were more of a financial burden to the School than an asset. The chief accomplishment during the past 60 years has been converting the School Forests from a sink to a source of money. One of the management objectives of the properties is the production of income for the School; virtually all of that must come from timber management. It is a policy that most management on the Yale Forests is done by students. The faculty members concerned with silviculture and forest management have traditionally been the directors. The present policy of paying a postgraduate fellow as the forest manager and students as apprentice foresters rather than special staff or consultants to manage the Forests is an attempt to combine the academic and education mission of the School with the commitment to demonstrate financial sustainability. This chapter describes the histories of management primarily focused on the two largest forestlands.

There are eight tracts of land in the Yale School Forest system (Figure 29.1). The largest parcel is the Yale-Myers Forest, established in the early 1930s through the generosity of alumnus, George H. Myers (Master of Forestry, 1902). Assembled by Myers from 1910 to 1920 from about 30 former farm holdings, Yale-Myers forest covers 7,860 acres (ac) (3,181 hectares (ha)) across four towns in northeast Connecticut. This forest has been managed for more than 80 years by the School as a multiple-use, working forest, and has the most activity in terms of education, research, and forest management operations. The forest is one of the largest privately held and professionally managed forest parcels in southern New England and is the largest physical possession of Yale University. The forest is comprised primarily of mixed second-growth hardwoods on glacial till soils. There is also a large component of eastern hemlock (*Tsuga canadensis*), several scattered white pine (*Pinus strobus*) stands mainly of old-field origin, and occasional red pine (*Pinus resinosa*) plantations started in the 1940s after field abandonment. There are numerous small ponds and most wetland areas have been created by beaver activity. The majority of this parcel is surrounded by large state (park and forest) and smaller private forest ownerships that together comprise more than 50,000 ac (20,235 ha) of open space within the region.

The second largest parcel is the Yale-Toumey Forest, located in southern New Hampshire near the city of Keene. Set up by Forestry School Dean James W. Toumey as a Research and Demonstration Forest in 1913, this is the oldest of the Yale School Forests. Toumey was the first forest ecologist at Yale and wanted a location for experiments exploring the growth and biology of trees. The basis of the parcel was several tracts donated by the same George H. Myers who had started land purchases in 1908, with most acquisitions completed by the School by 1938. An additional 450 ac (182 ha) was added in 2000 bringing the total land area to 1,930 ac (781 ha) in size. When the forest was established, it was virtually all in land that had recently cut over or in young stands of crooked old-field pine. Now, the forest is predominantly white and red pine, largely of plantation or old-field origin. This is the result of both large areas of glacial sandy outwash promoting coniferous forest, and large-scale experiments in growing and tending pure conifer stands. Nestled along the southern edge of the city of Keene, New Hampshire, this is the most urban of the Yale Forests. There is pressure to use some of the land for purposes other than forestry, and the Forest has taken on attributes of a suburban park for the surrounding towns.

Four smaller parcels in Vermont and New Hampshire comprise the rest of the Yale Forest System. These parcels see less use in regards to research and education due to their size and distance from New Haven, Connecticut. Crowell Forest

Forest Plans of North America. http://dx.doi.org/10.1016/B978-0-12-799936-4.00029-1

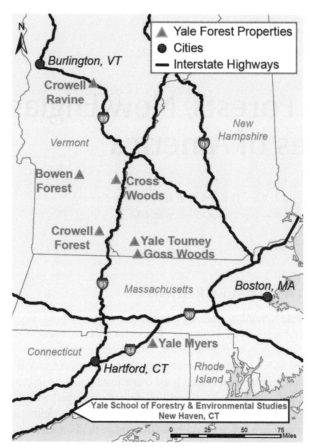

FIGURE 29.1 Location map of the Yale School Forests.

is 285 ac (115 ha) in size, and located in Dummerston, Vermont. This property consists of two tracts about 1 mile (mi) (1.61 kilometers (km)) apart. It is composed mainly of a northern hardwood forest with some old-field white pine stands. Crowell Ravine is 75 ac (30 ha) in size, and located in Duxbury, Vermont. This parcel consists of a northern hardwood forest that was cut over about 50 years ago. The land surrounds a steep ravine with water cascades at the bottom. Both properties were donated by Robert Crowell in 1983. The Bowen Forest covers 462 ac (187 ha), and is located in Mt. Holly, Vermont. Given to the School in 1924, it is the only Yale Forest that is deeded to be "kept as a forest" forever. This forest is mainly northern hardwood with white spruce (*Picea glauca*) plantations. It is located in a saddle near Okemo Mountain. Cross Woods is a 103 ac (42 ha) forest in West Windsor, Vermont that was given to the school as a gift from Mr. Gorham L. Cross in December 1982. It is a young hardwood forest that has been heavily cut over. Goss Woods is a 185 ac (75 ha) tract in Richmond, New Hampshire that was donated in 1986. Most of the land in this hardwood forest has been selectively cut over; its defining feature is a small mountain at one end.

Research on the School Forests has changed the practice of forestry in North America at different periods of time. James W. Toumey started research at the Yale-Toumey Forest in the early 1920s. He investigated plant-environment interactions of which his famous trenched plots (Toumey and Kienholz, 1931) were a part. This kind of work culminated during the 1940s in a series of forest ecology, forest soils, and silviculture textbooks that were the first of their kind in North America (Hawley, 1921, 1929, 1946; Lutz and Chandler, 1946; Toumey, 1929; Toumey and Korstian, 1937). The work was based in large part on the research work these authors did as faculty at the Yale Forests. Unlike Yale-Toumey, there was little ongoing research at the Yale-Myers Forest when this forest was donated as a series of parcels from 1926 to 1931. There were, however, a number of permanent plots that were set up by the U.S. Forest Service in the 1930s to monitor forest growth and health impacts from the chestnut blight (from the pathogenetic fungus *Cryphonectria parasitica*) and the gypsy moth (*Lymantria dispar dispar*).

The first complete description of land-use history, research, and management at the Yale-Myers Forest was published in 1945 (Meyer and Plusnin, 1945). During the Second World War and the post-war era (1940-1960), research at both forests was considerably curtailed. When David Smith was appointed as director in 1954 more research was gradually initiated.

Some of the first forest gap dynamics and controlled long-term regeneration studies in North America (1966–1968) were started at this time and continue today (Liptzin and Ashton, 1999; Smith and Ashton, 1993). Another focus that developed in the 1960s was the establishment of permanent plots in red and white pine plantations to evaluate growth and competition of differently spaced trees. This work eventually resulted in a better understanding of crown-growth allocation relationships that are now changing the ways foresters consider controlling growing space (Oliver and Larson, 1990; Seymour and Smith, 1987).

Research has been consistently increasing at the Yale-Myers Forest since around 1970, while it has remained somewhat steady at Yale-Toumey. This is in part related to time commitment and distance away from New Haven, but also due to the considerably richer land-use history and floristic complexity at the Yale-Myers Forest. This has provided a wider base upon which to explore a number of interesting social, biological, and physical questions. The first such research theme that developed at the Yale-Myers Forest was the documentation of patterns in stand development (Kelty, 1986; Kittredge, 1988; Oliver, 1978). This forest stand dynamics paradigm, conceived at the School Forests, has now become a way of thinking for many in North America (Ashton and Peters, 1999; Oliver and Larson, 1990; Smith, 1962; Smith et al., 1997; Toman and Ashton, 1996).

Since the mid-1980s, research has expanded significantly into four areas of work: (1) stand and regeneration dynamics of second-growth forests, (2) trophic food web complexity and controls on ecosystem stability, (3) aquatic ecology and population dynamics, and (4) soil and microbial ecology of forests. Work on forest gap dynamics continues with current investigations into patterns and interactions with soil moisture, soil nutrition, and radiation, and their effects on regeneration at microsite and landscape scales (Ashton, 1992; Ashton and Berlyn, 1994; Ashton et al., 1998; Frey et al. 2007; Kittredge and Ashton, 1990, 1995), and effects of fire on regeneration (Ducey et al., 1996; Moser et al., 1996). In addition, several studies investigated factors arresting successional dynamic in forest uplands (i.e., fernlands, white-tailed deer (*Odocoileus virginianus*) and forest understory interactions) (Kittredge and Ashton, 1995; Kittredge et al., 1992); more recent studies have been done on understanding the effects of disturbance on the herbaceous understory of forests (Aikens et al., 2007; Duguid et al., 2013; Ellum et al., 2010). Further, much work is now exploring trophic interactions and the role of biodiversity on ecosystem function at the Yale-Myers Forest. Most of this work centers on old fields and wetlands. Old-field systems studies have been examining the dynamics of ecological food chains and food webs comprised of grassland plants and insects and herbivore behavioral response to predators, including their effects on ecosystem dynamics and productivity (Abrams and Schmitz, 1999; Beckerman et al., 1997; Hawlena et al., 2012; Schmitz, 1994, 1997, 1998; Schmitz et al., 1997, 2000, 2013a,b; Uriarte and Schmitz, 1998). Work in wetlands has been focused on understanding patterns of species distribution and community composition, primarily of amphibians with special application to habitat conservation and management (Skelly et al., 2002, 2005, 2014; Wellborn et al., 1996). More recently, work has also expanded into soil ecology particularly into understanding the decomposition processes of forests in relation to site and disturbance history (Bradford et al., 2014; Covey et al., 2012; Keiser et al., 2014; Strickland et al., 2013a,b). Though these studies appear to be the major areas of research ongoing at the School Forests, there are many smaller shorter-term studies that are currently being conducted by faculty, doctoral students, and Master's students.

PLANNING ENVIRONMENT AND METHODOLOGY

The management of the Yale School Forests is designed around specific goals so that the properties provide the following:

- A hands-on, working (managed) forest laboratory for teaching
- A permanent, fully owned site for scientific research, especially that which extends over several decades
- An asset in the School's investment portfolio
- A dynamic example of the overall integrity and health of the forest ecosystem dynamic

To that end, the management plan of the School Forest was written with the balance of management, education, and research in mind, and specific objectives have been formulated as guidelines for faculty, staff, and students. The "working forest" criterion has been adopted for all School Forests. It clearly identifies economic self-sufficiency as a more realistic example of forest management with respect to private forest landowners. This contrasts with many other university forests, which are considerably subsidized, have their own endowments, or are primarily research forests. This working forest model promotes all operations that follow normal business and legal constraints for private nonindustrial forest owners who seek multiple benefits from their lands. Whenever possible, the management of the forest aims to avoid relying on special legal exemptions and unusual financial benefits generally granted to universities or nonprofit organizations. Taxes are paid in full along with all salaries, fellowships and internships, maintenance costs for facilities, roads and bridges, and normal costs of management operations and supplies. The Yale Forests serve as an example of sustainable and multiple-use management that is, in the long term, financially independent of the School. This policy stipulates long-term because unusual circumstances occur, particularly involving short-lived market conditions for certain products that merit capturing income at periodic intervals. This means selling products and services when stumpage markets are high and deferring sales when prices are low, rather than generating income on a continuous basis irrespective of market conditions.

The specific objectives for the Yale Forest System require regeneration and post-establishment treatments to the forest. The treatments for the Yale Forests aim to create a variety of stand conditions across a range of topographic and soil conditions. To meet these stand structure requirements across the whole forest, area regulation was adopted in 1991. In the past, management of the forest used volume regulation from a continuous forest inventory system conducted on a 10-year cycle starting in 1956. Silvicultural guidelines for area regulation are based on the most recent analysis of these plots and spatial analysis of stand age-class distribution. A variety of silvicultural treatments are used, and these consider the ecology and product value for each stand. Treatments are variable, but the different forests do have general guidelines to meet the goals and specific objectives of the School Forest. The majority of the silvicultural prescriptions seek to maintain the complex mixtures of New England hardwoods using knowledge of stand dynamics and the importance of securing advance regeneration. There are some exceptions, such as where management purposefully aims to suppress regeneration to create specific habitats, primarily, to maintain some early-seral species.

Using volume regulation, annual allowable cut for the entire School Forest system as of 2004 is calculated at 1 million board feet (MMBF) (2,360 cubic meters (m^3)) based on only the forest area defined for timber management. However, the annual cut rarely approaches or exceeds this number, and in some years barely achieved half this amount. When calculated on a 10-year basis, the mean annual cut for the period 1981–1990 was 800 thousand board feet (MBF) (1,888 m^3), for 1991–2000 it was 600 MBF (1,416 m^3), and for 2001–2010 mean annual cut was calculated to be 650 MBF (1,534 m^3). Annual cut will likely remain under 1 MMBF (2,360 m^3) in the future because of the other goals (research, system resiliency, and education) that have been identified for the management of the School Forests. Positive net annual growth (Table 29.1) has occurred not only because of cutting less than the estimated allowable cut, but also because of the age-class structure. Most forest stands fall in a narrow age range because of pre-Yale ownership land use history. Stands that have not been regenerated recently are relatively young (60–80 years) and continue to grow quickly. Therefore, current annual increment for the forests exceeds the mean annual increment. At this stage, it is expected that the annual growth will exceed the long-term allowable cut, but as the forest ages that annual increment will decrease.

Currently, the guidelines for the Yale-Myers Forest are to conduct crown thinning treatments on 100–200 ac (40–81 ha) per year for approximately 100–200 MBF (236–472 m^3) volume removed. Regeneration treatments (shelterwood or seed tree) are scheduled on 50–75 ac (20 ha) per year, with approximately half of the area as establishment treatments and the remaining area focused on release of established regeneration, harvesting roughly 150 MBF (354 m^3) of volume from each. The guidelines for Yale-Toumey Forest, Crowell Dummerston Forest, and Goss Woods are to conduct thinning treatments on 50–100 ac (20–40 ha) per year with an approximate volume removed of 100–150 MBF (236–354 m^3). Added to this are regeneration cuts on 25–50 ac (10–20 ha) per year yielding 50–100 MBF (118–236 m^3).

The remaining three forests (Bowen, Cross, and Crowell Ravine) have separate management plans approved by the State of Vermont. This timber harvest schedule for the Yale-Myers Forest was adopted in 1994, at the same time that the 7-year rotation of division inventories was established. Each division within the Yale-Myers Forest comprises about

TABLE 29.1 Standing Volume, Harvest Volume, and Net Change in Volume (All MMBF) from the Productive Portion of the Yale-Myers Forest over 20 Years, from 1984 to 2004

Species Group	Standing Volume			Harvest 1994–2004	Net change 1994–2004
	1984	1993	2004		
Oak	10.1	13.7	15.4	1.5	+1.7
Pine	6.1	8.7	11.6	1.1	+2.9
Hemlock	12.1	9.6	9.4	1.1	-0.2
Other hardwoods	5.8	7.4	6.0	0.4	-1.4
Total	34.1	39.4	42.4	4.1	+3.0

1,000 ac (405 ha), and each year management (detailed inventory and silvicultural treatments) is focused on one of these divisions. The Yale-Toumey forest works on a 5-year rotation of divisions (~500 ac (202 ha) per division), and includes Dummerston and Goss Woods.

Four hundred forest health and understory diversity plots were established in 1986 and are re-measured for floristic diversity, woody debris, forest structure, and tree species regeneration at 10-year intervals. These plots taken together assess current stand level conditions of groundstory herbaceous diversity, regeneration, vertical structure, and woody debris in relation to current silvicultural prescriptions and management regimes.

Regeneration measurements help guide management and assess regional differences in deer browse impact and susceptibility to regeneration failure related to forest light conditions, seed source, and soil type. Ground-level floristic diversity measures are used to assess and strategically plan for a sensitive areas network within the School Forests. Forest structure and composition and woody debris measures are used to gauge wildlife (bird, amphibian, and mammal) habitat suitability. In addition, wetlands throughout the forest have been cataloged through a combination of remotely sensed imagery and ground-collected data. Each year amphibians are quantified and related to local conditions, including hydroperiod, wetland area, water chemistry, and forest cover through a permanent network of plots that monitor water level fluctuation and amphibian breeding biology. The seral sequence set up by beaver activities, as well as the subsequent decay of their work, results in a highly dynamic mosaic of wetland environments, and these changes in wetland cover are also closely monitored. Taken together, these measures are used to quantitatively support landscape-level integration of sensitive and special areas (riparian systems and wetlands, biologically unique areas, older forest components, early-seral habitat, and recreational viewsheds) into a working forest landscape.

As outlined in the Management Guidelines for the Yale-Myers Forest, one of the three primary objectives of the forest is to provide a hands-on, working (managed) forest laboratory for teaching. The Yale School Forests provide a wide variety of ecosystems, ranging from pine plantations to wetlands, which can assist faculty in teaching courses by using the forests as an outdoor classroom as well as supervising students on projects on the School Forests. The educational and research opportunities on the Forests are often overlapping with the objectives of both uses met by one project. Many faculty have taken advantage of this opportunity and bring their students to the Yale Forests for field trips. In addition, the Yale-Myers Forest has the capacity for large groups to stay overnight on extended field trips. The stated goal of the Forests is to provide educational opportunities, and, in that way, it is important to have a direct faculty relationship to the management of the School Forests. The education objectives of the School Forests are currently met in 11 ways:

1. Field trips: The School Forests are used as an outdoor classroom. The Yale-Myers Forest is used more than the other properties for ecology, silviculture, and forest dynamics field trips because of its proximity to New Haven as compared to the Vermont and New Hampshire forests, and because of the facilities for overnight stays.
2. Exercises and demonstrations: Many workshops are held at the Yale-Myers Forest. These typically comprise topics such fire ecology and prescribed burning, harvesting operations, wetlands delineation, soil surveys, silvo-pastoral systems, and non-timber forest cultivation (maple syrup, shitake mushrooms, medicinal understory herbs).
3. Field modules: The Yale-Myers Forest is used each summer for the Ecosystem Measurement Module instructing incoming first-year Master's students on the basics of field measurement.
4. Student research: Doctoral students, Master's students, and research interns all use the School Forests. There are official dissertation and Master's research projects, as well as projects within the context of different courses such as economics and management.
5. Forest Apprenticeship Program: Each summer, 8–12 Master of Forestry students participate in an apprentice forester internship program at Yale-Myers Forest. The interns do all phases of forest management work, including continuous forest inventory, stand exams, mapping and photographic interpretation, timber sale layout, marking of silvicultural prescriptions, contract compliance and supervision, road drainage and maintenance, deed searches, and boundary marking.
6. Forest administration: The Forest Fellow and 4–6 student assistants run the day-to-day administration of the forest under the supervision of the School Forest Director. The policy of paying students, rather than special staff or consultants, to manage the Forests is a unique solution combining management with educational training.
7. Extension and demonstration areas for non-Yale visitors: Group meetings, workshops, and tours are conducted at the School Forests by students and faculty. Common events are those sponsored by the Society of American Foresters, Connecticut Forest and Park Association, Connecticut Forestry Extension, field visits by students from other universities, and field trips sponsored by the U.S. Forest Service. There are currently six unique demonstration areas that serve to illustrate our understanding of forest management to groups of professionals, students, and the public. About 40 events are held each year.

8. Summer research seminar and talks: Each summer, a research seminar series is held at the Yale-Myers Forest camp and features faculty, students, or other outside researchers. These seminars are open to both the surrounding community, natural resource professionals within the region, as well as those associated with Yale. The purpose is to demonstrate the nature of research and its management implications for New England forests.

9. Dissemination of published research: Papers resulting from research taking place on Yale Forest properties are listed on the Yale Forest Web site. In addition, summaries of both published and unpublished doctoral work relating to forest management are assembled into research bulletins geared toward environmental professionals. These bulletins are available online and distributed electronically to professional and public organizations across southern New England at regular intervals. There are more than 500 published studies on various topics that are stored in the archives and that have been completed on the Yale-Forests since their founding.

10. School Forests Newsletter and extension materials: An annual newsletter summarizing the last year's events and the plans for the future year is available to the public on a Web site, distributed to subscribers in the region, school forest alumni, and the School of Forestry and Environmental Studies faculty, students, and administration. In addition, periodic updates on demonstration areas, exercises, and educational opportunities are made available through electronic bulletins to subscribers.

11. The Yale School Forest Web site (http://environment.yale.edu/forests) and Facebook page serves to educate students and others along with disseminating information about the uses and activities (management, research, and education) conducted at the School Forests.

Recognizing the substantial challenges of maintaining a functional network of highly forest-literate landholders adjacent to the Yale-Myers Forest, the Yale School Forests is developing the Quiet Corner Initiative focused specifically on a Woodland Partnership. This Woodland Partnership began in 2010 when the Yale School Forests reached out to private forestland owners surrounding the Yale-Myers Forest to find neighbors interested in joining with us to enhance the pace, scale, and quality of forest conservation and management in the region. The Yale School Forest staff has worked to develop this network of landowners who are collaborating with one another and with the School Forests to provide education opportunities to students while improving the quality and coordination of land management in the area.

Since its inception, the partnership has grown to include more than 100 partners and 10,000 ac (4,047 ha) within three critical sub-watersheds of the upper Thames River, comprising the immediate surroundings of the Yale-Myers Forest. Participating landowner partners allow students the opportunity to hone forest management, forest planning, research, and land conservation skills. In return, landowners get high-quality, professional-grade work performed on their properties that helps them better manage these valuable assets. At a broader level, the partnership facilitates landowners working across traditional property boundaries to share costs, infrastructure, and information. In doing so, they work at larger economies of scale, ideally to gain preferred access to forest product markets. Whether landowners are interested in timber, recreation, conservation, or in developing non-timber products like maple syrup, Yale students and faculty help them to develop professional management plans tailored to their values and then to implement these plans. During implementation, students work with the landowners as well as forestry and conservation professionals from the area to again mix professional training with real, on-the-ground experience that delivers tangible results to the partners.

Plan Development Methods and Constraints

To carry out the objectives of the management plan, the School Forests have been zoned by land use. More than one-third of the land is zoned as part of a network of reserves demarcated as wetland and riparian protection areas, ecologically sensitive areas, research areas, early-seral meadows, and future late-successional forest areas. The remaining two-thirds comprise second-growth forest that is both managed for timber and increasing species, structural, and age-class diversity. The silvicultural methods for carrying out the area and volume-based cutting plan can be divided between treatments to conifers and hardwoods.

White pine now occupies mostly sandy and glacial outwash soils within the forest. Successful regeneration methods all rely upon natural regeneration through the seed-tree regeneration method. About 8–10 seed trees are left per ac (20–25 per ha), but in most cases reserves of smaller pine or other species are left as individuals or groups for increasing age-class structure, wildlife habitat, or timber value. Where regeneration does not satisfactorily establish, enrichment planting is practiced. In almost all circumstances, pine is the crop tree and is pruned and release cleaned at about 25 years (30–40 per ac, 74–99 per ha), followed by several crown thinnings. On most pine sites, rotations vary between 60 and 70 years.

Hemlock is managed in a diverse set of stand development pathways. Most hemlock stands comprise a scattered oak over-story with hemlock represented in all strata from the understory to the canopy. Selection systems that use groups or patches of 0.33 ac (0.13 ha) in size are the normal method to sustain hemlock on moist sites where the hemlock woolly adelgid (*Adelges tsugae*) is either less prevalent or where hemlock is more resistant. The gap size is enough to create a diversity of regeneration including oak, pine, and hemlock. In these circumstances, the cutting cycle is about 21 years, with 4–5 age classes represented within the stand. On north and northwest facing slopes, hemlock is managed through variants of the strip shelterwood method. Strip width to secure advance regeneration is about the height of the existing canopy. Such methods are new for the manage-ment of the forest but will be increasingly used under these circumstances and again where cool moist conditions allow for greater resilience of hemlock to adelgid infestation. On drier sites, hemlock stands are converted to oak-hardwood stands through uniform shelterwood systems (Figure 29.2), but group reserves of hemlock are left primarily as thermal cover and habitat structure for wildlife. These stands are considered adelgid susceptible because they are south facing or on upland sites.

There are two primary hardwood stand types: (1) upland and mid-slope oak stands and (2) valley or mesic maple-ash-tulip poplar (*Acer* spp.-*Fraxinus* spp.-*Liriodendron tulipifera*) stands. Oak stands are usually managed through shelterwoods with closer spacing (30–40 feet (ft), 9–12 meters (m)) between parent trees on drier sites as compared to mid-slopes (40–60 ft, 12–18 m) and valley sites (50–70 ft, 15–21 m). This is a reflection of the ability to secure oak regeneration relatively easier on drier sites that can persist for long periods of time in understory conditions (50% survival after ten years). Strong masting years throughout the forest are infrequent (once every 10 years) making valley sites very susceptible to poor oak regeneration establishment (few oak seedlings survive in understory conditions on such sites after 5 years). Oak shelterwoods on valley sites therefore need to be opportunistically planned immediately after heavy mast years with wide-open spacing to promote release from the shade tolerant competition of sugar maple (*Acer saccharum*) and black birch (*Betula lenta*).

Oak shelterwoods in almost all cases are planned as irregular systems that comprise 2- to 3-age classes with the regen-erating cohort the most dominant. Reserves of older age classes are arranged as single-tree and group reserves that serve to diversify species composition, increase wildlife habitat structure, or add additional pre-commercial or unrealized timber value. On mesic sites with sugar maple single-tree reserves of sugar maple are often part of oak shelterwood cuts. Also, in circumstances where prolific advanced regeneration is apparent, uniform and group shelterwood systems are bypassed in favor of one-cut shelterwoods with reserves. On drier oak sites with persistent mountain laurel (*Kalmia latifolia*) in the un-derstory, advance regeneration of hardwoods is often lacking. Under such circumstances, uniform shelterwoods are marked in tandem with site scarification and crushing of laurel or in certain circumstances with the use of prescribed groundstory fires. For maple-ash-tulip poplar sites, uniform shelterwoods are implemented with a relatively closer spacing (30–40 ft, 9–12 m) of masting maple, ash, and tulip poplar to encourage more sugar maple over shade-intolerant ash and poplar. Where adequate regeneration exists of sugar maple, one-cut shelterwoods are conducted with reserves of ash and poplar to supplement the regeneration release of maple.

Guiding Laws, Regulations, and Policies

The School Forests have been certified by a number of organizations including the American Tree Farm (since 1970), the Sustainable Forestry Initiative (2001–2010) and the Forest Stewardship Council (2001–ongoing). The State of Connecticut

FIGURE 29.2 A photograph depicting an area undergoing the establishment cut of a uniform shelterwood harvest for the development of an oak-hardwood stand. Most shelterwoods used at the Yale-Myers Forest are irregular, 2- to 3-age-class systems.

has regulatory oversight of management through the Connecticut Forest Practices Act, which mandates that the Director and the School Forest Manager be licensed foresters. Deer and turkey (*Meleagris gallopavo*) hunting (by permit and season) and public fishing are administered by the Connecticut Department of Energy and Environmental Protection. A 9 mi (14.5 km) portion of the Nipmuck Trail, maintained by the Connecticut Forest and Park Association, traverses the Yale-Myers Forest from south to north. This hiking trail is the only part of the Forest open to the general public for recreational use other than hunting and fishing. These recreational uses represent the extent of the "official" recreation program at the Yale-Myers Forest. However, activities such as hiking off of the blue trail, mountain-biking on the woods roads, horseback riding (by formal permission only), and cross-country skiing are tolerated as long as users are not causing damage or otherwise creating problems. These activities are permitted for individuals only. Non-invited groups are not permitted on the Forest. The above activities are allowed but are not actively promoted or encouraged because the School Forests Program has neither the human resources nor the physical presence at the Yale-Myers Forest to effectively manage large numbers of recreational users. Because recreation is not a primary management goal of the Forest, this is unlikely to change any time soon. That said, it is impossible and ultimately undesirable to exclude the public from using the Forest for recreational purposes, and the activities listed above are relatively low-impact and unlikely to cause problems or damage if user numbers are minimized. Yale is protected from liability under Connecticut's recreational-use statute as long as the public is not charged for entering the forest and Yale does not exhibit a "willful or malicious failure to guard or warn against a dangerous condition, use, structure, or activity." This protection applies both to instances when the public is merely allowed to recreate in the Forest and when they are specifically invited to do so. Entering into active timber harvests is strictly prohibited.

OUTCOMES OF THE PLAN

The plan can be considered to provide general, overarching guidelines to convert an even-aged second-growth forest to one with greater tree species composition, age-class diversity, and structure. There is enough flexibility that allows prescriptions to be unique and tailored to a stand on a yearly basis depending on the biophysical constraints faced and the economics of the market at that time. There are no mandates to treat each stand at a prescribed time, instead all stands are assessed at periodic intervals and prioritized for treatments that fulfill the annual cut and area mandates.

Regeneration harvests have now been ongoing since 1992. Nearly 20% of the production forest area has either started or been regenerated. This matches the estimated trajectory to create an all-aged forest. Surprisingly, the total growth volume increment of the forest continues to grow primarily because the actual annual harvests have been considerably lower than the allowable annual cut and because more than one-third of the forest is in some kind of reserve area.

DISCUSSION AND CONCLUSIONS

Learnings and Insights

The Yale School Forests are managed with a conservative and risk-averse approach. This focus is used primarily because of the unpredictability of accounting for external impacts unforeseen by following a more intensive program that has a focus on a higher yield, single-tree species, or single age-class. New England's forests are in and of themselves a reflection of lessons learned with the failure of agriculture in the 1800s and the past failures of intensification and single-species markets that could not be sustained on marginal soils. The forests are managed almost entirely through natural regeneration and include a diverse array of species and sites. This course of action is pursued to avoid the costs of planting and site control and to take advantage of markets that can change rapidly, are varied, and that are species specific.

Sustainability Issues

Because it is considered a working forest, the Yale School Forests are an example of sustainable and multiple-use management. The forests are an example of one of the most sustainably managed private lands in the region, owned and managed by the School, but with a demonstrated financial independence for more than 100 years. The Forests are a record of restoration with merchantable volume increasing from 10 MMBF (23,600 m³) in 1910 to 43 MMBF (101,480 m³) in 2004, and with over 50 MMBF (118,000 m³) harvested (mostly from thinnings) over the approximately 100-year management period.

Plan Implementation Challenges

Challenges to planning and implementation come from unforeseen events. To counter this, the approach is to have some broad generalized guidelines for creating a forest more capable of withstanding unpredictability, but also allowing yearly flexibility in meeting these guidelines. Past effects on management of the forests include impacts from hurricanes and disease, both beyond the control of management. The most notable was the death of American chestnut (*Castanea dentata*) in the early 1920s from the chestnut blight and the 1938 hurricane that leveled about 10 MMBF (23,600 m^3) of merchantable timber, dropping the standing volume from 23.8 MMBF (56,168 m^3) in 1937 to about 10 MMBF (23,600 m^3) in 1945. This period relied extensively on salvage logging operations and storage of logs in log ponds until depressed markets recovered from the flood of salvage timber. Today's challenge can be seen with the continuous impacts of exotic insects and diseases. The current impact that has been managed for the last 20 years has been the introduction to the forests of the hemlock woolly adelgid from Asia. New threats to ash (emerald ash borer (*Agrilus planipennis*)) and maple (Asian long-horned beetle (*Anoplophora glabripennis*)) exist. In addition, increased storms and droughts from the unpredictability of climate change continue to build.

REFERENCES

Abrams, P.A., Schmitz, O.J., 1999. The effect of risk of mortality on the foraging behaviour of animals faced with time and digestive capacity constraints. Evolutionary Ecology Research 1, 285–301.

Aikens, M.L., Ellum, D., McKenna, J., Kelty, M.J., Ashton, M.S., 2007. The effects of disturbance intensity on temporal and spatial patterns of herb colonization across a canopy opening of a southern New England mixed-oak forest. Forest Ecology and Management 252, 144–158.

Ashton, M.S., 1992. Establishment and early growth of advance regeneration of canopy trees in moist mixed-species forest. In: Kelty, M.J., Larson, B.C., Oliver, C.D. (Eds.), The Ecology and Silviculture of Mixed-Species Forests. Springer, New York, pp. 101–122.

Ashton, M.S., Berlyn, G.P., 1994. Species in different light environments. American Journal of Botany 81, 589–597.

Ashton, M.S., Peters, C.M., 1999. Even-aged silviculture in tropical rainforests of Asia: lessons learned and myths perpetuated. Journal of Forestry 97 (11), 14–19.

Ashton, P., Harris, P., Thadani, R., 1998. Soil seed bank dynamics in relation to topographic position of a mixed-deciduous forest in southern New England, USA. Forest Ecology and Management 111, 15–22.

Beckerman, A.P., Uriarte, M., Schmitz, O.J., 1997. Experimental evidence for a behavior-mediated trophic cascade in a terrestrial food chain. Proceedings of the National Academy of Sciences 94, 10735–10738.

Bradford, M.A., Warren II, R.J., Baldrian, P., Crowther, T.W., Maynard, D.S., Oldfield, E.E., Wieder, W.R., Wood, S.A., King, J.R., 2014. Climate fails to predict wood decomposition at regional scales. Nature Climate Change 4, 625–630.

Covey, K.R., Wood, S.A., Warren, R.J., Lee, X., Bradford, M.A., 2012. Elevated methane concentrations in trees of an upland forest. Geophysical Research Letters 39, doi:10.1029/2012GL052361, L15705.

Ducey, M.J., Moser, W.K., Ashton, P.M.S., 1996. Effect of fire intensity on understory composition and diversity in a Kalmia-dominated oak forest, New England, USA. Vegetatio. 123, 81–90.

Duguid, M.C., Frey, B.R., Ellum, D.S., Kelty, M., Ashton, M.S., 2013. The influence of ground disturbance and gap position on understory plant diversity in upland forests of southern New England. Forest Ecology and Management 303, 148–159.

Ellum, D.S., Ashton, M.S., Siccama, T.G., 2010. Spatial pattern in herb diversity and abundance of second growth mixed deciduous-evergreen forest within southern New England, USA. Forest Ecology and Management 259, 1416–1426.

Frey, B., Ashton, M.S., McKenna, J., Ellum, D., Finkral, A., 2007. Topographic-related patterns in seedling establishment, growth, and survival among masting species of Southern New England hardwood forests. Forest Ecology and Management 244, 31–45.

Hawlena, D., Strickland, M.S., Bradford, M.A., Schmitz, O.J., 2012. Fear of predation slows plant-litter decomposition. Science 336, 1434–1438.

Hawley, R.C., 1921. The Practice of Silviculture, with Particular Reference to Its Application in the United States. John Wiley & Sons, Inc., New York.

Hawley, R.C., 1929. The Practice of Silviculture. John Wiley & Sons, Inc., New York.

Hawley, R.C., 1946. The Practice of Silviculture. John Wiley & Sons, Inc., New York.

Keiser, A.D., Keiser, D.A., Strickland, M.S., Bradford, M.A., 2014. Disentangling mechanisms underlying functional differences in decomposer communities. Journal of Ecology 102, 603–609.

Kelty, M.J., 1986. Development patterns in two hemlock-hardwood stands in southern New England. Canadian Journal of Forest Research 16, 885–891.

Kittredge Jr., D.B., 1988. The influence of species composition on the growth of individual red oaks in mixed stands in southern New England. Canadian Journal of Forest Research 18, 1550–1555.

Kittredge, D.B., Ashton, P.M.S., 1990. Natural regeneration patterns in even-aged mixed stands in southern New England. Northern Journal of Applied Forestry 7, 163–168.

Kittredge, D.B., Ashton, P.M.S., 1995. Impact of deer browsing on regeneration in mixed stands in southern New England. Northern Journal of Applied Forestry 12, 115–120.

Kittredge, D.B., Kelty, M.J., Ashton, P.M.S., 1992. The use of tree shelters with northern red oak natural regeneration in southern New England. Northern Journal of Applied Forestry 9, 141–145.

Liptzin, D., Ashton, P., 1999. Early-successional dynamics of single-aged mixed hardwood stands in a southern New England forest, USA. Forest Ecology and Mangement 116, 141–150.

Lutz, H.J., Chandler, R.F., 1946. Forest Soils. John Wiley & Sons, New York.

Meyer, W.H., Plusnin, B.A., 1945. The Yale Forest in Tolland and Windham Counties. Yale University, New Haven, CT, Connecticut.

Moser, W.K., Ducey, M.J., Ashton, P.M.S., 1996. Effects of fire intensity on competitive dynamics between red and black oaks and mountain laurel. Northern Journal of Applied Forestry 13, 119–123.

Oliver, C.D., 1978. The Development of Northern Red Oak in Mixed Stands in Central New England. Yale University, New Haven, CT.

Oliver, C.D., Larson, B.C., 1990. Forest Stand Dynamics. McGraw-Hill, Inc., New York.

Schmitz, O.J., 1994. Resource edibility and trophic exploitation in an old-field food web. Proceedings of the National Academy of Sciences 91, 5364–5367.

Schmitz, O.J., 1997. Press perturbations and the predictability of ecological interactions in a food web. Ecology 78, 55–69.

Schmitz, O.J., 1998. Direct and indirect effects of predation and predation risk in old-field interaction webs. The American Naturalist 151, 327–342.

Schmitz, O.J., Beckerman, A.P., O'Brien, K.M., 1997. Behaviorally mediated trophic cascades: effects of predation risk on food web interactions. Ecology 78, 1388–1399.

Schmitz, O.J., Bradford, M.A., Strickland, M.S., Hawlena, D., 2013a. Linking predation risk, herbivore physiological stress and microbial decomposition of plant litter. Journal of Visualized Experiments 73, e50061. doi:10.3791/50061.

Schmitz, O.J., Hambäck, P.A., Beckerman, A.P., 2000. Trophic cascades in terrestrial systems: a review of the effects of carnivore removals on plants. The American Naturalist 155, 141–153.

Schmitz, O.J., Raymond, P.R., Estes, J.A., Kurz, W.A., Holtgrieve, G.W., Ritchie, M.E., Schindler, D.E., Spivak, A., Wilson, R.W., Bradford, M.A., Christensen, V., Deegan, L., Smetacek, V., Vanni, M.J., Wilmers, C.C., 2013b. Animating the carbon cycle. Ecosystems 17, 344–359.

Seymour, R.S., Smith, D.M., 1987. A new stocking guide formulation applied to eastern white pine. Forest Science 33, 469–484.

Skelly, D., Freidenburg, L., Kiesecker, J., 2002. Forest canopy and the performance of larval amphibians. Ecology 83, 983–992.

Skelly, D.K., Bolden, S.R., Freidenburg, L.K., 2014. Experimental canopy removal enhances diversity of vernal pond amphibians. Ecological Applications 24, 340–345.

Skelly, D.K., Halverson, M.A., Freidenburg, L.K., Urban, M.C., 2005. Canopy closure and amphibian diversity in forested wetlands. Wetlands Ecology and Management 13, 261–268.

Smith, D., 1962. The Practice of Silviculture. John Wiley & Sons, New York.

Smith, D.M., Ashton, P.M.S., 1993. Early dominance of pioneer hardwood after clearcutting and removal of advanced regeneration. Northern Journal of Applied Forestry 10, 14–19.

Smith, D.M., Larson, B.C., Kelty, M.J., Ashton, P.M.S., 1997. The Practice of Silviculture: Applied Forest Ecology. John Wiley and Sons, Inc., US.

Strickland, M.S., Hawlena, D., Reese, A., Bradford, M.A., Schmitz, O.J., 2013a. Trophic cascade alters ecosystem carbon exchange. Proceedings of the National Academy of Sciences 110, 11035–11038.

Strickland, M.S., McCulley, R.L., Bradford, M.A., 2013b. The effect of a quorum-quenching enzyme on leaf litter decomposition. Soil Biology and Biochemistry 64, 65–67.

Toman, M.A., Ashton, P.M.S., 1996. Sustainable forest ecosystems and management: a review article. Forest Science 42, 366–377.

Toumey, J.W., 1929. Foundations of Silviculture upon an Ecological Basis. John Wiley & Sons, New York.

Toumey, J.W., Kienholz, A.R., 1931. Trenched Plots Under Forest Canopies. Yale University, New Haven, CT.

Toumey, J.W., Korstian, C.F., 1937. Foundations of Silviculture upon an Ecological Basis. John Wiley & Sons, New York.

Uriarte, M., Schmitz, O.J., 1998. Trophic control across a natural productivity gradient with sap-feeding herbivores. Oikos 82, 552–560.

Wellborn, G.A., Skelly, D.K., Werner, E.E., 1996. Mechanisms creating community structure across a freshwater habitat gradient. Annual Review of Ecology and Systematics 27, 337–363.

Chapter 30

Bayfield County Forest, Wisconsin, United States of America

Jeff Barkley,[1] Jason Bodine,[2] Chris Hoffman,[3] Jeremy Koslowski,[4] Larry Stevens,[5] Joseph Schwantes,[6] and Jane Severt[1]

[1]Wisconsin County Forests Association, Rhinelander, Wisconsin, USA, [2]Bayfield County Forestry and Parks Department, Washburn, Wisconsin, USA, [3]Ashland County Forestry Department, Butternut, Wisconsin, USA, [4]Polk County Forestry Department, Balsam Lake, Wisconsin, USA, [5]Vilas County Forestry Department, Eagle River, Wisconsin, USA, [6]Wisconsin Department of Natural Resources, Madison, Wisconsin, USA

ABBREVIATIONS

BCF	Bayfield County Forest
EA	Environmental Analysis
IRMU	Integrated Resource Management Unit
WCFA	Wisconsin County Forests Association
WDNR	Wisconsin Department of Natural Resources
WisFIRS	Wisconsin Forest Inventory and Reporting System

MANAGEMENT SETTING AND BACKGROUND

Of Wisconsin's 72 counties, 29 have county forests. The Bayfield County Forest (BCF), located in extreme northern part Wisconsin, is the fourth largest of Wisconsin's county forests (Figure 30.1). Wisconsin's county forest program is unique in the United States and functions as a partnership between county governments and the Wisconsin Department of Natural Resources (WDNR). Counties direct the on-the-ground management within the sideboards of Wisconsin statutes 28.10 and 28.11. Locally, county forestry committees consisting of five to eight county board members provide guidance to professional county forestry staff responsible for the actual management of the forest. The statutes detail the purpose for management, planning requirements, responsibilities of the county and the WDNR, distribution of revenues, grant and loan availability, and entry and withdrawals from the county forest program. The WDNR provides technical assistance to each county, administers grants and loans, coordinates the forest certification certificates, and provides general oversight of the program's statutory requirements. Bayfield County employs a county forest administrator, an assistant county forest administrator, five foresters, one forest technician, an office manager, and receives approximately 3,395 hours of assistance from WDNR foresters annually.

Bayfield County's current long-range plan was developed in 2006 and covers the 15-year period from 2006 to 2020. Wisconsin law dictates both the planning duration and time frame for all Wisconsin county forests. The BCF originated in the late 1920s when Bayfield County obtained a large amount of land through tax delinquency. At the time, these lands were a burden on county governments who were reluctant to accept the "unwanted lands." Logged over and ravaged by fires, these lands presented no prospect for immediate financial return. Wisconsin's creation of the County Forest Program provided incentives for counties to retain these lands and manage them for the public good. Bayfield County officially became the sixth county to enter the program in 1932, with an initial entry of 52,832 acres (ac) (21,381 hectares (ha)). Since that time, additional lands have been entered bringing the area up to the current total of 169,394 ac (68,553 ha).

Forest Plans of North America. http://dx.doi.org/10.1016/B978-0-12-799936-4.00030-8

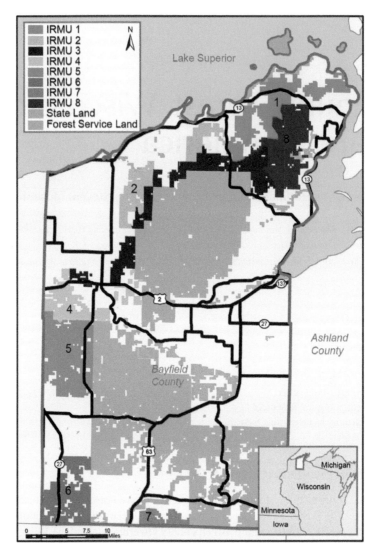

FIGURE 30.1 A map of the Bayfield County Forest integrated resource management units (IRMU).

The BCF is arranged into six *blocking boundary* units (administrative areas), depicted in the comprehensive land-use plan, which are further divided into 202 management compartments. More than 87% of the land within the BCF blocking boundary is county forest with the remaining 13% consisting of small private or industrial tracts. More than 94% of the land base is forested. Uplands are comprised of primarily aspen (*Populus tremuloides*), mixed hardwoods, oak (*Quercus* spp.), red pine (*Pinus resinosa*), white pine (*Pinus strobus*), and jack pine (*Pinus banksiana*), while fir-spruce (*Abies balsamea and Picea* spp.), lowland hardwoods, tamarack (*Larix laricina*), and northern white cedar (*Thuja occidentalis)* forest types occupy the forested lowlands. The remainder is classified as nonforested, including types such as open water, wetlands, rights-of-way, grass openings, shrubs, and bogs. Within the county forest boundaries there are 27 named and 104 unnamed lakes that have all or part of their shoreline under county ownership. An additional 30 named streams with more than 150 miles (mi) (241 kilometers (km)) of frontage on the county forest provide a wealth of aquatic habitat (Figure 30.2) and recreational possibilities. The topography of the forest and surrounding area is of glacial origin. Glaciers eroded hilltops and filled valleys, thus reducing relief. Elevations range from a low of 602 feet (ft) (183 meters (m)) above sea level at Lake Superior to 1,610 ft (491 m) on the western ridges of the Penokee Range in the southeast part of the county. Soils range from red clay and sand over clay in the northernmost portions of the forest to outwash sands in the central portions. Stony sandy loams and silt loams are found in the southernmost parts of the forest.

Bayfield County offers a rich mix of recreational opportunities infused within the management of the forest. Developed recreational opportunities include designated snowmobile, all-terrain vehicle, cross-country ski, mountain bike, walking

FIGURE 30.2   A Bayfield County Forest woodland stream.

and dog sled trails, picnic areas, boat landings, and fishing piers. Numerous regional and national events are held annually including the American Birkebeiner Ski Race, the Chequamegon Fat Tire Festival, and the Apostle Islands Sled Dog Race. More extensive opportunities including hunting, fishing, trapping, berry and mushroom picking, bird watching, hiking, mountain biking, and sightseeing are also present. These activities on the BCF contribute markedly to countywide tourism that annually generates more than $40 million (U.S. dollars (USD)) in direct spending and employs more than 600 people in Bayfield County (Wisconsin Department of Tourism, 2014). The BCF alone generates, on average, $2,380,450 USD timber stumpage each year, which bolsters an even larger regional forest industry, Wisconsin's second largest industry sector. Value-added economic impact of this timber is in excess of $57 million USD (Deckard and Skurla, 2011). The revenues from sustainable timber harvests directly reduce the local tax levy.

PLANNING ENVIRONMENT AND METHODOLOGY

From the mid-1800s until about 1910, Wisconsin forests were vital in providing raw material for the U. S. lumber industry. The demand for lumber and the forest management approaches practiced during these times resulted in overharvest of the forests and degradation of the landscape. People immigrated to Wisconsin in order to develop farms on the cleared areas. However, within just a few years, unsuitable soils, short growing seasons, and widespread fires, resulted in many of these lands being abandoned or tax delinquent. As a result, the Wisconsin County Forest program began with the acquisition of these tax-delinquent properties (Bayfield County Forestry and Parks Department, 2006).

County forestry staff engages with user groups and interested parties regularly on an informal basis. It is important to note the Bayfield County Forestry and Parks Committee, which provides oversight and guidance to the county forest staff, meet monthly in a public forum as well. These committee meetings are an opportunity for the public and interested user groups to provide feedback to this oversight committee. Additionally, the Bayfield County Tribal Relations Committee meets biannually with the Red Cliff Band of Lake Superior Chippewa to discuss, among other things, natural resource management on the County Forest located within Reservation boundaries. The Red Cliff Tribe has traditional use rights on the County Forest and owns adjacent lands. Ongoing interaction with user groups, the public and affected partners is crucial to plan acceptance and adaptive management.

Guiding Laws, Regulations, and Policies

A comprehensive county forest land-use plan (also called a *county forest plan*) needs to be prepared for all Wisconsin county forests, as directed by statute (the Wisconsin County Forest Law, WI. Stats 28.11(5)(a)). Annual work plans further define and supplement the comprehensive land-use plan and emphasize current needs of county forests and associated parks and trails programs. As noted in the Wisconsin County Forest Law (2011–12 Wisconsin State Statutes and Annotations. Updated through 2013 Wisconsin Act 380, Chapter 28 Public Forests, 28.11(5)(a), Accessed May 16, 2014), the BCF is managed for the

optimum production of forest products together with recreational opportunities, wildlife, watershed protection and stabilization of stream flow, giving full recognition to the concept of multiple-use to assure maximum public benefits; to protect the public rights, interests and investments in such lands; and to compensate the counties for the public uses, benefits and privileges these lands provide; all in a manner which will provide a reasonable revenue to the towns in which such lands lie.

Further, the Bayfield County Forest Plan is required by statute (§ 28.11(5)(a) WI. Stats) to

include land use designations, land acquisition, forest protection, annual allowable timber harvests, recreational developments, fish and wildlife management activities, roads, silvicultural operations and operating policies and procedures; it shall include a complete inventory of the county forest and shall be documented with maps, records and priorities showing in detail the various projects to be undertaken during the plan period.

This particular plan is the fifth iteration of the long-range planning required by statute. The law was revised in 1963 to create today's permanent program of county forests. Prior plans were written to cover a 10-year period, but the law was amended in 2005 to a 15-year time frame for long-range planning. Bayfield County Ordinances also codify policies specific to the county forest, in title 12, chapter 2. Included is the ordinance designating lands into county forest status and several ordinances pertaining to specific use policies on the BCF. For example, Section 12-2-6(a) describes a number of policies regarding permitted and prohibited recreational use of county forest lands, and Section 12-2-6(b) describes several allowable forest practices and harvesting operations, including stand improvement harvests, salvage harvests, and commercial timber sales.

Bayfield County also recognizes that the county forest is part of a diverse landscape and is managed not as an independent entity, but as part of a larger ecoregion as depicted in the National Hierarchical Framework of Ecological Units (Cleland et al., 1997). Management enhances and protects unique flora, fauna, and ecological areas. A Natural Heritage Inventory database (Wisconsin Department of Natural Resources, 2014a) is consulted on all ground-disturbing activities. When rare, threatened, or endangered species are indicated, the foresters have species-specific resources available to determine how best to avoid or mitigate negative impacts. Endangered species resource staff from the WDNR is available for consultation as well. Foresters are also attuned to forest conditions that are contrary to the norm and may flag areas for further inspection of the vegetation. All foresters are trained in the Forest Habitat Classification System (Wisconsin Department of Natural Resources, 2006b), which helps to interpret site capability (moisture and nutrients) based on ground vegetation. This helps in determining the silvicultural prescriptions on timber harvests and fitting species to the site's desired future conditions. It also sharpens the biotic eye of foresters and alerts them to potential rare herbaceous plants and invasive plant species. All activities employ Wisconsin's Forestry Best Management Practices for Water Quality (Wisconsin Department of Natural Resources, 2010) that have been proven to provide excellent protection for water resources and associated species. In addition, Wisconsin's Forestry Best Management Practices for Invasive Species are also applied to minimize spread of invasive species (Wisconsin Department of Natural Resources, 2009).

Methodology of Planning Process

As mentioned earlier, a comprehensive county forest land-use plan (also called a *county forest plan*) needs to be prepared for a 15-year time horizon. Mid-year 2002, at the onset of the planning process for the 2006–2020 county forest plans, a multidisciplinary ad hoc group was formed to aid in the plan development process and to ensure that all the county forests included all of the facets of the plan required by law. This was a statewide group and included five county forest administrators (local county-employed foresters), three WDNR foresters, a WDNR Team Leader, the WDNR County Forest program specialist, the Executive Director of the Wisconsin County Forests Association (WCFA), and WDNR program specialists covering endangered resources, wildlife, and water resources. The WCFA is a nonprofit organization that represents the collective forestry interests of all 29 county forests in Wisconsin. Over the next year and a half this group met several times. A statewide template plan was approved by both the WDNR and the WCFA by the fall of 2003. Included in the framework for the plan template were the following chapters with their chapter index number:

    100—Background
    200—General Administration
    300—Management Planning
    400—County Forest Blocking
    500—Land Management and Use
    600—Protection
    700—Roads, Trails and Access
    800—Integrated Resource Management
    900—Appendix and Maps

1000—Needs/Plan Implementation
2000—Annual Planning
3000—Accomplishments and Monitoring

Early in 2004, the template was distributed to all counties that then coordinated development of the county specific plan. At a minimum, all plans were required to include these elements in the template plan. All plans were also required to incorporate public input into the process. To that end, statewide public participation training was provided to all counties. The result of the training was Bayfield County holding three public meetings to identify issues and solicit comments for consideration prior to their plan development. Comments and issues raised at the time included:

1. The County was encouraged to continue multiple-use management of the forest.
2. Bayfield County was commended for seeking certification of the County Forest.
3. Comments were made indicating there is not adequate law enforcement on the County Forest. This was especially in reference to the use of motorized vehicles.
4. Citizens spoke against increased motorized use on the County Forest citing the noise, resource damage, and the impact on others trying to use the forest.
5. Citizens spoke for more motorized recreational use opportunities on the Forest, referencing a need for access to public land, use by older citizens who would otherwise be unable to access remote parts of the forest, and the economic benefits that recreational motorized use of public lands provide.
6. The County was encouraged to continue to acquire land to add to the County Forest.
7. The County was encouraged to continue management that maintains early-successional species and pine barrens.
8. The County was encouraged to maintain timber production as a priority on the Forest. Comments were also made about the importance of maintaining buffers around water resources and of incorporating aesthetics in harvest plans.

In response, Bayfield County elected to add two chapters (4000—Integrated Resource Management Unit (IRMU) summaries, and 5000—Monitoring) to the plan. In addition, the county elected to add sections pertinent to forest certification. In 2004, Bayfield County opted to commit to the Sustainable Forestry Initiative (SFI) certification scheme. Later, in 2007, it chose to also become certified under the Forest Stewardship Council (FSC). BCF has since been successfully audited on three different occasions by third-party independent auditors. After the initial draft was produced in late 2004, the 2006–2020 BCF plan was approved by the Bayfield County Forestry and Parks Committee and was made available for public comment. In January 2005, public meetings were held at four locations throughout the county. In addition, comments were solicited at monthly Forestry Committee meetings throughout the year. Public comments were incorporated and changes were made to the draft. The plan was formally approved by the Bayfield County Board in January of 2006. It then was submitted to the WDNR for approval.

Prior to the WDNR's final approval, Wisconsin law required that such plans also be evaluated for their environmental impact. Much the same as the original plan development, an overall environmental analysis (EA) was completed assessing the impacts of all 29 county forests (Wisconsin Department of Natural Resources, 2006a). Bayfield County then appended the EA with individual county data. The EA was made available through a public notice for comments in April of 2006, and given that there were no additional comments, was approved in May 2006 by the WDNR after a 30-day comment period. Public comments on the EA and later drafts of the actual plan were minimal. The process of soliciting and incorporating public concerns in advance of, and during the drafting of the plan, was instrumental in gaining public acceptance. This greatly streamlined the overall planning process. The public feedback, certification requirements, and local expertise helped modify the plan and steered the management direction within the sideboards of the statutory purpose for the forest. Formal approval of the overall BCF plan by the WDNR was granted in July 2006. The BCF plan is organized into the chapters previously identified. Much of the plan identifies aspects that apply forestwide. Laws and ordinances that apply to management on the forest are found primarily in chapters 100–400, 600, and 900 (appendix). Roles and duties for staff managing the forest are primarily in chapters 200 and 300. Policies and methodologies for on-the-ground management of the forest are found in chapters 500, 700, and 800. On-the-ground management strategies are further refined by zoning the forest with respect to important issues such as primary uses, access, aesthetics, and unique resources. IRMUs identified in chapter 4000 take the access, aesthetics, and unique resource decisions a step further. This information is brought forth into each respective IRMU, coupling it with the specific information on forest cover types, landforms, geology, soils, surface waters, cultural sites, and protection needs. This information then further refines the on-the-ground management in those areas.

Chapter 1000 (Plan Implementation) generalizes the anticipated levels of timber harvest by forest type and silvicultural prescription. It also projects the site preparation, release, and reforestation needs over the 15 years of the plan. This is determined by using the forester's field reconnaissance data and silvicultural prescription, including a recommended implementation date, for each stand of timber into the Wisconsin Forest Inventory and Reporting System (WisFIRS), a Web-based system (Wisconsin Department of Natural Resources, 2014b). The WisFIRS then generates sustainable long-term harvest goals using area control for the property. In doing so, it utilizes desired average property-specific harvest intervals and constraints. For example, on the BCF, the average harvest interval for even-aged aspen stands is 55 years and even-aged jack pine stands is 50 years. In application, however, some stands may be harvested early and some later than the target dates in order to best spread out the age classes for those types. This is important ecologically and operationally since many of the forest types originated at similar times after the cut-over period in the early 1900s.

The purpose of the county forest program implies importance to the production of forest products and the generation of income to offset local tax levies. To that end, all forested stands are initially considered eligible for timber harvest. However, the county forest law, SFI/FSC certification, and the public also demand a multiple-use management philosophy. Therefore, in addition to the forestwide aspects, Bayfield County divided the forest into zones to better organize and address key issues such as access and aesthetic management. It also identified *Special Management Areas* to better protect and enhance natural and recreational features. Forest certification requirements are embedded into the prescribed management laid out in the plan, rather than creating a separate management system within which to address specific certification requirements.

Access to the forest is perhaps the most controversial aspect of management, and chapter 700 was entirely devoted to this issue. An initial access management plan was developed in 1998 to address rising motorized recreational use. Environmental damage and conflicts with silent sports users were occurring. That plan was incorporated and modified into chapter 700 of the 2006–2020 plan. In October 2013, chapter 700 was amended to address the ever-changing access demands being put on the land base. Stakeholders were identified at the start of this process and kept abreast of all discussions. Feedback was solicited throughout the entire process, including an open house in May 2013. In addition, comments were solicited at every monthly Forestry Committee meeting and during the County Board meeting prior to final approval. While some uses are permissible in all parts of the forest, the conflict between motorized and nonmotorized use caused the county to create motorized use zones. Roads and trails are closed to all forms of motorized vehicle use unless posted open or designated open as per the motorized use classifications. These zones are used as a general guideline although there are some exceptions to each that are then signed accordingly. Table 30.1 illustrates the forestwide breakdown of motorized access areas and permitted uses.

The BCF is an actively managed forest, yet aesthetics are very important to many of the public user groups. Chapter 500 establishes four aesthetic zones. Zone A identifies those areas where scenic values are most important. The emphasis on scenic values decreases in Zones B and C. Zone D identifies aesthetic management specific to special management areas identified in chapter 800. There are some restrictions on permitted uses and seasonal use based on the aesthetic zone designation. Forestry staff members apply aesthetic modifications when establishing timber for sale in Zones A and B. Identification and management of special management areas is also separated out and referenced in chapter 500. These areas contain unique resources such as rivers, lakes, bogs, as well as unique geological, historical, and archeological features. A globally rare pine barrens area of nearly 10,000 ac (4,047 ha) is also individually identified for unique and prescriptive management in the plan (Bayfield County Forestry and Parks Department, 2013b).

Eight IRMUs subdivide the BCF (Figure 30.1) to aid in refining the forestwide policies and procedures applied on the forest. Soil types were used as the primary criteria in delineating these units. Each IRMU is organized similarly with the following information:

- Compartment numbers and areas
- Forest cover types: Existing and desired
- Landforms, geology, and soils
- Surface water resources inventory
- Recreation uses
- Historical and cultural sites
- Special management areas
- Protection needs
- Access management, roads, and trails
- Management issues, concerns, and opportunities

Foresters establishing practices in these respective IRMUs consult this information and adjust the standard management practices accordingly to address the special concerns or opportunities identified for these areas.

The forest plan is dynamic and adaptive, and is amendable as needed. Plan amendments also require approval by the full Bayfield County Board and the WDNR. Much of the detailed work planning is included in an Annual Bayfield County Forest Work Plan (Bayfield County Forestry and Parks Department, 2013a). This document is an annual amendment to

TABLE 30.1 Summary of Permitted Motorized Use by Type, Per Area Classification

Area Classification	Acres	Highway Vehicles		Off-Highway Vehicles		Snowmobiles		Nonmotorized		Total
		(mi)	(%)[a]	(mi)	(%)[a]	(mi)	(%)[a]	(mi)	(%)[a]	(mi)
High motorized	47,430	354	95	368	99	368	99	5	1	373
Moderate motorized	84,697	148	25	488	82	590	99	106	18	594
Low motorized	37,267	3	1	6	2	143	60	233	98	239
Total	169,394	505	42	862	72	1,101	91	344	28	1,206

[a]The distance for each type of use in each area classification divided by the total distance for each area classification.

chapter 2000 in the overall plan. Annual harvests are identified via WisFIRS by compartment and stand number, summed, and included in the annual plan. In addition, work goals for all the forest reconnaissance, reforestation, wildlife, access, recreation, and land acquisition for that particular year are included. The Annual BCF Work Plan identifies the management anticipated to occur given the budget that the County Board has approved for the Forestry Department for the upcoming year. It also acts as a measuring stick for year-end accomplishments reported to the County Board and the WDNR at year's end. Accomplishments are included in chapter 3000 each year. Ongoing, running totals of the long-term plan accomplishments are included in chapter 5000—Monitoring.

OUTCOMES OF THE PLAN

A number of measureable outputs are realized in managing the BCF. For example, with respect to forest products, the markets for Bayfield County timber are typically robust, and nearly all timber established for sale is sold. Dual forest certification under the SFI and FSC programs aids in making Bayfield County timber appealing to forest industry. This helps the County to attain plan goals for a sustainable harvest of all major forest types (Figure 30.3). The base BCF plan generally prescribes the management philosophy for each of the various forest types, recreation, cultural activities, integrated pest management, and special uses. Approximately 4,494 ac (1,819 ha) are evaluated for harvest annually, resulting in an average of 3,850 ac (1,558 ha) established for sale. Similarly, needs for release, site preparation, and reforestation are identified by year for the term of the plan. On average, 1,002 ac (406 ha) of site preparation, 344 ac (139 ha) of timber stand improvement, and 739 ac (299 ha) of planting or seeding are targeted each year.

FIGURE 30.3 Area by cover type on the Bayfield County Forest.

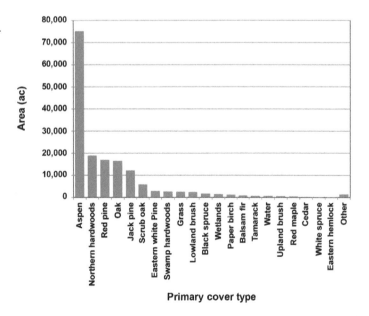

Thus far (2006–2013), annual harvests average 3,317 ac (1,342 ha) 62,000 cords (about 224,690 m³), and 1,167 MBF (2,754 m³) of sawtimber. More than $718 USD of stumpage is generated from each acre harvested. This generates, on average, $2,382,450 USD of annual timber revenue, 90% of which is retained by the county, with the remaining 10% distributed to the local townships. While most forest regeneration is secured naturally, approximately 739 ac (299 ha) are artificially reforested through planting or direct seeding annually. Cultural activities such as site preparation, release, timber stand improvement, and forest reconnaissance updating are also accomplished each year. Timber stumpage revenue more than offsets the costs of vegetative management and those incurred for the recreation program and markedly reduces the overall tax levy. Actual timber revenues have exceeded budgeted revenues by $667,916 USD per year on average during the plan period.

Natural succession of many even-aged types such as aspen, red oak (*Quercus rubra*), paper birch (*Betula papyrifera*) and jack pine to later-successional forest types is occurring statewide. This is particularly true on private lands and National Forest lands that have been less aggressive in managing these types. Bayfield County recognizes that these types are important ecologically, particularly from a wildlife perspective. They strive to maintain them and also to balance the age-class distribution within each type. Some forest types, primarily northern hardwoods, are typically managed on an uneven-aged basis. These comprise approximately 13% of the forested area and total nearly 20% of the harvested area.

The BCF plan provides for a diverse array of recreational opportunities which are key to a very important local and regional tourism industry. The plan recognizes that not all uses are compatible on every piece of the forest but opportunities exist somewhere on the forest for nearly every form of outdoor recreation. Designated recreational trails are supported on the BCF, and these include trails designed for all-terrain vehicles (38 mi, 61 km), cross-country skiing (32 mi, 51 km), dog sledding (44 mi, 71 km), hiking (15 mi, 24 km), mountain biking (21 mi, 34 km), and snowmobiling (95 mi, 153 km). The latter of these uses does not include trails located on town roads or those trails that are located on private lands. Notable inclusions to this system of recreational trails are parts of the North Country National Scenic Trail and the Birkebeiner Ski Trail, which hosts the annual American Birkebeiner Ski Race. Chequamegon Area Mountain Bike Association Trails, which are affiliated with the International Mountain Bike Association, are also very popular. In addition, a great number of undesignated trails are available for users consistent with the use zoning and seasonal restrictions in chapter 700 of the plan. This includes over 1,000 mi (1,609 km) of primary and secondary roads. A wide variety of dispersed recreation is available forestwide including hunting, trapping, fishing, camping, and picnicking.

Projects for game and non-game wildlife species are carried out in conjunction with the local WDNR Wildlife program. Forest game management (white-tailed deer (*Odocoileus virginianus*), ruffed grouse (*Bonasa umbellus*), American black bear (*Ursus americanus*), wild turkey (*Meleagris gallopavo*), snowshoe hare (*Lepus americanus*), and numerous furbearers) is centered on maintaining early-successional species such as aspen, jack pine, paper birch, and scrub oak (*Quercus ellipsoidalis* and poor quality *Q. rubra*). Upwards of 120 neotropical migrant songbird species spend a portion of each year in Wisconsin. Bayfield County management is cognizant of the diversity of habitat needed for these and tries to provide a wide variety of habitat types and age classes. From 2009 through 2011, the Bayfield County Forestry and Parks Department initiated a complete breeding bird survey over the entire County Forest. As a result, 1,200 survey points were collected over every major cover type. Data collected from the survey will provide a better foundation for assessing the impacts of forest management on breeding birds as well as provide additional tools in the development of forest management prescriptions.

Bayfield County has identified five areas in particular that have unique water, ecological, geologic, historic, or archeological attributes. These areas are identified in the plan and are managed to enhance and protect their unique features. The total area containing these resources is nearly 15,000 ac (6,070 ha). The largest and most noteworthy is the Barnes Barrens Management Area, an 11,500 ac (4,654 ha) block of globally imperiled barrens habitat with its own management plan nested within the larger plan. Through prescribed fire and unique harvest prescriptions, this area is maintained in a specific mix of open barrens, brush prairie, savanna barrens, and pine barrens. A number of plant and animal species are dependent on this habitat type, including the federally endangered Kirtland's warbler (*Setophaga kirtlandii*).

The above-listed mix of outputs has proven effective in balancing the mix of economic, ecological, and social issues present locally and regionally. The forest plan has evolved into an effective and adaptable document over the nearly 50 years that long-term plans have been in place on the forest.

DISCUSSION AND CONCLUSIONS

The Bayfield County Forest Plan has proven to be an effective framework for successful public land management and adapting to change. Bayfield County remains committed to working with others to implement its forest plan and provide multiple forest-related benefits for the citizens of Bayfield County and the public as a whole.

Learnings and Insights

The Bayfield County Forest, as a forest that is aligned with many others based on Wisconsin statutes, with a similar over-arching purpose and goals, benefited from having a basic plan template from which to work and from conducting an EA that began at a larger scale. These two aspects of the planning process were very helpful in providing consistency and in streamlining the local planning and assessment process. Public input prior to plan initiation was very helpful in providing a sound understanding of the public sentiment, in identifying user groups and controversial issues, and was instrumental in framing the local plan. A prime example of this was the input received on motorized versus nonmotorized recreation, which identified the need to do further analysis on where each might be most appropriate on the forest.

During the planning process, consideration of other local planning efforts such as Comprehensive Outdoor Recreation Plans, State Land Master Plans, adjacent National Forest Plans, Town and County Comprehensive Plans, and Land and Water Resources Plans was vital to avoiding later conflicts. The BCF shares borders with private individuals, private industry, State lands and National Forest lands. Bayfield County benefits from staff engaging in these planning efforts on neighboring lands at the onset of periodic planning and when new planning efforts arise.

Ongoing public feedback is also critical in adaptive management and maintaining buy-in from user groups. The plan must be amendable with limited processes in order to address unanticipated changes (e.g., windthrow, insect and disease, social initiatives) and to make changes based on unexpected results from prior years. Bayfield County amends the plan yearly with its Annual Work Plan. This amendment, and other more sporadic amendments, must receive a Forestry Committee recommendation and approval by the full County Board. The WDNR must then also approve the amendment. These processes provide management stability and minimize the potential for an influential interest group or individual to impart a personal agenda on the forest. This level of scrutiny seems to offer just enough oversight to ensure changes do not come on a whim, yet it is not such a hurdle that it deters the county forest from bringing forth justified amendments.

Sustainability Issues

In the BCF plan, it is noted that the *County Forest will be sustainably managed at a level of operation that considers the health of the forest and the products and amenities desired by the public in accordance with the goals identified.* The BCF plan was therefore designed to support sound ecological, social, and economic management. The plan addresses issues important to the public, provides for management that returns a reasonable income to the county and to towns, supports the forest and tourism industries, facilitates adaptive management, and meets the requirements of the County Forest law. Sustainable forest management is a prevalent theme (chapter 800) and guides the integrated management approach that has been adopted by the county.

Plan Implementation Challenges

The BCF and the other county forests are working forests. Sustainable forest management is at the core of achieving management objectives for wildlife, recreation, the ecosystem, and maximizing the public benefit of the forest. Stumpage revenues are substantial and very important in off-setting costs for recreation and infrastructure on the forest. They also help to compensate the townships for road maintenance expenditures. A number of over-arching issues and associated challenges lie at the heart of the continued success of the BCF and other county forests in northern Wisconsin.

First, the forest industry of the Lake States has undergone dramatic changes in the past 25 years. The forest industry is now truly a global marketplace. Pulp and paper mills, once a local fixture, have been acquired by larger companies. Some have been reconditioned and repurposed while others have been dismantled and shuttered. Economies of towns long dependent on *the mill* have been decimated by these mill closures. While Wisconsin's forest industry remains the second largest state industry at $18 billion USD annually, it is in a constant battle to retain market share. For BCF and the other county forests to be successful, markets for the timber must be retained. Maintaining a successful wood products industry is an ongoing challenge for the state, the industry, elected officials, and the public as a whole.

A second challenge, which is a trickle-down effect of the evolution of the forest industry, is related to changes in the logging community. Wisconsin's loggers are well-trained, well-organized, and ever-more true businessmen. Costs of machinery are high, hours are long, and, without stable markets, many have been forced out of the profession. From 2003 to 2010, the number of logging businesses decreased by 20%. It is a difficult business and new entrepreneurs are few. The average age of a Wisconsin logger is 52 years, and less than 11% are under the age of 35 (Rickenbach and Vokoun, 2013). Groups like the Great Lakes Timber Professionals Association have been very helpful in representing the profession and maintaining the logging force. Programs such as the Wisconsin Master Logger Certification program have national acclaim and are helping loggers address forest certification standards on Wisconsin's public lands. The Lakes States region is a stronghold for forest certification in the United States, and while this has likely helped to retain industry it also imposes additional requirements on landowners, consuming mills, and logging professionals. Retention and recruitment of loggers remains an ongoing challenge and is essential if county forests are going to be able to market their forest products and generate revenue.

A third challenge for continued success in selling and marketing timber is maintaining the road and railroad infrastructure necessary for moving the products to markets. Townships, and to a lesser extent, Bayfield County and the U.S. Forest Service, have responsibility for local road maintenance. In addition to receiving road management assistance from the Wisconsin Department of Transportation, townships receive compensation from the BCF that amounts to 10% of the stumpage revenue. This averages approximately $238,000 USD each year. Townships also receive $0.30 per ac ($0.74 per ha) as a payment in lieu of taxes on an annual basis from the WDNR, a figure that has not been adjusted since 1989. Bayfield County allocates an additional 1% of annual stumpage revenues ($23,800 USD) toward town road projects, providing access to the BCF if towns provide matching funds. The cost of road maintenance is substantial, and these payments are sometimes not acknowledged or deemed insufficient by local town boards. Some perceive the logging industry as the root of all road maintenance issues and seek to impose tight vehicle weight restrictions on local roads. Some go so far as to require liability agreements with individual loggers. This discourages loggers from bidding on timber in those townships, lessens timber stumpage revenue, and makes it more difficult on the loggers. Maintaining open lines of communication with town officials is crucial. Legislative efforts to increase aid payments to towns and efforts to educate and appease town board members are ongoing.

An additional aspect of the road infrastructure issue is addressing weight limits on roads leading to markets. Heavier weights allow haulers to move larger amounts of product to mills and reduce transportation costs per unit of wood. However, heavier loads can also exponentially increase damage to roadways leading to expensive repair and reconstruction projects. The forest products industry has made great strides in recent years to lessen road damage on rural and secondary roads by adding axles and lowering axle weights. While these actions have been successful in helping mitigate increased transportation costs, hurdles still remain in educating local officials on the differences between gross and axle weight. Deterioration of aging bridges remains problematic.

Other Interesting Issues Related to the Plan

The BCF and other public lands are increasingly being pressured to provide access to all forms of recreational pursuits. This poses challenges for the future. The recreation industry is constantly evolving and coming up with new forms of motorized and non-motorized outdoor recreational opportunities. Demographics of forest users are also changing, as are their expectations from public lands. The challenge is to offer trails or places to recreate for all user groups, while minimizing use conflicts or creating unsafe recreational environments. For example, the side-by-side utility vehicles (UTV) and fat-tired mountain bikes are two recreational vehicles that are relatively new to the BCF. Incorporating these new vehicles into existing networks while avoiding conflicts with existing uses will be a major challenge. UTVs are heavier, wider, and until recently were not accompanied by registration fees that could be used to help support trail use or development. Many existing trails and bridges are not wide enough to accommodate the UTVs, and safety issues are heightened due to increased speed and size. Fat-tired mountain bicycles, while seemingly harmless as a nonmotorized form of recreation, pose use conflicts with other pursuits. New regulations and policies will need to be developed to accommodate this recreational opportunity. It remains important to provide for outdoor pursuits on our public lands but it also remains a work in progress. Good communication and adaptive forest plans will be keys to success.

Lastly, our public lands need to be promoted and advocated for, if they are going to continue to be a part of the public domain. The value of public lands is frequently questioned by some taxpayers and politicians who narrowly focus on the tax base of lands in their jurisdiction. They fail to recognize the higher cost of services to access and accommodate private lands and the spinoff values associated with nearby public lands. Further, payments in lieu of taxes for public lands need to be at a level that local private landowners and taxpayers do not incur a disproportionate tax burden for services that are valued by other people from the local community and beyond. Educating the public and elected officials on these tradeoffs will be an ongoing challenge going forth.

ADDITIONAL READING AND RESOURCES

This chapter represents a synthesis of the 2006 County Forest Plan for Bayfield County, Wisconsin. To view the plan itself, please visit this Internet site, which was available on March 30, 2014:

http://www.bayfieldcounty.org/250/Forestry-Plan-2006-2020

If in the future the link to the site appears broken, search the Internet using the title of the plan and the keywords provided.

REFERENCES

Bayfield County Forestry and Parks Department, 2006. Bayfield County Forest Comprehensive Land Use Plan 2006–2020. Bayfield County Forestry and Parks Department, Washburn, WI. 212 p.

Bayfield County Forestry and Parks Department, 2013a. Bayfield County Forestry & Parks Department Annual Work Plan, January 1 through December 31, 2014. Bayfield County Forestry and Parks Department, Washburn, WI. 13 p.

Bayfield County Forestry and Parks Department, 2013b. Barnes Barrens Management Plan. Bayfield County Forestry and Parks Department, Washburn, WI. 11 p.

Cleland, D.T., Avers, P.E., McNab, W.H., Jensen, M.E., Bailey, R.G., King, T., Russell, W.E., 1997. National hierarchical framework of ecological units. In: Boyce, M.S., Haney, A. (Eds.), Ecosystem Management Applications for Sustainable Forest and Wildlife Resources. Yale University Press, New Haven, CT. pp. 181–200.

Deckard, D.L., Skurla, J.A., 2011. Economic Contribution of Minnesota's Forest Products Industry—2011 Edition. Minnesota Department of Natural Resources, Division of Forestry, St. Paul, MN. 18 p.

Rickenbach, M., Vokoun, M., 2013. Wisconsin Loggers: Then and Now. Blogging Logging 5—Business Demographics (May 2013). University of Wisconsin, College of Agriculture and Life Sciences, Madison, WI. http://notcountingtrees.fwe.wisc.edu/taxonomy/term/60?page=1 (Accessed March 30, 2014).

Wisconsin Department of Natural Resources, 2006a. Environmental Analysis and Decision on the Need for an Environmental Impact Statement (EIS). Wisconsin Department of Natural Resources, Washburn, WI. 28 p.

Wisconsin Department of Natural Resources, 2006b. Silviculture Handbook, Chapter 12, Forest Habitat Type Classification System. Wisconsin Department of Natural Resources, Madison, WI. http://dnr.wi.gov/topic/forestmanagement/documents/24315/12.pdf (Accessed April 13, 2014).

Wisconsin Department of Natural Resources, 2009. Wisconsin's Best Management Practices for Invasive Species. Wisconsin Department of Natural Resources, Madison, WI. 56 p.

Wisconsin Department of Natural Resources, 2010. Wisconsin's Forestry Best Management Practices for Water Quality. Wisconsin Department of Natural Resources, Division of Forestry, Madison, WI. 162 p.

Wisconsin Department of Natural Resources, 2014a. Wisconsin's Natural Heritage Inventory (NHI). Wisconsin Department of Natural Resources, Madison, WI. http://dnr.wi.gov/topic/nhi/ (Accessed April 13, 2014).

Wisconsin Department of Natural Resources, 2014b. Wisconsin Forest Inventory and Reporting System—WisFIRS. Wisconsin Department of Natural Resources, Madison, WI. http://dnr.wi.gov/topic/ForestManagement/wisfirsIntro.html (Accessed April 30, 2014).

Wisconsin Department of Tourism, 2014. County Total Economic Impact. Wisconsin Department of Tourism, Madison, WI. http://industry.travelwisconsin.com/Research/Economic-impact (Accessed March 30, 2014).

Chapter 31

Chattahoochee-Oconee National Forest, Georgia, United States of America

Pete Bettinger,[1] Krista Merry,[1] Erika Mavity,[2] Dick Rightmyer[2] and Ron Stevens[2,†]

[1]Warnell School of Forestry and Natural Resources, University of Georgia, Athens, Georgia, USA, [2]Chattahoochee-Oconee National Forest, Gainesville, Georgia, USA

MANAGEMENT SETTING AND BACKGROUND

The Chattahoochee-Oconee National Forest is located in north and central Georgia (Figure 31.1), and is composed of two national forest tracts of land administered as a single unit (U.S. Department of Agriculture, Forest Service, 2004a). The Chattahoochee portion of the national forest was once part of Nantahala (North Carolina) and Cherokee (Tennessee) National Forests and in 1936 became separate from these. The Oconee portion of the national forest became national forest-land in 1959 through a proclamation of President Eisenhower, and the combined forest was then formed (U.S. Department of Agriculture, Forest Service, 2013a). The total land area at the time of the development of the 2004 forest plan was about 866,000 acres (nearly 350,500 hectares), with about 87% of the land in the more northern Chattahoochee portion of the national forest. Nearly 91% of the national forest arose through acquisitions or exchanges in association with the Weeks Act (36 Stat. 961) or its amendments. The current organization of the national forest consists of a supervisor's office and four districts. Within the four districts are 10 congressionally designated wilderness areas that comprise nearly 14% of the area of the forest. Two of these recent wilderness designations occurred after the 1985 Forest Plan was completed and were brought forward primarily by public interests.

The Chattahoochee-Oconee National Forest is heavily recreated; it currently has 33 campgrounds, 28 day use areas, 7 overlooks, 6 beaches, 5 shooting ranges, and 2 visitors' centers (U.S. Department of Agriculture, Forest Service, 2013b). Approximately 730 miles (mi) (about 1,177 kilometers (km)) of trails have been developed for hiking purposes, 248 mi (about 400 km) of trails for equestrian uses, 141 mi (about 227 km) of trails for biking purposes, and 112 mi (about 181 km) of trails for off-highway vehicle use. The stream systems of the Chattahoochee-Oconee National Forest are situated within five major river basins (Tennessee, Savannah, Coosa, Oconee, and Ocmulgee Rivers). Within the Chattahoochee portion of the forest resides an estimated 2,763 mi (about 4,456 km) of perennial cold-water streams. Within the Oconee portion of the forest resides an estimated 393 mi (about 634 km) of warm-water perennial streams. According to the U.S. Department of Agriculture, Forest Service (2013b), the main recreational uses of the forest are relaxation, hiking or walking, fishing, and camping.

The majority (60%) of the Chattahoochee portion of the national forest is composed of deciduous forests (Figure 31.2), although pine (25%) and mixed (15%) forests are also prevalent. On the Oconee portion of the national forest, pine forests dominate (69%), while hardwood forests comprise about 28% and mixed forests account for 3% of the forest area. The 2004 forest age-class distribution indicates that more than two-thirds of the land area contains forests that were on average between 60 and 100 years old (Figure 31.3). The current forest plan emphasizes the restoration, maintenance, and enhancement of forest ecosystems, while servicing the public demand for an array of uses and products (U.S. Department of Agriculture, Forest Service, 2004a). The plan emphasizes native plant communities, wildlife, and fish populations and their habitats, old-growth forest communities, and forest health, along with other nature-based recreational opportunities. Forest health concerns include oak decline, nonnative diseases (e.g., Dutch elm disease), native insects (e.g., southern pine

---

[†]Retired.

Forest Plans of North America. http://dx.doi.org/10.1016/B978-0-12-799936-4.00031-X

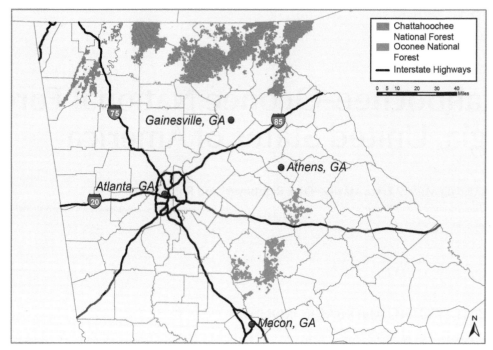

FIGURE 31.1 A map of the land area of the Chattahoochee-Oconee National Forest.

FIGURE 31.2 A view of Lake Winfield Scott in the fall. *Photo courtesy of Becky Vaughters.*

beetle, *Dendroctonus frontalis*), nonnative insects (e.g., hemlock woolly adelgid, *Adelges tsugae*), and nonnative plant species (e.g., kudzu, *Pueraria lobata*) (U.S. Department of Agriculture, Forest Service, 2013b).

PLANNING ENVIRONMENT AND METHODOLOGY

The Forest Service indicates that adaptive management guided the development of the 2004 plan (U.S. Department of Agriculture, Forest Service, 2004b). A planning process described in the National Forest Management Act and its implementing regulations was followed, which included the following steps:

- Identify the purpose and need: issues, concerns, and opportunities
- Develop planning criteria
- Collect information and inventory data
- Analyze the management situation
- Formulate alternatives
- Estimate the effects of alternatives

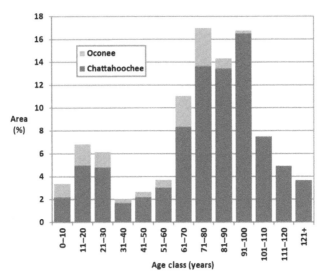

FIGURE 31.3   An age-class distribution of the forest resources on the Chattahoochee-Oconee National Forest.

- Compare alternatives
- Select a preferred alternative
- Approve and implement the plan
- Monitor and evaluate the plan

At the onset of the planning process, public comment was solicited in order to understand the planning issues. Information regarding the planning process was communicated to the public through a number of methods, including press releases, web site content, public meetings, and the newsletter "*Eco21*." Further, an assessment of the current management situation helped determine the suitability and feasibility of the forest for meeting social needs (U.S. Department of Agriculture, Forest Service, 2004c). The open meetings that were held, and the progression of public input that was provided, allowed the Forest Service to develop a range of possible management alternatives for the national forest.

Guiding Laws, Regulations, and Policies

A number of federal statutes, regulations, and executive orders guided the development of planning criteria for the forest plan. The statutes included the Clean Air Act, the Clean Water Act, the Wilderness Act, the Endangered Species Act, the National Historic Preservation Act, the National Forest Management Act, and the Roadless Area Conservation Rule. Goals and objectives from the Forest Service strategic plan, along with agency policy, also guided the development of the planning criteria. The plan also suggests that the plans of other agencies, state and local governments, and Indian tribes may have been used to develop the planning criteria.

The analysis of forest resources was conducted separately for the Chattahoochee and the Oconee portions of the national forest, given historical differences in products produced and stumpage values and due to differences in topography. Timber supply was analyzed with regard to local markets, recent sales, forest growth rates, logging operability, tree stocking, average tree size, and tree quality. The analysis suggested that the national forest had greater significance as a holder of inventory (e.g., high-quality hardwood) rather than a general producer of timber, that the rate of growth was far greater than the recent rate of harvest, and that the average annual mortality exceeded the average annual harvest (U.S. Department of Agriculture, Forest Service, 2004c).

In the initial planning stages, a land classification was performed to understand the suitability of land for the sustained yield of timber production (U.S. Department of Agriculture, Forest Service, 2004c). Through this analysis, unsuitable lands were identified using National Forest Management Act criteria and, thus, defined as those that were:

- Not forest
- Withdrawn by Congress or other authority above the Regional Forester
- Not capable of producing timber
- Subject to irreversible damage to soil productivity or watershed conditions given current technology
- Unable to be restocked after 5 years
- Had inadequate response information

These lands are not capable for timber production because to do so they would be subjected to irreversible damage, or they are generally unable to be restocked after 5 years, thus were grouped into the category "not appropriate for timber production" in Table 31.1. After the land classification process, the land was further divided into strata, which were initially identified by ecological classification units (e.g., Blue Ridge, Piedmont, etc.), forest type group, age, productivity, and condition class. Upon inspection, it was determined that this set of strata would require more than 26,000 yield tables. Therefore, the decision was made to remove the productivity and condition class from the stratification process, as it was assumed

TABLE 31.1 Land Classification and Management Prescription Allocation for the 2004 Chattahoochee-Oconee National Forest Plan

	Chattahoochee (ac)	Oconee (ac)
Total land area	750,770	115,215
Total suitable forest land	367,196	93,899 [a]
Total unsuitable forest land	383,571	21,316 [a]
Nonforest land	2,126	2,216
Withdrawn from timber production	155,001	5,673
Not appropriate for timber production	222,117	13,427
Inadequate information to classify	4,327	0
Management area prescriptions [b]		
Wilderness areas	117,436	0
Recommended wilderness areas	8,094	0
Wild, scenic, and recreational river areas [c]	11,084	4,854
National scenic areas	7,122	0
National recreation areas [d]	25,689	0
Dispersed recreation areas	96,732	9,368
Remote backcountry recreation areas	28,261	0
Appalachian National Scenic Trail corridors	16,655	0
Natural areas	17,943	0
Scenic byways and corridors	29,017	0
Other scenic areas	41,352	0
Experimental forests	0	9,364
Watershed protection and restoration areas	27,179	0
Management, maintenance, and restoration of plant associations	172,718	35,006
Outstandingly remarkable stream areas	17,868	4,730
Old-growth forests	28,657	1,617
Mix of successional forest habitats	68,323	0
Forest interior, mid- to late successional forests	23,693	0
Red-cockaded woodpecker habitats	0	47,108
Other areas	12,947	3,168

[a] *Slightly different than in USDA Forest Service (2004c), perhaps due to rounding error.*
[b] *A riparian management prescription is embedded in most of these on perennial and intermittent streams.*
[c] *Designated and recommended.*
[d] *Designated and proposed.*

(through an analysis of the forest inventory data) that the national forest contained land and forests of predominantly average quality (U.S. Department of Agriculture, Forest Service, 2004c). As a result, the actual number of strata was fewer than 1,000 for each forest plan alternative. The growth of the forest resources was projected using the Forest Vegetation Simulator, or FVS (Dixon, 2013). U.S. Department of Agriculture Forest Inventory and Analysis (FIA) measurement plots were then associated with each stratum, however, some strata had few or no plots assigned to them, and adjustments were made for these instances. FVS was then used to project potential timber volumes for each stratum using associated plots. In assessing potential thinning options in FVS, minimum volumes per unit area were required as were minimum residual basal area levels. At most, about 20 silvicultural regimes were developed through FVS for each stratum. The potential harvest volumes and net present value of each alternative were estimated using the SPECTRUM linear programming matrix generator (U.S. Department of Agriculture, Forest Service, 2008) in conjunction with the C-WHIZ commercial solver (Ketron Optimization, 2000). For strata that might accommodate thinnings, regimes that included up to three thinnings were available to the forest planning model (SPECTRUM). Ultimately, the potential number of strata x harvest options for each alternative modeled approached 200,000.

In terms of the management regimes recognized in the SPECTRUM analysis, final harvests, shelterwood harvests, and uneven-aged management harvest entries were modeled as the removal of a percentage of the standing inventory. Minimum regeneration harvest ages ranged from 40 to 140 years based on ecological community type. While the potential regimes acted as inherent constraints on the development of the plan, the overall objective was to maximize the net present value subject to these types of resource constraints and constraints on early-successional habitat development. Other constraints included those related to wood-flow (nondeclining yields), riparian areas, and management requirements. Scheduled harvest volumes estimated through the SPECTRUM analysis were then adjusted downward to account for the mortality caused by the southern pine beetle. The time horizon modeled was 200 years, with time periods represented by decades (U.S. Department of Agriculture, Forest Service, 2004c).

In sum, SPECTRUM was used to examine how timber management could be used to achieve the goals and objectives of each forest plan alternative. Management prescriptions were assigned to land allocations, and thus SPECTRUM was used to model some management constraints and to determine the most efficient way to meet the objectives through silvicultural prescriptions. Only the benefits and costs associated with activities of the timber program were included in the SPECTRUM analysis, however, recreation and wildlife (viewing, etc.) values were included externally in the overall analysis of forest plan alternatives. The effects of activities on other resources were also addressed outside of the SPECTRUM analysis, based on results generated from the SPECTRUM analysis. For example, the recreation opportunity spectrum, watershed cumulative effects, and other economic and social impacts were modeled externally to SPECTRUM using the guidance provided with the SPECTRUM-generated results of each forest plan alternative.

OUTCOMES OF THE PLAN

Of the plan alternatives that were assessed, the selected alternative was assumed to provide the best balance between the social, physical, biological, and economic environment. The desired resource conditions that land managers should attain were identified in the forest plan, and goals, objectives, and management prescriptions are used to close the gap between current and desired conditions (U.S. Department of Agriculture, Forest Service, 2004b). The preferred alternative had a net present value of nearly $2.3 billion U.S. dollars, mostly as a result of recreation and wildlife values associated with the alternative, although these were not assumed to be directly received by the forest (U.S. Department of Agriculture, Forest Service, 2004c). The preferred alternative emphasized restoration and maintenance of forest and aquatic ecosystems, the production of high-quality water resources, the sustainability of a variety of old-growth forest conditions, and the development and maintenance of nature-based recreation opportunities. As an example, the 2004 forest plan guides the development and maintenance of a network of medium to large old-growth forest patches for a variety of potential old-growth communities. On the Chattahoochee portion of the national forest, 23% of the forest area either has old-growth characteristics or is where management prescriptions have been applied that should lead to old-growth conditions. For the Oconee portion of the national forest, 7% of the forest area has old-growth characteristics or has assigned management prescriptions that should lead to old-growth conditions (U.S. Department of Agriculture, Forest Service, 2004b).

The 2004 forest plan documents are not prescriptive with regard to the timing and placement of management activities. Many of the outcomes were estimated through several stages in the analysis, yet were influential in the estimation of a few quantifiable products such as scheduled timber harvest volume. For example, rare communities and old-growth forest areas were identified and projected at various stages of the analysis. In addition, riparian areas were implied in each management area emphasis (Table 31.1). In lieu of rigid prescriptions, objectives and standards were identified for each management area. For example, in the *mix of successional forest habitats* areas, the objective is to have a minimum of 50% of the land in mid- to late-successional forests, a minimum of 20% of the land just in late-successional forests, and 4–10% of the land in early-successional forest. As suggested, the plan is not spatially explicit as to where the activities will be placed, and the forest supervisor retains discretion on the rate of plan implementation, which is influenced by budgets and regional/national forest priorities. In sum, the plan allows, but does not mandate the implementation of management activities; site-specific analyses will ultimately determine the activities that can be implemented. In terms of timber harvest volume scheduled, the plan adheres to the National Forest Management Act requirements of a nondeclining even-flow of timber harvest volume and long-term sustainable yields. The long-term sustained yield was estimated to be 880 MMBF (about 2.08 million m^3) per decade, while the scheduled standing (green) timber harvest volumes for the first decade were projected to be 600 MMBF. This included 116 MMBF of hardwood sawtimber and 352 MMBF of softwood sawtimber per decade. In addition, another 214 MMBF of salvage harvest volume was estimated to be produced during the first decade (U.S. Department of Agriculture, Forest Service, 2004d). However, timber production is essentially an outcome or byproduct (not a goal) of the forest plan, and, in practice, the production of wood generally occurs through forest health projects, projects aimed at the reintroduction of native trees species, or projects that address mast production or oak decline.

DISCUSSION AND CONCLUSIONS

The Chattahoochee-Oconee National Forest plan addresses strategic issues for two very different units of the national forest. The Chattahoochee portion is located in the northern part of the state of Georgia and is covered extensively with deciduous forests. The area contains a significant amount of topographic relief, and the issues (recreation, forest health) are much different than the issues of the Oconee portion of the national forest. The Oconee portion is located in the central Piedmont region of the state and contains a significant amount of coniferous forests. Management in this area is often guided by concerns for red-cockaded woodpecker (*Picoides borealis*) habitat development and maintenance. Across the forest, examples of the recent silvicultural activities include thinnings in pine stands to address forest health concerns and regeneration harvests on mountain ridges of the western side of the Chattahoochee portion of the national forest to promote the development of longleaf pine (*Pinus palustris*).

Learnings and Insights

Nearly all of the planning issues encountered during the development of the 2004 plan centered on questions of *how much* (amount of activity or forest type) and *where* (distribution of activity or forest type). For each issue that the plan addressed, management prescriptions were integral to the answers. One of the important outcomes of the plan was the general impression by the planning team that decisions made during the previous planning process (in the 1980s) could not be undone or revisited, and that the more recent issues brought forth by the public were to be folded into the previous plan to create the new plan. Critics of the plan seemed to rally around management direction that involved the *restoration* of natural systems, yet agreement on the methods by which restoration would be achieved was difficult to obtain. However, increased public recognition of the dynamic nature of forests and the benefits of certain practices (e.g., prescribed fire) used to develop and maintain wildlife habitat values have been observed.

Sustainability Issues

Clean water, visual quality, and sustainable diversity of plant and animal species were all suggested by the public as principles that should guide the management of the forest, and these are inherent in the plan. The National Forest Management Act also required the forest planning staff to examine the suitability and sustained yield of timber production, even though timber production outcomes are inherently tied to forest health and restoration issues. Although planning rules for national forests have since been amended, the 2004 plan was also structured to address the *viable populations of existing native and desired nonnative vertebrate species* mandate of the National Forest Management Act by facilitating the development and maintenance of a very large and complex assessment of species and habitats. This *viability standard* arose from the 1982 planning rule (47 FR 43037), which guided the development, amendment, and revision of national forest plans. By not fully implementing the plan objectives, the national forest runs the risk of not meeting this objective. Monitoring could reveal the trajectory of the forest (in both an amount and distribution) with regard to this concern, which could become an issue with the development of subsequent plans.

Plan Development Challenges

During the strategic planning process, which extended over a period of about 8 years, more than 300 public meetings were held and attended by constituencies with a variety of expectations of the national forest. The planning staff attempted to be inclusive and transparent during these meetings, although they were occasionally contentious. The end result was a relatively slow and expensive planning process, although some of the delays were budget-related. A regional decision also directed the planning team to coordinate work in phases consistent with five other eastern national forests, and this extended the time required to develop the 2004 plan. Finally, the strategic plan provides guidance for, but does not prescribe, actual management activities. A separate project-level planning process is employed for individual management actions. In this case, the public is again involved, an environmental analysis is performed, and proposed alternatives are offered.

In assessing the costs of activities, different sources of data were necessary, each with varying levels of detail. One of the main challenges involved the estimation of road management activities. Determining the amount of new roads needed to address each alternative plan required an exploration of several data-intensive methods, some involving analyses of the terrain, and most proving to be too detailed for the level of information needed for the strategic plan. Ultimately, an analysis of unroaded areas was performed, and an assessment was made of the roads that might need to be developed. Scheduled harvests in these management areas were then charged a cost for potential road development in the harvest scheduling analysis. Other factors, such as increasing National Environmental Policy Act requirements, steadily increasing upward reporting requirements through automated tracking programs, and reductions in personnel all had an effect on the management costs associated with the development of the plan (U.S. Department of Agriculture, Forest Service, 2004c).

Plan Implementation Challenges

As mentioned earlier, the rate of implementation of the plan's management actions is subject to budget conditions and other priorities of the forest and the region within which it resides. Project-level planning processes further require the national forest to work with the public when examining specific management alternatives for specific areas of land.

Other Interesting Issues Related to the Plan

One important aspect of the 2004 forest plan was the development of a monitoring and evaluation program intended to inform the forest managers of the progress in meeting the goals and objectives. The information generated through this program can enable the forest managers to respond to changing conditions, emerging trends, the introduction of new technology and information, and evolving social concerns. Unfortunately, every goal, objective, and standard cannot be monitored independent of the others (U.S. Department of Agriculture, Forest Service, 2004b) due to budget considerations and other administrative issues. Therefore, monitoring priorities will be developed that are relevant to current issues, that are important with respect to legal and agency policies, and that can result in scientifically credible direction. A number of monitoring approaches were described in the plan, ranging from the assessment of trends in habitat conditions to the estimation of wildlife populations and their demographics. Some of these efforts are to be performed on 5-year intervals (e.g., trends in the conditions of rare communities, area of old-growth by community class) or are performed annually (e.g., trends in red-cockaded woodpecker populations, trends in prairie warbler (*Setophaga discolor*) occurrence in early-successional habitat), while others are to be continuously assessed (e.g., amount of vegetation management in riparian areas, conditions and trends of forest fuels) or assessed as needed or as detected (e.g., presence of nonnative plant species, or insect and disease problems). This 2004 forest plan is required by regulations to be revised on a 10- to 15-year cycle, or perhaps sooner if resource conditions change significantly (U.S. Department of Agriculture, Forest Service, 2004a).

ADDITIONAL READING AND RESOURCES

This chapter represents a synthesis of the 2004 Land and Resource Management Plan for the Chattahoochee-Oconee National Forest in Georgia. To view the plan itself, please visit this Internet site, which was available on March 13, 2014:

http://www.fs.usda.gov/detailfull/conf/landmanagement/planning/?cid=stelprdb5413247&width=full.

If in the future the link to the site appears broken, search the Internet using the title of the plan and the keywords provided.

REFERENCES

Dixon, G.E., 2013. Essential FVS: A user's guide to the Forest Vegetation Simulator. U.S. Department of Agriculture, Forest Service, Forest Management Service Center, Fort Collins, CO. 226 p.

Ketron Optimization, 2000. C-WHIZ Linear Programming Optimizer. Ketron Optimization, Sterling, VA. http://www.ketronms.com/documentation/KMS_CWHIZ_brochure.pdf (Accessed February 9, 2014).

U.S. Department of Agriculture, Forest Service, 2004a. Final Environmental Impact Statement, Appendix B. U.S. Department of Agriculture, Forest Service, Southern Region, Atlanta, GA. 81 p.

U.S. Department of Agriculture, Forest Service, 2004b. Appendix F: Timber Suitability, Total Timber Sale Program, Vegetation Management Practices. U.S. Department of Agriculture, Forest Service, Southern Region, Atlanta, GA.

U.S. Department of Agriculture, Forest Service, 2004c. Summary Final Environmental Impact Statement and Land and Resource Management Plan Chattahoochee-Oconee National Forests. U.S. Department of Agriculture, Forest Service, Southern Region, Atlanta, GA, Management Bulletin R8-MB 113 D. 36 p.

U.S. Department of Agriculture, Forest Service, 2004d. Record of Decision for the Final Environmental Impact Statement of the Land and Resource Management Plan Revision for the Chattahoochee-Oconee National Forests. U.S. Department of Agriculture, Forest Service, Southern Region, Atlanta, GA, Management Bulletin R8-MB 113 C. 30 p.

U.S. Department of Agriculture, Forest Service, 2008. Spectrum User Guide. U.S. Department of Agriculture, Forest Service, Ecosystem Management Coordination, Planning & Analysis Group, Washington, DC.

U.S. Department of Agriculture, Forest Service, 2013a. Chattahoochee-Oconee National Forest history. U.S. Department of Agriculture, Forest Service, Washington, DC. http://www.fs.usda.gov/detail/conf/learning/history-culture/?cid=fsm9_029299 (Accessed February 9, 2014).

U.S. Department of Agriculture, Forest Service, 2013b. Chattahoochee-Oconee National Forests 2012 Quick Facts. U.S. Department of Agriculture, Forest Service, Chattahoochee-Oconee National Forest Supervisor's Office, Gainesville, GA. 11 p.

Chapter 32

City of San Francisco, California, United States of America

Jon Swae

City & County of San Francisco Planning Department, San Francisco, California, USA

ABBREVIATIONS

DPW   Department of Public Works
FUF   Friends of the Urban Forest

MANAGEMENT SETTING AND BACKGROUND

The city of San Francisco is home to 812,826 residents with a daytime population of almost 1,000,000 people (U.S. Census Bureau, 2011). San Francisco has one of the smallest tree canopies of any major U.S. city (13.7%), less than Chicago (17%), Los Angeles (21%), and New York City (24%) (San Francisco Planning Department, 2014). The size of San Francisco's urban forest is estimated at approximately 669,000 trees (Figure 32.1), with a capital value of $1.7 billion U.S. dollars (USD) (Nowak et al., 2007). The city and county of San Francisco government has jurisdiction over approximately 236,000 trees, including 105,000 street trees (Figure 32.2) and 131,000 trees in parks and open spaces. Canopy cover within the city varies between neighborhoods (Figure 32.3). The Department of Public Works (DPW) has jurisdiction over all trees and greening in the public right-of-way. The DPW prunes street trees, responds to tree emergencies, performs tree inspections, conducts tree-related sidewalk repair, and issues permits for the planting and removal of street trees. Although the DPW has the ultimate authority over all trees within the public right-of-way, the agency is responsible for maintaining only about 40% (or 40,000) of these trees. The responsibility for the remaining 60% falls to adjacent private property owners. The Recreation and Park Department is responsible for 131,000 trees on 4,196 acres (ac) (1,698 hectares (ha)) of parkland. These include trees in city parks, identified natural areas, and public golf courses.

Other significant stands of trees within the city are managed by a variety of state and federal agencies, such as those within the Golden Gate National Recreation Area and the Presidio. Trees located on private residential, institutional, and commercial properties account for another significant portion of San Francisco's urban forest and are managed by a mix of institutions and private property owners. The local nonprofit organization Friends of the Urban Forest (FUF) carries out the majority of street tree planting in San Francisco (Friends of the Urban Forest, 2014). Since 1981, FUF and its volunteers have planted more than 48,000 new and replacement trees within the city.

Unlike cities with naturally occurring forests, San Francisco's original landscape had very few trees. Prior to European arrival in the early 1800s, San Francisco's environment was characterized by sand dunes, grasslands, wetlands, riparian areas, and coastal scrub areas that supported primarily low-lying vegetation. Small, scattered stands of native trees grew near creeks, canyons, and on the city's less windy eastern side. Trees native to the area include coast live oaks (*Quercus agrifolia*), bay laurel (*Umbellularia californica*), willows (*Salix* spp.), and California buckeye (*Aesculus californica*). The lack of expansive native tree cover reflects San Francisco's microclimate, windy conditions, and sandy and serpentine soils.

Today, San Francisco is a vibrant city with a highly altered natural environment. Most of the natural landscape has been transformed by urbanization. Creeks, wetlands, and parts of San Francisco Bay have been filled to accommodate urban development. Massive tree planting efforts over the last 150 years have created an urban forest where none existed prior. San Francisco's streets and parks contain more than 200 species of trees. Many of these are introduced species from places

Forest Plans of North America. http://dx.doi.org/10.1016/B978-0-12-799936-4.00032-1

FIGURE 32.1 A map displaying the distribution of street tree maintenance responsibilities in San Francisco.

as far away as Australia, Asia, and Africa that are well adapted to thrive in the city's Mediterranean climate. Open spaces, parks, and natural areas still retain significant native landscapes and habitats. These support diverse local plant and wildlife communities. Efforts have been made to protect and restore these areas.

San Francisco's urban forest (Figure 32.2) is primarily the result of human determination and ingenuity. The first large-scale tree plantations were introduced in the late 1800s. By 1879, approximately 1,000 ac (405 ha) of shifting sand dunes had been stabilized by the planting of 155,000 trees to create Golden Gate Park, the city's largest public park. In 1883, the U.S. Army initiated the planting of 350,000 trees to reduce wind and visually isolate its military base at the Presidio. Trees such as Monterey cypress (*Cupressus macrocarpa*), Monterey pine (*Pinus radiata*), and blue gum eucalyptus (*Eucalyptus globulus*) were the primary species planted. These species demonstrated fast growth and a tolerance of salty air, poor soil conditions, and dry summers.

By the mid-1900s, only a few of the city's parks, boulevards, and large pieces of land had been transformed by trees. Photographs from this time show many neighborhoods devoid of tree cover. In 1955, the DPW established a Tree Planting Division. Thousands of street trees were planted and public interest in trees grew. The environmental movement of the 1960s and 1970s further strengthened interest in urban tree planting. City-sponsored plantings expanded throughout San Francisco. During this period, many new species were introduced, such as Indian laurel fig (*Ficus microcarpa*) and blackwood acacia (*Acacia melanoxylon*), that would later be discovered to cause sidewalk disruption and other problems.

In 1981, the city's tree planting program was eliminated. Severe municipal budget cuts dramatically altered the city's tree planting and forestry programs. The halt of city-sponsored tree planting led to the creation of the nonprofit organization, FUF. Today, the majority of street tree plantings in the city continue to be coordinated and carried out by this group. Since its inception, FUF has planted 48,000 new and replacement street trees and engaged thousands of volunteers in

FIGURE 32.2   San Francisco's 105,000 street trees are a major component of its urban forest.

growing and caring for the urban forest. By the 1990s San Francisco's municipal tree crews had shifted their focus and limited resources to tree maintenance along major streets and corridors.

The global economic recession of 2008 further reduced the scale of the San Francisco's urban forestry program. The DPW was forced to eliminate key maintenance positions that provided critical services for street trees. These reductions have extended the average pruning cycle from 5 years to 12 years per street tree. In response to repeated budget cuts, the DPW was forced to initiate a Tree Maintenance Transfer Plan in 2011. Under that plan, the DPW has begun to transfer responsibility for thousands of trees previously under its care to adjacent private property owners. This has made property owners responsible for services previously provided by the city, including tree pruning and sidewalk repair. The controversial program has raised concerns among many city residents who do not wish to or are unable to take on maintenance responsibilities. It has concerned environmental policy makers who feel this strategy may leave the urban forest in a high state of risk, neglect, and damage. These ongoing urban forestry financing challenges were a major driving force behind the creation of San Francisco's Urban Forest Plan.

PLANNING ENVIRONMENT AND METHODOLOGY

The San Francisco Urban Forest Plan arose out of the need to create a strategy to ensure the ongoing health and sustainability of the city's trees including long-term funding for maintenance. The Plan will be divided and carried out in three separate phases.

1. *Phase 1 (Street Trees).* The plan's first phase discusses the overall urban forest with a primary focus on street trees. The plan highlights the benefits of trees and landscaping within San Francisco. It also recommends increasing the street tree population and developing a comprehensive approach to street tree management.

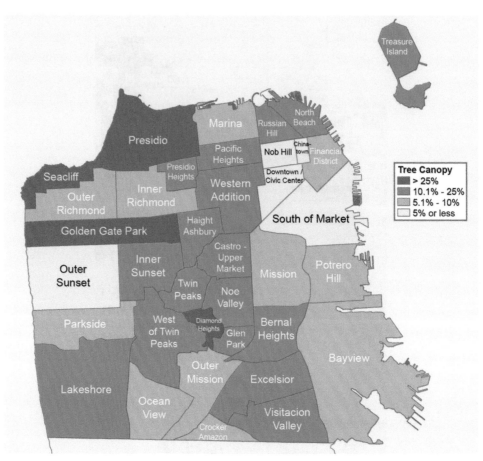

FIGURE 32.3   A map showing the variation in tree canopy cover by neighborhood within San Francisco.

2. *Phase 2 (Trees in Parks and Open Spaces).* This plan's second phase will create a specific vision and strategy for trees in parks and open spaces that addresses the policy, management, and financing needs of park trees.

3. *Phase 3 (Trees on Private Property and Greening Buildings).* The third phase of the plan will include efforts aimed at greening buildings and private properties. Recommendations and guidelines will be developed for trees and landscaping on private property and living architectural elements on buildings such as green roofs and walls.

The Urban Forest Plan for Street Trees (Phase 1) was developed over a 2-year period by the San Francisco Planning Department (2014) and the DPW in collaboration with the city's Urban Forestry Council and FUF. Funding for the plan was made available through a grant from the State of California Strategic Growth Council's Urban Greening Planning Program. Street trees were prioritized for Phase 1 because of the urgent issues related to the decline in maintenance funding. The plan identifies policies and strategies to proactively manage and grow the city's street tree population with the goal of creating an expanded, healthy, and thriving urban forest now and for the future. Content for the plan was informed by a series of meetings, workshops, public forums, and think tanks with residents, city agencies, landscape professionals, and urban forestry specialists.

The plan was also shaped by the results of the Street Tree Census and Street Tree Financing Study. The Planning Department commissioned an economic consultant to conduct a finance study (AECOM, 2013) that evaluated potential costs and funding mechanisms associated with the establishment of a municipal street tree program, whereby the city would take responsibility for maintaining 100% of San Francisco's street trees. As part of the Street Tree Census, data on tree age, location, species and condition was collected for 25,000 of the city's 105,000 street trees. This data provided a snapshot of major issues affecting street trees throughout the city and helped inform the plan's strategies to improve street tree management and care.

The Urban Forest Plan (Phase 1) addresses the following threats to the long-term health of the city's street trees:

1. *Harsh growing environment.* Streets are a difficult place for trees to take root and flourish. Small growing spaces, compacted soil, and neglect make it hard for urban trees to survive and reach optimal health.

2. *Insufficient tree canopy*. San Francisco lags behind other major cities in tree canopy coverage. Thousands of potential planting spaces remain empty. The lack of tree cover prevents the city from realizing the full benefits of urban trees as valuable "green infrastructure."

3. *Fragmented maintenance system*. San Francisco's fragmented approach to street tree maintenance makes achieving a coordinated and standard level of care difficult to achieve. Although the DPW has ultimate authority over all trees within the public right-of-way (streets and sidewalks), the agency is responsible for maintaining only about 40% of the city's street trees with the responsibility for the rest (60%) falling to a confusing mix of private property owners and other public agencies. This discontinuous maintenance patchwork has resulted in an inefficient and costly tree maintenance program.

4. *Inadequate resources*. The city's urban forestry budget has decreased dramatically over the last 8 years. This has restricted the DPW's ability to plant and care for trees. As a result, maintenance responsibility for thousands of trees is being transferred to private property owners. Widely unpopular with the public, this approach has put San Francisco's urban forest in a high state of risk for neglect, and damage. Identification of stable maintenance funding sources is essential to ensure the health of San Francisco's urban forest.

The Urban Forest Plan (Phase 1) is divided into four sections containing: (1) a vision for the urban forest, (2) a policy framework, (3) key recommendations, and (4) an implementation strategy. The plan provides the following vision for the city of San Francisco's trees: *Our urban forest will be a healthy, well-maintained and sustainably financed collection of trees and greenery that improves the city's ecological function and brings enormous benefits to the people of San Francisco* (San Francisco Planning Department, 2014, p. 38).

The plan's policy framework is based on five goals. Each goal is accompanied by a series of strategies and actions required to achieve it. The goals include the following:

- *Grow* the urban forest through new planting to maximize the social, economic, and environmental benefits of trees and urban greening.
- *Protect* the urban forest from threats and loss by preserving the city's existing trees.
- *Manage* the urban forest through coordinated planning, design, and maintenance to ensure its long-term health and sustainability.
- *Fund* the urban forest by establishing a dedicated funding stream for the city's trees.
- *Engage* residents, public agencies, and the private sector in caring for the urban forest and developing a deeper connection to nature.

The plan identifies four key recommendations that summarize the many policies and strategies contained in the plan document.

1. *Maximize the benefits of urban trees*. San Francisco's trees provide a wide range of important social, economic, and environmental benefits (estimated at $9.4 million USD annually) (San Francisco Planning Department, 2014). Some of these benefits include air filtration, storm water absorption, carbon storage, habitat creation, and improved public health. The plan recommends maximizing the benefits of urban trees by identifying species that provide high levels of ecosystem services. This will provide forest managers and property owners with the information needed to more carefully select and plant trees based on the benefits most desired. In addition, the plan promotes making the benefits of trees more visible to the public and policymakers through signage, quantification of benefits, and a public information campaign.

2. *Grow the street tree population by 50%*. The plan recommends the planting of 50,000 new street trees on San Francisco's streets over the next 20 years. This would expand the city's street tree population by half from 105,000 street trees (2014) to 155,000 street trees (2034). These new trees are needed to stem the decline of the urban forest, increase the city's tree canopy, and expand the benefits of trees to more of the city's neighborhoods. They will also result in a more equitable distribution of tree canopy and help reduce greening inequities throughout the city.

3. *Establish and fund a citywide street tree maintenance program*. Cities recognized as urban forestry leaders typically hold responsibility for the management and maintenance of all city street trees. The plan proposes halting the practice of transferring maintenance responsibility to private property owners. Alternately, the plan recommends centralizing maintenance responsibility for 100% of San Francisco's street trees under the DPW through a newly created municipal street tree program. Under such a program, homeowners would be relieved from the responsibility of maintaining trees fronting their property and making tree-related sidewalk repairs. Creation of a citywide street tree maintenance program would require the establishment of a dedicated long-term funding source to finance the program. The Street Tree Finance Study (AECOM, 2013) identified a variety of funding options for consideration by decision makers, including an assessment district, parcel tax, and general obligation bonds. The plan recommends further evaluation of these tools to determine their feasibility and potential to achieve the plan's goals.

4. *Manage trees throughout their entire life cycle.* The plan recommends managing trees throughout their entire life cycle including the processes related to their growth, maintenance, and disposal in order to create a more sustainable resource flow. Components of a street tree life-cycle management program include the creation of a street tree nursery, a citywide tree removal and succession strategy, and a wood re-use program to create second-life products from dead or removed street trees. Life-cycle management can help minimize waste, reduce travel distances, and provide second-life opportunities for locally grown urban wood.

The plan's final component is an implementation strategy. This section of the plan document includes a table assigning responsibility for every plan action to various agencies, departments, and institutions and timelines for completion (1–20 years).

OUTCOMES OF THE PLAN

At the time of this writing, a number of positive outcomes associated with the city's urban forest have occurred as a result of the plan. First, a ballot measure to establish a citywide street tree maintenance program is under consideration. The Urban Forest Plan proposes identifying stable and dedicated funding to centralize maintenance responsibility for street trees under the DPW. This will require a significant increase in urban forestry funding over what is currently available from existing sources. The ballot measure under consideration would put before San Francisco voters the option of approving a parcel tax on properties within the city to generate the necessary funding to create a citywide street tree maintenance program. Success of the ballot measure will require a two-thirds majority vote in favor. If the ballot measure is successful, property owners would be relieved of the burden and expense of maintaining trees fronting their property and making tree-related sidewalk repairs.

Second, an urban forest wood re-use study has been completed. To advance the plan's tree life-cycle management recommendation, a wood re-use study was conducted through a partnership between city agencies and a local educational institution. The study examined the potential for various end-of-life uses for street and park tree wood including lumber, engineered wood, chips/mulch/compost, biochar, hog fuel, ethanol, and paper products. The study provided recommendations for the city to increase recycling and re-use of wood from dead or removed trees through an enhanced wood utilization program.

Third, a citywide street tree census is being completed. As a result of the plan, the citywide street tree census will be conducted for all of San Francisco's 105,000 street trees. Comprehensive data on location, species, age, condition, and structural conditions will be collected. This data will result in the creation of the first comprehensive database of the city's street trees. The data will be made accessible to the public through an online local urban forest map Internet site (urbanforestmap.org).

Finally, urban design strategies for street trees are being developed. Some of San Francisco's most visually memorable streets and urban places are shaped by trees (Figure 32.2). These streets employ unique species palettes to achieve dramatic effects. Many of the city's neighborhoods and streets, however, feature less intentional plantings and an uncoordinated patchwork of trees. Consistency and variation in tree form, color, and seasonal display can be used to create dynamic and harmonious streetscapes. A follow-up study is being carried out to identify urban forest design strategies and to determine how to increase the public and private realm's capacity to accommodate more trees.

DISCUSSION AND CONCLUSIONS

The City of San Francisco's Urban Forest Plan has catalyzed a larger discussion regarding the definition and role of infrastructure within the city. Capital planning and investment discourse has largely been dominated by the construction, repair, and maintenance of so-called "gray" infrastructure systems. These include streets, public transit systems transportation, and energy and water delivery systems. Highly engineered infrastructure systems have historically overshadowed the important role and function of "green" infrastructure—the natural and living systems within the city. Public agencies and decision makers are beginning to recognize the many benefits and ecosystem services provided by the city's trees, parks, and watersheds. These natural elements fulfill important functions such as managing storm water, filtering air pollution, decreasing habitat fragmentation, and creating more livable neighborhoods. San Francisco's ecological infrastructure, including its urban forest, holds great potential to help the city achieve a variety of goals including adaptation to climate change and improved public health. The plan strives to elevate the importance of the natural systems, such as the urban forest, to the same level of importance and attention of "gray" infrastructure systems. It makes the case for recognizing trees as valuable components of the city's infrastructure and advocates for increasing investment to ensure their long-term health.

Sustainability Issues

The plan advances many of San Francisco's sustainability policies and programs. These include aggressive targets for reducing waste and greenhouse gases, supporting biodiversity, and creating a more equitable distribution of trees. San Francisco has set a goal of achieving "zero waste" by the year 2020 by diverting 100% of the city's solid waste from landfills. Increased recycling and composting are an essential component of this strategy. The plan advocates for increased in-house recycling and re-use of wood waste to create second-life products such mulch, compost, finished lumber, and artisan wood products. The city's Climate Action Strategy (San Francisco Department of the Environment, 2013) establishes targets for greenhouse gas reduction, including a 25% decrease from 1990 levels by 2017 and a 40% decrease by 2025. The plan proposes growing the urban forest by 50,000 trees, significantly increasing its capacity to act as a carbon sink. In addition, by promoting the creation of high-value wood products such as usable lumber and furniture from city trees, the plan aims to halt the release of carbon back into the atmosphere from decomposing wood waste. The plan also supports protecting and enhancing urban biodiversity by encouraging the planting of local wildlife supportive trees and other plants. The plan promotes social equity by emphasizing tree planting in neighborhoods with a disproportionate lack of trees, especially within disadvantaged and underserved communities.

Plan Development Challenges

The initial scope of the Urban Forest Plan envisioned a much larger and expansive effort. The limited availability of resources, however, would not allow for the comprehensive examination of urban forestry resources originally planned. In response, the plan's initial scope was divided into three phases that could be carried out independently and be combined at a later time. As a result, the plan's first phase only addresses street trees. Trees in parks and on private properties are to be the subject of future plan phases. Outreach and involvement in citywide policy issues such as the urban forest also posed a challenge to plan development. Residents are often more likely to engage in planning efforts that directly address their neighborhood. Citywide policy issues such as street tree maintenance can seem abstract. Neighborhood organizations and members of FUF have been helpful in elevating urban forestry issues and making them relevant to local stakeholders.

Plan Implementation Challenges

Implementation of the Urban Forest Plan and its recommendations will require a significant increase in funding. The plan proposes establishing and funding a citywide street tree maintenance program. It also recommends the planting of 50,000 new street trees. The cost of these activities is estimated at $22–31 million USD annually (AECOM, 2013). This represents a drastic increase in funding from current levels. The city's ongoing funding obligations such as public safety, education, and healthcare have historically taken priority over urban forestry in the city's budgeting process. Without new fees or taxes, it is unlikely all the plan's recommendations can be realized. A ballot measure is under consideration that would put before voters the option of creating dedicated funding for a citywide street tree maintenance program. However, it will require significant political support and a two-thirds majority vote by San Francisco voters.

Other Interesting Issues Related to the Plan

The plan supports the use of technology as a way to connect a broader range of stakeholders to the urban forest. The San Francisco Urban Forest Map (urbanforestmap.org) makes the city's existing database of trees available to residents through an interactive Internet-based map. Visitors to the site can explore information about trees near their homes as well as throughout the city. They can also upload additional information about specific trees by locating them on the map or adding ones that are missing. The "crowdsourcing" of tree data by residents can provide forest managers with additional information and identify where new maintenance needs may be emerging. The Urban Forest Map also quantifies the real-time benefits of trees, including gallons of storm water filtered, pounds of air pollutants captured, kilowatt-hours of energy conserved, and tons of carbon dioxide removed from the atmosphere. This map will be updated with the results of the citywide Street Tree Census and will eventually be made available to a wider audience through the creation of a mobile phone app.

ADDITIONAL READING AND RESOURCES

This chapter represents a synthesis of the 2014 Urban Forest Plan for the city of San Francisco, California. To view the plan itself, please visit this Internet site, which was available on July 11, 2014:

http://www.urbanforest.sfplanning.org/

If in the future the link to the site appears broken, search the Internet using the title of the plan and the keywords provided.

REFERENCES

AECOM, 2013. Financing San Francisco's Urban Forest: The Costs + Benefits of a Comprehensive Municipal Street Tree Program. AECOM, San Francisco, CA, 54 p.

Friends of the Urban Forest, 2014. Greening San Francisco. Friends of the Urban Forest, San Francisco, CA. http://www.fuf.net/ (Accessed June 6, 2014).

Nowak, D.J., Hoehn III, R.E., Crane, D.E., Stevens, J.C., Walton, J.T., 2007. Assessing Urban Forest Effects and Values: San Francisco's Urban Forest. U.S. Department of Agriculture, Forest Service, Northern Research Station, Newtown Square, PA, Resource Bulletin NRS- 8. 22 p.

San Francisco Department of the Environment, 2013. San Francisco Climate Action Strategy. San Francisco Department of the Environment, San Francisco, CA. http://www.sfenvironment.org/cas (Accessed June 6, 2014).

San Francisco Planning Department, 2014. San Francisco Urban Forest Plan, Phase One: Street Trees. San Francisco Planning Department, San Francisco, CA.

U.S. Census Bureau, 2011. American FactFinder. U.S. Department of Commerce, Census Bureau, Suitland, MD. http://factfinder2.census.gov/faces/nav/jsf/pages/index.xhtml (Accessed June 6, 2014).

Chapter 33

Forest Management Unit 13—Forest Management License (FMU) 3, Manitoba, Canada

Kevin Crowe[1] and Laird Van Damme[2]

[1]Lakehead University, Thunder Bay, Ontario, Canada, [2]KBM Resources Group, Thunder Bay, Ontario, Canada

ABBREVIATIONS

CIAC	Communities of Interest Advisory Committee
FML	Forest Management License
FMU	Forest Management Unit
FLI	Foreign Lands Inventory
HEC	Habitat Element Curves
L-P	Louisiana-Pacific Canada, Ltd.
OSB	Oriented Strand Board
SAC	Stakeholders' Advisory Committee

MANAGEMENT SETTING AND BACKGROUND

Forest Management License 3 (FML 3) is located in west-central Manitoba, Canada. The total area of FML 3 is 2.6 million hectares (ha) (6.42 million acres (ac)). Within this vast area are multiple administrative boundaries of land zoned for non-forestry objectives. These include a provincial park that dominates the area, privately owned land (primarily used for agriculture), multiple protected areas that vary in size from large Wildlife Management Units to small voluntarily protected private areas, and five First Nations reserves and land claims. Forest Management Unit 13 (FMU 13) (Figure 33.1) is comprised of about 376,000 ha (929,096 ac), 83% of which is productive forest, and 39,000 ha (96,369 ac) of which is a provincial park area (Duck Mountain Provincial Park). The park has two zones where harvesting is prohibited: backcountry zones and recreation zones. Selective logging of white spruce (*Picea glauca*) began in the early 1900s on portions of FMU 13 near farmlands for the use of settlers in the region. Softwood lumber and some hardwoods have been industrially logged in the Duck Mountains since the 1950s. Louisiana-Pacific Canada, Ltd. (L-P) began harvesting hardwoods in 1996. Thus, FMU 13 has a relatively short legacy of recorded anthropogenic disturbance.

The managed forest of FML 3 is within the boreal biome and is comprised of hardwood, mixedwood, and conifer stands (Figure 33.2). The hardwood stands are dominated by trembling aspen (*Populus tremuloides*), balsam poplar (*Populus balsamifera*), and white birch (*Betula papyrifera*). Hardwoods comprise 47.5% of the productive forest in FML 3. The mixedwood and conifer forest is dominated by black spruce (*Picea mariana*), with a smaller representation of white spruce and jack pine (*Pinus banksiana*). The managed forest is slow-growing with merchantable harvests typically occurring in stands only over the age of 70 years, yielding on average, 150 cubic meters (m^3) per ha (1,429 cubic feet (ft^3) per ac) at the age of rotation. The forest also has a short growing season with a mean daily temperature below 0 °C (32 °F) between October and March. The managed forest's primary natural disturbance comes from fire. Forest fires create landscape mosaics of large and small openings, with irregular boundaries leaving burned and unburned trees standing.

Forest Plans of North America. http://dx.doi.org/10.1016/B978-0-12-799936-4.00033-3

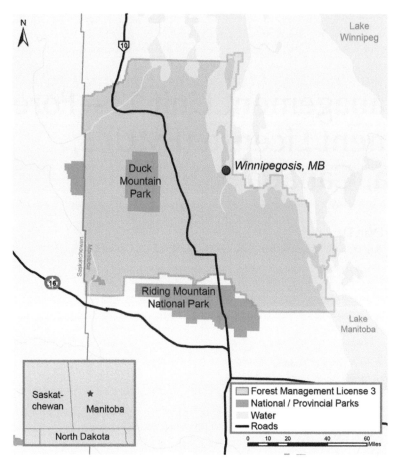

FIGURE 33.1 A map of Forest Management Unit 13, within Forest Management License 3.

FIGURE 33.2 Coniferous forests within Forest Management Unit 13 of Forest Management License 3.

Historically, FML 3 included a mixture of open prairie grasslands and a combination of hardwoods, softwoods, and mixedwoods. In the twentieth century, much of the grassland and forest was cleared for agricultural use. Today, approximately 100,000 people live in the region, and a large portion of the economy depends upon agriculture. The forest within FML 3 was not used for industrial purposes until the 1950s; and therefore, it contains an abundance of old forest cover, a legacy of active fire suppression, and low industrial use.

In 1996, L-P began operations in a new mill located in Swan Valley, Manitoba, to produce oriented strand board (OSB). L-P is one of the leading OSB manufacturers in North America, and its new mill (Swan Valley) required a sustainable supply of hardwood from the region. In 1996, L-P also received approval on its first 10-year management plan

(1996–2006) to begin harvesting from the government-owned forest in FML 3. Approval was given by the landowner, the Province of Manitoba, through its administrative branch, Manitoba Conservation. The approved timber supply consists of approximately 629,000 m³ (22.2 million ft³) of hardwood and 176,000 m³ (6.2 million ft³) of softwood per year from FML 3. Another 229,000 m³ (8.1 million ft³) of hardwood is available in a volume agreement in the Porcupine Mountain Provincial Forest immediately north of FML 3. The combined volume of hardwood timber supplied to the OSB mill in Swan Valley is up to 900,000 m³ (31.8 million ft³) per year, including purchased wood from private land.

PLANNING ENVIRONMENT AND METHODOLOGY

In 2006, L-P's first 10-year management plan for FML 3 was due to expire. L-P's second management plan for FML 3 was to cover a period of 20 years (2006–2026). FML 3 is divided into three distinct forest management units (FMUs): FMUs 10, 11, and 13. For diverse administrative reasons, planning is performed on each management unit separately. Therefore, in this chapter, we will review only the 20-year management planning of FMU 13 and the Duck Mountain Provincial Forest contained within it. To prepare for planning, L-P had to complete the following six activities:

1. Create an ecological forest lands inventory.
2. Identify watersheds.
3. Identify areas eligible and ineligible for harvest.
4. Establish reliable ecological yield curves.
5. Establish participation in and communication of planning activities.
6. Collect and interpret data from monitoring, surveying, and research.

A new forest lands inventory (FLI) was completed early in 2003 for FMU 13. The forest inventory and sampling program was completed under the joint direction of L-P and Manitoba Conservation and took 2 years to complete. Two sets of aerial photographs at a scale of 1:15,000 were used: (i) black and white near infrared, and (ii) color infrared "leaf-off" photos. The "leaf-off" photographs enabled identification of understory conifers that would normally be masked by the leaves of the hardwoods.

Stratification of ecological boundaries into polygons was determined first, and then stand, soil, and topographic attributes were interpreted for each polygon. The FLI also distinguished multiple tree canopy layers to describe vegetation at different heights, a phenomenon common in mixedwood stands. The interpretation of photographs was validated by 1,429 sampling plots that were measured in a statistically rigorous sampling design with a great depth of ecological resolution and detail.

The 2003 FLI revealed that there was 57,771 ha (142,752 ac) of nonforested land area (e.g., water, wetlands) in the FMU, along with 6,356 ha (15,706 ac) of nonforested uplands. The productive forest area was approximately 83% of the FMU, although about 12% was unavailable (closed) to active forest management. Areas were removed for the following reasons:

1. *They were considered heritage sites.* A 50 m (164 ft) buffer (i.e., zone where harvesting is excluded) was placed around sites containing archaeological evidence that the site holds relevance to the heritage of First Nations peoples.
2. *They were Provincial roads.* For visual quality objectives, a 100 m (328 ft) buffer was placed along all provincial roads.
3. *They were riparian buffers.* The buffer widths required by Manitoba Conservation for FMU 13 range from 50 m (164 ft) to 200 m (656 ft) on selected lakes, streams, and rivers; 50 m (164 ft) buffers were applied on all remaining water bodies.
4. *They included steep slopes.* Areas with slopes greater than 40% were considered to be inoperable and excluded from harvest consideration.
5. *They were located within a Provincial park.* Provincial parkland within FMU 13 is closed to harvesting.
6. *They contained non-merchantable productive forests.* Lowland spruce stands, owing to their small-diameter distribution and low productivity, were considered non-merchantable and were closed to harvesting.

Within the productive areas of the FMU that were open to active management, the stand types were identified as hardwood (43%), mixedwoods (26%), and softwood (31%). The hardwood areas consisted of pure trembling aspen (25% of the total forest area), aspen, balsam poplar, and white birch (18%). The mixedwoods forests were either hardwood-dominant (15% of the total forest area) or black spruce-leading (11%). The softwood forests contained upland black spruce (12% of the total forest area), lowland black spruce (8%), white spruce (6%), and jack pine (5%).

The age-class structure of the FMU 13 (Figure 33.3) indicates an abundance of area in older stands. It should be noted that the range of ages within which the harvesting is profitable is narrow. For example, for trembling aspen, the mean annual increment is maximized when the rotation age selected is 70 years and the operability range is approximately between 60 and 75 years. For white spruce, the optimal rotation age is around 80 years, and its range of operability is between 70 and

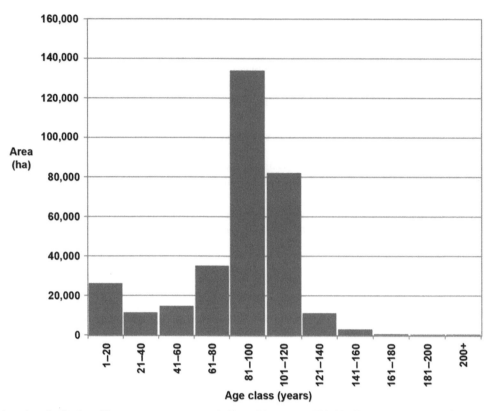

FIGURE 33.3 Age-class distribution of forest open to management in Forest Management Unit 13 of Forest Management License 3.

90 years of age. Operability in stands over 120 years is a challenge in this forest. This is because the proportion of rot in the stem increases dramatically after 120 years, a symptom of the tree's senescence.

Annual variation in peak flows of rivers and streams is a natural phenomenon in FMU 13, but there exists the potential to increase peak flows through the removal of forest cover in a watershed. The standard definition of a watershed is an area of land from which surface run-off drains into a common body of water. Watershed boundaries can be drawn at multiple spatial scales (e.g., sub-basins, drainage basins, etc.). In FMU 13, L-P, in cooperation with the provincial government's Manitoba Conservation and the federal government's Department of Fisheries and Oceans, divided FMU 13 into 16 sub-basins and 77 distinct watersheds. The Department of Fisheries and Oceans required that no more than 30% of a distinct watershed (at sub-drainage scale) can be in a harvested state. For L-P planners, an obvious question to address, given this constraint, is what are the recovery rates of harvested areas in FMU 13? To answer this question and plan harvesting with minimal disturbance to the watershed, analysts at L-P developed an impact assessment approach for each watershed and sub-basin using the WRENNSS® water yield model (U.S. Department of Agriculture, Forest Service, 1980). In addition, since each watershed has a unique set of properties, L-P used the WRENNSS® model to identify watersheds that were at higher risk than others.

The yield curves for the long-term timber supply analysis were developed from volume sampling conducted in 2001 and 2002. L-P and Manitoba Conservation jointly determined the sampling stratification and the number of sample plots to be established. A total of 1,470 plots were sampled from both FMU 13 and the adjacent FMU 14 to provide data for developing the yield curves. A total of 11 different yield curves, or "cover classes," were developed, and each polygon in the forest was assigned to one cover class with a yield curve predicting the volume of merchantable timber (by species) that can be harvested (in m³ per ha) from that cover class based on stand age.

The new yield curves based on temporary sample plots were demonstrated to be valid, insofar as they could be used to predict the merchantable volume of a cover class over time, but their ability to predict natural succession in unharvested stands proved problematic. Predicting natural succession in mixedwood stands has been a persistent problem in the management of boreal forests across Canada due to the inherent variability of mixedwood stand dynamics over time. For the pragmatic purposes of timber supply modeling, the Province of Manitoba made the decision that a death age would be assigned to each cover class. The death age is the age at which the overmature trees in the modeled stand are forecast or expected, on average, to die naturally. At the death age, both the stand volume and stand age would assume to reset to zero, and the original forest cover type would regenerate.

Analysts at L-P were dissatisfied with the death age assumption made on succession in the new yield curves. Therefore, in 2004, L-P commissioned the development of new empirical yield curves. In addition to the data collected from the 1,470 temporary sample plots in 2001–2002, L-P included the analysis of Canadian Forest Service 1,484 permanent sample plots. The plot data consisted of measurements within mixedwood stands, in the nearby Riding Mountain National Park, which featured periodic measurements over the last 50 years. To develop a model of stand development that could be used to account for forest successional trends over 100 years into the future, work by Dr. Norm Kenkel (University of Manitoba Ecology) was also integrated into the new yield curves. Dr. Kenkel quantitatively determined successional trends in boreal mixedwood forest from permanent sample plots, established and remeasured in the Riding Mountain area, from 1946–1969, and a sub-set of 286 permanent sample plots were remeasured by LP in 2002. These successional trends, in stands aged 90 to 200 years old, were utilized to guide development of new yield curves for older stands. Rather than resetting to age zero, older stands showed a 20-year decline, followed by a 20-year increase in a sine wave pattern. This pattern was designed to mirror the observed gap phase dynamics detected by Kenkel's work.

Given the quality and quantity of permanent sample plot and temporary sample plot data, priority was given to construct a set of yield curves that not only forecasted changes in standard forest volume over time, but also described linkages between key stand attributes and habitat elements (e.g., snags, coarse woody debris, and shrub cover). Thus, as the data and the purpose evolved to track habitat elements (in addition to growth and yield), the name Habitat Element Curves (HEC) was adopted in place of yield curves. The completed sets of HECs were designed to forecast expected changes in the forest due to aging (natural succession) and management. Manitoba Conservation's Forestry Branch requested that the HECs be validated. Therefore, during the spring and summer of 2005, the HECs were tested to validate model components, using independent data sources, including sets of temporary and permanent sample plot data provided by L-P. Validation exercises included rigorous sets of statistical procedures that compared predictions from the HECs with actual observed data. This work was completed by August of 2005.

The first step in establishing participation in planning activities was to assemble a planning team. The planning team was comprised of individuals from L-P and many other organizations, including Provincial and Federal government agencies, Canadian Universities, and consultants. L-P's intention in forming the planning team was to bring together the appropriate environmental, regulatory, and academic capacity to provide the resources and information required to develop the plan. The second step taken was developing a document containing a set of guiding principles and strategies for public communication and participation. The third step involved executing the strategies contained in this document.

Public participation in planning was accomplished through a set of meetings that occurred between February 2003 and May 2006. The meetings fell into three categories, a local Stakeholders' Advisory Committee (SAC), a broader Communities of Interest Advisory Committee (CIAC), and the general public. The Stakeholders' Advisory Committee consisted of a cross-section of individuals with various interests in the management of forestlands in FML 3. Membership was drawn from more than 45 First Nations, local associations, or conservation groups including, for example, membership from organizations such as the Intermountain Snowmobile Club, Pine Creek First Nations, Manitoba Naturalist Society, Duck Mountain Trappers' Association, Manitoba Conservation, Ducks Unlimited, Swan Valley Sport Fishing Enhancement, University of Manitoba Department of Botany, and the University of Winnipeg. The Advisory Committee has been an active part of L-P's planning process since its inception in 1994. The Advisory Committee met with L-P 20 times between 2003 and 2006. The CIAC had a similar cross-section, but included leaders of various organizations focused on urban input from the City of Winnipeg, whereas SAC provided local (Swan River) and mostly rural input.

L-P also hosted public events that involved both SAC and CIAC members during the development of the management plan. The most significant public events planned were a set of planning workshops where different planning scenarios for FMU 13 were shown to members of the public and their input was sought. Three such workshops took place between November of 2004 and March of 2006. These workshops were surprisingly valuable to the L-P planning team. The feedback from these workshops induced the planning team to make multiple changes in designing its preferred scenario. The changes incorporated into the management plan included strategies or targets for managing:

- Biodiversity
- Water quality and quantity
- Soil resources
- Recreation opportunities
- Esthetics
- Economics

The valuable changes resulting from these workshops are authentic representations of public participatory planning, and are recorded in detail in the documentation of the Forest Management Plan. Strategies for public communication of planning activities included the hosting of open house meetings, in both Swan River and Winnipeg (the province's capital city), the publication and distribution of 12 editions of a newsletter, and the maintenance of a comprehensive Internet site, www. SwanValleyForest.ca (Louisiana-Pacific Canada Ltd., 2014).

The final strategy by which L-P sought to communicate the quality of its management activities was through pursuing certification. The Sustainable Forest Initiative (SFI) program is one of the world's most widely applied standards and voluntary systems. It provides companies with the ability to demonstrate to the public their commitment to sustainable forest management. L-P began its certification efforts in 2001, and in December of 2002, after a two-phase audit process, was certified by the SFI. SFI certification was successfully renewed by L-P Swan Valley in 2003, 2004, 2005, and it has been maintained ever since.

Many of the results from surveying and monitoring activities were summarized in a document entitled *Report on Past Forest Operations*, which was Chapter 2 of the 2006 Sustainable Forest Management Plan (Louisiana-Pacific Canada Ltd., 2014). This report acts as a point of reference for linking past and proposed activities. It summarized the operations, monitoring, and surveying that have occurred since 1996, when the first 10-Year Management Plan was approved. Elements of the report include summaries of all pre-harvest surveys, roads, harvest blocks, fiber procurement opportunities, independent logging contractors, trucking statistics, applicable Crown fees, stream crossing surveys, forest regeneration, and insect and disease problems. A pre-harvest survey is a site-specific ecosystem assessment of a scheduled harvest block conducted one year prior to logging. Because the OSB mill in Swan Valley procures its wood from many sources (it holds multiple licenses to harvest from government land and maintains multiple sources of private wood), the planning team reviewed information regarding the role FMU 13 plays in the bigger picture of fiber procurement. L-P's OSB mill can use only hardwood trees; therefore, L-P did not need exclusive access to wood in the FMU 13. In the prior plan, L-P entered into multiple timber purchase agreements whereby the harvesting of mixedwood stands saw hardwood stems sent to L-P and softwood stems sent to Spruce Products Ltd. and other quota holders. Before planning, the relationships with other quota holders and the local demand for their softwood stems were reviewed. Crown fees for wood harvested are submitted, on a quarterly basis, to Manitoba's Minister of Finance. These fees are for three distinct purposes: (a) stumpage fees for the wood harvested from government-owned land, (b) a forest renewal and stand management Trust Fund, and (c) a forest protection charge. The planning team was particularly interested in the second category of fees, because it placed an approximate upper bound on the total annual cost of regeneration activities it could plan for while developing its strategic plan.

With regard to regeneration, the removal of the overstory in hardwood stands promotes natural root suckering of aspen and balsam poplar, while white birch regenerates primarily from stump coppicing. Regeneration surveys on hardwood stands were performed 5 years after harvest. In the prior plan, all hardwood regeneration met or exceeded provincial standards. On mixedwood and softwood sites, conifer seedlings are planted and regeneration surveys are performed at 7 years post-harvest. In the prior plan, all mixedwood and softwood sites met or exceeded provincial standards. Finally, information on insects and disease is collected during pre-harvest surveys or as observed by staff in the field. In FMU 13, the major sources of fiber loss from disease are armillaria root rot (*Armillaria mellea* species complex) and western gall rust (*Endocronartium harknessii*). This information is important in tactical planning, because the primary strategy used to control these diseases is accelerated harvest of infected areas.

In addition to the contents of the *Report on Past Forest Operations*, the planning team also reviewed the many research projects completed in FML 3. Results and data collected from these projects were incorporated, where possible, into the Forest Management Plan. In Swan Valley, L-P was engaged in extensive research and monitoring programs. Many of these programs were in response to conditions for the Environment Act license. A summary is given below in terms of the research priority and the number of research projects executed under that priority.

- Ecosystem management (3 projects)
- Economic and social dimensions of forests (2 projects)
- Forest renewal (13 projects)
- Knowledge transfer (2 projects)
- Human-caused environmental effects (17 projects)
- Natural environmental effects (8 projects)
- Implementing sustainable forest management (17 projects)
- Forest function (19 projects)

Strategic Planning Model

For forestlands owned by the province of Manitoba, it is required that the allowable annual cut, and its supporting analysis, be determined by the Forestry Branch of Manitoba Conservation. The licensee, L-P, is required to constrain its harvest levels and operations to the annual allowable cut calculated by the Forestry Branch. The major steps in preparing for strategic planning have been described above. Therefore, in this section, we limit ourselves to describing the wood supply model, the objectives, constraints, and results of the timber supply analysis.

The wood supply model used by the Forestry Branch was the widely used Woodstock model (Remsoft, 2014). Woodstock is a linear programming model of the strategic harvest scheduling problem. In these models, the decision variables are typically of the form x_{ijk}, where x_{ijk} represents the number of hectares to harvest from cover type i, within age class j, during time period k. Note that the decision variable does not indicate, on a map, where the harvest is to occur. Woodstock provides what is often called an "aspatial solution." The solution from a strategic linear programming model is typically allocated in the form of cut-blocks, on a map of the forest, during tactical planning. This spatial allocation of a strategic model's solution can occur using another optimization model in a second and separate step.

The Forestry Branch of Manitoba completed its wood supply analysis in 2004 and published a *base case* scenario, or a scenario that should be considered as a baseline for further sensitivity analysis. The primary objective in the base case scenario was to determine the sustainable harvest level of wood fiber for hardwood, softwood, and mixedwood strata in FMU 13. The objective function used in the model was to maximize total merchantable volume of hardwood and softwood that can be harvested from the FMU 13, over a 200-year planning horizon, with 5-year planning-periods. The objective function was subject to the following constraints:

1. Maintain an even-flow of hardwood and an even-flow of softwood harvested over the planning horizon (i.e., 0% volume flow fluctuation).
2. Ensure that total operable growing stock of the forest does not decline in the last 50 years of the planning horizon.
3. The harvested softwood or softwood-leading cover types are to be regenerated with the same cover types after a 5-year time lag. Hardwoods and hardwood-leading cover types receive the same treatment without a regeneration delay.
4. The minimum allowable harvest age is the lower age in a cover type's operability range. The operability requirement reflects current operating practices by L-P on this land base. Hardwood-dominated strata must have at least $75\,m^3$ per ha ($1,072\,ft^3$ per ac) of merchantable hardwood volume to be eligible for harvest. Softwood-dominated strata must have at least $50\,m^3$ per ha ($715\,ft^3$ per ac) of softwood volume to be eligible for harvest.
5. The death age occurs across all cover types. Since the wood supply analysis was completed in 2004, and L-P had not completed its new yield curves until 2005, the older yield curves were used in the timber supply analysis conducted by the Forestry Branch; therefore, since data were not available on succession in older stands, the analysis assumed that death, not succession, occurs in all cover types. The ages of death ranged from 140 to 180 years. At the age of death, the stand is assumed to return to age zero, and continue aging.

The timber supply analysis also assumes that there will be no loss to the standing volume of the forest in the event of a catastrophic natural disturbance. The policy of the Forestry Branch is that should a natural disturbance arise that depletes more than 5% of the standing volume, the analysis will be performed again using the depleted inventory.

The base case scenario was also evaluated by the following measures:

- Total harvest volume by cover type over time
- Growing stock on harvestable land base over time
- Harvest area by cover type over time
- Harvest volume per ha over time
- Mean age of harvested stands over time
- Area of forest in each age class over time
- Mortality (measured in m^3) over time
- Habitat area for elk and moose over time

Tactical Planning Model

The tactical planning model used was by L-P was Patchworks® (Spatial Planning Systems, 2009). The Patchworks® software formulates the tactical planning problem as an integer goal programming model solved using a metaheuristic algorithm. There are three significant consequences arising from the selection of Patchworks® as a planning tool. First, since the decision variables in Patchworks® are integer (binary, to be precise), the solution produced is spatially explicit (i.e., the optimal schedule of harvesting and road-building activities can be mapped directly from the solution). The second consequence of using Patchworks® was that it is a goal programming model. In Patchworks®, many of the constraints one might have in a linear programming formulation of the problem can be turned into goals, and the objective function in a goal programming model is to minimize a solution's total deviation from the target values assigned to these goals. Each goal, in the objective function, can also be weighted; therefore, trade-off analysis between conflicting goals can be facilitated by applying different weights (reflecting different goal priorities) to different goals. Thus, the structure of the model dictates how some analyses can be performed. A third significant consequence of selecting Patchworks® as a planning model is that analysts were not restricted to looking at tactical planning problems alone. Because it is a spatially explicit model, the L-P analysts have used Patchworks® to address strategic, tactical, and operational planning questions. The analysts at L-P have used Patchworks® in this manner because they wanted to select a solution with a spatial harmony and long-term continuity between the strategic, tactical, and operational levels of planning.

A coarse-filter approach to conserving biodiversity assumes that the majority of the habitat conditions in a forest can be conserved if a natural forest structure is maintained. A natural forest, in this context, can be measured by three criteria: a natural species composition, a natural age-class structure, and a natural patch size distribution of stands across the forest. To plan for this first objective, L-P analysts used their new (2005) yield curves (which included successional events occurring in stands 120–200 years of age), the annual allowable cut values as targets in the goal programming model, targets in their goal programming model to keep the area in each forest cover class stable across the planning horizon (of 200 years), targets to maintain an age-class structure within the bounds of natural variation, and targets to achieve a distribution of disturbance patch sizes across the forest. The values for these patch sizes were intended to emulate the natural disturbance pattern of fires in the boreal forest. Patch sizes were divided into six categories based on area: 0–25 ha (0–62 ac), 25–50 ha (62–124 ac), 50–100 ha (124–247 ac), 100–250 ha (247–618 ac), 500–1,000 ha (617–2,471 ac), and 10,000+ ha (over 2,471 ac). To aid in evaluating the output of this coarse-filter strategy, the L-P analysts used SLAMS®, a Spatial Landscape Assessment Model (Rempel and Donnelly, 2010), to evaluate the landscape resulting from the above set of harvesting goals. Of course, the targets on patch size distributions violated the Province's constraint of a maximum opening size of 100 ha (247 ac); therefore, L-P treated this target as an exploratory scenario to be used in conjunction with SLAMS®.

Implementation of Goals and Objectives

The goal of conserving ecosystem condition and productivity is addressed by limiting the length, duration, and densities of forest roads. A target was placed on road density, which is measured in Patchworks® as the ratio of linear road length to the area that is being accessed. A density target was selected such that the optimal solution would contain a cluster of road-building and harvesting activities over time rather than a broad dispersal of such activities across the entire forest and tactical planning horizon.

Under the goal of conserving biodiversity, L-P analysts constrained the harvesting of any rare ecosites. For example, rare ecosites on wetlands include open bogs, exposed marshes, and open rich fens. On uplands, rare ecosites include jack pine and black spruce on dry sandy sites, and hardwoods and white spruce on moist, sandy sites. In addition to this constraint, a goal was set to minimize the proximity of any harvesting or road-building activities within a certain distance of these rare ecosites. To accomplish this, all stands and candidate road segments in the forest were assigned a proximity index to rare ecosites. A target of low proximity was set in the goal programming model, and a solution that minimized the deviation from this target was found.

Pest and disease problems were managed primarily by accelerated harvest and preferentially scheduled for harvest. In the past, these decisions were made at the operational level of decision making, but in this plan, disease and pest management have been integrated into the objectives of strategic planning. Patchworks® is used to schedule these operational level interventions, and the forecast outcomes are then fully integrated into the strategic objectives on forest structure and composition. Hence, pest management provides an example of integrating operational aspects into strategic planning made possible by the use of spatially explicit forest management planning.

To maintain the representation of the current range of wildlife habitat associations, a synthesis of the many research activities occurring in FML 3 was conducted and multiple indicators of wildlife habitat associations were established. In strategic planning, the scheduling of harvest blocks was constrained to ensure that the full representation of these indicators persisted through time.

The Department of Fisheries and Oceans required that, as a condition of L-P's Environment Act license, no more than 30% of a distinct watershed (at sub-drainage scale, termed second-order stream) can be in a harvested state (0–5 years for hardwood, 0–14 years for softwood). Planned harvesting, at the strategic level, was constrained by setting targets in Patchworks® for the maximum amount of equivalent clear-cut area allowed in any sub-basin, in any given period.

Measures of carbon sequestration were derived from L-P's research program. Carbon yield curves were utilized in Patchworks® as directly resulting from harvesting and regeneration activities forecast over time. Since FMU 13 contains a legacy of older stands, constraints were not applied to maintain the current stock of carbon in the forest. Instead, the measures of carbon sequestration were used as indicators. Carbon stability was projected to occur around year 50 of the plan.

All of the above objectives were derived and verified through three scenario planning workshops attended by the general public and both SAC and CIAC members. At these planning workshops, input was sought through the presentation of six different scenarios defined as follows:

1. The base case scenario developed by the Forestry Branch.
2. A natural disturbance scenario.
3. A scenario that maximizes biodiversity.
4. A scenario that divides FMU 13 into different zones of management intensity.
5. A scenario based on watershed management.
6. A scenario that maximizes fiber production.

From the feedback given at these workshops, L-P refined its objectives and constraints to produce a selected scenario. The last workshop was held in March of 2006 and featured the objectives described above.

OUTCOMES OF THE PLAN

The 20-year allowable annual cut for FMU 13 determined from the province of Manitoba's base case was 146,280 m³ (5.17 million ft³) per year of softwood and 348,823 m³ (12.32 million ft³) per year of hardwood. L-P followed the standard practice of tactical planning by scheduling a set of blocks to be harvested and roads to be built over the next 20 years. The allocation of blocks was not to exceed the 20-year allowable cut determined by the Forestry Branch, limited to cut-block sizes of 100 ha (247 ac), and constrained against harvesting adjacent blocks for a 10-year period.

In addition to standard tactical planning designed to deliver the most economical wood supply possible, L-P analysts decided to address a host of ecosystem components, such as carbon stocks. Many of these additional ecosystem values were not required by Provincial regulators. L-P chose to pursue an analysis of both tactical and strategic-level questions using Patchworks®. L-P analysts wished to pursue these questions as part of its corporate goal to plan for conservation of biodiversity at the landscape scale. The results from the three course-filter objectives were satisfied in the selected planning scenario. First, the species composition remains relatively stable over the long run because of the practices of understory protection and planting on mixedwood sites. Second, the objective of maintaining an age-class structure within the bounds of natural variation was also satisfied. Table 33.1 illustrates the changes in the age classes over the first 20 years of the plan, and reveals that age classes of forest stands younger than 60 years increase from 29.4% of the total area to 36.8% by the end of the planning period on the managed land base. Similarly, trees of age class 120 years and older will increase from 4.3% of the total area to 20% by the end of the 20-year time frame. The harvesting activity is concentrated on the 60–100 year age class. Hence, the representation of these areas decline.

The results from the third objective of the coarse-filter strategy (maintaining a natural distribution of patch sizes across the landscape) reveal that the smallest three size classes (harvest areas of 0–100 ha (0–247 ac)) made up an average of 92% of all the disturbances over the 20-year plan. In comparison, the two largest size classes (harvest areas of 500 ha (1,236 ac) or more) make up an average of 0.3% of all disturbances over the same time period. In emulating the natural disturbance pattern of fire in the boreal forest, such a distribution comes closest to the ideal distribution created by nature.

TABLE 33.1 Managed Area (% of Total Area) by Age Class and Period Over the 20 Years of the Plan of Forest Management Unit 13 in Forest Management License 3, Manitoba

	Age Class (Years)								
Year	0–20	21–40	41–60	61–80	81–100	101–120	121–140	141+	Total
0	9	4	5	12	43	23	3	1	100
5	15	2	5	9	39	25	4	1	100
10	21	3	5	5	25	31	9	1	100
15	25	5	5	4	19	29	11	2	100
20	26	9	4	5	10	26	17	3	100

DISCUSSION AND CONCLUSIONS

L-P submitted the Management Plan to the Provincial Government in 2006. This plan, like other forest management plans in the province of Manitoba, was never formally approved by the government. A dispute on whether to follow a harvest schedule derived under the Forestry Branch's base case or the tactical plan presented to the public by L-P was never resolved. Nonetheless, L-P's annual operating plans, which followed the tactical plan, were approved each year by Manitoba Conservation. By 2016, the 10-year allocations of the unapproved tactical plan will have been harvested and regenerated according to specifications of the tactical plan generated by L-P. Operations staff members at L-P have reported that the allocations in the tactical plan were both economical and practical. The tactical plan also made the annual operating plans easy to develop and to defend to stakeholders.

Learnings and Insights

The L-P plan had several unique features relative to many forest management plans in Canada. First, the base data was current and of exceptionally high quality. All of the research programs, developed to meet and exceed environmental license conditions, produced ideas and data that drove the setting of objectives through successive scenario planning workshops. Second, extraordinary measures were taken to engage stakeholders. This produced a strong endorsement of the plan and strengthened L-P's social license to operate in the region. Third, the main decision support tool (Patchworks®) created a strong link between strategic objectives and tactical planning. This link protected biodiversity and water conservation goals while providing an economical harvest and renewal solution.

Sustainability Issues

The coarse-filter approach to conserving biodiversity was the fundamental strategy used in meeting the goal of environmental sustainability, but it was not the only strategy used. For example, computer simulations of future forest conditions, using song-bird habitat diversity as an indicator of biodiversity, supported the coarse-filter assumptions by indicating that biodiversity is likely to be sustained under the proposed management strategy and tactical plan. The strength of these types of analyses, combined with constraints on harvesting levels in watersheds, adjacent blocks, and rare ecosites, enabled L-P to address the question of its sustainable practices in a public forum and receive certification of sustainability under the SFI program.

Plan Development Challenges

Although the plan itself was not approved by the regulatory agency in charge of forest management over technical issues that were never resolved (yield curves and forest succession patterns), the supporting annual operating plans were approved. Hence, the plan was approved incrementally, an unusual circumstance in Canada.

Plan Implementation Challenges

Operational staff at L-P found the plan was relatively easy to implement on the ground and that it was defensible to public stakeholders. This outcome is a result of the investment in good base data, scientific studies, public engagement, and strong linkages between strategic and tactical plans.

REFERENCES

Louisiana-Pacific Canada Ltd., 2014. Swan Valley Forest Management Plan Website. Louisiana-Pacific Canada Ltd, Swan River, MB. http://www.swanvalleyforest.ca/ (Accessed April 29, 2014).

Rempel, R.S., Donnelly, M., 2010. A Spatial Landscape Assessment Modeling Framework for Forest Management and Biodiversity Conservation. Sustainable Forest Management Network, Edmonton, AB. 36 p.

Remsoft, 2014. Remsoft Solution Suite. Remsoft, Fredericton, NB. http://www.remsoft.com/technology.php (Accessed April 29, 2014).

Spatial Planning Systems, 2009. What is Patchworks? Spatial Planning Systems, Systems, Deep River, ON. http://www.spatial.ca/products/index.html (Accessed April 29, 2014).

U.S. Department of Agriculture, Forest Service, 1980. An approach to Water Resources Evaluation of Non-point Silvicultural Sources (A Procedural Handbook). U.S. Environmental Protection Agency, Office of Research and Development, Environmental Research Laboratory, Athens, GA. EPA-600/8-80-012. http://water.epa.gov/scitech/datait/tools/warsss/rrisc_handbook.cfm (Accessed April 29, 2014).

Chapter 34

Fort Wainwright, Alaska, United States of America

Dan Rees and Adam Davis

DPW Environmental Division, Fort Wainwright, Alaska, USA

ABBREVIATIONS

INRMP   Integrated Natural Resource Management Plan

MANAGEMENT SETTING AND BACKGROUND

Headquartered just east of Fairbanks, Alaska, the Department of the Army installation Fort Wainwright (Figure 34.1) is responsible for managing more than 1.6 million acres (ac) (0.648 million hectares (ha)) of land for military use, most of which is withdrawn from the public domain. Fort Wainwright shares responsibility for forest vegetation management with the U.S. Department of Interior's Bureau of Land Management (BLM) on withdrawn public domain lands. The total land area available for forest management at Fort Wainwright is 173,247 ac (70,112 ha). Forests of the area encompassed by this military installation include white spruce (*Picea glauca*), black spruce (*Picea mariana*), paper birch (*Betula papyrifera*), balsam poplar (*Populus balsamifera*), and quaking aspen (*Populus tremuloides*) tree species.

Army forest management is required to support and enhance the immediate and long-term military mission while meeting natural resource stewardship requirements set forth in federal laws and the Army's Environmental Strategy. It is Fort Wainwright policy to maintain, restore, and manage its forestlands on an ecosystem basis. The harvesting of forest products, including other consumptive and nonconsumptive activities that take advantage of the forest environment, are allowed and encouraged when conducted consistently with protecting and maintaining a viable, self-sustaining forest ecosystem. Revenues generated from the commercial harvesting of forest products will be used to maintain, improve, and, as necessary, restore previously degraded forest ecosystems. Forest ecosystem management strategies should be broad-based to optimize overall natural resource benefits, and should not be focused on a single management objective such as the maximization of timber production. Forest ecology and management are to be an integral part of the master planning process and review.

Three vegetation rights categories exist on Fort Wainwright lands: (1) Department of Defense fee simple (Army owns vegetation rights), (2) public domain lands withdrawn for military use where withdrawals specify BLM vegetation management authority (BLM controls vegetation rights), and (3) public domain lands withdrawn for military use where vegetation management authority is not specified (Army controls vegetation rights). On lands in which the Army has vegetation rights, any sale of timber would be processed through the Army's forest management system. If the BLM has vegetation management authority for the forest vegetation resources at the project site, any vegetation manipulation must be performed after BLM approval. Proceeds from BLM sales are deposited into the general U.S. Treasury and would not be processed through the Army's forest management system. Further, if vegetation rights management authority belongs to the BLM, consultation is required with the designated BLM forestry representative.

Within the forest management plan, seven management areas are identified:

- Fort Wainwright Main Post
- Tanana Flats Training Area
- Yukon Training Area
- Donnelly East Training Area

Forest Plans of North America. http://dx.doi.org/10.1016/B978-0-12-799936-4.00034-5

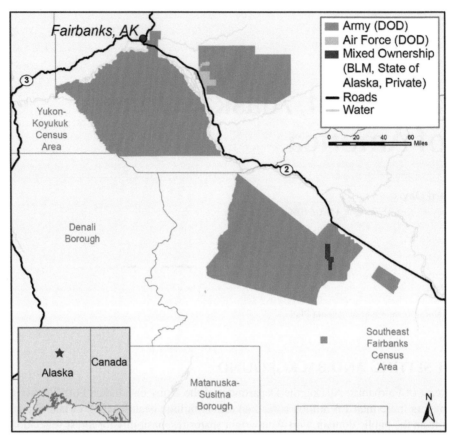

FIGURE 34.1 A general vicinity map of the Fort Wainwright lands.

- Donnelly West Training Area
- Gerstle River Training Area
- Black Rapids Training Area

The Fort Wainwright Main Post covers a total of 15,558 ac (6,296 ha) of upland forests that include birch and aspen, mixed hardwood-white spruce, and white spruce forests on relatively well-drained, warm sites. Forested areas cover 39% of the Main Post. Here, a high level of human-caused disturbance was experienced in the early 1900s, and fire suppression activities began around the 1950s. While there are significant areas of regeneration at the Main Post (Figure 34.2), the age-class distribution is dominated by the 60–120 year categories. The regeneration class consists of forests with tree diameters between 1.0 and 5.9 inches (in) (2.5–15.0 centimeters (cm)). Lowland forests (30%) of the Main Post include balsam poplar, mixed balsam poplar-spruce, and white spruce stands. Mixed birch-spruce stands also occur, especially on older lowland sites. Lowland sites are subject to a variety of natural disturbances—erosion, flooding, and ice damage near active river channels, fire, insects and disease, windthrow, and thermokarsting (land surface characterized by small hollows and hummocks formed as ice-rich permafrost thaws). The total volume of wood on the Main Post, at the time of development of the forest management plan, was about 4 million cubic feet (ft³) (about 113,250 cubic meters (m³)).

The Tanana Flats Training Area contains 655,985 ac (265,473 ha) of land that is 41% forested and composed of about 13% commercial forests. At the time of the development of the forest management plan, the total timber volume in this training area was about 163 million ft³ (4.6 million m³), most of which is white spruce, birch, and aspen sawtimber and poletimber. The poletimber class consists of trees with diameters between 6.0 and 8.9 in (15.1–22.8 cm) for softwoods and between 6.0 and 10.9 in (15.1–27.9 cm) for hardwoods. The sawtimber class consists of trees with diameters greater than 9 in (22.9 cm) for softwoods and 11 in. (27.9 cm) for hardwoods. From about 1900-1940, extensive harvesting occurred in lowland sites, especially along the Tanana River. Mining also disturbed lowland forests. These disturbances were typically smaller scale than the large upland fires, and they created a complex mosaic of stand types and ages. The Yukon Training Area (Figure 34.3) contains approximately 259,969 ac (105,208 ha) of land, most (69%) of which is forested. The estimated total timber volume in this area was about 84 million ft³ (2.4 million m³), mainly consisting of white spruce, birch, and aspen sawtimber and poletimber.

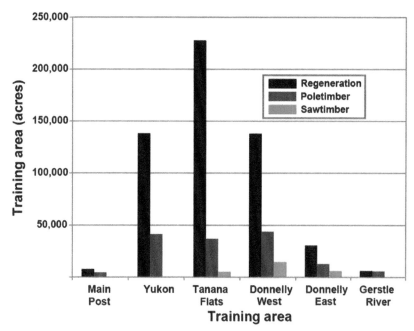

FIGURE 34.2   Timber size-class distribution by management area on forested lands within Fort Wainwright.

FIGURE 34.3   Typical forested terrain found in Fort Wainwright's Yukon Training Area.

The Donnelly East Training Area (east of the Delta River) contains approximately 101,662 ac (41,142 ha) of land, and forests occupy about 49% of the land area. In the Donnelly East Training Area, the total timber volume was estimated to be more than 25 million ft^3 (about 707,814 m^3), most of which is commercial sawtimber and poletimber. The Donnelly West Training Area (west of the Delta River) contains about 531,141 ac (214,500 ha) of land, and about 37% of these are considered forest. Here, the total timber volume was estimated to be about 103 million ft^3 (2.9 million m^3), most of which is commercial sawtimber and poletimber. The Black Rapids Training Area contains approximately 3,373 ac (1,365 ha) of forested and nonforested lands, with forests occupying 14% of the land area. Here, the total timber volume was estimated to be about 55,740 ft^3 (1,578 m^3), all of which is commercial sawtimber and poletimber. The Gerstle River Training Area covers 20,737 ac (8,392 ha) of forest and non-forest lands, with forests occupying 60% of the land area. In 1994, the Hajdukovich fire burned 11,320 ac (4,581 ha), which was classified as non-forest with reed grass (*Calamagrostis* spp.), willow (*Salix* spp.), and other hardwood regeneration covering most of the burned area. Parts of the burned area are now considered forest regeneration areas. Only approximately <1% of this training area contains forests with sawtimber-size trees, yet approximately 27% of the training area contains poletimber-sized trees. Here, the total timber volume was estimated to be more than 2.5 million ft^3 (70,781 m^3), all of which is commercial sawtimber and poletimber.

PLANNING ENVIRONMENT AND METHODOLOGY

The primary objective of the management of lands at Fort Wainwright is to support the U.S. military training mission. Another major objective of the forestry program at Fort Wainwright is to promote a healthy ecosystem capable of supporting mission and conservation requirements. The forest management plan is a required component of a larger Integrated Natural Resource Management Plan (INRMP). Therefore, the forest management plan must consider the military mission, preservation of habitat, and recreation for all of the forested areas. Harvests of forest products from Fort Wainwright lands are permitted, but are not mandatory. In 2007, Fort Wainwright developed a forest management plan that outlines short- and long-term forestry objectives. The plan addressed the following goals and objectives:

- Manage forest resources for long-term stewardship in order to maintain productive forest ecosystems and forest health and vitality, and to meet the future needs of the mission
- Institute practices to protect the natural resources of the installation (soil, water, fish and wildlife, natural vegetation, recreation, and esthetics)
- Balance the ecological, social, and economic values of forested parcels
- Maintain healthy forest ecosystems through active management

Although not the subject of this chapter, an Urban Forest Management Plan is also an integral part of the INRMP and includes professional standards (e.g., National Arborist Association), technical specifications, training, certification, and requirements for activities impacting the planting, growth, and survival of trees in the urban forest ecosystem of the installation.

Guiding Laws, Regulations, and Policies

Laws, instructions, regulations, and guidance that must be complied with when performing forest management activities at Fort Wainwright include, but are not limited to, the Endangered Species Act of 1973, as amended; National Environmental Policy Act; Sikes Act (16 U.S.C. 670a-670o, as amended by the Sikes Act Improvement Act 1997); National Historic Preservation Act of 1966, as amended; Defense Finance and Accounting Service Indianapolis Center (DFAS-IN) Regulation 37-1; Chapter 14-Finance And Accounting Policy Implementation, January 2000; 32 CFR Part 651, Environmental Analysis of Army Actions; Army Regulation 405-80, *Management of Title and Granting Use of Real Property*; 10-Oct-97; and Army Regulation 405-90, *Disposal Of Real Estate*, 10 May 1985.

Forest management projects must comply with the applicable laws and an INRMP developed in accordance with the National Environmental Policy Act (NEPA) that addresses the impact, if any, on the composition, structure, and function of natural communities and biological diversity. The installation must also comply with 36 CFR 800, or the "Protection of Historic Properties, Native American Graves Protection and Repatriation Act," and other applicable statutes as outlined in Army Regulation 200-1, coordinate with the U.S. Fish and Wildlife Service, and consider the potential effects of activities on significant archeological resources. The broader INRMP establishes best management practices to be conducted during the timber sales. For example, a two-level buffer is required. Timber harvest activity is not allowed within 50 ft (15.2 m) immediately adjacent to an anadromous fish stream or high-value resident fish water bodies. Between 50 and 100 ft, a 50% minimum retention of trees must occur.

A Memorandum of Understanding is necessary to define the expectations, terms, and conditions of forest management and timber disposal between the installation and the supporting U.S. Army Corps of Engineers District. Generally, the installation is responsible for timber management while the district is responsible for timber disposal. The Memorandum of Understanding identifies the roles and responsibilities of both parties in relation to administration of timber sales and forest management. The installation develops specific reports of availability in advance with the disposal agency in order to maximize market potential for timber. The use of an Army Environmental Database on Environmental Quality tracks how well the Army's environmental compliance status meets statutory and regulatory requirements and partially responds to a requirement to perform environmental compliance assessments, as defined by Army Regulation 200-1.

To facilitate annual work planning requirements, Fort Wainwright and the Alaska Corps of Engineer District, develop an annual report of timber availability that is consistent with the forest management plan and that describes a list of each planned timber sale by management unit. This annual report needs to comply with the forest management plan, the INRMP, and federal, state, and local environmental compliance standards. Finally, the state of Alaska is entitled to 40% of the net revenue from forest product disposal, as noted in 10 U.S.C. 2665, "Sale of Certain Interests in Land; Logs (Reimbursable Forestry)" and the Department of Defense Financial Management Regulation. Surplus funds from the sale of forest products are collected and stored in the Department of Defense Forestry Reserve Account. These funds are made available for military departments for improvements to forest lands, unanticipated contingencies, and other management actions that support established plans, including:

- Fish and wildlife habitat improvements or modifications
- Range rehabilitation where necessary to support wildlife
- The control of off-road vehicle traffic
- Specific habitat improvement projects and related activities and adequate protection for species of fish, wildlife, and plants considered threatened or endangered

To support forest management and silvicultural activities that directly relate to the management of the forest ecosystem, some expenses can be supported by the Conservation Reimbursable Forestry Account, however, the fund fluctuates from year to year. Forest management activities that directly support mission landscape requirements and environmental stewardship in the forest management plan section of the INRMP are eligible for Conservation Reimbursable Forestry Account funds.

The broader INRMP may be viewed as a higher-level planning effort. A lower and more specific level of planning produces forest land-use plans for specific activities, and these are prepared prior to commercial sales of forest products. These plans include sale boundaries, cruised volume, silvicultural prescription to be employed, road layout, best management practices for prevention of soil erosion and sedimentation, water quality considerations, cultural resources protection, wildlife considerations, harvest method(s), scaling requirements, slash disposal, site preparation, and regeneration requirements. Other laws and regulations guide the development of forest land-use plans for specific activities.

Silvicultural Activities Considered

Silvicultural treatments used on Army lands are designed to improve military mission areas, and where possible, facilitate multiple-use and sustained yield timber management objectives while enhancing watersheds, wildlife habitats, and esthetic values along scenic corridors. To ensure appropriate regeneration of forests, clear-cut activities are employed in all forest types, seed tree and group selection harvests are employed in white spruce, paper birch, and mixed species forests, and shelterwood harvests are used in white spruce, aspen, balsam poplar, and mixed forest areas.

Thinnings, sanitation harvests, and timber stand improvement activities are all employed in various timber types to meet the objectives of the forest management plan. Timber stand improvement activities are implemented to improve the quality of forest stands, to support military training activities, and to improve wildlife habitat for species such as spruce grouse (*Falcipennis canadensis*) and Townsend's warbler (*Setophaga townsendi*). These activities can include thinning, spacing, chemical injection, chipping, prescribed burning, and others, all designed to improve species composition, improve tree quality, improve growth rates, or reduce hazardous fuel levels. The military needs to train personnel under certain environmental conditions, and this may require activities such as thinnings to facilitate maneuverability in certain areas.

Regeneration of forests can be made through planting seedlings, planting sprigs, coppice cuts, or seeding. Scarification and prescribed fire may be necessary for site preparation efforts. When developed, an operational reforestation plan would outline the objective of the regeneration project and additional treatments needed if the objective is not being met and would define the site preparation and regeneration methods and the target species and density.

OUTCOMES OF THE PLAN

The sustained yield timber base was defined as those forested areas where periodic harvests could be regulated and continued for a long period of time—possibly even indefinitely. Generally, only productive timberland is included in the sustained yield timber base. Through observations of some Fort Wainwright lands, it has been determined that forests containing poor stocking levels or containing black spruce generally occur on relatively unproductive sites, and, therefore, these were excluded. In addition, areas containing recently burned timber are excluded from the sustained yield timber base. These areas are excluded because the productivity status of the site may change as result of the fire and subsequent changes in species composition and stocking levels. Harvest levels for white spruce and hardwoods (Table 34.1) were based on a simple area cut method that divides the total productive forest land by the rotation age. The result of this computation provides the area that can be harvested each year. Rotation ages were determined to be 120 years for white spruce and 80 years for hardwoods, not including a 10-year window of time for ensuring adequate regeneration. The majority of the volume projected to be produced from these activities falls in the poletimber and sawtimber classes.

TABLE 34.1 Estimated Harvest Levels at Fort Wainwright

Area	White Spruce		Birch-Aspen	
	Potential Total Harvest Area (ac)	Estimated Annual Harvest Area (ac)	Potential Total Harvest Area (ac)	Estimated Annual Harvest area (ac)
Fort Wainwright Main Post	1,470	11	2,290	25
Tanana Flats Training Area	74,629	574	7,761	86
Yukon Training Area	18,653	143	23,062	256
Donnelly East Training Area	12,870	99	3,011	33
Donnelly West Training Area	41,584	320	18,399	204
Gerstle River Training Area	3,568	27	1,584	18
Black Rapids Training Area	60	<1	–	–

DISCUSSION AND CONCLUSIONS

Forest resources are protected on Fort Wainwright lands through the Integrated Natural Resources Management Plan as well as through local and Army regulations. The INRMP and associated forest management plan establish best management practices to be used during timber sales, clearing, or construction activities to protect surrounding forest resources, wetlands, surface waters, and wildlife. If construction activities cannot avoid clearing of forest resources, these activities must follow correct procedures to minimize impact and effectively utilize forest products as specified in the INRMP's forest management plan section.

Learnings and Insights

The development of Fort Wainwright's forest management plan involved the integration of an understanding of military training requirements on the landscape and the silviculture of boreal tree species. The military utilizes forests for foot and vehicular maneuver training. Forests provide value to military trainers by providing obstacles to maneuver and horizontal and vertical concealment for soldiers and equipment (Hale et al., 1999). An understanding of growth form and habits of boreal tree species was incorporated into the needs of military training. An example of this integration of forest management activities for military training is areas thinned for vehicular maneuvers. Boreal tree species tend to be shallow-rooted due to cold soils, and they do not tolerate excessive compaction around the roots. In some areas, where the forest was thinned for vehicle maneuvering training exercises, trees are dying. A solution to the problem was to use a clumped retention approach to better protect the roots of trees from vehicle traffic. The clumps of trees were also found to provide better concealment options for military equipment than were individual trees.

Sustainability Issues

To determine the suitability of areas for forest management activities, each type of activity will be considered separately, taking into account potential environmental impacts, the season of use, and opportunities to balance seasonal use with other uses of the land. Installation conservation personnel determine technical feasibility of management activities for achieving landscape requirements and integrate environmental compliance (e.g., endangered species protection), stewardship, and sustainability requirements. The environmental and related impacts of forest management activities will be assessed according to Army Regulation 200-1. Coordination with adjacent private and public landowners and managers will be included in the assessment process. Coordination must be made to ensure all local, state, and federal requirements are met.

Army lands may be designated for one or more types of forest management activities in response to a demonstrated need, provided that sufficient suitable areas are available. Lands that may not be designated for forest management activities use are as follows:

- Areas restricted for security or safety purposes, such as explosive ordnance impact areas
- Areas containing geological and soil conditions, flora or fauna, or other natural characteristics of fragile or unique nature, which would be subject to excessive or irreversible damage
- Areas that are considered key fish and wildlife habitat, which are primarily along streams, and where older-aged trees, snags, and coarse woody debris are to be maintained
- Areas that contain archeological sites, historic sites, petroglyphs, or pictographs, or areas set aside for their scenic value

Plan Development Challenges

The Sikes Act requires the development of an INRMP and the forest management plan falls within the guise of the INRMP. One challenge to the development of the plan was locating adequate funding for various projects. Interior Alaska has a small forest products industry, and the majority of forest products are utilized exclusively within the local area. Fort Wainwright's forests have limited high-value sawtimber material. Fort Wainwright's forest products therefore generate limited revenue and are mostly utilized as firewood in the local communities. As a result, forest products do not provide enough funding for the plan implementation. While the forest plan and the forestry program support and enhance military training opportunities, the forest plan outlines a program that utilizes forest management practices to support multiple requirements at Fort Wainwright, such as forest clearings for construction projects, wildlife habitat enhancement projects, and military training area improvement projects. Funding for the forest plan and program is provided by the many clients served by the forestry program. Army regulation 200-1 requires salvaging of forest products, yet resistance to the forest plan development stems from the salvage requirement within clearing projects. The high cost of salvaging wood increases construction costs for projects, and the wood salvaged has limited value. To address the funding issue, during the plan development process an idea was brought forward that the forestry program would try to salvage the usable forest products from a construction site before clearing processes begin. If conditions were not feasible for the salvage of all of the wood, or if the site was not accessible for salvage operations, the wood could be purchased by the project proponent and disposed of in the most economically feasible method. The incorporation of multiple options for adhering to the salvage requirements for forest products into the forest plan brought wide support for the plan within the Fort Wainwright community.

Plan Implementation Challenges

Individual forest management projects must support the mission of the installation, and projects cannot encumber land needed to conduct mission operations. Natural resource managers must coordinate with mission operators to identify opportunities that can improve long-term mission access to land, increase training realism, and improve training flexibility.

Maintaining good forest health is also a challenge to plan implementation, from both control and funding perspectives. Prevention of outbreaks is the primary approach to insect and disease control in intensively managed sites, and the approach consists of silvicultural practices that enhance the natural control of insects and diseases and the removal of infested trees. Insects are active and significant components of Alaska's ecosystems, and arctic and boreal insects are represented by few species with large populations. Bark beetles (e.g., the spruce beetle (*Dendroctonus rufipennis*)) are one of the most important disturbance agents in mature white spruce stands in interior Alaska. They respond quickly to large-scale windthrow, fire damage, and flood injuries. The Alaska Department of Natural Resources estimates that 30–50% of the forest stands older than 150 years are infected in the Fort Wainwright area (Department of Natural Resources, 2001). A species of engraver beetle (*Ips* spp.) is also found throughout Alaska, but it is most prevalent in the Interior. These engraver beetles favor areas with accumulation of forest debris (slash), and infestations occur mainly along river floodplains and areas disturbed by erosion, spruce top breakage (e.g., from snow-loading), harvest, or wind. Mortality caused by engraver beetles has stimulated research into management tactics that utilize semiochemicals such as pheromones and tree bark volatiles. Stress caused by drought and aspen leaf miner (*Phyllocnistis populiella*) activity have significantly affected the condition of aspen forests in the interior of the state (U.S. Forest Service, 2012).

Other Interesting Issues Related to the Plan

A wildland fire management plan was developed in close conjunction with the forest management plan. Wildfires have burned significant portions of Fort Wainwright within the past 10 years. The availability of timber resources is volatile and changing whenever new areas are burned. The forest plan incorporates this variability in timber availability by focusing on burned areas for forest management activities such as firewood sales. Firewood is the dominant forest product sold by Fort Wainwright and the public prefers to obtain this product from standing dead wood. Dead, standing, burned trees retain their value as firewood for several years after a fire, allowing ample time to salvage the wood. Forest management activities in the plan also focus on protecting military infrastructure and modified landscapes designed for maneuvering activities from wildfire. Because of fire danger during summer months, military training opportunities can be limited through restrictions placed on pyrotechnic use within the ranges. The forest plan identifies and targets these areas for thinning and hazard fuel reduction projects. The landscape around the ranges are treated to make the forest less susceptible to wildfire and to increase military training opportunities by reducing hazardous fuels.

REFERENCES

Department of Natural Resources, 2001. Tanana Valley State Forest Management Plan. Department of Natural Resources, Division of Forestry, State of Alaska, Anchorage, AK. 157 p.

Hale, T., White, S., Burns, D., Palmer, D., Jones, D., Skoglund, M., Michaels, K., 1999. Tactical Concealment Area. U.S. Army Engineer Research and Development Center, Champaign, IL. NSN 7540-01-280-5500. 93 p.

U.S. Forest Service, 2012. Forest health conditions in Alaska 2011. U.S. Department of Agriculture, Forest Service, Alaska Region, Anchorage, AK. FHP Protection Report R10-PR-25. 68 p.

Chapter 35

Green Diamond Resource Company, California, United States of America

Jim Hawkins,[1] Gary Rynearson[1] and Kevin Boston[2]

[1]Green Diamond Resource Company, Korbel, California, USA, [2]Department of Forest Engineering, Resources and Management, Oregon State University, Corvallis, Oregon, USA

ABBREVIATIONS

CEQA	California Environmental Quality Act
THP	Timber Harvesting Plan
WDR	Waste Discharge Requirement

MANAGEMENT SETTING AND BACKGROUND

Green Diamond Resource Company is a family-owned company established in 1890 by Solomon Simpson. Its California operations occupy roughly 390,000 acres (ac) (157,831 hectares (ha)) of commercially managed timberlands in Del Norte and Humboldt Counties (Figure 35.1). The bulk of its ownership is within 20 miles (mi) (32 kilometers (km)) of the Pacific Ocean, and the most easterly portion is approximately 30 mi (48 km) inland. Like most large managed timberland properties in California, the estate is a collection of individual parcels that range from 20 ac (8 ha) to 100,000 ac (40,469 ha). Green Diamond's core ownership of redwood forest (Figure 35.2) is located along the west coast of the coastal and Klamath mountains and includes about 355,000 ac (143,667 ha) of land. These privately owned and commercially managed timberlands are part of a larger landscape consisting of private nonindustrial forests ownerships, rural residences, the Six Rivers National Forest (managed by the U.S. Department of Agriculture, Forest Service), Redwood National Park, and various state and county parks. Other commercially managed timberland ownerships include lands managed by Sierra Pacific Industries, Humboldt Redwood Company, and Soper-Wheeler. The surrounding Tribal lands include the Yurok lands along the Klamath River and Hoopa lands in the interior part of the state. Historically, all of these lands were primarily managed for timber production, including Redwood National Park, which was commercial timberland until the later part of the last century when it was created through two separate Congressional actions using eminent domain powers.

There are three ecoregions on the property: Northern California Coast, Northern California Coast Range, and the Klamath Mountains. The majority of Green Diamond's lands are within the Northern California Coast ecoregion. These include mountains, hills, and surrounding valleys that have a climate that is greatly modified by the Pacific Ocean. The predominate forests types in this ecoregion include Sitka spruce (*Picea sitchensis*), Bishop (coastal) pine (*Pinus muricata*), redwood (*Sequoia sempervirens*), Douglas-fir (*Pseudotsuga menziesii*), and tanoak (*Lithocarpus densiflorus*). The precipitation varies significantly from 20 inches (in) (508 millimeters (mm)) to 120 in (3,048 mm) per year. In the summer fog region, the growing season is between 225 and 310 days per year. As the distance from the coast increases, the summer fog decreases, resulting in a shortening of the growing season to between 80 and 250 days primarily due to lack of moisture. Redwood is unique among tree species in that it can absorb water directly from the air through the stomata, increasing water content by up to 11% (Limm et al., 2009); thus, summer fog is an important element in the growth of redwood.

The area within and around the Green Diamond properties contains numerous species that are listed as threatened under both federal and state laws, including chinook salmon (*Oncorhychus tshawytscha*), coho salmon (*Oncorhychus kisutch*), steelhead trout (*Oncorhnychus mykiss*), marbled murrelet (*Brachyramphus marmoratus*), and northern spotted owls (*Strix*

Forest Plans of North America. http://dx.doi.org/10.1016/B978-0-12-799936-4.00035-7

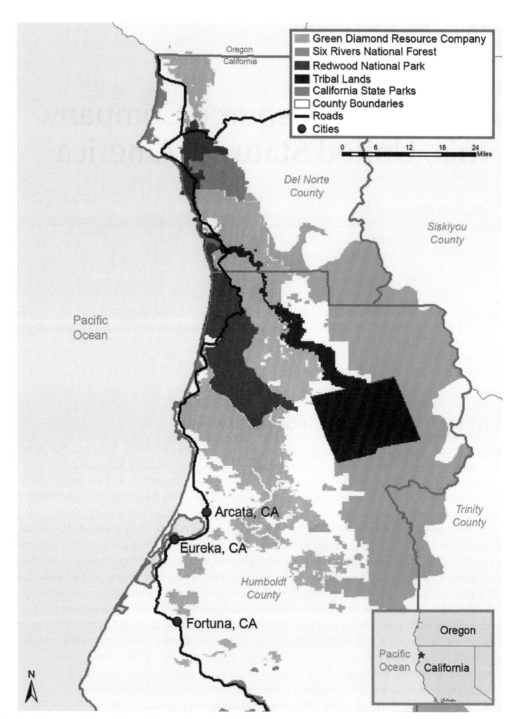

FIGURE 35.1 Location map of Green Diamond forests in northern California.

occidentalis caurina). Thus, Green Diamond must consider the potential impacts of its forest management actions on these species or their habitat in the context of the regulatory goals of these laws.

PLANNING ENVIRONMENT AND METHODOLOGY

Guiding Laws, Regulations, and Policies

The major federal laws that impact the management of the Green Diamond properties are the Endangered Species Act (16 U.S.C. 1531–1544) and the Clean Water Act (33 U.S.C. 1251–1376). Additionally, Green Diamond must consider its

FIGURE 35.2 Second-growth redwood trees on the Green Diamond property.

management actions in light of the Migratory Bird Treaty Act (16 U.S.C. 703–712) and National Environmental Protection Act (42 U.S.C. § 4321 *et seq.*). California has a well-deserved reputation for having the most stringent state forest practices and intensely regulated private timberlands in the nation. Green Diamond must manage its timberlands in compliance with these state laws that include: the California Forest Practice Act and Forest Practice Rules, the California Environmental Quality Act (CEQA), the Porter-Cologne Water Quality Control Act, the California Fish and Game Code, and the California Endangered Species Act. Additionally, Green Diamond has developed a corporate compliance program to monitor and ensure that the company remains in compliance with this multitude of programs.

The California Z'berg-Nejedly Forest Practice Act is much more prescriptive than most other state's forest practice rules. It influences forest management through multiple rules and regulations that are created by the State Board of Forestry and Fire Protection. The Act states that timberlands are to furnish high-quality timber "while providing watershed protection, and maintaining fisheries and wildlife" (California Public Res. §4512). Thus, the option of fiber farming that occurs in much of the country is considered to be inconsistent with the goals for forest management in California.

In addition to the Forest Practice Act, any projects that may alter the environment require a state permit and an analysis under the CEQA. If adverse impacts to the environment *may* occur, *the project proponent is required* to prepare an Environmental Impact Report (EIR). The EIR document must include an assessment, a description of the local and cumulative effects, and must propose mitigation measures to reduce these impacts. A timber harvesting plan (THP) satisfies the EIR requirement and is at the heart of the permitting process for California timber harvest. THPs are prepared by registered professional foresters and are reviewed by various state agencies before being approved. A THP is considered functionally equivalent to an EIR and satisfies both the Forest Practices Act and CEQA. It is through this THP permit that the majority of the state regulations are enforced for forestry.

The forest practice rules have many substantive regulations that impact forest planning in California. The rules specify minimum rotation ages for even-aged silvicultural systems: 50 years for Site I, 60 years for Site II and III, and 80 years for Site IV and V lands (CCR 913.1). Additional constraints limit the size of harvest units using even-aged silvicultural techniques ranging from 20 ac (8 ha) for ground-based systems to 30 ac (12 ha) for cable and helicopter, with a maximum of 40 ac (16 ha). The largest opening requires special approval from the Director (CCR 913.1(2)) based on unique circumstances. This is smaller than the requirements found in the Pacific Coast guidelines of the Forest Stewardship Council (FSC) standards, which limit the clear-cut opening to 60 ac (24 ha). Additionally, large ownerships, defined as those that exceed 50,000 ac (20,235 ha), are required to develop a Sustained Yield Plan, or an "Option A" plan, to address the maximum sustained production of high-quality timber products (CCR 913.11). When Green Diamond developed its Option A plan, it was to satisfy the five following requirements:

1. Produce timber products while accounting for limits on productivity due to the constraints imposed from the consideration of other forest values including, but not limited to recreation, watershed health, wildlife, and range.
2. Balance growth and harvest over time.
3. Ensure adequate site occupancy of the species to be managed.
4. Maintain good stand vigor.
5. Ensure adequate regeneration.

Additionally, the company's forest management practices have been certified by both the FSC and Sustainable Forestry Initiative (SFI) programs. This imposes additional voluntary duties on the property. Both programs require the development of a forest management plan that involves some public inputs. The programs do have their differences; however, as SFI requires that the management plan be developed using the best available science, while FSC requires additional public involvement, interaction with local Native Americans, and significant protection of high-value conservation forests. FSC also forbids the use of genetically modified organisms and bans the use of certain herbicides.

Planning Process and Silvicultural Direction

The beginning of the forest planning process involved the creation of the desired future conditions. These include the following four major ideas:

1. Create a mosaic of age classes (Table 35.1) in a dynamic pattern across the landscape using even-aged silvicultural prescriptions. This involves creating openings that are between 1 and 40 ac (0.4 and 16 ha), with an average opening size of approximately 15 ac (6 ha).
2. Provide for the retention of older stands along watercourses within a matrix of regenerating younger forests. Riparian issues are a major concern for the property as approximately 25% of the landscape is contained in the riparian management zones (RMZs). The implementation of selection silvicultural practices in the RMZs will result in the average tree age increasing from 44 to 94 years during the next 50 years. Selection harvesting or no harvest zones that are implemented around site specific habitat retention areas, unstable areas, or other site specific areas will add to the mosaic of young and old stands across the landscape.
3. Retain key habitat elements within the even-aged openings, such as snags, large hardwoods, legacy trees, and large woody debris.
4. Promote commercial thinning of stands prior to even-aged harvest to concentrate stand growth on the best available crop trees.

The silvicultural prescriptions have been designed to achieve these goals and vary with stand age, stand condition, site class, and species composition. Silvicultural prescriptions have multiple goals, including promoting the maximization of sustained yield. Others have been created to maintain, alter, or create various types of habitat. Green Diamond's even-aged management practices closely resemble the variable retention silvicultural prescription as defined in the California Forest Practice Rules. The prescription was designed to promote the retention and development of structural elements and biological features in the stand such as large trees, snags, and down wood to create a more complex post-harvest stand into the next rotation. Furthermore, to promote replanting and to reduce the risk of fire, biomass operations have

TABLE 35.1 Forest Age Class Percentage by Hydrographic Planning Area Within the Green Diamond Property in California, United States

Age Class (Years)	Smith River	Coastal Klamath	Korbel	Humboldt Bay
0–9	10.1	5.4	12.7	17.1
10–19	15.7	4.3	8.4	21.6
20–29	7.0	18.9	14.0	22.5
30–39	23.6	35.1	10.0	8.0
40–49	20.6	21.4	20.1	4.3
50–59	13.8	8.2	14.1	15.2
60–69	2.7	1.0	8.1	4.3
70–79	1.2	0.2	3.9	1.4
80+	1.5	2.6	4.9	4.5
Nonforest	3.8	2.9	3.8	1.1

been deployed to assist in slash disposal following clear-cutting. The goal is to treat approximately 1,000 ac (405 ha) per year in this manner.

The yields are estimated for these multiple prescriptions using the Forest Projection and Planning System (Forest Biometrics Research Institute, 2014) for both inventory tracking and growth and yield modeling that has been exclusively calibrated for the Green Diamond property. The growth model has been calibrated with a system of large, 0.75 ac (0.3 ha) permanent plots located throughout the property. The results are used to estimate projected volumes and many of the habitat features.

Green Diamond has engaged in a number of conservation planning efforts on the property. One includes a northern spotted owl habitat conservation plan (HCP) that was developed in consultation with the U.S. Fish and Wildlife Service. This HCP has been approved for a 30-year term, with a comprehensive review after 10 years. The agreement provides Green Diamond with an incidental take permit for northern spotted owls. As part of the northern spotted owl HCP monitoring program, Green Diamond has performed extensive research on its property with regard to the habitat use of the owl. It has determined that, within the area studied, the dusky-footed woodrat (*Neotoma fuscipes*) is the owl's primary prey. On Green Diamond lands, the northern spotted owl benefits from habitat that includes mature timber that is adjacent to younger forests. The owls appear to associate with a suitable nesting habitat that is in close proximity to their prey, and they tend to flourish in these young stands. This finding has been used to create the desired future conditions on the property that promote the mosaic of small, even-aged patches of young seral forest harboring prey that are adjacent to older patches of forest in riparian zones and core areas for nesting owls. Furthermore, recent research has demonstrated that the larger and more aggressive barred owl (*Strix varia*) is a threat to the northern spotted owl, and experimental measures are being applied under the cooperative studies undertaken by the U.S. Fish and Wildlife Service, U.S. Forest Service, Green Diamond, and the Hoopa Tribe to promote the recovery of the northern spotted owl. These studies include issuing take permits under the Migratory Bird Treaty Act to remove barred owls from some areas while maintaining other forest stands as control sites. Initial results indicate that where the barred owls have been removed, the spotted owls have reoccupied the site. One result of the continual research on the impact of forest management on wildlife species is the preparation of the new forest HCP that is being prepared by Green Diamond covering a wide range of terrestrial species that will include both the northern spotted owl and the Pacific fisher (*Martes pennanti*).

There is a common set of constraints associated with the many rivers and streams that flow through the Green Diamond ownership. The aquatic systems contain many species listed under both the California and Federal Endangered Species Act. Additionally, there are rivers and streams, as well as segments that do not meet their designated water quality standards under the Clean Water Act, and thus are subject to total maximum daily load limits primarily for both sediment and temperature under section 303(d) of the Clean Water Act. Thus, the combination of the constraints being applied due to water quality and the listing of various species of fish has led Green Diamond to develop an aquatic HCP.

In 2007, the U.S. Fish and Wildlife Service and National Marine Fisheries Service approved an aquatic HCP for Green Diamond's timberlands in Northern California. This HCP targets aquatic species and provides substantial protection for riparian forest stands. The result is that timber harvesting is very limited within riparian areas on the property. The five main goals of the aquatic HCP are as follows:

1. Maintain cool water temperatures that are consistent with the requirements of the individual species.
2. Minimize and mitigate human-caused sediments being deposited into water bodies.
3. Provide the recruitment of large woody debris into all classifications of stream channels.
4. Improve populations of amphibian species in the plan area through the minimization of harvest-related impacts.
5. Implement a monitoring program that will allow the plan to adapt to new information as it becomes available.

The HCPs were developed to ensure compliance with the federal Endangered Species Act, but there were still issues surrounding sediment discharges under state regulations promulgated under California's Porter-Cologne Act and enforced by the regional water quality control boards. Thus, Green Diamond has applied for, and was recently issued, two separate waste-discharge requirement (WDR) permits to satisfy the standards developed by the North Coast Regional Water Quality Control Board. A road management WDR was acquired first to allow for a landscape approach to road maintenance and road upgrading issues. This was followed by a forest-wide WDR to allow for a programmatic approach to addressing harvest-unit-related sediment sources. These permits are valid for the same term as the aquatic HCP and create a comprehensive landscape approach for managing sediment from harvesting operations and forest roads. The goal of the aquatic HCP and roads WDR is to treat 50% of all identified controllable road-related sediment within the first 15 years of the permit. Non-road related sediment sources associated with harvest units, such as legacy skid trail crossings, will be inventoried and treated concurrently with THP preparation and operations.

In addition to the landscape-based water quality permits, Green Diamond worked with the California Department of Fish and Wildlife to obtain a long-term permit for the management of the roads and the associated drainage system called a Master Agreement for Timber Operations. This permit also required an environmental analysis under CEQA in the form of a *Mitigated Negative Declaration*. This permit was approved prior to the Roads WDR, and the environmental analysis supported the issuance of the Roads WDR. The Master Agreement for Timber Operations allows annual planning for the installation and upgrading of culverts and bridges both associated with THPs and with the general road upgrade programs. The term of the Master Agreement for Timber Operations is the same as the aquatic HCP.

Green Diamond believes that the landscape approach is superior to obtaining individual WDR permits that are attached to individual THPs. It allows the company to prioritize the repair of road-related sediment sources across a watershed and upgrade or abandon the worst sites first. Sites that have a lower risk of failure can be upgraded or abandoned over an extended schedule and can be fixed outside of the THP process. The landscape approach has two components. The first is the road maintenance and inspection program that identifies sites that needed to be treated. The second component is the road implementation plan. This plan uses a systematic approach to road upgrading or decommissioning of the road system.

Finally, Green Diamond forestry practices have been certified to both SFI and FSC forest certification system standards. Both systems require the development of a forest management plan. FSC Principle 7 requires that a forest management plan be prepared at the appropriate scale and intensity of the operation. Under the FSC program, high-value forests such as old-growth forests shall be managed using the precautionary approach, a risk-adverse approach that requires significant evidence of a future benefits before an action can occur. Because of the potential to negatively impact the high-conservation forests, credible outside review of the assessment is required.

The SFI program requires that a management plan be prepared using the best scientific information available. Forest management plans must be developed that promote the long-term productivity of the forest and include a consideration of biological diversity. It requires the areas of special significance to be managed in a manner that protects their integrity and unique qualities, but does not create the high standard of the precautionary principle used by the FSC program. Thus, both sets of principles of sustainable forestry are included in constraints encountered by Green Diamond. In addition to the opportunity for public comment on proposed projects that are provided under the federal and state regulations, Green Diamond has committed to further engage its stakeholders through the two most common forest management certification systems in the United States. Thus, Green Diamond has placed a significant emphasis on developing and enhancing its social license to operate in a highly litigious environment.

Planning System

A multiple-stage forest planning system is used by Green Diamond to plan its forest operations. Additionally, these plans are used to ensure that the operations are in compliance with the many permits required to operate in California. The long-term system considers activities throughout the California estate. The planning horizon is 100 years as required by the Option A plan. This constrains the harvest so that the average annual projected harvest level during any 10-year rolling period cannot exceed the predicted long-term sustained yield. This harvest level is tracked by both Green Diamond and California Department of Forestry and Fire Protection, and the harvest level can be challenged by the public if there is evidence that the harvest exceeds growth.

The second level in the multiple-stage harvest planning system is the *Harvest Stand Availability Forecast*. This is a proprietary system that is used to plan the mid-time frame activities. It produces a 10-year forecast of available harvest volume. This process uses a combination of automated harvest-unit scheduling software and forester's guidance to create a mid-term harvest forecast that includes the various state regulations governing clear-cut size and green-up periods. Thus, it creates a spatially feasible harvest schedule. The forecast is used to guide harvest-unit planning to ensure that the landowners desired short-term harvest levels can be attained while complying with long-term sustained yield targets for volume. This forecast is updated at least every 3 years and may be updated more frequently if the forecast becomes out-of-sync with on-the-ground operations. Because this is a mid-term forecast, allowances are made for some variances in harvest levels due to changes in the markets. However, the planned harvest levels remain below the level required from the Option A plan.

The third level is the annual operational plan and it uses the *Log Production Projection System*. This is a rolling 1-year plan that is updated monthly. This system is a database harvest-unit planner that is used to create a detailed harvest schedule for individual logging crews for a 1-year period. It reports the projected harvest volume for each harvest unit by species, as well as projected production rates for each harvest unit and summary reports of projected and actual harvest volumes. This system included numerous additional reporting features that are used by the logging operations department, the forestry department, and the accounting department to project and track weekly, monthly, and annual harvest operations.

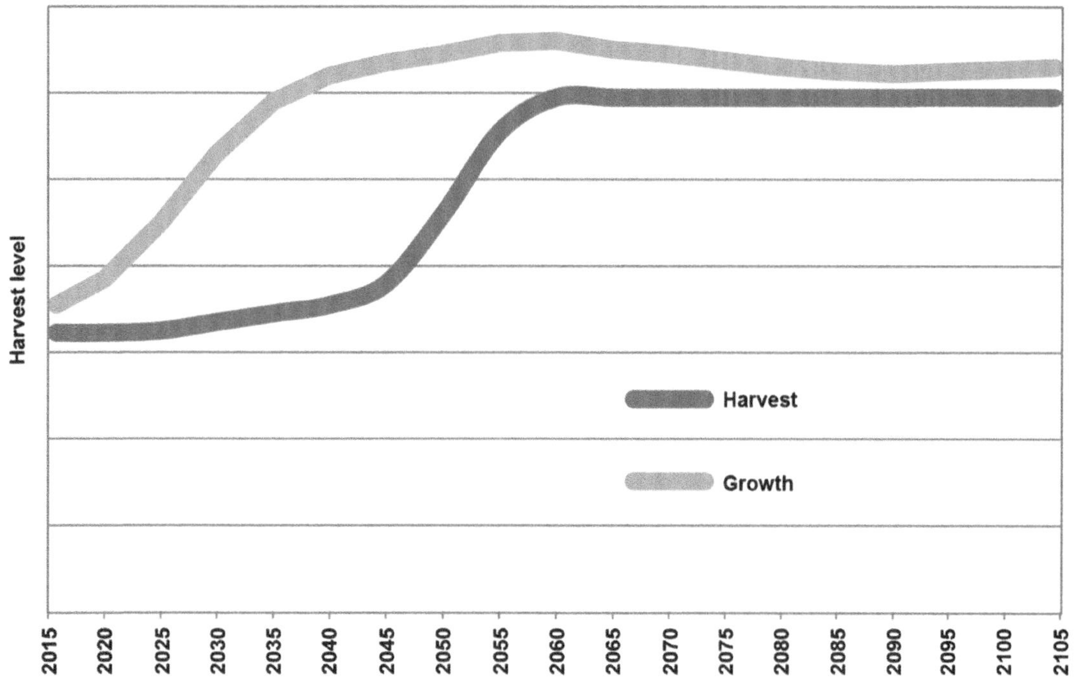

FIGURE 35.3   Relative conifer harvest and growth by 5-year time period, 2015–2105 for the Green Diamond.

OUTCOMES OF THE PLAN

The forest planning process includes a significant set of constraints created from the various environmental agreements engaged by the company. The result is a harvest schedule where growth exceeds harvest during the first 50 years of the 100-year planning horizon. This is also a result of the dedicated decision to manage stocking levels to increase growth and yield of regenerated stands to allow for future harvests (Figure 35.3). After 50 years, the harvest rapidly increases and approaches growth after 55 years.

Additionally, with designated protection areas within the results from the planning models, Green Diamond has been provided more "regulatory certainty," where THPs are designed to be consistent with landscape-level programmatic permits, and agency review and approval of THPs is conducted within the framework of these approved permits. The move toward a landscape-level planning effort has also streamlined some aspects of THP preparation processes. Specifically, the cumulative impact analysis included in each THP has been made more robust and easier to prepare by integration of the landscape-level impact analysis conducted for the programmatic permits.

DISCUSSION AND CONCLUSIONS

Learnings and Insights

One result of this planning effort has been the development of data to support a wide range of permits that have enhanced the regulatory certainty of Green Diamond's actions. These are often multiple-decade permits, and they allow Green Diamond the maximum amount of flexibility at implementing actions as it does not need to wait for the piecemeal approval required by a myriad of state and federal agencies. The ability to incorporate these landscape or property agreements into the THP process also allows for a consistent documentation of the cumulative effects analysis required by the THP process.

The process of documentation has provided Green Diamond with the opportunity to inform the public about the detailed forest management activities being performed by Green Diamond. Furthermore, the combined certification to both SFI and FSC standards, with mandatory monitoring programs and an internal corporate accountability program, should provide Green Diamond with the maximum transparency of its actions, which should lead to rebuilding the social license that may have been harmed during the controversies of the 1990s.

Sustainability Issues

Sustainability is a key aspect of the management of Green Diamond lands in California, and each of the agreements associated with the management plan requires some level of monitoring. Additionally, both certification systems used by Green Diamond require monitoring as part of the adaptive management requirements to ensure that sustainable forest management principles are central to the management of the land. The monitoring program developed by Green Diamond for the aquatic HCP has several levels. The first is *rapid response monitoring*, which provides the early warning to ensure that the biological goals of the aquatic HCP are being met. It includes measurement of water temperatures on Class I and Class II watercourses and monitoring of changes in abundance of coastal tailed frogs (*Ascaphus truei*) and southern torrent salamanders (*Rhyacotriton variegatus*) across the landscape. It includes implementation and effectiveness monitoring of road management measures that are designed to reduce sediment delivery into the streams. Long-term responses are monitored to document the achievement of conservation goals in the aquatic HCP. This process includes both Class I channel monitoring as well as sediment monitoring on the Class II streams. The long-term trend monitoring has a research component imbedded in it, and is the basis for knowledge accumulation necessary to practice adaptive management. There is an understanding that it may take many years to accumulate the knowledge relevant to propose extensive set of research projects. However, there is a proposed research project to modify the shade canopy along streams to allow for increasing sunlight to reach the stream. The project will attempt to determine if increased sunlight reaching a stream leads to increased fish size or fish numbers by increasing primary productivity in the streams. The final stage of monitoring includes four watersheds that have been judged to be representative of the geology and physiographic provinces within the aquatic HCP. These watersheds are identified as experimental watersheds that will allow for additional research and monitoring to be conducted with the goal of better understanding the interaction between forest management activities and riparian and aquatic ecosystems.

Plan Development Challenges

The software used to prepare the 100-year plan (the Option A plan) is the Forest Projection and Planning System, which is the flagship product of the Forest Biometrics Research Institute (2014). The computer program has three primary components or modules: forest inventory, growth and yield, and forest planning. A big challenge in preparing the forest plan alternatives involved using the program. A significant amount of time was required to develop various alternatives, to analyze the output, and to adjust issues that arose (i.e., processing certain stands in certain ways). Green Diamond also relied on site visits by Forest Biometrics Research Institute to ensure that certain subroutines within the software were used correctly and to help analyze the output from the many alternatives that were generated.

The 100-year Option A plan has to be approved by the California Department of Forestry and Fire Protection. The Department has a specialist in Sacramento, the Sustained Yield Forester, who carefully reviews each and every Option A plan submitted to the agency. The forester's job is to ensure that the harvest levels proposed in these plans are sustainable and based on sound inventory methods and reliable growth and yield models. The Sustained Yield Forester conducts one or more field visits and randomly selects stands to field-check to make sure the inventory in the system matches the current condition of the forest. Green Diamond has been through this process three times since 1995, with the most recent approval of its current Option A plan coming in 2009.

The Harvest Stand Availability Forecast is the mid-term plan that Green Diamond relies upon for management 10 years into the future. It takes several iterations of the planning process to develop the Harvest Stand Availability Forecast. To begin, Green Diamond's GIS staff prepares a set of maps that show the Chief Foresters all of the harvest units that are mature and ready to cut. The Chief Foresters pore over the maps, adjust harvest-unit boundaries as needed, double-check the inventory information, and then start assigning a harvest year to each harvest unit. The challenge is to be very mindful of California's adjacency rules so that the proper green-up period passes before an adjacent block of timber is scheduled for harvest. When the foresters have completed their work, the maps are given back to the GIS staff to enter all this new information into the system, update the appropriate databases, and then prepare a new Harvest Stand Availability Forecast. The updated forecast is then checked to make sure it projects a suitable flow of logs by tree species. If not, the Chief Foresters make adjustments as needed and the Harvest Stand Availability Forecast is updated once again.

Plan Implementation Challenges

Unfortunately, the Harvest Stand Availability Forecast can become "out of date" quickly. Unforeseen issues usually arise soon after the plan is completed. Units scheduled for harvest in the first year are often delayed for one reason or another. For example, a northern spotted owl might be found in a harvest unit, a harvest unit might not contain the type of logs currently needed by Green Diamond's sawmill customers, or there might be a problem getting a THP approved in a sensitive watershed. That is why it is important to update the Harvest Stand Availability Forecast frequently (generally every 3 years).

Communication between the company's departments is also critical for successful implementation of the forest plan. Required road work, in most instances, must occur one year ahead of the planned harvesting operations to give the roads a chance to "set up" and harden. Areas that are selected for harvesting must be surveyed for wildlife and rare plants. If the surveys are not completed in a timely manner, harvesting operations cannot begin. Perhaps most importantly, once harvesting operations are completed they must be immediately scheduled for reforestation so that adjacent stands with mature timber can be available for harvest as soon as the state-mandated green-up period has passed. To facilitate communication, frequent meetings are scheduled so that key supervisors can exchange information to ensure that all operations proceed in an orderly fashion.

REFERENCES

Forest Biometrics Research Institute, 2014. Forest Projection and Planning System (FPS). Forest Biometrics Research Institute, Portland, OR. https://www.forestbiometrics.com/software/fps/ (Accessed March 3, 2014).

Limm, E.B., Simonin, K.A., Bothman, A.G., Dawson, T.E., 2009. Foliar water uptake: a common water acquisition strategy for plants of the redwood forest. Oecologia 161, 449–459.

Chapter 36

Northwest Oregon State Forests, United States of America

Robert Nall[1] and John Sessions[2]

[1]Oregon Department of Forestry, Salem, Oregon, USA, [2]College of Forestry, Oregon State University, Corvallis, Oregon, USA

ABBREVIATIONS

BOF	Oregon Board of Forestry
FPA	Forest Practices Act
GPV	Greatest Permanent Value
IPs	Implementation Plans
NW FMP	Northwest Oregon State Forest Management Plan
OAR	Oregon Administrative Rule
SBM	Structure Based Management

MANAGEMENT SETTING AND BACKGROUND

The Northwest Oregon State Forest Management Plan (NW FMP)[1] was adopted in 2001 and revised in 2010. It covers approximately 610,193 acres (ac) (246,942 hectares (ha)) of forests managed by the State Forests Division of the Oregon Department of Forestry for the Oregon Board of Forestry (BOF) and the State Land Board. Most of these forests occur in the northwest corner of the state and are divided into three designated state forests (Clatsop (22%), Tillamook (60%), and Santiam (8%)) and are administered by four districts, while the remaining lands (10%) are composed of scattered parcels administered by two districts (Figure 36.1).

The lands owned by the BOF (98% of the planning area) were acquired from counties beginning in the late 1930s and continuing through 1950s. The counties had originally acquired the lands during or shortly after the Great Depression of the 1930s, primarily due to default of taxes. At that time, the counties did not have the resources to maintain or manage the lands, so state legislation was passed that enabled the counties to deed these lands over to the BOF in exchange for a portion of all future revenues derived from these lands. Legislation requires that the lands acquired by the BOF be managed to "secure the greatest permanent value (GPV) of such lands to the state" (ORS § 530.050).

The lands owned by the State Land Board (2% of the planning area) were acquired by the state from the federal government through the Oregon Admission Act of 1859 (11 Stat. 383) for the purpose of supporting public education. These lands were placed under the jurisdiction of the State Land Board, which is comprised of the State Governor, State Treasurer, and Secretary of State. The SLB's objective for these forests is to provide income to the State's Common School Fund consistent with sound land management techniques. To meet these objectives, the State Land Board has contracted with the Oregon Department of Forestry to manage the forests lands under an agreement that all revenue derived from these lands go to the Common School Fund, and the State Land Board reimburses the Oregon Department of Forestry for the cost of management.

---

1. The federal "Northwest Forest Plan" covers the federal lands in a larger overlapping region, and it is sometimes confused with the NW FMP. These two plans are completely separate; they apply to different lands and have no policy interactions.

Forest Plans of North America. http://dx.doi.org/10.1016/B978-0-12-799936-4.00036-9

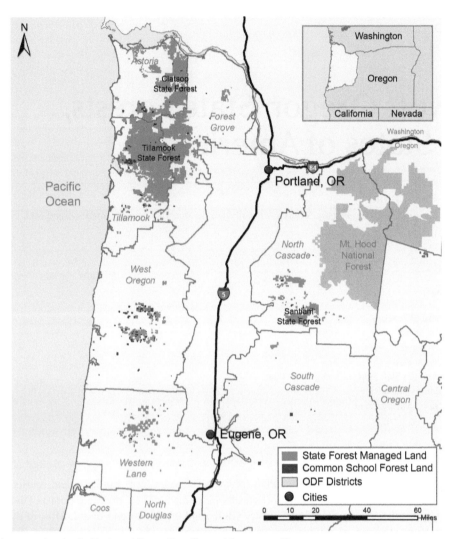

FIGURE 36.1   Lands managed under the Northwest Oregon State Forests Management Plan.

The BOF in Oregon is composed of a seven member board whose members are appointed by the governor and approved by the State Senate. The BOF supervises all matters of forest policy within Oregon, appoints and supervises the State Forester, and adopts rules regulating forest practices (Private Forests Division) and wildfire protection (Protection from Fire Division) on state and private lands. The BOF provides policy direction for the State Forests Division and adopts forest management plans as Oregon Administrative Rules (OARs), but it is prohibited by statute from setting annual timber harvest objectives.

The State Forests Division is a self-supporting government entity with 98% of its funds derived from timber sale receipts on BOF lands (36.25% of the net revenue) plus reimbursement of costs from the State Land Board (Oregon Department of Forestry, 2012). The remaining funds are obtained through competitive grants, licensing fees for off-highway vehicles, campground fees, and sales of miscellaneous forests products, such as ferns, moss, and other greens for the florist. Today, the State Forests Division harvests approximately 240 MMBF (566,400 m³) per year state-wide with net revenue of $65–$75 million U.S. dollars (USD) (after accounting for logging and road construction costs). From these revenues, the Oregon Department of Forestry distributes about $37–$46 million USD per year to the beneficiary counties.

Under the NW FMP, the State Forests Division uses its share of the revenue to plan and prepare harvest operations on 10,187 ac (4,123 ha) including surveys for threatened and endangered species at a cost of approximately $2,000,000 USD per year, manages 4,459 miles (mi) (7,175 kilometers (km)) of road, plants 2,495,500 seedlings, operates 16 campgrounds, and manages 618 mi (994 km) of recreational trails annually (Table 36.1).

Most of the forests are located in the Coast Range Mountains, between the Willamette Valley and the Pacific Ocean, with elevations ranging from a few hundred feet to more than 4,000 feet (ft) (1,219 meters (m)). The Coast Range is a relatively

TABLE 36.1 Key Resource Statistics and Activities by District for Northwest Oregon State Forests

Resource or Activity	District						
	Astoria	Forest Grove	North Cascade	Tillamook	Western Oregon	Lane	Total
Area (acres)	136,835	114,987	47,626	250,462	35,224	25,059	610,193
Standing inventory (MMBF)[a]	3,875	3,160	1,154	5,200	931	832	15,152
Riparian Zones (acres)[b]	13,018	9,886	3,527	32,271	2,486	1,318	62,506
NSO areas (acres)[c]	7,803	5,107	10,521	26,429	7,707	14,583	72,150
MM management areas (acres)[d]	1,737	–	–	5,385	3,952	1,610	12,684
Volume harvested annually (MMBF)[a]	73	61	19	47	12	6	218
Trees planted annually	659,000	508,000	166,750	925,500	79,000	90,750	2,429,000
Road system (miles)	968	878	480	1,676	318	139	4,459
Campgrounds	3	5	4	4	–	–	16
Off-highway vehicle trails (miles)	15	71	5	400	5	–	496
Nonmotorized trails (miles)	21	60	10	21	10	–	122

[a] *Volume in Scribner Board Feet.*
[b] *Riparian Zones are based on the Stream Bank and Inner Zone as defined in the NW FMP and extend from the stream edge to 25, 50, or 100 ft (7.6, 15.2, or 30.5 m) depending on the individual streams attributes.*
[c] *Northern spotted owl areas are state forest lands that are within the provincial circle (either 1.5 or 1.2 mi (2.4 or 1.9 km) radius) around the activity center of northern spotted owl pair or resident single, so "take-avoidance" policies specific to this species apply.*
[d] *Marbled murrelet management areas contain stands where surveys have determined to be occupied by marbled murrelet, so "take-avoidance" policies specific to this species apply.*

young geologic formation with steep slopes that are unstable in many locations. The region experiences mild, wet winters, with annual precipitation exceeding 200 inches (in) (5,080 mm) in some locations, while the summers are warm and generally dry. The Santiam State Forest is located east of the Willamette Valley in the foothills of the volcanic Cascade Mountains. The elevations of this forest range from 1,000 ft (305 m) to over 5,000 ft (1,524 m).

Logging of the lands that would become state forests began around 1900, but after the lands were logged they had little value to the landowner, although some were converted to small homesteads. Other areas lost their value due to wildfires. As we noted earlier, during the Great Depression of the 1930s, many properties became tax-delinquent and were foreclosed on by the counties. These lands were eventually transferred to the BOF as described under Management Setting.

The largest single parcel (255,000 ac (103,197 ha) or 42% of the planning area) of the state forests covered by the NW FMP came from an area called the "Tillamook Burn" that was created by a series of four overlapping fires that occurred every 6 years from 1933 through 1951 covering a total of 364,000 ac (147,309 ha). A total of 7.5 billion board feet (17.7 million m³) of fire-killed timber was salvage logged between 1933 and 1955 with much of it used to meet the lumber demands of World War II.

After the "Tillamook Burn" came into state ownership, Oregonians passed a bond measure in 1948 to reforest the burn. Between 1948 and 1973, contractors, inmates, and volunteers (including school children) planted 72 million Douglas-fir (*Pseudotsuga menziesii*) seedlings, and 36 tons of Douglas-fir seeds were aerially spread by helicopter. While most of the lands were burned over or under-productive when acquired, they did contain some mature forest that could be harvested, so the State Forests Division started developing forest plans in the early 1950s that included an *allowable cut* calculation based on the current forest inventory. These early plans were very timber-oriented with little recognition of other forest resources. In 1971, Oregon enacted the Forest Practices Act (FPA) that regulated harvesting and other forest management activities on state and private lands. The management plans for state forests grew in complexity from 1971 onward, with extensive economic analysis of rotation lengths and reforestation activities. These plans explicitly recognized protection measures for streams, wildlife habitat, and other areas deserving of special conservation measures. These plans also saw the first use of a computer simulation model to determine harvest objectives.

The listing of the northern spotted owl (*Strix occidentalis caurina*) and marbled murrelet (*Brachyramphus marmoratus*) as threatened and endangered species under the federal Endangered Species Act (ESA) in the 1990s, along with an increase in public interest in state forest management, resulted in the need for a far more complex forest planning process addressing many more resources than just timber.

PLANNING ENVIRONMENT AND METHODOLOGY

Simultaneous to embarking on development of a new forest management plan for Board of Forestry lands in northwest Oregon, the BOF determined that the complexities of modern forest management required a clearer policy foundation than a single sentence in statute (i.e., "secure the GPV of such lands to the state"). The BOF spent several years developing and adopting a series of OARs to provide that policy foundation. The most important of these is known as the GPV Rule adopted as ORS 629-35-20 in 1998 that elaborates on the statement "secure the GPV of such lands to the state" to mean "healthy, productive, and sustainable forest ecosystems that over time and across the landscape provide a full range of social, economic, and environmental benefits to the people of Oregon." The GPV Rule directs the State Forester to secure the GPV through "active management in a sound environmental manner to provide sustainable timber harvest and revenues to state, counties, and local taxing districts."

The GPV Rule has been described as placing the management goal and priorities of state forests somewhere between those of private industrial landowners and those of federal lands. Based on the GPV Rule, a number of guiding principles were identified for developing the NW FMP that includes direction that the plan will:

- Be a comprehensive, integrated forest management plan that will cover many forest resources ranging from air quality and cultural resources to water quality and wildlife habitat, and will also meet the requirements of the state and federal ESA
- Recognize the importance of ecosystem restoration and watershed health
- Be developed within the context of managed forests and will contribute to the timber supply for present and future generations
- Recognize the financial interests of the counties and the State Land Board
- Examine opportunities to achieve goals through cooperative efforts with other agencies and groups
- Seek input from a diverse group of stakeholders during its development
- Commit the State Forests Division to using monitoring and research as well as adaptive management to ensure that the best available knowledge is used to manage forest resources

Constraints and Opportunities for the Planning Area

There are a number of constraints that affect the planning area. Some constraints are related to the forest itself and its composition, some are related to economics and the costs of conducting forest management in the Coast Range, and some are related to the requirements for environmental protection (discussed under Regulatory Environment).

When the NW FMP was initially adopted in 2001, the forests had an age-class distribution that made it difficult to meet the state's management goals, with almost 56% of the forest area between the ages of 26 and 55, while 21% of the area was between 56 and 85 years old, and only 5% of the area over age 85. This age-class distribution would have been relatively easy to implement a timber management strategy based on the culmination of mean annual increment, however, in the short-term it was not well suited for an integrated forest management plan where ecological restoration and the development of habitat for late-seral species were goals. However, the relatively young forest presented a great opportunity where silvicultural interventions (partial cutting) could be performed to produce revenue and to develop more complex forest structure.

In addition to the forest being relatively young, with most of the stands occurring in a very narrow range of ages following reforestation efforts associated with the Tillamook Fires, this massive reforestation effort was only marginally successful in some portions of the forest. Significant areas in the western half of the Tillamook State Forests are either poorly stocked or are stocked with low-value species.

The topography of the Coast Range is a significant constraint on conducting management activities on the lands covered by the NW FMP. About one-quarter of all state forest lands are on slopes greater than 65%, while almost half of the Tillamook District slopes are greater than 65%. As a result, more than 70% of the planning area must be harvested using the more expensive cable logging systems, and about 8% of the area cannot be logged at all primarily due to topographic considerations (i.e., roads cannot be built to access the necessary landings). The steep slopes also drive up road construction costs.

Guiding Laws, Regulations, and Policies

Management of state forests is influenced by two levels of external regulations: state-wide statutes and rules (e.g., Oregon Forest Practices Rules (FPA) and Oregon Endangered Species Act), and federal statues and rules (e.g., the federal ESA and Clean Water Act). The FPA was adopted in 1971 and regulates forest management activities on all state and private forest lands in Oregon, including logging, road construction, reforestation, and herbicide applications. The FPA gives exclusive authority to the BOF to develop and enforce rules affecting forest management activities, while consulting with other state and local governments concerned with the forest environment. The State Forests Division is responsible for administering the FPA and associated rules on state forests lands with advice from the Private Forests Division. In some cases, such as road construction and maintenance, the State Forest Division's management strategies rely heavily on the FPA, but for other resources, such as aquatic and riparian habitat, the Division develops its own more conservative management strategies in order to attain the "GPV." One of the most operationally significant portions of the FPA is its limit on the size of clear-cut harvest units. Clear-cut harvest units cannot exceed 120 ac (49 ha) until the trees planted in the adjacent clear-cut unit are 4 years old or 4 ft (1.2 m) tall. If this condition is not met, then a 300 ft (91 m) strip of uncut trees must be left between the units.

The FPA is the method by which the forest activities in Oregon comply with the federal Clean Water Act. The OARs based on the FPA also provide for the protection of certain sensitive species, including the northern spotted owl, however, the rules clearly state that they are site protection strategies and landowner and operators should contact the U.S. Fish and Wildlife Service or National Marine Fisheries Service regarding compliance with the federal ESA. Since the FPA does not provide rules as an avenue for federal ESA compliance, the State Forests Division develops its own policies for complying with the law. Currently, the state forests covered by the NW FMP contain two threatened birds, the northern spotted owl and the marbled murrelet, as well as several threatened anadromous fish species (salmon (*Oncorhynchus* spp.) and steelhead (*Oncorhynchus mykiss*)).

The State Forests Division started conducting surveys for the northern spotted owl in 1989 and for the marbled murrelet in 1992. There are currently 90 northern spotted owl sites on or adjacent to state forests, affecting management and contributing to owl protection on approximately 72,150 ac (29,199 ha). There are 166 designated marbled murrelet management areas contributing to murrelet protection on approximately 12,684 ac (5,133 ha). The State Forests Division has internal policies for addressing the Section 9 requirement that nonfederal landowners "avoid take" of listed species. The strategies reserve sufficient habitat around sites occupied by either of these species to avoid disturbance or harm to them. Guidelines for protecting these sites are established through policy, but they are applied to specific sites through a Biological Assessment written by one of the Division's wildlife biologists. The Biological Assessments for the northern spotted owl sites and marbled murrelet management areas are reviewed by wildlife biologist at the Oregon Department of Fish and Wildlife, and then they are sent to the U.S. Fish and Wildlife Service for review. "Take-avoidance" policies for salmon and steelhead are achieved through the riparian management strategies written directly into the Division's FMPs.

The BOF and State Forests Division have explored forests certification through Forest Stewardship Council and Sustainable Forestry Initiative several times over the last 15 years. Analyses to date have not shown that certification will improve the environmental, social, and economic outcomes on state forests. There are several primary reasons that led to this conclusion:

- The State Forests Division already bases its management strategies on the best available science, so certification through either Forest Stewardship Council or Sustainable Forestry Initiative is unlikely to change or improve management strategies or improve environment outcomes
- There is a cost to participate in these certification programs that would reduce the Division's discretionary budget and reduce funds available to invest in activities such as stream enhancement projects and recreation facilities
- Certification does not appear to result in an increase in stumpage values in northwest Oregon, so there would be no economic gain
- Certification of state forests has not been a priority for the stakeholders and interest groups of state forest lands
- Staff time and resources have been dedicated to helping the Board of Forestry consider different management strategies for state forest lands instead of certifying existing strategies

In addition to the GPV Rule, the BOF adopted a Forest Management Planning Rule that directs the State Forester to develop forest management plans that are based on the "best available science" and contain the following elements: guiding principles, description and assessment of state forest resources, resource management goals, and resource and asset management strategies. The forest management plans also contain guidelines for research and monitoring and adaptive management. A significant requirement of the rules is that the BOF must make a determination that the forest management plan meets "the obligation to secure the GPV" as defined in the GPV Rule prior to adopting the proposed forest management plan as an OAR. Since the FMP is formally adopted as administrative rule, it must undergo an extensive process that includes an economic analysis, a public comment period, public hearings, and a "hearings officer" report addressing the issues and concerns identified by the public. The Forest Management Planning Rule requires the BOF to review forest management plans at least once every 10 years.

In addition to the Forest Management Planning Rule's requirement for broad, strategic long-range forest management plans, it also requires short-term Implementation Plans (IPs) and Annual Operations Plans. The IPs are specific to each district and describe how the strategies of the FMP will be applied to that specific district and the forest management activities (harvesting, reforestation, recreation, etc.) the district intends to carry-out to achieve the goals of the FMP. The annual harvest objective, represented as volume and area, for the district is identified through the development of the IPs. The draft IPs undergo a 45-day public comment period. The State Forester then approves the IPs after determining that they achieve the goals of the FMP.

The Annual Operations Plans are, as the name implies, prepared annually and describe the specific forest management operations that will occur during the fiscal year and how they will achieve the objectives of the IP. The Annual Operations Plans describe the specific timber operations to be prepared and sold, roads to be constructed and maintained, sites to be planted, trails to be built, surveys to be conducted, and all other forest management activities. The plans also undergo a 45-day public comment period. The district forester signs the Annual Operations Plans indicating that he has found that activities described are consistent with all statutes, rules, plans, and policies. The State Forests Division has a number of other plans and policies that serve a variety of functions:

- The northern spotted owl and marbled murrelet policies described in the "Regulatory Environment" are designed to address the ESA and other laws.
- There are policies for *Aquatic Anchors* and *Terrestrial Anchor Sites* that are designed to fulfill the requirements of the NW FMP.
- There are many policies designed to assist in the conduct of forest activities including policies for reforestation and young stand management, forest roads and engineering, cultural resources, recreation, and invasive species.

NW FMP Development Methods

Prior to the adoption of the GPV Rule, state forests were under what could be called a timber management plan that was based on even-aged silvicultural systems where the final clear-cut harvest age was set at the culmination of mean annual increment, and most other forest resources were managed according to the FPA. With the adoption of the GPV Rule, the State Forest Division needed to develop an entirely new forest management approach. Although development of the GPV Rule had just begun, scoping for a new management strategy for state forests began in 1994 with a series of public meetings in northwest Oregon to begin development of the guiding principles upon which the plan would be based. Over the next 6 years, the Division continued a very strong commitment to public involvement in the planning process that included numerous public meetings, field tours, a full color newsletter, a web site featuring the planning information, and a citizen advisory committee.

As the plan's strategies developed, it became apparent that obtaining an Incidental Take Permit and Habitat Conservation Plan (HCP) from the U.S. Fish and Wildlife Service may be beneficial to the achievement of the plan's goal for achieving a balance of economic, social, and environmental outcomes. The major benefit of obtaining the HCP was greater predictability and stability in planning activities, rather than needing to respond to each new listing of a species under the ESA or identification of an occupied site. In 1996, the Division initiated the development of the *Western Oregon State Forests Habitat Conservation Plan*. Development of this plan continued for more than 10 years before it was determined by the BOF that the plan would not improve the environmental, social, and economic outcomes on state forests and, thus, was discontinued.

The State Forests Division worked closely with the Oregon Department of Fish and Wildlife and the U.S. Fish and Wildlife Service in the development of the HCP. In 1998, the Division also established an 11-member Public Interest Committee to define and attempt to resolve the scientific, technical, and policy issues associated with the draft HCP. The members represented various interest groups including recreation, environmental, fishing, timber, and counties with forest trust lands. In addition, the Department also initiated an Independent Scientific Review (Hayes, 1998) of the draft HCP and draft NW FMP to receive a critique from a diverse group of scientists on the scientific underpinnings of the strategies identified for the plans. The Independent Scientific Review team consisted of 27 scientists from a variety of universities, institutes, and research stations across the United States.

As the goals and strategies of the draft NW FMP were coalescing in 1999, the State Forests Division determined it needed a harvest scheduling model to analyze the environmental, social, and economic effects of implementing the plan's strategies and to compare those effects with the outcomes of alternative forest management scenarios. The State Forests Division partnered with Oregon State University to develop this model. The harvest scheduling model was developed using goal programming with a spatially explicit heuristic program that projected outcomes in twenty 10-year periods (Sessions et al., 2006). Goal programming was used to avoid infeasibilities during solution and facilitate sensitivity analysis. The spatially explicit heuristic model was necessary because of the 120 ac (49 ha) clear-cut rule in the FPA and the size of the problem (the largest district contained over one million binary decision variables). Eventually, nine management scenarios were analyzed and presented to the BOF: five variations of the proposed plan, two industrial scenarios (based on the FPA), a reserve-based scenario, and a "grow-only" scenario.

OUTCOMES OF THE PLAN

In January 2001, the BOF adopted the NW FMP after finding that it met the obligation of attaining GPV. The final plan is 580 pages in length, including appendices. The NW FMP contained management strategies for 14 different resources. Some of these management strategies have very site-specific requirements, but do not affect very much area (e.g., all of the cultural resource sites in the planning area do not accumulate to 100 ac (40.5 ha)), while others have led to an evolution in management practices (e.g., broadcast burning of clear-cuts has been largely replaced by other methods of slash disposal to reduce impact on air quality). The four strategies that direct management under the NW FMP include: Structure Based Management (SBM), Landscape Design, Aquatic and Riparian, and Species of Concern.

The primary management strategy of the NW FMP can be described as a SBM approach that was developed based on the concepts presented by Oliver and Larson (1996). SBM is a forest management strategy that focuses on developing the historic array of stand structures across the landscape through active management (thinning and clear-cut harvest). The intent SBM was to develop an integrated management plan that emphasized the compatibility of forest values over time and across the landscape as opposed to the "either-or" debate that has been dominating the management of federal lands in the Pacific Northwest (Bordelon et al., 2000) for the past 20 years.

The SMB approach describes five structure types and, over time, will progress from the simple to the more complex structure types. In addition, the NW FMP set long-term goals as a percentage of the landscape for each of these stand structures; however, it did not set a specific timeline for achieving the goals.

1. Regeneration—Young, early serial stands generally less than 10–15 years old (long-term goal: 15–25%).
2. Closed Single Canopy—Young stands where the crowns of the trees have grown together and are preventing sunlight from reaching the forest floor resulting in little understory vegetation. These stands are generally 15–40 years old, but they can be older (long-term goal: 5–15%).
3. Understory—As stands grow and mature, the crowns open up due to a reduced number of trees (either through natural mortality or harvesting) and the understory vegetation, including tree seedlings and saplings, begins to develop and occupy the site (long-term goal: 30–40%).
4. Layered—Older, more complex forest stands that have multiple layers of tree canopies and well-developed understory vegetation. Many variables play a role in how quickly layered stands develop, but in the planning area, stands generally develop into this structure classification between ages 60 and 90 (long-term goal: 15–25%).
5. Older Forest Structure—This is the most complex of the structures and serves SBM's surrogate for "Old Growth" forest. In addition to having a complex, multi-layered forest canopy, these stands contain at least eight trees per ac (20 trees per ha) that have a diameter at breast height that exceeds 32 in (81.3 cm), large snags, and a large volume of down wood (long-term goal: 15–25%).

Figure 36.2 is a view of an understory stand where the overstory composed of Douglas-fir, has been commercially thinned in order to promote layering by releasing the understory of western hemlock (*Tsuga heterophylla*). In addition to advancing the complexity of the stand, the operation provides significant revenue to the beneficiaries.

The plan also includes a Landscape Design Strategy that requires the districts to identify where they intend to create these structures in the forest in order to create a variety of patch sizes and connectivity corridors benefiting native wildlife. This landscape design mapping focused on the Layered and Older Forest Structure (generally referred to as "complex" structure), because they are in the shortest supply, require the most time and effort to create, and provide habitat for the threatened northern spotted owl and marbled murrelet. The mapping process was completed through the development of the district IPs, another requirement of the NW FMP. The district IPs would also identify a more refined goal for each of the stand structures based on each district's forest, wildlife, topographic, and other site-specific conditions.

The Landscape Management Strategies require maintaining or creating structural components for wildlife habitat across the landscape. The structural habitat components include: maintaining a diversity of trees species, leaving at least two snags

FIGURE 36.2 Photo of commercial thinning to promote layering within areas of the Northwest Oregon State Forests.

per ac (5 per ha) across the landscape, retaining five green trees per ac (12.4 per ha) in clear-cuts, and retaining 600–900 ft^3 per ac (42-63 m^3 per ha) of recent down wood in clear-cuts. The NW FMP also prohibited the cutting of any remaining old-growth stands or individual trees, where operationally feasible.

The Aquatic and Riparian Strategy is the third major component of the NW FMP, which has the explicit goal of providing "properly function aquatic habitats" as required by the GPV Rule. While this strategy is more restrictive than the FPA riparian rules, it is not nearly as restrictive as the riparian management strategies applied on federal lands in the region. The Aquatic and Riparian Strategy creates six stream classifications with three management zones. The Stream Bank Zone (0–25 ft (0–7.6 m) from the stream) is a no-harvest zone on five of the six classifications. The Inner Zone (25–100 ft (7.6–30.5 m)) is designed for limited active management and has the goal of creating a mature forest condition along all fish bearing streams and along the medium and large non-fish-bearing streams. The Outer Zone (100–170 ft (30.5–51.8 m)) is an area of active management designed to provide some additional protection, especially on stream segments where the Inner Zone is deficient of trees.

The Aquatic and Riparian Strategies provide protection standards for lakes, ponds, and wetlands of various sizes and prohibit harvesting on the inner gorges of streams. The Aquatic and Riparian Strategies include a strong emphasis on aquatic habitat restoration projects. The projects consist of placing logs in streams to create fish habitat, and are carried out under the direction of a fish biologist from the Oregon Department of Fish and Wildlife. Other projects include removal of legacy roads and the replacement of road culverts that block fish passage. A large part of the Aquatic and Riparian Strategies include the development and implementation of a comprehensive watershed assessment process.

The Species of Concern components in the NW FMP are intended to address "individual species of concern including salmonids, northern spotted owls, marbled murrelets and other sensitive species." The NW FMP, as adopted in 2001, did not contain the details of these strategies, but referred to the draft HCP that was being developed at the time. However, the NW FMP did state "if the HCP is not adopted, this forest management plan will be expanded to include further detail on strategies for managing the habitats of threatened endangered or sensitive species." Negotiations with the U.S. Fish and Wildlife Service and National Marine Fisheries Service on the draft HCP continued for several more years. These negotiations were delayed for several years in order to work on revising an HCP for the Elliott State Forest and to develop a more advanced harvest model to analyze the outcomes of adopting or not adopting the HCP. After additional analysis, it was finally determined by the BOF that adoption of the HCP would not improve the environmental, social, and economic outcomes on state forests. The decision to not adopt the HCP led to the NW FMP being revised in April 2010, to the development of Aquatic Anchor strategy to address the needs of salmonids, and to the Terrestrial Anchor Site strategy to address the habitat needs of northern spotted owls, marbled murrelets, and other late-seral dependent species.

DISCUSSION AND CONCLUSIONS

The 2001 and 2010 revisions of the forest plan for state forests in northwest Oregon need to be managed in a way that achieves the GPV to the people of the state. The revisions incorporated new strategies for the management of terrestrial and aquatic resources of concern for certain wildlife and fish species.

Learnings and Insights

After a 6-year process and a significant public involvement process, the BOF finally adopted the 2001 NW FMP by consensus. The long process enabled the Division to build broad support for the plan among a variety of stakeholders. However, to build a plan with broad support, the plan had to be written with very broad, high-level goals and strategies. The plan did not include any firm direction on the most controversial issue of any forest planning project: harvesting. As a result, many of the stakeholders were able to read the same plan and see something that fit their objectives. Some stakeholders saw a plan that provided high harvest volumes and adequate resource protection, while others saw a plan that provide high resource protection and some harvest volume. The broad support for the plan rapidly eroded when the harvest levels where identified in the draft IPs. Both groups—those wanting high harvest levels and those wanting high resource protection—were disappointed by the results. After the IPs were approved, some stakeholders were so displeased with the results that they attempted to change the management of state forests through the legislature and through the ballot box. However, none of those attempts succeeded. Since the approval of the IPs in 2003, the Division has been in almost constant discussions with the stakeholders regarding the appropriate harvest levels.

The broad strategies in the plan were also designed to provide the field foresters with a high level of flexibility to implement the NW FMP using sound forest management principles and professional judgment. While this flexibility had many benefits for the implementation of the plan, it also had some drawbacks and often led to situations where the intent of the plan was unclear to the foresters. For example, the NW FMP has a goal of creating complex forest structures on a significant portion of the landscape, but it does not provide a clear goal for those stands that are not destined for complex structure. As a result, the silvicultural pathway is unclear: should a stand be thinned once or twice, and should it be clear-cut at age 40, 60, or 80 years?

Sustainability Issues

The Division has been implementing the NW FMP for 12 years with some variations in the annual harvest objectives and other policies as new information becomes available. Given current information and modeling, the harvest objectives appear to be sustainable over the long term. It also appears that the NW FMP's long-term structure goals are attainable; however, it is anticipated to take between 60 and 100 years to achieve the goals for Older Forest Structure. Figure 36.3 illustrates the changes in the estimate of stand structure over the planning area between 2003 and 2013. The figure shows significant increases in the Layered and Understory structures with a decrease in the Closed Single Canopy structure. These changes are largely due to an aggressive partial cut harvest objective carried out since the implementation of the

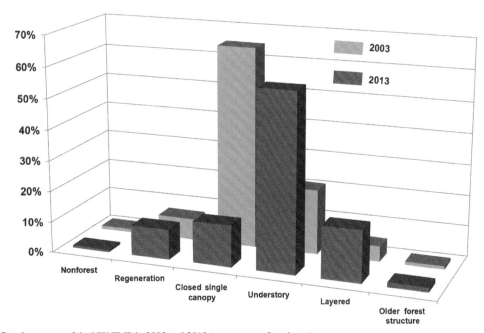

FIGURE 36.3 Stand structure of the NW FMP in 2003 and 2013 (percentage of total area).

NW FMP. There has been little change in the estimated amount of Older Forest Structure, but a 10-year period is too short of a time period for the development of Older Forest Structure. The data that this chart is based on should be used with caution, because the forestry inventory system and the method of identifying stand structure changed between 2003 and 2013.

Adaptive management is one of the key strategies of the NW FMP to ensure that the plan is sustainable over the long term. This strategy asks the questions "are the goals of the plan being achieved?" and "if the goals are not being achieved, what needs to change in order to achieve those goals?" Since plan adoption in 2001, there have been a number of projects developed to monitor the implementation and effectiveness of the NW FMP. These projects are:

1. *Second Party Assessment of the NW FMP*: The study conducted by an external consulting firm sought to evaluate the effectiveness of the plan's strategies to meet its goals and their implementation. The general findings of the study include: Division staff is using the strategies in the NW FMP; management is moving the plan's stand structure goals and meeting the riparian area guidelines; and some improvement could be made regarding road and recreational trail impacts to water quality (Rockwell et al., 2006).
2. *Implementation Monitoring Report*: This internal study examined the implementation of the strategies within the NW FMP were being properly understood and applied. The study showed that staff did understand the intent of the strategies; there was broad compliance with the upland and riparian strategies with a few areas of over- or under-achievement; and the plan's goals will require significant investment to be achieved (Hayes, 2010).
3. *Riparian Effectiveness Study* ("Ripstream" study): This study included a Before-After-Control-Impact design and included several years of pre- and postharvest data. Stream temperature data was the first data set to be analyzed. Results indicate that the state forests strategies are effective at meeting Department of Environmental Quality water quality standards for temperature (Dent et al., 2008; Groom et al., 2011).
4. *Commercial Thinning and Wildlife Study*: This study examined the hypothesis that thinning would create complex forest structure and lead to the colonization of these stands by birds associated with these structures (the "field of dreams" hypothesis). Findings generally indicate support for the hypothesis (Cahall et al., 2013).

There is one significant question regarding sustainability of the NW FMP and that is the financial sustainability of the State Forests Division under this plan. Until the Great Recession of 2009, the Division was operating with a substantial financial reserve. During the recession, the reserve was lost and the State Forests Division reduced staff by approximately 20% and made drastic cuts to its inventory, recreation, and research programs. Since the recession, the Division's finances have stabilized to some extent, but they are still fragile. The State Forests Division has been able to fill some of the vacant positions and increase investments in the forests, however, it is estimated that the State Forests Division is underfunded by $4–$5 million (USD) a year. In addition, the Division only has a 6-month operating reserve. The State Forests Division is currently undertaking several projects to address this issue, including exploring potential adaptations to the current NW FMP that would result in it being more financially viable.

Plan Development Challenges

The State Forests Division faced four main challenges in developing the NW FMP: simultaneous development of administrative rules; significant public interest in the plan; identification of plan strategies that are supported by science; and shifting the culture of the Division's staff from a primary timber focus to broader forest resource focus. Clearly, developing the OARs, especially the GPV Rule, that governs the forest management plan at the same time the plan was being developed showed the whole process down. First, from a basic staff workload perspective, and because changes to the draft GPV Rule required changes to the forest management plan strategy. The strong public interest in the plan and the Division's commitment to involving the public in the planning process required significant staff time to meet with stakeholders and to negotiate the details of the GPV Rule and the NW FMP goals and strategies with them.

When it was proposed, the SBM strategy was a new concept for managing a large forest, so many aspects of the strategy had to be developed and measured against existing science, where available. The Aquatic and Riparian Strategies incorporated into the NW FMP required significant scientific review. Both of these evaluated through an independent scientific review that was described earlier. Even today, some people consider SBM and the NW FMP's Aquatic and Riparian Strategies to be experimental.

The final challenge to developing and implementing the GPV Rule and the NW FMP was to change the culture of the State Forests Division's staff. Prior to the development of SBM and the NW FMP, the Division's staff had a common forest management vision that was based on meeting the FPA and on clear-cutting stands at a specific rotation age. The staff was well trained and experienced to achieve this vision. The GPV Rule and the strategies of the NW FMP constituted a major change in direction for the State Forests Division. It took several years of training, field tours, meetings, and conferences to ensure that State Forests Division staff fully understood, accepted, and supported the new direction.

Plan Implementing Challenges

It can be said that the approval of the NW FMP was the end of the beginning for planning on state forests in northwest Oregon. The approval of the NW FMP led to a great deal of additional work to be completed:

- The IPs with landscape designs and harvest objectives were drafted and approved
- A new forest inventory was developed that collected the additional data necessary to implement SBM, such as information on understory vegetation, snags, and down wood
- Watershed assessments were conducted across most of the land base
- Silvicultural prescriptions were developed and evaluated for the potential to move stands to more complex structures

These projects were explicit, or at least anticipated requirements of the forest management plan and took years to implement. The IPs were not approved until March 2003, while it took until late 2004 to collect enough new inventory data for it to be widely useable. Work on watershed assessments continued through 2007. The evaluation of the silvicultural prescriptions for developing stand structure is still in progress.

The adoption of the NW FMP also resulted in some unanticipated projects, such as the development of definitions of harvest types, numeric definitions of stand structures, and the development of a more robust harvest model.

The SBM strategy results in harvesting where the line between clear-cuts and partial cuts becomes very blurred. After a couple of years of implementing the plan, the State Forests Division received comments from representatives of both forest industry and conservation groups that requested that the State Forests Division develop a set of harvest type definitions that provides a better description of the harvest activities taking place on state forests. The State Forests Division developed definitions for three types of regeneration harvests and six types of partial cuts that describe the harvest activities based on the residual stand condition. These definitions only describe the harvests; they do not prescribe where the harvests can occur. The NW FMP provided good functional definitions of SMB's five stand structure types. However, those definitions were inadequate for field identification at the stand level and were impossible to apply to a forest inventory. The Division developed, over a period of several years, a numeric algorithm for identifying the stand structures through the Division's newly develop forest inventory. The most difficult aspect is the identification of the layering component associated with the Layered and Older Forests Structure. After several revisions, this process is now providing consistent and credible results.

The identification of the appropriate harvest objective was the most difficult aspect of implementing the NW FMP. The initial IPs did not use the harvest model developed with Oregon State University in 1999, because that model was designed to compare alternative forest management plans rather than identify specific harvest objectives. Instead, the harvest objectives in the IPs were based on a simple area control harvest calculation. These harvest objectives generated a great deal of controversy from various groups. The counties, who are the primary financial beneficiary of these lands, and forest industry thought the harvest objectives were too low, while various environmental groups thought the harvest objectives were too high.

When the State Forester approved the initial IPs in 2003, he initiated a separate 3-year project to develop a more robust harvest model for state forests with the goal of providing a definitive method of identifying harvest objectives and assisting in the analysis of the proposed HCP. This project became known as the *Harvest and Habitat Model Project* and published its final report in 2006 (Oregon Department of Forestry, 2006). This project continued model development with Oregon State University and resulted in a number of significant improvements, including:

- More accurate yield tables with a wide array of silvicultural prescriptions generated by the Forest Vegetation Simulator (Dixon, 2013) and based on the Division's new forest inventory
- Incorporation of realistic harvest units within the model and coordination of thinning in the Riparian Zone of a harvest unit when the upland area is harvested
- Integration of a transportation system into the model where log hauling, road maintenance, and road construction is tied directly to the harvesting of specific units
- Inclusion of three northern spotted owl population scenarios to inform the decision on adopting the draft HCP

The new model proved to be very useful in establishing harvest objectives for the districts that will result in a desired amount of complex structure (Layered and Older Forest Structure). The analysis on adoption of the draft HCP proved inconclusive, largely due to the highly variable results produced by the northern spotted owl population scenarios. A third-party assessment of the model found that model was useful for identifying harvest objectives, but a number of improvements would be required before it produced reliable estimates on habitat (Marmorek et al., 2006). The State Forests Division incorporated many of the recommendations from the third-party assessment into the model and has been using it to conduct analysis of various forest management plans and policies, including a proposed HCP, a revision to the NW FMP, revisions to five IPs, evaluation of species of concern policies and landslide hazard policies.

REFERENCES

Bordelon, B., McAllister, D., Holloway, R., 2000. Sustainable forestry Oregon style. Journal of Forestry 98, 26–34.

Cahall, R., Hayes, J., Betts, M., 2013. Will they come? Long-term response by forest birds to experimental thinning supports the "Field of Dreams" hypothesis. Journal of Forest Ecology 304, 137–149.

Dent, L., Vick, D., Abraham, S., Shoenholtz, S., Johnson, S., 2008. Summer temperature patterns in headwater streams of the Oregon Coast Range. Journal of the American Water Resources Association 44, 803–813.

Dixon, G.E., 2013. Essential FVS: A User's Guide to the Forest Vegetation Simulator. U.S. Department of Agriculture, Forest Service, Forest Management Service Center, Fort Collins, CO. 226 p.

Groom, J., Dent, L., Madsen, L., 2011. Response of western Oregon stream temperatures to contemporary forest management. Forest Ecology and Management 262, 1618–1629.

Hayes, I., 2010. State Forests Division Implementation Monitoring Report AOP Years 2002–2006. Oregon Department of Forestry, Salem, OR.

Hayes, J., 1998. An Independent Scientific Review of Oregon Department of Forestry's Proposed Western Oregon Habitat Conservation Plan. Oregon State University, Department of Forest Science, Corvallis, OR.

Marmorek, D., Murray, C., Droessler, T., Dunsworth, G., Monserud, R., Kiester, R., Northway, S., 2006. Scientific Peer Review of H&H (Harvest and Habitat) Model Project Review Results. ESSA Technologies Ltd., Vancouver, BC.

Oliver, C.D., Larson, B.C., 1996. Forest Stand Dynamics. John Wiley and Sons, New York, NY.

Oregon Department of Forestry. 2006. Harvest and Habitat Model project final report. Report presented to the Oregon Board of Forestry on March 8, 2006. Oregon Department of Forestry, Salem, OR.

Oregon Department of Forestry. 2012. State Forest Financial Viability Work Group final report. Report presented to the Oregon Board of Forestry on December 7, 2012. Oregon Department of Forestry, Salem, OR.

Rockwell, W., Ebel, F., Rochelle, J., 2006. Forest Management Assessment Report for the Oregon Department of Forestry's Northwest and Southwest Forest Management Plans. Strategic Resource Systems, Saint Johns, MI.

Sessions, J., Overhulser, P., Bettinger, P., Johnson, D., 2006. Linking multiple tools: an American case. In: Shao, G., Reynolds, K. (Eds.), Computer Applications in Sustainable Forest Management, Including Perspectives on Collaboration and Integration. Springer, New York. pp. 223–238.

Chapter 37

Revelstoke Community Forest—Tree Farm License (TFL) 56, British Columbia, Canada

Randy Spyksma,[1] Cam Brown,[1] Del Williams[2] and Kevin Bollefer[2]

[1]Forsite Consultants Ltd., Salmon Arm, British Columbia, Canada, [2]Revelstoke Community Forest Corporation, Revelstoke, British Columbia, Canada

ABBREVIATIONS

AAC	Annual Allowable Cut
BC	British Columbia
RCFC	Revelstoke Community Forest Corporation
TFL	Tree Farm License
THLB	Timber Harvesting Land Base

MANAGEMENT SETTING AND BACKGROUND

Situated 40 kilometers (km) (25 miles (mi)) north of Revelstoke, British Columbia (BC), Canada, the Revelstoke Community Forest Corporation owns and operates Tree Farm Licence (TFL) 56. A TFL is an area-based tenure on provincial Crown land, which grants rights to manage the land and harvest timber. Approximately 15% of the annual harvest in BC is associated with this type of tenure. TFL 56 covers an area of 119,820 hectares (ha) (296,075 acres (ac)) north of the town of Revelstoke. The TFL is bounded on the west by the Lake Revelstoke Reservoir, on the east by the height-of-land of the Selkirk Mountains, on the north by the Goldstream River, and on the south by the Downie-Carnes height-of-land (Figure 37.1). The land is extremely rugged and dominated by two valleys that run roughly east-west along Downie Creek and Goldstream River, and one north-south valley, that of the Columbia River (Lake Revelstoke Reservoir). The elevation ranges from 573 meters (m) (1,880 feet (ft)) at reservoir level to 3,050 m (10,007 ft) at Carnes Peak. The terrain is dominated by glaciated valley bottoms rising steeply to high alpine (Figure 37.2). Most harvesting is confined to valley sidewalls and valley bottoms. The remaining *high country* is too rugged or does not support marketable timber. The forested land base is a relatively small proportion of total area, and the timber harvesting land base (THLB) is even a smaller proportion still (22,575 ha (55,783 ac), or 19% of the total area). Table 37.1 provides area statistics for the TFL.

The ruggedness of the area has minimized human use; hence, there are no settlements, little private land, and until recently, little recreation use. Recreation now consists of a canoe route on the Goldstream River, mountain biking and hiking trails into the alpine areas, hunting, snowmobiling, and backcountry skiing. A world-class helicopter ski operation also has two backcountry lodges located adjacent to the TFL that use alpine areas and some of the forested terrain in the TFL for ski runs. One highway (Highway 23 N) traverses the TFL. Vehicular traffic is light and dominated by logging trucks and other industrial traffic.

Wildlife utilizes the TFL area extensively. Some species of special interest include mountain caribou (*Rangifer tarandus caribou*), grizzly bears (*Ursus arctos horribilis*), wolverines (*Gulo gulo*), and rare bats. The TFL is also home to many common big game species such as moose (*Alces alces*), mule deer (*Odocoileus hemionus*), mountain goats (*Oreamnos americanus*), black bears (*Ursus americanus*), and wolves (*Canis lupus*), which are hunted by big game outfitters and residents. Mountain caribou have been a major management concern since the 1990s and recent federal government initiatives under the *Species at Risk Act* have increased the focus on this species. The creation of younger forest age classes in the region has enhanced forage availability for moose, leading to expanding populations, which in turn has led to increased wolf populations and increased predation of caribou.

Forest Plans of North America. http://dx.doi.org/10.1016/B978-0-12-799936-4.00037-0

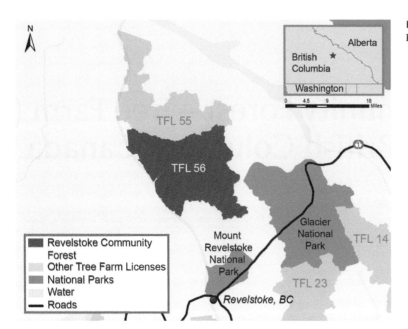

FIGURE 37.1 Location Map of Revelstoke Community Forest (TFL 56).

FIGURE 37.2 Steep-sided valleys and dense forests found in the Revelstoke Community Forest (TFL 56).

The forests of the TFL are dominated by western hemlock (*Tsuga heterophylla*), Engelmann spruce (*Picea engelmannii*), western red cedar (*Thuja plicata*), subalpine fir (*Abies lasiocarpa*), and Douglas-fir (*Pseudotsuga menziesii*). Other species that occur less commonly include western white pine (*Pinus monticola*), black cottonwood (*Populus trichocarpa*), paper birch (*Betula papyrifera*), quaking aspen (*Populus tremuloides*), and lodgepole pine (*Pinus contorta*). The amount of area by leading tree species in the THLB is:

- Engelmann spruce: 6,862 ha (16,956 ac)
- Western red cedar: 6,822 ha (16,857 ac)
- Western hemlock: 5,330 ha (13,170 ac)
- Douglas-fir: 2,061 ha (5,093 ac)
- Balsam poplar: 1,016 ha (2,511 ac)

The age-class structure over the entire productive land base is shown in Figure 37.3. The forested area is distributed across a wide range of age classes, with most of the THLB area falling within 0–100 and 230–320 year age classes.

From 1955 to 1992, the area was part of a larger Tree Farm License (TFL 23) last owned by Westar Timber. In 1992, the southern portion of TFL 23 was sold to Pope and Talbot Ltd. In 1992, Westar Timber sold the remaining TFL area to Evan's Forest Products Ltd. Due in large part to concerns identified by the citizens of Revelstoke, a revised deal was negotiated that resulted in the area being split into two TFLs (55 and 56). As a result, TFL 55 was sold to Evans Forest Products, and TFL 56 was sold to the city of Revelstoke (Revelstoke Community Forest Corporation, 2009).

TABLE 37.1 Landbase Area Summary of the Revelstoke Community Forest (TFL 56) in British Columbia

Factor	Total Area (ha)	Total Area (ac)	Effective Area (ha)[a]	Effective Area (ac)[a]	Percent of Total Area	Percent of PFLB[b]
Total Tree Farm License area			119,823	296,083	100.0	
Less						
Nonforest/nonproductive land			58,822	145,349	49.0	
Existing roads, trails, and landings			1,146	2,832	1.0	
Total Crown Forested Land Base (CFLB)			59,855	147,902	50.0	100.0
Less						
Inoperable/inaccessible areas	23,770	58,736	23,770	58,736	19.8	39.7
Environmentally sensitive/unstable terrain areas	1,822	4,502	1,741	4,302	1.5	2.9
Low productivity sites	3,540	8,747	360	890	0.3	0.6
Nonmerchantable forest types	2,503	6,185	1,725	4,262	1.4	2.9
Riparian reserves	1,492	3,687	1,124	2,777	0.9	1.9
Wildlife habitat areas	2	5	2	5	0.0	0.0
Downie saltlick	19	47	14	35	0.0	0.0
Mountain caribou reserves	10,611	26,220	7,984	19,728	6.7	13.3
Isolated stands	123	304	123	304	0.1	0.1
Site-specific inoperable areas	669	1,653	437	1,080	0.4	0.7
Timber Harvesting Land Base (THLB)			22,575	55,783	18.8	37.7
Less other removals						
Estimate of future roads, trails, landings			459	1,134	0.4	0.8
Wildlife tree patches			388	959	0.3	0.6
Old growth management areas			355	877	0.3	0.6
Effective long-term Timber Harvesting Land Base			21,373	52,813	17.8	35.7

[a]*Effective netdown area represents the area that was actually removed as a result of a given factor. Removals are applied in the order shown above, thus areas removed lower on the list do not contain areas that overlap with factors that occur higher on the list. For example, lake buffers netdown does not include nonforested area.*

[b]*Productive forest land base (PFLB) refers to the forest crown land area within the FMA. A subset of this area is suitable for timber harvesting.*

The Revelstoke Community Forest Corporation (RCFC) was formed in 1993 for managing and operating TFL 56. The corporation is owned by the RCFC Holding Company Ltd., which is owned by the city of Revelstoke. Three local forest industry companies helped finance the purchase of the licence. The city owns the Holding Company while the industry partners have timber removal rights for a portion of TFL 56's annual allowable cut (AAC). The city's sawlog allocation (50% of the AAC at the time of Management Plan #4, now sitting at 60%) is sold through a log sort yard, and prices are determined through a competitive bidding process or through direct negotiations. The industry partners' sawlog allocation (the other 40–50%) is provided at cost (averaged annually), with tree species and log grades that are representative of the wood harvested. Pulpwood from TFL 56 is sold under a separate contract, and the proceeds of these sales are factored back into the cost of sawlogs (Revelstoke Community Forest Corporation, 2009).

The RCFC and the Holding Company are governed by a board of directors that is composed of the mayor, two city councillors, the city administrator, and three people appointed from the community. A staff of three employees manages the day-to-day business. The industry partners provide input through a management advisory committee. Most forest management, construction, logging, and silviculture services are obtained from contractors and consultants, with the goal

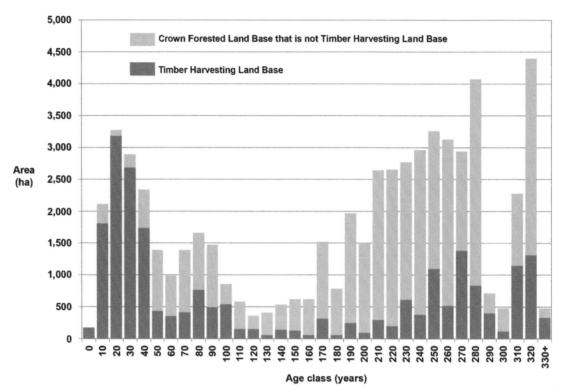

FIGURE 37.3 Revelstoke Community Forest (TFL 56) forest age-class distribution in 2008.

of maximizing employment and economic benefit within the local community. The corporation is funded through the proceeds of log sales because a commitment was made to city tax payers that they would not be called upon to fund the venture (Revelstoke Community Forest Corporation, 2009).

PLANNING ENVIRONMENT AND METHODOLOGY

Key considerations in the management of TFL 56 are:

1. *Sustainable forest management practices* that provide: (a) timber for local mills and other purchasers through regular sales; (b) suitable habitat to sustain wildlife populations and maintain biodiversity; and (c) values and experiences important to local residents and tourists (e.g., recreation, viewscapes).
2. *Profitability*: The TFL must be self-sustaining based on sales of harvested timber. The high proportion of low value pulp-grade logs in the TFL make economics challenging, particularly when pulp markets are poor.
3. *Economic opportunities for local businesses*: Suppliers, contractors, consultants, etc.

Guiding Laws, Regulations, and Policies

The Province of BC's *Forest Act* requires holders of TFLs on Crown land to develop a management plan and a timber supply analysis. At the time that TFL 56 Management Plan #4 was completed, the requirement for updates to the management plan and timber supply analysis were every 5 years. This time frame has been extended to 10 years since that time. The management plan's goal is to identify and propose for approval to BC's Chief Forester the management objectives and strategies for meeting timber and non-timber objectives, and an associated AAC. The resource management objectives to be addressed are identified in BC's *Forest and Range Practices Act* ([SBC 2002] chapter 69) (Queen's Printer, 2014), higher level plan orders, and *Government Action Regulation* orders.

The Revelstoke area is subject to the *Revelstoke Higher Level Plan Order* (Government of British Columbia, 2001), which reflects significant government and local efforts spent on land use planning in the 1990s. The Order establishes resource management zones and associated management objectives; plus, it gives direction on monitoring and ongoing review of the Order. It defines the planning requirements for biodiversity, old and mature forests, caribou, and grizzly bears in the Revelstoke area.

In May 2002, the Committee on the Status of Endangered Wildlife in Canada designated woodland caribou (including the mountain ecotype) within the Southern Mountain Population as *threatened*. Under the federal *Species at Risk Act*, the BC Government was responsible for preparing a recovery plan that would provide direction in the recovery of the mountain caribou, and the province's Species at Risk Coordination Office designated the Revelstoke habitat team to develop a recovery plan (British Columbia Ministry of Environment, 2007). This plan took effect in December 2008 and included the selection of habitat reserves necessary to get the population back to pre-1996 numbers. A Government Action Regulation Order was developed to legally define the spatial habitat reserves (Government of British Columbia, 2009).

The province's Chief Forester asked to approve the plan and determine an Annual Allowable Cut (AAC) based on the factors described in Section 8 of the *Forest Act*, including the social and economic objectives of the Crown and all other relevant information.

Management Plan Development

The development of Management Plan #4 benefited from significant investments in planning that occurred in the 10 years preceding plan development. Specifically, *total chance* harvest planning was completed on the potentially operable land base to identify the full extent of potential road systems and harvest areas (cutblocks). This was accomplished using office review of stereo photos and land base information that included 5 m (16.4 ft) contour mapping, followed by ground-truthing of potential roads and harvest areas. The objective was to ensure that any investment in road building would best serve the TFL in the long term and minimize the potential for isolating timber from harvest. Because of the TFL's steep slopes, cable logging is the predominant harvesting system, and the location of roads and landings plays a critically important role in determining how much area can be harvested and the extraction costs. Placing the roads in the correct locations was seen as a very important objective for long-term viability of the TFL. Areas that could not be accessed at reasonable cost became default reserve areas. As a result of this planning, a solid information base was available to build a long-term spatial plan *from the bottom up*. This level of tactical planning is not typical of planning in BC, where a harvest level is usually determined for a unit and then effort is spent working out how to implement it on the ground (often referred to as *top-down* planning).

The 20-month management planning process that was used to create Management Plan #4 followed the steps listed below:

1. *Government review of previous management plan.* Ministry staff conducted a review of the existing management plan and provided comments on the plan, the licensee's performance in respect to the plan, and a list of relevant guidelines in effect.
2. *Development of the timber supply information package and draft management plan.* The timber supply analysis information package is submitted for approval to the Ministry, which includes the information, assumptions, and approach the licensee proposed in conducting the timber supply analysis. A draft management plan is developed and its purpose is to propose management objectives and strategies for achieving those objectives within the TFL, and to include details on the status of various inventories. The draft management plan also provided an opportunity for the public and other agencies to review and comment on the proposed objectives and strategies. The licensee provides an opportunity for public review of the draft management plan by publishing notices locally.
3. *Development of a timber supply analysis and a 20-year plan.* The purpose of the timber supply analysis is to present information regarding the timber supply forecasts for consideration by the Chief Forester in the determination of the AAC under Section 8 of the *Forest Act*. This analysis establishes a *base case* timber supply forecast that reflected existing practice in the TFL and explored sustainable harvest levels over 250 or more years. It also uses sensitivity analyses to assess the impacts of potential changes to modeling assumptions, and gain further understanding of the dynamics at work in the base case forecast. The RCFC does not have its own growth and yield program but does cooperate with the BC provincial program. The Ministry staff then provides notice to the licensee on the acceptability of the analysis—included strengths and weaknesses of the analysis. The 20-year plan (a time interval requirement at the time the plan was developed) is a nonoperational plan that requires local Ministry acceptance and consideration by the Chief Forester in the AAC determination. The plan is a tool to help confirm availability of areas for harvest given constraints reflected in the timber supply analysis. There is no requirement to follow the 20-year plan after management plan approval. In this case, the bottom-up planning approach ensures that the plan is both reasonable and achievable.
4. *Approval of the management plan and determination of the AAC.* If deemed suitable, the Chief Forester will approve the management plan and determine an AAC. A determination considers the technical work completed in the timber supply analysis, the social and economic objectives of the Crown, and all other relevant information. The reasons for the AAC determination are documented in a rationale statement. The AAC is generally applicable for a 10-year time frame. The Chief Forester also outlines any conditions for the approval of the management plan in a letter to the licensee.

A range of non-timber resource values are present on the TFL and required consideration in the plan. The following values were specifically identified within the plan and incorporated into the timber supply analysis modeling work were relevant to timber supply:

- Recreation resources—A series of designated trails, helicopter ski runs and backcountry huts are located within the TFL, with two helicopter ski lodges located immediately north of the TFL
- Visual resources—Significant viewscapes and visual resources exist that are being managed in order to maintain scenic quality
- Wildlife and Fish resources—Many wildlife species occur within the TFL including large carnivores (e.g., grizzly bear, black bear, and cougars), ungulates (e.g. moose, deer, mountain goats) and a number of species at risk
- Archaeological resources—There is very little historic use by First Nations evident in the area; therefore, no specific archaeological resources management strategies were incorporated into the management plan, and these resources will be managed if found on the TFL
- Water licenses—There are close to a dozen commercial and domestic water licenses within the TFL

The timber supply analysis completed for TFL 56 was completed using Patchworks™ (Spatial Planning Systems, 2009). Patchworks is a spatial forest estate model that can incorporate real-world operational considerations into a strategic planning framework. It utilizes a goal-seeking approach and an optimization heuristic to schedule activities across time and space in order to find a solution that best balances the targets and goals defined by the user. Targets can be applied to any aspect of the problem formulation.

OUTCOMES OF THE PLAN

The TFL 56 Management Plan #4 included a series of commitments or management strategies for the TFL, along with a recommended timber harvest level. A series of management strategies were identified within the management plan, with commitments for implementation. These included the following:

1. *Harvesting.* As a part of the timber supply analysis, supported by the total chance harvest plan, a plan was developed that identified proposed roads and cutblocks in the next 20 years. This plan was to be used within the context of current market conditions to prioritize harvesting. A range of harvesting systems and silviculture systems used are aligned with the prevailing slopes, landform, ecology, and reforestation objectives for a given site.
2. *Roads.* The goal here is to provide safe, efficient, and environmentally appropriate transportation corridors from the forest stands to the public highway. Road maintenance activities are conducted in order to permit safe operation of logging trucks, to provide safe access to the public, and to prevent environmental damage.
3. *Reforestation.* The goals of the silviculture program are to: (a) regenerate all logged areas within a maximum of 3 years of logging being completed (the average time is expected to be less than 2 years); (b) conduct an aggressive brush control program to maximize stand vigor and health; (c) comply with the Forest Stewardship Plan (Queen's Printer, 2014) in satisfying free-growing stocking standards; (d) use regeneration techniques that will increase productivity; (e) establish regeneration with mixtures of two or more species ecologically suited to the growing site; and (f) establish silvicultural trials where knowledge of an activity or treatment is inadequate.
4. *Forest Health.* To address this concern, annual flights are conducted over the TFL to assess windthrow and pest conditions. Cooperation with government pest specialists is necessary to ensure pooling of knowledge and exchange of data. The area is closely monitored so that known disease problems will be detected. Results are used to prepare action plans and treatments. Surveys of infected areas are conducted to monitor pest activity, and where present, an attempt is made to maintain pests at endemic levels by preventing the conditions that favor build-up and spread.
5. *Fire Protection.* The goal is to minimize damage from fire in the forested land base and to maximize the timber salvage from fire damaged stands. In doing so, some fire breaks are created by using broadcast burning in logical areas to prepare for climate change and the increased fire activity predicted in the area.
6. *Recreation Resources.* Access to important recreational areas and trail heads is maintained. Activities are jointly implemented with the Ministry of Forest, Lands and Natural Resource Operations to maintain or enhance existing recreation sites and trails and to identify and manage potential recreation sites and trails. Commercial recreation firms also maintain commercial recreation opportunities.
7. *Visual Resources.* Harvesting units are designed with basic visual principles in mind, including the shape and configuration of cutblocks.

8. *Streams, Lakes, and Wetlands.* These concerns relate to maintaining and protecting the productive capacity of fish habitat, maintaining streamside vegetation and the integrity of stream channels, preventing unnatural stream bank erosion, sedimentation and introduction of woody debris, and maintaining the integrity of wetlands.

9. *Wildlife and Biodiversity.* These concerns relate to developing an ongoing retention strategy that incorporates required reserves with operational limitations and desirable wildlife reserves, utilizing the retention of mature forest, managing the patch size distribution to mimic natural disturbance, retaining wildlife trees, and managing access to support wildlife and biodiversity management.

10. *Public, First Nations, and Stakeholders.* To address this concern, a series of consultation opportunities will be provided with the public, including an annual public meeting, annual report, frequent advertisements and an "open door policy" for public consultation. In doing so, meaningful consultations are conducted with non-timber tenure holders including guide-outfitters, trappers, commercial recreation operators, and water users with specific additional consultation for key situations. Opportunities are also provided for information sharing with First Nations people.

The management strategies are incorporated into a timber supply analysis to develop a long-term sustainable timber supply for the TFL. The timber supply modeling was completed for a minimum of 300 years for each scenario to confirm that the harvest and growing stock levels remain stable. Results showed a harvest level of 88,000 m³ per year (3.11 million ft³ per year) for the first 100 years and then a rise to a long-term level of 101,000 m³ per year (3.57 million ft³ per year). The spatial reserves used to address old-seral objectives and caribou habitat protection proved to impose the largest impacts on timber supply.

DISCUSSION AND CONCLUSIONS

The 20-year plan that was developed for TFL 56 has subsequently supported forest development in the years since Management Plan #4 was completed. It will be updated with newer management objectives over time and is a method for recording these changing practices and objectives. Over time, it is essentially an evolving document. The annual report that is generated for the management plan has proven valuable for tracking forest management activities; these contain information on volume produced from the community forest, silvicultural activities implemented, and other issues pertinent to the management of the forest (Revelstoke Community Forest Corporation, 2009).

Sustainability Issues

As was noted earlier, the TFL needs to be a self-sustaining endeavor that is based on timber sales, and two of the important considerations in the management of the TFL include implementing sustainable forest management practices and providing for a predictable wood production level for local mills and other purchasers. Further adding to the complexity of the management plan is the notion that suitable habitat to sustain certain wildlife populations and to maintain biodiversity is important, and that recreational and viewshed experiences of local residents and tourists are important.

Plan Development Challenges

Challenges to the development of the management plan include the accuracy of the forest inventory for the TFL, timber related constraints (e.g., pulp log prices, high harvesting costs), and a wide range of non-timber resource values (e.g., visual resources and caribou). A clear understanding of these timber and non-timber factors needed to be understood and balanced in ensuring a viable plan.

Plan Implementation Challenges

A key challenge to the implementation of Management Plan #4 has been the further development and refinement of habitat conservation and management strategies within the TFL. Recent (2013) regulatory changes have occurred that have required resource management zones and habitat conservations requirements for caribou to be modified. This has resulted in changes to the available land base and will ultimately influence future operations throughout the TFL. A second key and ongoing challenge for operating within the TFL under Management Plan #4 is the market values of pulp logs within the region and province. Given the large component of pulp logs across the TFL, pulp prices will dictate the level of harvest activity within the TFL from year to year.

The terrain within TFL 56 is steep, rugged, and mountainous. This combined with a challenging timber profile (a high pulp log component) makes the area a challenge to operate in. In order to address these conditions within Management Plan #4, detailed operational planning was completed in the form of a total chance harvest plan that fully identified future roads and harvesting throughout the TFL. This detailed operational plan was completed to ensure that the timber supply analysis and supporting 20-year plan were realistic.

ADDITIONAL READING AND RESOURCES

This chapter is based on the TFL 56 Management Plan #4 (2009) completed by RCFC in BC, Canada. To view the plan itself, please visit the following Web site, which was available on May 18, 2014:

http://rcfc.bc.ca/wordpress/wp-content/uploads/2012/10/Management_Plan_4-May_25_09.pdf.

If in the future the link to the site appears broken, search the Internet using the title of the plan and the keywords provided.

REFERENCES

British Columbia Ministry of Environment, 2007. Mountain Caribou Recovery. British Columbia Ministry of Environment, Victoria, BC. http://www.env. gov.bc.ca/wld/speciesconservation/mc/#resources (Accessed May 28, 2014).

Government of British Columbia, 2001. Revelstoke Higher Level Plan Order. Government of British Columbia, Victoria, BC. http://archive.ilmb.gov. bc.ca/slrp/lrmp/cranbrook/revelstoke/files/plan_order.pdf (Accessed May 28, 2014).

Government of British Columbia, 2009. Order—Ungulate Winter Range #U-4-010, Mountain Caribous—Kinbasket Planning Unit. British Columbia Ministry of Environment, Victoria, BC. http://www.env.gov.bc.ca/wld/documents/uwr/u-4-010_order_09Dec09.pdf (Accessed May 28, 2014).

Queen's Printer, 2014. Forest and Range Practices Act. Government of British Columbia, Victoria, BC. http://www.bclaws.ca/Recon/document/ID/ freeside/00_02069_01 (Accessed May 28, 2014).

Revelstoke Community Forest Corporation, 2009. Management Plan #4. Revelstoke Community Forest Corporation, Revelstoke, BC. http://rcfc.bc.ca/ wordpress/wp-content/uploads/2012/10/Management_Plan_4-May_25_09.pdf (Accessed May 28, 2014).

Spatial Planning Systems, 2009. What is Patchworks? Spatial Planning Systems, Deep River, ON. http://www.spatial.ca/products/index.html (Accessed April 29, 2014).

Chapter 38

South Puget Planning Unit, Washington, United States of America

Cathy Chauvin, Angus Brodie, Heather McPherson and Abu Nurullah
Washington Department of Natural Resources, Olympia, Washington, USA

ABBREVIATIONS

DNR   Washington State Department of Natural Resources
WDFW   Washington Department of Fish and Wildlife

MANAGEMENT SETTING AND BACKGROUND

The South Puget Habitat Conservation Plan (HCP) Planning Unit's boundary (Figure 38.1) encompasses farms, parks, reservations, a major military base, cities, and thousands of acres of state trust lands. Five of the larger blocks of state trust lands are designated as state forests (Green Mountain, Tahuya, Tahoma, Elbe Hills (Figure 38.2), and Tiger Mountain). Most of the state trust lands in this unit were deeded to Washington by the federal Enabling Act of 1889 (25 U.S. Statutes at large, c. 180 p. 676), while others were acquired through land transactions or transferred to the state from the counties for reforestation and revenue production. The South Puget planning unit also includes DNR-managed natural area preserves and natural conservation areas, which are lands taken out of trust status to protect native ecosystems, rare plant and animal species, and other natural features. Preserves and conservation areas do not generate revenue but contribute to the DNR's ecological objectives.

This planning unit is situated in the most densely populated area of Washington State, the Puget Sound. More than half of the unit's area is urbanized, supporting the greater Seattle area as well as Tacoma, Olympia, and numerous smaller but growing cities. Because of its proximity to major urban areas, this unit attracts thousands of visitors eager to enjoy what state trust lands have to offer: a rustic experience close to home with 450 miles (mi) (724 kilometers (km)) of trails, 30 developed recreation sites, and many acres (ac) of dispersed recreation opportunities. The DNR estimates that Tiger Mountain State Forest alone attracts more than 375,000 visitors per year, with Tahuya State Forest a close second at 250,000 visitors. In addition, state trust lands are highly visible to the traveling public from roads that cross the planning unit. U.S. Interstate Highway 90, a designated national scenic byway called the "Mountains to Sound Greenway," carries traffic past Tiger Mountain State Forest and into Seattle. State Highway 18, a cut-off route between Interstates 5 and 90, runs along Tiger Mountain's eastern edge, and State Highway 706 carries a stream of visitors each year past the Elbe Hills and Tahoma state forests to the Nisqually entrance of Mount Rainier National Park. Traffic at the Nisqually entrance totaled 197,829 cars in 2013 (National Park Service, 2014).

Most forests in this planning unit are low-lying and wet. Approximately 80% of state trust lands are situated below 3,000 feet (ft) (914 meters (m)) in elevation, and 50% are below 1,000 ft (305 m). The Olympic Mountain rain shadow influences the climate, and precipitation (usually rain) averages 20–70 inches (in) (510–1,780 millimeters (mm)) per year. Forests at these elevations are dominated by Douglas-fir (*Pseudotsuga menziesii*) with western hemlock (*Tsuga heterophylla*) and western red cedar (*Thuja plicata*) as other primary species. The conifer forest mosaic is interspersed with hardwood tree species such as bigleaf maple (*Acer macrophyllum*) and red alder (*Alnus rubra*). The remaining 20% of state trust lands in this unit are located at elevations up to 5,000 ft (1,524 m), where precipitation averages 50–140 in (1,270–3,560 mm) per year and often falls as snow. Conifer forests at these elevations are similar to those found at lower elevations, although Pacific silver fir (*Abies amabilis*) and noble fir (*Abies procera*) are also present, mostly at higher elevations.

Forest Plans of North America. http://dx.doi.org/10.1016/B978-0-12-799936-4.00038-2

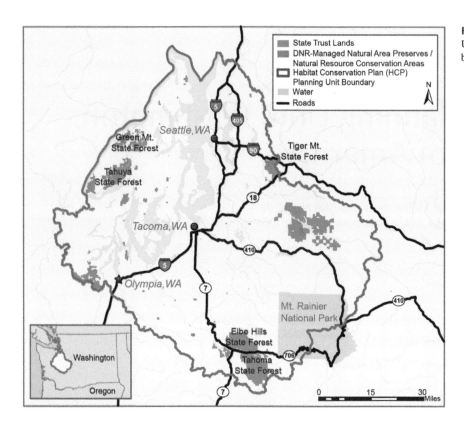

FIGURE 38.1 South Puget HCP Planning Unit. State trust lands located outside the unit boundary are not shown.

FIGURE 38.2 Elbe Hills State Forest with Mount Rainier in the background and a variable retention harvest in the foreground.

Forest conditions in this unit are the legacy of natural disturbances and forest management actions conducted under previous policies and laws. At the time of plan development, forest stands tended to be young and often overstocked. Using a stand development stage classification system based on the work of many authors in the Pacific Northwest (Carey, 2007; Carey et al., 1996; Franklin et al., 2002; Oliver and Larson, 1996; Van Pelt, 2007), the DNR classified 20% of state trust lands in this planning unit as being in a *competitive exclusion* stand development stage, during which tree densities are high and competition is intense for light, nutrients, and growing space. Only 2% of state trust lands were classified as being in mature stages of development (*niche diversification* and *fully functional* stages). In these stages, forests have multiple canopy layers, snags, and other elements of structural complexity. The remainder of state trust lands was classified as being in the *ecosystem initiation* (27%), *understory development* (40%), or *biomass accumulation* (11%) stages. With regard to stocking, 22% of state trust lands were considered overstocked (with a relative density between 70 and 100) and 3% were considered extremely overstocked (relative density over 100). These forest conditions have implications for northern spotted owls (*Strix occidentalis caurina*), as will be discussed in this chapter.

PLANNING ENVIRONMENT AND METHODOLOGY

The purpose of forestland planning is to translate the DNR's policies into practical guidance for managing a planning unit on a day-to-day basis. The DNR manages nine HCP planning units in Washington. When this project began, forestland planning had not been completed for any of them.

Guiding Laws, Regulations, and Policies

State trust lands are managed within the framework of federal and state laws, which include the federal Endangered Species Act (16 U.S.C. 1531 et seq.) and Washington's State Environmental Policy Act (43.21c RCW). State trust lands also are managed under Washington's Multiple Use Concept (RCW 79.10.120), which allows multiple uses on state trust lands when those uses are in the best interest of the state and its citizens and consistent with the DNR's trust management responsibilities. Those responsibilities are clearly laid out in the 1984 landmark decision *County of Skamania v. State of Washington* (102 Wn.2d 127, 685 P.2d 576), commonly referred to as the *Trust Mandate*. Per Washington's Supreme Court, a trustee (DNR) must act with undivided loyalty to the trust beneficiaries to the exclusion of all other interests, and manage trust assets prudently. In addition, the state legislature directs the DNR to manage state trust lands to generate revenue for trust beneficiaries in perpetuity, in other words, *forever*. Doing so requires responsible management with an emphasis on long-term sustainability.

State trust lands are also managed under the DNR's policies. The *Policy for Sustainable Forests* is a key policy document that governs the 2 million ac (about 0.81 million hectares (ha)) of forested state trust lands in the DNR's care. This document directs the DNR to maintain forest health, protect areas of special ecological concern, and target 10–15% of each planning unit for older forest conditions based on structural characteristics. This document also includes guidance for social values including mitigation of visual impacts and public access and recreation. The DNR also has responsibilities under its *State Trust Lands HCP*. Authorized under the Endangered Species Act, the *State Trust Lands HCP* is a long-term management plan that includes a suite of conservation strategies and associated objectives for the restoration and maintenance of habitat for federally listed species in conjunction with forest management activities. Although the *State Trust Lands HCP* is a multispecies plan, it concentrates on northern spotted owls, marbled murrelets (*Brachyramphus marmoratus*), and salmon.

Objectives for the Analyzed Area

The DNR's goal for the South Puget planning unit was a forestland plan that would address local issues; reduce uncertainties involved in meeting revenue, ecological, and social objectives; and guide the location of timber sales. The deliverables of the planning process included the following:

1. *An environmental impact statement (EIS) prepared under the State Environmental Policy Act.* The preferred alternative in the final EIS would function as the management plan; no separate planning document would be produced. This was a strategic decision. Planning needed to be dynamic and responsive to changes in forest conditions, policies, and other influences. Key components of the plan would be kept separate and updated as needed.
2. *A harvest schedule generated from a forest estate model.*
3. *A spatial digital tool.* This tool would enable foresters to select an area on a digital map and bring up information such as forest stand development stage, habitat status, and relative density.
4. *Management strategies and their associated procedures.* These items would be written to address the specific challenges of managing this unit on a day-to-day basis.

For this planning effort, the DNR preferred to stay within the bounds of current DNR policies, since planning is designed to implement, not change, those policies. However, the DNR knew that some changes would be unavoidable. One example was the current sustainable harvest level. To regulate harvest levels, the DNR assigns state trust lands across the state to one of 20 sustainable harvest units based on county boundaries, administrative regions, and other considerations. Each sustainable harvest unit is assigned a sustainable harvest level that is updated each decade. In the South Puget planning unit, state trust lands belong to one of seven sustainable harvest units: Federal Grant, King County, Kitsap County, Lewis County, Mason County, Pierce County, or Thurston County. The DNR understood that addressing all major management issues through this planning process might change the current harvest level in one or more of these units. The planning process was therefore viewed as an opportunity to refine these levels, with the understanding that changes to sustainable harvest levels would require approval by the DNR's Board of Natural Resources.

Plan Development Methods

The steps of the DNR's forestland planning process included understanding the issues, developing and analyzing management alternatives, selecting a final alternative, and implementing the plan. The plan was developed jointly by the DNR's planning team located at the agency headquarters in Olympia, Washington and personnel in the DNR's South Puget region office in Enumclaw, Washington, with extensive input from the public and stakeholders.

The project began with meetings between the planning team and region personnel to understand the day-to-day challenges of managing this unit. The DNR also conducted extensive public outreach, which included a series of workshops and scoping meetings to gather input on key issues. In these efforts, three issues rose to prominence: visual impacts, recreational use, and northern spotted owl habitat.

Visual impacts were important due to the visibility of state trust lands from major roads, the proximity of nearby communities, and the sheer number of recreational visitors. Specific areas for visual impact mitigation were identified during the public workshops. Most of these areas were already mitigated through existing procedures, including riparian and wetland buffers and protection of potentially unstable slopes. To address impacts in other visually sensitive areas, the DNR's foresters needed greater flexibility in the number of leave (residual, live) trees they were allowed to retain in stand replacement harvests.

Recreational use was already high on South Puget lands and was expected to increase. Populations in some counties in the planning region (King, Kitsap, Thurston, and Pierce) were anticipated to grow by more than 580,000 people by the year 2030 (Washington State Department of Transportation, 2014). The DNR anticipated that many of these new residents would seek out recreation opportunities. Issues with high recreation use identified in meetings with the public and region personnel included vandalism, trespass on adjacent private lands, illegal trail use, and resource damage. A related issue was public awareness of logging. The DNR needed to disperse forest management activities across the South Puget planning unit and across decades to remind recreational visitors and surrounding communities that these are working forests, not parks. Such awareness is critical to maintaining the DNR's intangible social license to operate.

With regards to northern spotted owls, the DNR's conservation objective was to provide habitat that makes a significant contribution to demographic support, maintenance of species distribution, and facilitation of dispersal (Washington State Department of Natural Resources, 1997). Dispersal is the movement of young, juvenile, or adult owls from one sub-population to another (Miller et al., 1997). The South Puget HCP planning unit contains the majority of designated dispersal management areas on state trust lands managed under the *State Trust Lands HCP*. However, the DNR's definition of *dispersal habitat*, as presented in the *State Trust Lands HCP*, did not include requirements for structural diversity (such as snags or down wood), nor did it place upper limits on the number of trees per acre. As a result, many stands classified as owl dispersal habitat were not functioning well in that role. Structurally simple and densely stocked, these stands were difficult for owls to fly through, contained few perches for roosting or foraging, and provided little habitat support for the owls' main prey species, the northern flying squirrel (*Glaucomys sabrinus*). This habitat definition was written for the *State Trust Lands HCP* in 1997 using existing biological information and was meant to be replaced when more information became available (Washington State Department of Natural Resources, 1997).

Since the signing of the *State Trust Lands HCP*, new scientific information had been published regarding habitat use by dispersing northern spotted owls (Miller et al., 1997), spotted owl demography during the dispersal phase (Forsman et al., 2002), and the deficiencies of Washington dispersal habitat definitions in meeting life requirements of dispersing owls (Buchanan, 2004). Based on this new information and understanding of northern spotted owl dispersal requirements, as well as the DNR's assessments of habitat conditions, the question was posed whether the DNR could improve its northern spotted owl conservation efforts by altering its definition of dispersal habitat.

Once all of the issues were understood (such as northern spotted owl habitat, recreation, and visual impacts), the DNR developed management alternatives to address them. The alternatives reflected oral and written comments from the general public and stakeholders and conversations with region personnel. The *no action alternative* represented current management practices. The *exploratory alternative* was designed to stretch, but not break, the boundaries of existing policy, and the *preferred alternative* included management strategies to solve the issues identified in the first phase of the project. Both the preferred and exploratory alternatives included new northern spotted owl dispersal habitat definitions (refer to the Outcomes section later in this chapter).

To test and refine the alternatives, the DNR used a forest estate model. The model was developed with the Spatial Planning System, a commercial software package developed by Remsoft, Inc. The model required four types of inputs: land classification, yield tables, objectives, and constraints.

1. The DNR classified forested state trust lands by unique combinations of 13 attributes or themes. These themes included watershed, surface and timber ownership groups, forest type, site class, size class (based on quadratic mean diameter), stocking class (based on relative density), silvicultural status (for example, thinned, unthinned, or regenerated, administrative areas, spotted owl management units, deferrals land class (uplands, riparian, or general management areas), rain-on-snow sub-basins, and road access.

2. To develop the yield tables, the DNR first stratified the planning unit into 306 strata using combinations of forest inventory parameters (forest type, site class, stocking class, and size class). The DNR then used the United States Forest Service's (USFS) Forest Vegetation Simulator to model each stratum under 12 silvicultural pathways. Potential harvest methods used in the model included uniform and variable density thinning, and variable retention harvest, in which structural elements of the existing stand (snags, down wood, leave trees, and other elements) are retained from one rotation to the next in order to promote structural diversity across the landscape.
3. The DNR's revenue objective was represented in the model as the objective function, which was to maximize net present value.
4. The DNR represented its other management objectives as hard and soft constraints on the model. Hard constraints are absolute values that must be met. Examples of hard constraints include ensuring that stand replacement harvests are followed by regeneration activities, or that standing volume in upland areas is maintained within 90% of the previous decade. Soft constraints are not required, but incur a penalty to the objective function if they are not met. Examples include maintaining hydrologic maturity in specific watersheds, a certain percentage of northern spotted owl habitat, or continuous forest cover on potentially unstable slopes.

Within the model, each management alternative was represented as a scenario. Each scenario was subsequently solved by the model in a linear programming formulation. In this formulation, for each scenario the model locates a solution (when, where, and by what method to harvest) that maximizes net present value over 10 decades, meets all hard constraints, and incurs the fewest possible penalties for not meeting soft constraints (ecological objectives). If the model's solution to a scenario did not effectively balance the DNR's revenue and ecological objectives, the DNR adjusted the management alternative's parameters and asked the model to solve the scenario again. For example, in two of its alternatives the DNR adjusted northern spotted owl habitat thresholds until it identified those thresholds that could be attained with minimal or no change in cost to the trust beneficiaries (habitat thresholds will be discussed in the Outcomes section of this chapter).

Once the alternatives were finalized, the DNR used the model to produce a database called the *state of the forest file*, which is a projection of how forest stand conditions may change over 10 decades as a result of implementing the model's solution. The DNR analyzed the state of the forest file data for each alternative in order to identify probable significant adverse environmental impacts. This quantitative analysis was performed using the *criteria and indicator* concept of the Montréal Process (Working Group on Criteria and Indicators for the Conservation and Sustainable Management of Temperate and Boreal Forests, 1995). Criteria are broad concepts, such as forest health, that are measured with a set of indicators. For example, the criteria *forest health* was evaluated using the measurable indicator *stand density*. In this example, the DNR analyzed the state of the forest file data to determine if, over 10 decades, stand density increased to a point that would cause a probable significant adverse environmental impact to forest health. Criteria, indicators, and the methods used to evaluate them were based on DNR policies, applicable laws, professional judgment, and recommendations in the scientific literature.

To analyze potential environmental impacts to northern spotted owls, the DNR evaluated the state of the forest file data with northern spotted owl dispersal assessment tool stand-level models (Gordon et al., 2014). Built in close cooperation with the Washington Department of Fish and Wildlife (WDFW) and the USFS, these models assign forest stands a score based on their ability to support movement, roosting, and foraging (Figure 38.3). The DNR analyzed these scores to assess changes in habitat quality over time.

OUTCOMES OF THE PLAN

The DNR's decision maker adopted the preferred alternative as the best balance of the DNR's objectives. While the harvest level and net present value were projected to drop slightly under the preferred alternative, gross revenue was projected to increase, and the percent of the planning unit in older forest condition also was expected to increase (Table 38.1). This alternative included a number of strategies and their associated procedures, such as the even apportionment of timber harvest volume (referred to as *even flow*) between the South Puget's four administrative units. Under even flow, each decade's harvest volume in each unit must be within 15% of the volume harvested during the first decade of the plan. This strategy has resulted in a more equitable apportionment of the harvest volume between the administrative units and helped the DNR to maintain a continued forest management presence. This alternative also included modifications of the leave tree strategy. To provide foresters greater flexibility in mitigating visual impacts, the DNR increased the number of trees that could be left standing within a stand replacement harvest. In visually sensitive areas, foresters could leave up to 16 trees per ac (40 per ha) in clumped or dispersed patterns. In the Nisqually River visual impact corridor, which is within the Elbe Hills State Forest (visible from Highway 706), that limit was raised to 20 trees per ac (49 per ha).

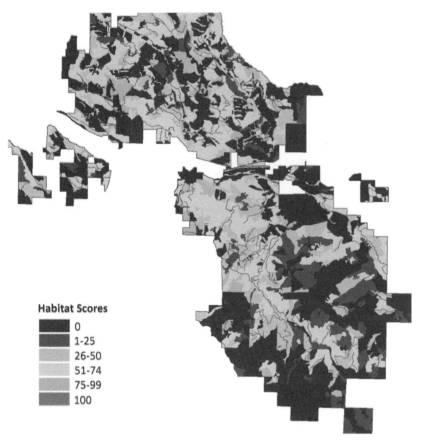

FIGURE 38.3 Sample results for foraging, northern spotted owl data assessment tool stand-level model. Scores of 50 or above indicate the forest stand supports foraging.

Habitat Scores
- 0
- 1-25
- 26-50
- 51-74
- 75-99
- 100

TABLE 38.1 Projected Outcomes of Each Alternative Examined in the Final EIS

Alternative	Harvest Level, Decade 1 (MMBF)	Gross Revenue, Decade 1 ($ Million USD[a])	Cumulative net Present Value After 100 Years ($ Million USD[b])	Percent of Planning Unit in Older Forest Condition[b]	Date Northern Spotted Owl Dispersal Management Areas Reach 50% for South Puget Movement Habitat (Decade)
No action	374	95	178	16	Does not reach
Preferred	367	106	171	21	2,047
Exploratory	410	126	179	26	2,037

[a]U.S. dollars.
[b]Niche diversification and fully functional stand development stages.

The DNR adopted a separate strategy for visual impacts within the Tiger Mountain State Forest. Previously, this forest had been managed under a separate management plan. Approved in 1986, the plan stipulated a minimum rotation age of 60 years and restricted the harvest in each watershed administrative unit to no more than one-sixth of its area in any given decade. The purpose of these restrictions was to create a mosaic of forest stands of different age classes across the forest. Through this planning effort, the DNR integrated the 1986 Tiger Mountain plan into the South Puget forestland plan and lifted both of these restrictions. The rotation length is now based on current stand conditions and each stand's productive potential, and the DNR can harvest up to one-sixth of the available harvest base. In addition, the clustering of timber harvests is not allowed. For visually sensitive areas, the DNR developed a variety of leave tree strategies, with no one strategy dominating. These changes provided foresters greater flexibility for mitigating visual impacts and spread the harvest more evenly across the unit.

The DNR's management of recreation resources did not change substantially as a result of this planning process. However, through this process, the DNR realized the importance of site-specific recreation planning to determine the types and levels of recreation the land can sustain, how much recreation activity can be supported without negatively affecting trust management responsibilities, and how much activity can be supported financially. The DNR also recognized the potential impacts recreation can have on elements of the environment. This thinking led to subsequent, separate recreation planning efforts for the Green Mountain and Tahuya State Forests. The development of the Snoqualmie Corridor Recreation Plan, which includes the Tiger Mountain State Forest, is in progress.

One of the most significant contributions of this planning project was a new way to define and manage northern spotted owl dispersal habitat. Under its existing strategy, the DNR designated spotted owl management units within each dispersal management area. Fifty percent of each spotted owl management unit was maintained as dispersal habitat per the requirements of the *State Trust Lands HCP*. This strategy had two areas for improvement. First, the habitat definition needed to be changed. Second, the spotted owl management units were small and difficult to track.

The process to change the habitat definition for this planning unit began in 2006 with numerous meetings and field visits with experts from the DNR and the USFS, WDFW, and United States Fish and Wildlife Service (USFWS). Working closely with these agencies, the DNR developed a *South Puget movement* habitat definition. This definition amended the original definition with a stand density requirement (no more than 280 trees per ac (692 trees per ha) greater than or equal to 3.5 in (8.9 centimeters (cm)) diameter at breast height (DBH)). The DNR also determined that a second habitat type would be needed to meet the owls' needs. This habitat type was called *movement, roosting, and foraging*. Its definition incorporates forest structural components needed for movement (tree density, cover, and canopy layering), foraging (snags and coarse woody debris), and roosting (canopy layering). This definition also included a minimum snag size of 15 in (38 cm) DBH and a minimum tree height of 15 ft (4.6 m). Changing the habitat definitions resulted in many stands being disqualified as northern spotted owl habitat. In these stands, foresters were previously unable to thin below a relative density of 48, per the DNR's policies that stated that areas considered *dispersal habitat* could not be taken out of that condition. Once declassified, stands would be available for more intensive treatments that would allow them to develop into habitat more quickly.

In addition, the DNR aggregated existing spotted owl management units into larger areas called *dispersal management landscapes*. Within each landscape, the DNR would restore and maintain habitat to the following thresholds: 35% as movement, roosting, and foraging habitat, and 15% as South Puget movement habitat. This change, as well as the new habitat definitions, required concurrence by the Federal Services (USFWS and National Oceanic and Atmospheric Administration Fisheries) to amend the *State Trust Lands HCP*.

The DNR anticipated that these changes would improve habitat conditions over time. Under the preferred alternative, projections showed that the DNR would reach the 50% northern spotted owl habitat threshold in dispersal management landscapes by the fourth decade of the forestland plan.

As a result of proposed changes in management, the DNR determined that harvest levels would need to be altered in two sustainable harvest units. The harvest level for the Pierce County sustainable harvest unit would increase from 12 to 33 million board feet (MMBF) (28,320–77,880 m³), while the harvest level for the Kitsap County sustainable harvest unit would drop from 25 to 17 MMBF (59,000–40,120 m³). These recommendations were approved by the Board of Natural Resources.

Another key outcome of this planning process was the harvest schedule. The DNR developed a harvest schedule in spatial database (map) form for the preferred alternative for the first decade of the plan (2009–2018) to help guide timber sales planning. Developing these maps was a cooperative process. Because of the complexity of the DNR's GIS data, the first set of maps included numerous small areas that were financially infeasible to harvest. To solve this problem, the DNR worked with region staff to combine some harvest units and eliminate others. The result was a harvest schedule that balanced the DNR's objectives and was practical and feasible to implement.

DISCUSSION AND CONCLUSIONS

In many ways, the South Puget forestland plan was an overwhelming success. The DNR's staff expressed the satisfaction of providing practical solutions for complex management challenges. The even-flow strategy has been helpful in maintaining public awareness of working forests, and the new northern spotted owl habitat definitions have been workable solutions. In addition, the final EIS has been praised as a highly readable document that institutionalizes key agency knowledge about the South Puget planning unit. The final EIS also has been effective in training new foresters and is consulted regularly by region managers.

Sustainability Issues

The success of the plan and its implementation was evident in a recent audit by the Forest Stewardship Council. An internationally recognized certification organization, the Forest Stewardship Council found that the DNR's sustainable resource management for the South Puget planning unit met its certification standards for environmental impacts, the maintenance of high conservation value forests, the monitoring and assessment processes, the benefits of the forest (including generation of trust revenue), the forestland planning process, and other standards (Bureau Veritas, 2013). The South Puget planning unit is the only DNR planning unit to be certified under both the Forest Stewardship Council and the Sustainable Forestry Initiative. Certified forests are grown to an approved set of standards for environmentally responsible, socially beneficial, and economically prosperous management (Forest Stewardship Council, 2014).

Plan Development Challenges

Though successful, the development of the forestland plan was not without its challenges. The planning effort required large investments of time and personnel in both the Olympia and region offices. The effort to develop a detailed, accurate model was far more complex than anticipated. It also was difficult to maintain agency focus over the long period of time required to complete the forestland plan. From initiation to completion, the process spanned over 5 years.

Plan Implementation Challenges

Implementation was challenging due to a lack of resources. Immediately following release of the final EIS, the planning team in Olympia was absorbed into another, urgent forestland planning process. Because no Olympia staff was available to assist with plan implementation, the entire burden of implementation was placed on region personnel. Although region personnel were given a harvest schedule, the model itself was archived and the spatial digital tool envisioned at the start of the project was never developed. Other subsequent events, such as litigation that affected management of northern spotted owl habitat on all state trust lands, also prevented the plan from being implemented as fully as envisioned.

Learnings and Insights

The most prevalent area of improvement for future plans is to dedicate time and resources to assist with plan implementation. The final EIS could also be distilled into a concise field guide containing strategies, procedures, and other relevant information. Along with the envisioned spatial digital tool, the field guide could be loaded onto the personal electronic devices (tablets, smart phones) that many new foresters prefer. These items could be highly useful in the event of staff turnover, and the field guide could be shared with stakeholders to increase transparency. Finally, it would be valuable to dedicate time and resources at 5- or 10-year intervals to examine the plan—what is working, what is not, whether objectives are being met as anticipated, and what adjustments may need to be made.

Overall, forestland planning for the South Puget planning unit was a valuable learning experience, providing the DNR its first opportunity to develop strategies for an individual planning unit and produce a harvest schedule through forest estate modeling. Given its success, the South Puget forestland plan has been a model for subsequent planning efforts within the DNR.

REFERENCES

Buchanan, J.B., 2004. In my opinion: managing habitat for dispersing northern spotted owls—are the current management strategies adequate? Wildlife Society Bulletin 32, 1333–1345.

Bureau Veritas, 2013. FSC Certification System, Recertification Public Report, Forest Management Certification, Washington State Department of Natural Resources South Puget Sound HCP Planning Unit. Bureau Veritas, Paris, France.

Carey, A.B., 2007. Aiming for Healthy Forests: Active, Intentional Management for Multiple Values. U.S. Department of Agriculture, Forest Service, Pacific Northwest Research Station, Portland, OR, General Technical Report PNW-GTR-721.

Carey, A.B., Elliot, C., Lippke, B.R., Sessions, J., Chambers, C.J., Oliver, C.D., Franklin, J.F., Raphael, M.G., 1996. Washington Forest Landscape Management Project—A Pragmatic, Ecological Approach to Small-Landscape Management. Washington Department of Natural Resources, Olympia, WA. Washington Forest Landscape Management Project Report Number 2. 110 p.

Forest Stewardship Council, 2014. Mission and Vision. Forest Stewardship Council U.S, Minneapolis, MN. https://us.fsc.org/mission-and-vision.187.htm (Accessed March 18, 2014).

Forsman, E.D., Anthony, R.G., Reid, J.A., Loschl, P.J., Sovern, S.G., Taylor, M., Biswell, B.L., Ellingson, A., Meslow, E.C., Miller, G.S., Swindle, K.A., Thrailkill, J.A., Wagner, F.F., Seaman, D.E., 2002. Natal and breeding dispersal of northern spotted owls. Wildlife Monographs 149, 1–35.

Franklin, J.F., Spies, T.A., Van Pelt, R., Carey, A.B., Thornburgh, D.A., Berg, D.R., Lindenmayer, D.B., Harmon, M.E., Keeton, W.S., Shaw, D.C., Bible, K., Chen, J., 2002. Disturbances and structural development of natural forest ecosystems with silvicultural implications, using Douglas-fir forests as an example. Forest Ecology and Management 155, 399–423.

Gordon, S.M., McPherson, H.M., Dickson, L., Halofsky, J., Synder, C., Brodie, A.W., 2014. Wildlife habitat management. In: Reynolds, K.M., Hessburg, P.F., Bourgeron, P.S. (Eds.), Making Transparent Environmental Management Decisions. Springer-Verlag, Berlin, Germany.

Miller, G.S., Small, R.J., Meslow, E.C., 1997. Habitat selection by spotted owls during natal dispersal in western Oregon. Journal of Wildlife Management 61, 140–150.

National Park Service, 2014. Mount Rainier NP. National Park Service, Washington, DC. https://irma.nps.gov/Stats/SSRSReports/Park%20Specific%20 Reports/Traffic%20Counts?Park=MORA (Accessed March 18, 2014).

Oliver, C.D., Larson, B.C., 1996. Forest Stand Dynamics. John Wiley & Sons, New York. 520 p.

Van Pelt, R., 2007. Identifying Mature and Old Forests in Western Washington. Washington State Department of Natural Resources, Olympia, WA. 104 p.

Washington State Department of Natural Resources, 1997. Final Habitat Conservation Plan. Washington State Department of Natural Resources, Olympia, WA.

Washington State Department of Transportation, 2014. Population Growth in Relation to the State's Counties. Washington State Department of Transportation, Olympia, WA. http://www.wsdot.wa.gov/planning/wtp/datalibrary/population/popgrowthcounty.htm (Accessed March 18, 2014).

Working Group on Criteria and Indicators for the Conservation and Sustainable Management of Temperate and Boreal Forests, 1995. The Montreal Process on Criteria and Indicators for the Conservation and Sustainable Management of Temperate and Boreal Forests. Montréal Process Liaison Office, Tokyo, Japan. http://www.montrealprocess.org/documents/publications/techreports/1995santiago_e.pdf (Accessed March 18, 2014).

Chapter 39

Weyerhaeuser, North Carolina, United States of America

Bud Bigelow, Robert A. Ewing and Venkatesh Kumar

Weyerhaeuser Company, Timberlands Strategic Planning, Federal Way, Washington, USA

ABBREVIATIONS

HBU	Higher and Better Use
HYF	High Yield Forestry
NCDWQ	North Carolina Division of Water Quality
RSPS	Remsoft® Spatial Planning System
USGS	U.S. Geologic Survey
WREDCO	Weyerhaeuser Real Estate Development Company

MANAGEMENT SETTING AND BACKGROUND

In the Coastal Plain region of North Carolina, Weyerhaeuser holds approximately 540,000 acres (ac) (218,535 hectares (ha)) of land (Figure 39.1) in fee simple ownership (absolute legal title). Loblolly pine (*Pinus taeda*) plantations (Figure 39.2) constitute the majority of the forests in these timberlands, with a mixture of pines and hardwoods occupying the streamside management zones (SMZs) and bottomlands. Weyerhaeuser North Carolina timberlands are also certified through the Sustainable Forestry Initiative (SFI®). The North Carolina property is a part of the Weyerhaeuser business segment that owns and manages about 7 million ac (2.83 million ha) of commercial forestland in the United States and Uruguay. The company also holds long-term license agreements on approximately 13.9 million ac (5.63 million ha) of provincial land in Canada. Weyerhaeuser is a forest products company with business segments that currently include timberlands, wood products, cellulose fibers, and real estate. All of Weyerhaeuser's timberlands, including those in North Carolina, are managed as profit centers to generate returns over the near and long term. Returns are generated primarily by focusing on the selling of timber and logs to wood products manufacturing customers. Some of the key ideas that guide the management of Weyerhaeuser lands include practicing silviculture that results in strong financial returns, best fit with the biological potential of the land, and managing resources so that harvests align with markets. The organization operates at different scales to maintain cost advantages and manages land and resources sustainably to meet customer and public expectations. A last key idea reflecting the management of Weyerhaeuser lands is an anticipation of future opportunities through flexibility and experience.

Weyerhaeuser is a timberlands real estate investment trust (REIT). A timber REIT is a U.S. tax structure that allows qualified entities the ability to deduct from taxable income the dividends it pays to its shareholders. The tax efficiency of a REIT addresses the double-taxation problem of traditional C-corporations in the United States. The corporation pays corporate taxes and shareholders must pay an additional tax on distributed dividends. A timber REIT provides a liquid and tax-efficient structure for investor participation in timberland-related investments.

Forest Plans of North America. http://dx.doi.org/10.1016/B978-0-12-799936-4.00039-4

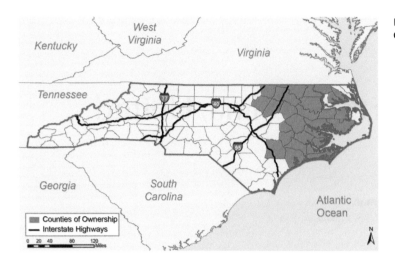

FIGURE 39.1 A location map of the counties in North Carolina that contain Weyerhaeuser holdings.

FIGURE 39.2 Loblolly pine plantations in the Weyerhaeuser North Carolina woodlands.

In North Carolina, Weyerhaeuser operates lumber mills in Greenville, New Bern, and Plymouth, and they are proximate to Weyerhaeuser's timberlands holdings. These mills can produce more than 600 MMBF (over 1.4 million m³) of high-quality lumber products, including Weyerhaeuser Framer Series™ Lumber. The Weyerhaeuser pulp mill at New Bern processes coniferous tree species to produce approximately 310,000 air dry metric tons of specialty and fluff pulp annually.

Weyerhaeuser Real Estate Development Company (WREDCO), a branch of Weyerhaeuser's Timberlands Unit, also influences land management activities. WREDCO's goal is to identify specific land holdings for North Carolina timberlands that show characteristics of higher and better use (HBU) and develop these properties for real estate use. Over the last 30 years, the North Carolina operations unit has successfully developed small and large planned communities in and around the town of New Bern, beginning with the evolution of waterfront and small community developments. The following communities were developed in North Carolina:

- Greenbrier—a small residential community with a golf course
- Cypress Landing—a combined semi-private golf and marina community
- Taberna—a residential community and private golf course
- River Dunes—a marina community
- Carolina Colours—a golf course community

Other smaller communities were also built: Sandy Point, Waterways Edge, and Windsong. WREDCO is working on its newest project, Craven Thirty located near New Bern. WREDCO has long-term plans to continue to identify and develop properties that show HBU characteristics.

Weyerhaeuser's forestry practices are based on High Yield Forestry (HYF) principles and financial attractiveness of even-aged silviculture regimes. The components of HYF include regeneration and genetics, intensive stand management, and final harvest at a financially optimal age. Scientifically based HYF is practiced to maximize timber growth and value, while conserving soil productivity, protecting water quality and biodiversity, and providing wildlife habitat. Silviculture practices in North Carolina include site preparation for planting, use of superior seedlings, stocking control, pruning, vegetation and nutrient management. Soil is prepared using a variety of site preparation methods. For example, chemical or mechanical methods are used to improve the survival and growth of seedlings after planting on the harvested site. Raised beds are the predominant method of site preparation in North Carolina to overcome drainage problems. Genetically superior pine seedlings are planted in rows by hand or machine. Trees are thinned and pruned at a certain age to reduce competition among trees and improve wood properties. A proprietary fertilizer (Arborite®, a phosphate coated urea that includes phosphorus, nitrogen, and boron) is applied to trees at a recommended dosage and frequency to help trees to achieve the desired rate of growth. Competition from non-crop trees and other vegetation for nutrients and water is managed as required. Stands are targeted for harvest at financial maturity and then replanted.

HYF practices in North Carolina produce a wide variety of ecosystem outputs and services across the landscape. Some of these services already produce products and services with market value; others provide opportunities for additional revenue or marketing potential in the future as the markets develop; and some, although not measurable in dollars, illustrate the range of values that accompany our managed forests. Weyerhaeuser collects, tracks, and reports a broad list of ecosystem services so that customers, communities, and other interested stakeholders may have a better understanding of the diversity across our landscape.

Sustainable forest management means meeting the forest resource needs and values of the present without compromising the similar capability of future generations. This includes a land stewardship ethic that integrates reforestation, managing, growing, nurturing, and harvesting of trees for useful products with the conservation of soil, air and water quality, wildlife and fish habitat, and aesthetics. Weyerhaeuser developed a sustainable forest management policy that incorporates the above principles. Significant research and scientific resources are committed in the laboratory and the field to understand how forest practices can improve management opportunities. This information often reinforces the Best Management Practices (BMPs) that are used, knowledge of hydrology and water management, and biodiversity and habitat management in general. Third-party certification programs are then employed to demonstrate that forests are responsibly managed.

PLANNING ENVIRONMENT AND METHODOLOGY

Harvest scheduling within Weyerhaeuser is a business-driven activity based on timberlands strategic direction, changes in inventories, existing and future market conditions, and product demand of customers. Harvest planning uses a combination of commercial and proprietary software to determine sustainable harvest levels, to maximize financial returns, and to maintain environmental sustainability. The outcomes are used in annual, near-term (2–5 years), and long-term planning processes. The key assumptions used in the analysis are prices, costs, merchandizing parameters, and silviculture treatments. The forest inventory for each management unit (~100 ac or 40 ha) is extracted from the resource inventory system and entered into a proprietary growth and yield model to generate yield curves based on a range of silviculture options. Analysts use the Remsoft® Spatial Planning System (RSPS) to deploy a linear programming model capable of evaluating outcomes from applying different silviculture options, changing harvest timing, and complying with adjacency and other constraints due to legal and voluntary certification requirements. Multiple RSPS model scenarios are developed to address a range of management options such as assessing trade-offs between near-term cash flow and long-term value creation, meeting various customer demands, and evaluating alternative business strategies. The output from each scenario includes projected harvest flows, which are used to generate near-term and long-term financial projections. Business leaders review these scenarios and select one as the official scenario. The official scenario ties into near-term (2–5 years) operational plans, is used to set target harvest levels at the regional scale, and produces volume estimates and financial projections for internal and external use.

Weyerhaeuser maintains its timber inventory in an integrated resource inventory and geographic information system (GIS). The resource inventory component of the system is proprietary and is largely based on internally developed technologies, including growth and yield models developed by the Weyerhaeuser Research & Development group. The GIS component leverages ESRI GIS software. The inventory system is the database of record and contains GIS layers that describe owned and leased lands Weyerhaeuser manages. It is a major source of information that is reported annually, both internally and externally. The system supports the full range of operational management opportunities, from regeneration and silviculture treatment activities through harvest planning and logistics. It also provides the key data, which supports long-term harvest planning, land adjustment activities, strategic planning, and portfolio management. The timber inventory data collection and verification techniques include the use of industry-standard field sampling procedures as well as proprietary remote sensing technologies in some geographical locations where they generate improved estimates. The data is collected and maintained at the timber stand level and is updated on a predefined sampling cycle or when stand conditions change.

Guiding Laws, Regulations, and Policies

As mentioned above, SMZs and other harvest boundaries are delineated in agreement with current North Carolina BMPs, riparian buffer rules, and the North Carolina Coastal Area Management Act. North Carolina BMPs specify forest practices to be implemented to protect water quality. These BMPs supplement mandatory, statewide performance standards (Forest Practice Guidelines, N.C. Administrative Code 15A NCAC 01I .0100–.0209). BMPs include provisions for SMZs along intermittent and perennial streams and other perennial waters. Ephemeral streams are also protected by minimizing disturbances and maintaining groundcover adjacent to the ephemeral stream area. Operations are permitted in forested wetlands, but all applicable BMPs must be followed in order to limit soil and hydrologic disturbance. All water quality standards and forest practice guidelines apply to all wetlands, and jurisdictional wetlands are subject to the authority of the U.S. Corps of Engineers.

Proper implementation depends on the identification of intermittent and perennial streams. However, classification of streams can be difficult in the low-gradient stream systems that are commonly found in the eastern portion of the state. The *Methodology for Identification of Intermittent and Perennial Streams and Their Origins* (North Carolina Division of Water Quality, 2010) is used to determine stream system type. In select eastern North Carolina river basins, any stream that appears on a U.S. Geologic Survey (USGS) 1:24,000 (7.5 min) quadrangle topographic map or on a U.S. Department of Agriculture, Natural Resources Conservation Service (NRCS) soil survey map, is considered a *blue line* stream and subject to buffer rules unless a further determination under the North Carolina Division of Water Quality (NCDWQ) identification process demonstrates otherwise. The USGS blue line streams were drawn to topographic standards that do not represent stream function or exact location, and field determination by the NCDWQ process often shows discrepancy. The BMP guidelines provide operational detail but can be summarized as:

1. Maintain an SMZ width adequate to control sediment, reduce nutrient input, stabilize the stream bank, and protect wildlife and aquatic biota. The width of the SMZ is generally 50 feet (ft) (15.2 meters (m)) on either side of the stream.
2. Operations can occur within the SMZ, but no more than 50% of the canopy can be removed and no more than 20% of the ground can be exposed as bare soil.
3. No equipment operation or road construction may occur within 10 ft (3 m) of a stream.
4. Roads require detailed planning and construction, and guidance is found in the North Carolina BMP manual.
5. Forested wetlands are regulated by the U.S. Army Corps of Engineers, and additional assessment and operational provisions are necessary in these areas.

Much of the land managed by Weyerhaeuser North Carolina Timberlands is drained with field ditches and collector canals. Beginning in 1766, canals were developed to drain the forested wetlands for agricultural purposes. In 1909, the North Carolina Drainage District law (Article 11 § 156-135) promoted leveeing, ditching, draining, and reclamation of wet and overflowed lands, and, subsequently, large blocks of land were purchased and drained to sell as smaller family farms. In 1973, Open Ground Farms was drained, creating a 45,000 ac (18,211 ha) block—the largest in the eastern United States. Drainage continued to be encouraged until wetlands regulations stopped the installation of new canals. Although current regulations prohibit new drainage, other than what is considered minor drainage in the Clean Water Act, existing structures may be maintained to the original specifications. Weyerhaeuser performs minimal ditch maintenance approximately once per timber rotation. Drainage reduces the impact of equipment on soil and extends sustainable operational conditions.

A water management study was initiated by Weyerhaeuser in 1986 and conducted in conjunction with scientists from North Carolina State University. The study was designed to examine the effects of water table management on the growth of loblolly pine plantations in poorly drained soils and on water quality. The study was installed on a 185 ac (75 ha) tract in Carteret County called Carteret 7. The tract was divided into three equal-area plots, drained by parallel ditches at 330 ft (100 m) spacing. The ditches were isolated by plot and flow routed through moveable V-notch weirs to control the water table (Amatya et al., 1996; McCarthy et al., 1991). Although the original purpose of the study was improving tree productivity through water table manipulation, water quality and quantity became the focus of the study. A prominent agricultural model was extended to forested systems and is in wide use today (McCarthy and Skaggs, 1992), and much other research has continued to expand the understanding of forest operations in poorly drained soils. The results demonstrate that BMPs are effective in controlling sediment and nutrients in artificially drained areas and the water quality of outflow from ditched areas resemble that from unditched areas (Lebo and Herrmann, 1998; Lebo and Hughes, 1998). Current research at the Carteret site has expanded to include more intensive practices, including forest-based bioenergy systems, and will guide future practices (Nettles et al., 2011).

In the southeastern United States, practices relative to conservation of biological diversity are largely voluntary, except for federally protected species (Endangered Species Act) in some cases. Weyerhaeuser manages for biodiversity using coarse-filter and fine-filter approaches. At the coarse filter, it is assumed that normal silvicultural activities at the landscape scale promote habitat diversity, which provides habitat conditions for a diversity of species. This includes different aged plantations, set-aside areas (e.g., SMZs) of mature forest stands, differential application of silvicultural treatments (e.g., variety of site preparation options that are widely employed) across the ownership, and diversity in vegetation communities resulting from similar silvicultural treatments due to differences in soil types, site productivity, and so forth. This assumption is well supported by research that clearly demonstrates the value of managed forests in the southeastern U.S. for conservation of biological diversity. To track habitat diversity at this scale, Weyerhaeuser annually assesses habitat conditions across southern timberlands.

It is also recognized that certain species of conservation concern, or communities of exceptional conservation value, benefit from a more focused approach. In these cases, specific management actions are taken, as appropriate, to provide habitat for these species or conserve these communities. This approach is particularly applied to species listed as federally threatened or endangered. Across Weyerhaeuser's southern ownership, these species include red-cockaded woodpecker (*Picoides borealis*), gopher tortoise (*Gopherus polyphemus*), Red Hills salamander (*Phaeognathus hubrichti*), American burying beetle (*Nicrophorus americanus*), and American bald eagle (*Haliaeetus leucocephalus*). Unique ecological communities include such features as ephemeral wetlands or unique plant communities. In North Carolina, current management includes consideration of foraging habitat for red-cockaded woodpecker colonies located on the Croatan National Forest (adjacent to Weyerhaeuser property), and identification and management of wetlands, particularly ephemeral wetlands, which are critical habitat features for many amphibian species. Additionally, endangered red wolves (*Canis rufus*) use Weyerhaeuser land in eastern North Carolina and Weyerhaeuser is engaged in research projects involving this species.

Although not a species of concern, the American black bear (*Ursus americanus*) is a socially and economically important game species in eastern North Carolina. In the early 1970s, concern existed about the effects of large-scale establishment of intensively managed pine stands in eastern North Carolina on black bears. For the past 40 years, Weyerhaeuser has worked cooperatively with the North Carolina Wildlife Resources Commission to understand ecology of black bears on industrial forest landscapes and manage the species. Today, black bears are very abundant on Weyerhaeuser-owned timberlands in North Carolina, and the cooperative relationship between Weyerhaeuser Company and NCWRC has been very successful in conservation of this species.

A key element of Weyerhaeuser's management of biological diversity is an internal research program. Although this research addresses issues south wide, most of the research broadly applies to Weyerhaeuser's forestry practices, including eastern North Carolina. The primary goal of this program is to provide technical information regarding effects of forest management activities on wildlife habitat, species, and communities. The program also provides data to support science-based recommendations for cost-effective integration of wildlife habitat goals with forestry operations. The program has been active since the early 1970s and, over time, has shifted from primarily a game species focus to a nongame species focus, continually adapting to address emerging issues.

Forest certification recognizes and confirms responsible forest management. It achieves this by setting out guidelines and standards for how forestland should be managed to protect cultural, environmental, and social values of the land. Credible standards are based on science and developed with input from social, environmental, and economic interests. Certified forestland owners and managers verify that they meet the standards through independent, third-party audits. Weyerhaeuser's southern timberlands are certified under the SFI® program. This certification sets standards for forest management, including management of biological diversity (Objective 4 of the SFI® standard). Weyerhaeuser uses SFI® standards to help set expectations for biodiversity management on company timberlands. Weyerhaeuser's timberlands and products are certified to third-party standards. This system allows it to offer customers a reliable supply of high-quality products certified to a consistent, globally recognized standard. To meet the needs of the many family forest owners who provide up to 60% of its wood fiber, Weyerhaeuser supports the American Tree Farm System as the most effective forest management standard for certifying small forest landowners in the United States. When sourcing from noncertified landowners, certification by the SFI® Certified Sourcing standard requires the use of BMPs, qualified logging professionals, and education to support protection of other resources such as water quality, wildlife habitat, biodiversity, and unique cultural sites. Together, the SFI® and American Tree Farm standards provide effective, complementary programs for promoting sustainable forestry.

Modeling Process

Weyerhaeuser evaluates forest management plans against key financial and environmental performance indicators. Key financial indicators include net cash flow, earnings before interest, taxes, depletion/depreciation and amortization, net present value, and internal rate of return. Key environmental indicators include clear-cut size and adjacency/green-up constraints. The key assumptions used in the analysis include prices, costs, merchandizing parameters, and silviculture treatments.

RSPS is used to construct a linear programming model capable of evaluating outcomes from options that apply different silviculture options, harvest timings, and compliance with adjacency constraints. Plans are evaluated financially by first developing a maximum NPV scenario. This is a theoretical exercise where the model uses a maximum NPV objective, ignoring key constraints such as harvest flows. This scenario represents the theoretical maximum value that can be obtained from managing the property, given the other assumptions used. The financial metrics for subsequent scenarios (considerations on volume flows to customers or swapping near-term versus long-term cash flow) are compared to the theoretical maximum case, and against each other, to assess the financial tradeoffs. Multiple RSPS model scenarios are developed to address a range of management options such as: trade-offs between near-term cash flow and long-term value creation, customer demands, and alternative business strategies. The output from each scenario includes projected harvest flows, and these are used to generate near-term and long-term financial projections.

OUTCOMES OF THE PLAN

As an example of the types of outcomes generated during the development of the forest plan for the North Carolina timberlands, Figure 39.3 compares the harvest flows from the Base Case Scenario and an alternative labeled Scenario 1. The Base Case harvest flow was developed to achieve high near-term cash flows and to align activities with expected customer demands. The disciplined application of silvicultural practices over time increases the sustained harvest flow in the long-term plan over near-term levels. Scenario 1 represents a more traditional nondeclining even-flow objective. The Scenario 1 plan was not preferred as it produced poorer financial results, did not align as well with customer demands, and delayed the achievement of higher long-term harvest levels.

Region planners use the projections from RSPS for their annual and near-term planning efforts. The planners obtain an initial selection of the harvest candidate tracts from the early model results and refine the selection using the resource inventory system to target certain stand conditions in the current management strategy. With the aid of inventory details, such as maps, aerial photos, and field visits, the initial harvest candidate list is modified to comply with green-up and adjacency constraints, natural heritage areas, Environmental Protection Agency guidance zones, and other operational issues. The revised sequence of management units is used in the final RSPS runs. In the woods, candidate stands are prepared for harvesting in accordance with Weyerhaeuser's forest management principles, which include:

1. Roads, ditch conditions, and boundary-line definitions for the selected tracts are evaluated.
2. Environmental managers develop the stream system exemptions (a drainage reserve rule in North Carolina) and reviews for threatened and endangered species, such as American bald eagles and red-cockaded woodpeckers, and submit these to state and federal agencies for approval.
3. Forest planners develop plans for sensitive areas according to the Weyerhaeuser Southern Timberlands Aesthetics Guidelines and submit these to the environmental managers for review.

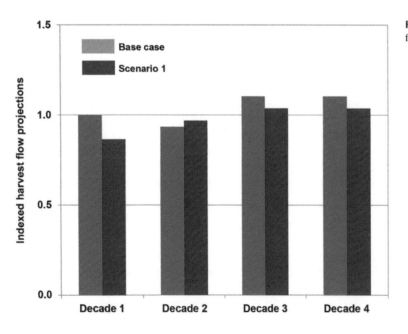

FIGURE 39.3 A comparison of projected harvest flows from two modeled scenarios.

4. Stand-level wildlife habitat elements are created within the company's Resource Management System (RMS) in accordance with the Weyerhaeuser Southern Timberlands Stand Level Biodiversity Guidelines.
5. SMZs and other harvest boundaries are delineated in agreement with current state BMPs, riparian buffer rules, and the North Carolina Coastal Area Management Act.
6. Stand maps for each planned harvest setting are developed to highlight harvest boundaries and applicable sensitive areas. These are produced using the updated inventory database.
7. Harvest plan notebooks containing tract maps, photographs, sensitive area plans, and other supporting information are provided to the harvesting functional team for execution.

DISCUSSION AND CONCLUSIONS

The long-lived nature of timberland investments necessitates taking a similar long view to guide forest management decisions. It is important for forest plans to be able to anticipate the potential effects of uncertain events by building in flexibility and robustness. By evaluating outcome sensitivity against market, customer, and biophysical parameters, it is possible to identify those plan attributes that are required to hold up well under the most plausible set of circumstances. Weyerhaeuser uses its own proprietary research and shared information from forestry cooperatives as a basis for understanding the fundamental dynamics of the forests it manages.

Models are deployed to simulate how timberlands grow and respond to management actions. A combination of commercial and proprietary software helps determine harvest levels, optimization of log and timber sales, financial projections, and environmental outcomes. In the end however, Weyerhaeuser forest planning is an interactive process conducted by planners and operating teams driven by business goals and the need to manage the forest environment in a sustainable way.

Plan Implementation Challenges

Weyerhaeuser operates both timberlands and mills in North Carolina as separate profit centers. Changes in lumber markets affect the demand for logs. Adapting to those changes in log demand from internal and external customers can create challenges to plan implementation. During the recession period of 2009–2011, plans were modified to adapt to the severe market downturn.

Weather events ranging from rainstorms to hurricanes can disrupt harvesting and manufacturing operations and also damage forest inventory. The size and extent of such problems varies with the storm event. Typically, such events have only minor, short-term effects. Fortunately, no severe weather events have affected current plan implementation.

Despite the challenges, Weyerhaeuser's planning tools and processes position the company to profitably adapt harvest plans to changes in the marketplace going forward.

ACKNOWLEDGMENTS

The authors gratefully acknowledge the contributions of Jami Nettles and Darren Miller (Southern Timberlands R&D), Edie Sonne Hall (Sustainable Forestry), Scott Dahlquist (WREDCO), Murray Johnson and Earl Birdsall (Timberlands Strategic Planning), and Jianping Shan (North Carolina Timberlands).

REFERENCES

Amatya, D.M., Skaggs, R.W., Gregory, J.D., 1996. Effects of controlled drainage on the hydrology of drained pine plantations in the North Carolina coastal plain. Journal of Hydrology 181, 211–232.

Lebo, M.E., Herrmann, R.B., 1998. Harvest impacts on forest outflow in coastal North Carolina. Journal of Environmental Quality 27, 1382–1395.

Lebo, M., Hughes, J., 1998. Silvicultural BMP effectiveness during harvest in coastal North Carolina. In: Watershed Management: Moving from Theory to Implementation. Water Environment Federation, Alexandria, VA. pp. 393–400.

McCarthy, E.J., Skaggs, R.W., 1992. Simulation and evaluation of water management systems for a pine plantation watershed. Southern Journal of Applied Forestry 16, 48–56.

McCarthy, E.J., Skaggs, R.W., Farnum, P., 1991. Experimental determination of the hydrologic components of a drained forest watershed. Transactions of the ASAE 34, 2031–2039.

Nettles, J., Youssef, M., Cacho, J., Grace, J., Leggett, Z., Sucre, E., 2011. The Water Quality and Quantity Effects of Biofuel Operations in Pine Plantations of the Southeastern USA. IAHS Press, Wallingford, Oxfordshire, UK. pp. 115–120.

North Carolina Division of Water Quality, 2010. Methodology for Identification of Intermittent and Perennial Streams and Their Origins, Version 4.11. North Carolina Department of Environment and Natural Resources, Division of Water Quality, Raleigh, NC. 41 p.

Chapter 40

Yakama Reservation, Washington, United States of America

Markian Petruncio and Steve Andringa

Yakama Nation Forestry, Toppenish, Washington, USA

MANAGEMENT SETTING AND BACKGROUND

The 1.37 million acre (ac) (about 557,265 hectare (ha)) Yakama Reservation includes 675,811 ac (273,497 ha) of forest and woodlands (Figure 40.1) in portions of Yakima, Klickitat, and Lewis counties in south-central Washington State. When the Yakama Reservation was established, much of the area had forests with open canopies in park-like conditions, which were dominated by scattered old-growth ponderosa pine (*Pinus ponderosa*) (Yakama Nation and the Bureau of Indian Affairs, 2005). The Yakama Reservation is a magnificent resource that provides water, food, medicine, spiritual values, employment, and revenue for the Yakama Nation. The Yakama people and their ancestors based their economy on salmon fishing, root, berry, and nut gathering, hunting, and other intertribal commerce. Inevitable disturbances, including climate change, will require proactive management strategies to restore forest health and achieve the sustainability mandates of the National Indian Forest Resources Management Act (P.L. 101-630), Code of Federal Regulations, and the Indian Affairs Manual (53 IAM). With respect to the Yakama Reservation forests, the Yakama Nation provides both short-term and long-term direction for the forestry program. In addition to the protection of watershed and stream flows, the National Indian Forest Resources Management Act also requires reservation lands to be managed for permanent forest production in accordance with sustained-yield principles (Bureau of Indian Affairs, Yakama Agency, 2005). The tribe employs people in professional, technical, and labor positions under an Indian preference policy. The Bureau of Indian Affairs (BIA) provides technical recommendations upon request, develops and maintains the Geographic Information Systems (GIS) data, and assists in developing silvicultural prescriptions.

Since the signing of the Treaty of 1855 (12 Stat. 951), the environmental conditions of the Yakama Reservation have been greatly changed by livestock grazing, timber harvesting, insects, diseases, wildland fires, and fire exclusion policies. Livestock grazing, particularly sheep grazing in the early 1900s, changed the amount and species composition of ground vegetation. The reduction of fine ground fuels prevented surface fires from burning across the landscape, as they did historically. Suppression of natural fires also prevented fires from performing important ecosystem functions such as regulating tree species composition, reducing forest stand densities, and cycling nutrients. Selective timber harvesting removed large ponderosa pines, and grand fir (*Abies grandis*) and Douglas-fir (*Pseudotsuga menziesii*) often regenerated in place of the pines.

In terms of fish and wildlife species of concern, northern spotted owls (*Strix occidentalis caurina*) can be found in some parts of the forest, along with goshawks (*Accipiter gentilis*), steelhead (*Oncorhynchus mykiss*), and bull trout (*Salvelinus confluentus*). The northern spotted owl has been the primary listed species that has affected management on Yakama lands. The BIA, as trustee of Yakama lands, is mandated to follow federal laws, including the Endangered Species Act. Listed species, under this law, are regulated by the U.S. Fish and Wildlife Service, which develop the mandatory terms and conditions for the protection of those species. The Yakama Nation has a desire to protect all species and generally does not view one species as more important than another. The Yakama Nation Forestry Program is therefore striving to meet present tribal needs and values while providing a desirable range of forest management options for future generations. Forest health is perhaps the main concern; however, social and cultural issues are important as well. A tremendous amount of forest restoration work is being accomplished while the underfunded and understaffed natural resources programs are struggling to meet unfunded federal mandates.

Forest Plans of North America. http://dx.doi.org/10.1016/B978-0-12-799936-4.00040-0

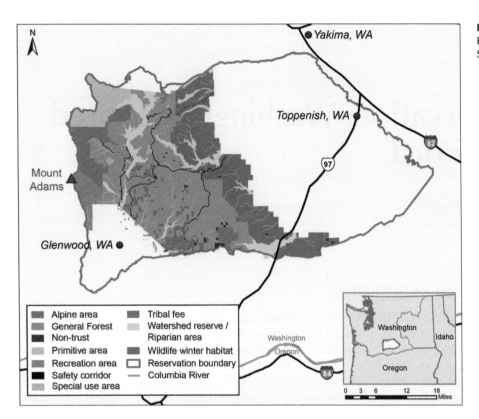

FIGURE 40.1 A map of the Yakama Reservation in south-central Washington State.

The pronounced elevation and precipitation gradients across the reservation greatly affect the distribution and growth of vegetation. Annual precipitation ranges from 7 inches (in) (17.8 centimeters (cm)) along the Yakima River in the east to 100 in (254 cm) along the Cascade Crest in the west. Elevation ranges from 700 feet (ft) (213 meters (m)) along the Yakima River to 12,276 ft (3,742 m) at the summit of Pahto (Mount Adams). The 1934 and 1997 forest inventories and cover type maps provide references for how the forest has changed in recent times. The four forest cover types that comprise most of the forested area today include ponderosa pine (35%), pine-fir (23%), mixed conifer (21%), and true fir (subalpine fir, *Abies lasiocarpa*)/mountain hemlock (*Tsuga mertensiana*) (19%). Pure stands of lodgepole pine (*Pinus contorta*) occur on about 2% of the land area. In general, the ponderosa pine, pine-fir, and mixed conifer forest types represent a gradient from open-canopy, low-density, ponderosa pine stands to closed-canopy, high-density mixed conifer stands. A comparison of forest type acres shows that ponderosa pine stands decreased from 58% of the area in 1934 to 26% in 1997. As Douglas-fir became established beneath ponderosa pine, many of the pine stands converted to pine-fir while, at the same time, some of the pine-fir stands converted to mixed conifer stands. Mixed conifer forest types increased from 8% of the area in 1934 to 32% in 1997. The 1934 forest cover type map also showed 36,380 ac (14,723 ha) that burned in the early 1900s. Many of the burned areas regenerated naturally and are now classified as true fir-mountain hemlock timber types. The 1934 forest inventory indicated that the standing timber volume was about 4 billion board feet (9.43 million cubic meters (m³)). Sixty-two years later, after 6 billion board feet (14.15 million m³) was harvested, the standing volume had increased to 11 billion board feet (25.94 million m³). The net volume growth over the period was approximately 13 billion board feet (30.66 million m³). The 2005 inventory showed a standing volume of 9 billion board feet (21.23 million m³).

The highly regarded management goals of the Yakama Reservation were setback by a western spruce budworm (*Choristoneura occidentalis*) epidemic that lasted two decades. As a result of the changes in species composition and increased stand densities, competition increased in many areas and tree vigor decreased, and forests became susceptible to outbreaks of defoliators, such as the western spruce budworm and the Douglas-fir tussock moth (*Orgyia pseudotsugata*), both of which are native insects that prefer to forage on grand fir and Douglas-fir foliage. Budworm populations began to increase in 1985 in the southwest area of the reservation, with a corresponding increase in defoliation that resulted in a high mortality rate. Many Douglas-firs were defoliated for several consecutive years and the stress of defoliation made them susceptible to attack by the Douglas-fir beetle (*Dendroctonus pseudotsugae*). In 2000, budworm defoliation of susceptible trees was occurring on approximately 206,000 ac (83,367 ha). A mountain pine beetle (*Dendroctonus ponderosae*) epidemic further resulted in extensive mortality of lodgepole pine. Lodgepole pine occurs in pure stands and across large areas as a component of the true fir-mountain hemlock and subalpine fir timber types. Most of the beetle-affected areas have been salvaged for pulp. An appropriate strategy will be to manage lodgepole pine on shorter rotations (70 years); the trees will be harvested before they become susceptible to bark beetles. The resulting high mortality generated from these events has greatly increased the wildland fire hazard, especially large stand replacing fires, which is at its highest during late summer when fuels are dry and lightning storms move across the region.

Timber harvesting on the Yakama Reservation began in 1944 when 9 million board feet (MMBF) (21,240 m³) of ponderosa pine was harvested because of a bark beetle epidemic. Harvested timber volumes increased as railroad and truck access improved and forestry staffing levels increased. The annual timber harvest level peaked at 226 MMBF (533,360 m³) in 1999 in response to the budworm epidemic and then decreased in the following years. During the budworm timber sales, the amount of pine harvested, as a percent of total volume harvested, decreased while the amount of Douglas-fir and grand fir harvest increased. This trend reflected the increased growth of Douglas-fir and grand fir over the last century, as well as the deliberate removal of these species because they are the budworm's preferred food. The species mix of timber harvests is changing now as a result of the successful conversion to early-seral species, and timber harvest volumes are decreasing.

PLANNING ENVIRONMENT AND METHODOLOGY

The first forest plan for the Yakama Reservation was developed in 1942, and subsequent plans have been developed about every 10 years (Yakama Nation and the Bureau of Indian Affairs, 2005). The first plan proposed a rotation age of 120 years for even-aged stands, and nearly all of the harvesting activities were directed toward ponderosa pine stands that were heavily attacked by western pine beetle (*Dendroctonus brevicomis*). The first forest-wide inventory was developed for the 1962 forest management plan. The 1974 management plan suggested uneven-aged management of ponderosa pine, pine-fir, and mixed conifer forest types, while lodgepole pine and true fir forest types were to be managed as even-aged forests. The most recent plan was completed in 2005 (Yakama Nation and the Bureau of Indian Affairs, 2005) and uneven-aged management was again noted as the preferred silvicultural system (except in lodgepole pine and true fir forests). The Yakama Nation's Land and Natural Resources Policy Plan states that a forest management plan shall design new management strategies to address changes in forest conditions that occurred over the last century. On the Yakama Reservation, natural resource planning occurs at two scales: reservation-wide landscape (the scope of the forest management plan) and project. The reservation-wide planning process identifies and prioritizes projects, analyzes the relationship of each project to other projects, and provides direction with respect to a desired range of future conditions. The project-scale process involves specific actions, the management directions, silvicultural prescriptions, road plans, and other requirements that are important for site-specific projects. A number of issues, concerns, and opportunities arose during the development of the forest management plan, including (in alphabetical order):

- Big-game habitat
- Forest health
- Old-growth forests
- Revenue and employment
- Threatened and endangered species
- Water quality

As described in the 2005 forest management plan, some of the expected future conditions of the Yakama Reservation include:

1. Landscapes should closely resemble those that were created by the activities of historic disturbance agents, such as fires, wind events, insect and disease outbreaks, and animals.
2. Air quality should be maintained at high levels.
3. Culture, tradition, and practices commonly employed by people living on the reservation should remain.
4. Populations of native and desired non-native fish species, and their habitat, are viable and maintained.
5. Habitat conditions for desired native and non-native flora and fauna are suitable and exist in order to maintain biodiversity.
6. Year-round, traditional use opportunities are diverse and are provided with an emphasis on resource utilization and protection.
7. Short-term and long-term economic stability for the Yakama Nation membership is viable.
8. Surface and ground waters are maintained in sufficient quantity and distribution to sustain the water, soil, and other resources.

Guiding Laws, Regulations, and Policies

A number of published works guided the development of the 2005 forest management plan for the Yakama Reservation. For example, an Environmental Assessment was completed in 2005, which listed five alternatives from which the Yakama Nation Tribal Council selected a preferred alternative (Bureau of Indian Affairs, Yakama Agency, 2005). An archaeological overview (Uebelacker, 1984) was completed for a previous planning process and was still considered applicable, providing information on how to locate, evaluate, preserve, and enhance prehistoric and historic resources. Best management practices were developed to address water quality, soil, and fisheries objectives. These are required by the Clean Water Act (P.L. 80-845, 33 U.S.C. §§ 1251-1387) on federally administered lands and represent policies, guidelines, and practices designed to improve water quality. The objectives of the best management practices address road and drainage system design and maintenance, timber harvests, stream crossings, landing locations, dust abatement, and other issues and practices affecting water quality. Best management practices were also developed for aesthetic resources, and management activities need to conform to Clean Air Act requirements (Yakama Nation and the Bureau of Indian Affairs, 2005). Other federal laws that are applicable to the forest management plan include the Antiquities Act of 1906 (P.L. 59-209), the National Historical Preservation Act of 1966 (P.L. 89-665), the National Environmental Policy Act of 1969 (P.L. 91-190), and the National Indian Forest Resources Management Act of 1990 (mentioned earlier), among others.

The Yakama Nation utilizes Tribal Council Resolutions to transact business. These are developed in accordance with a General Council Resolution that was passed in February 18, 1944. Tribal Council Resolutions that related to the development of the 2005 forest plan included those that established the goals, policies, and appropriate uses of land (T-92-87), harvest policies (T-67-88), adoption of plan alternatives (T-021-04), and management plan approval (T-159-05) among others.

Planning Process

In developing the management plan, the forests of the Yakama Reservation were divided into management emphasis areas (Table 40.1). The majority of the Yakama Reservation is open to tribal members but closed to the general public. The primary exception is the *Tract D Recreation Area*. The Recreation Area, encompassing approximately 21,000 ac (about 8,500 ha), is located on the southwestern portion of the Yakama Reservation on the slopes of Mount Adams (Figure 40.2). A predominantly natural alpine meadow environment is found at the base of Mount Adams, and rugged, steep glaciers along its slopes characterize the area. Ownership of this area has been in dispute dating back to the Treaty of 1855 and the disappearance of the 1855 map of the reservation boundaries. An 1890 survey of the Reservation, accepted by the General Land Office, did not include the Tract D area. The original treaty map, which included the Tract D area, was found in 1930. In the meantime, 98,000 ac (39,660 ha) of the Glenwood Valley had passed into private hands. Another 21,000 ac

TABLE 40.1 Management Emphasis Areas of the Yakama Trust Forest (Revised April 2014)

Management Emphasis Area	Acres	Hectares
General forest	248,544	100,584
Wildlife winter habitat	132,514	53,628
Riparian	97,459	39,441
Canyon	54,129	21,906
Alpine	43,027	17,413
Primitive	31,157	12,609
Woodlands	23,000	9,308
Roads	22,734	9,200
Tract D recreation	13,954	5,501
Safety corridor	7,321	2,963
Traditional use	2,332	944
Total	675,811	273,497

FIGURE 40.2   Mount Adams, as viewed from inside the Yakama Reservation.

(about 8,500 ha) were part of the Gifford Pinchot National Forest, which is administered by the Department of Agriculture. However, in 1972 Richard Nixon authorized the return of the 21,000 ac of National forestlands to the Yakama Nation by Executive Order 11670.

The Yakama Nation will continue to keep this area open to the general public, subject to tribal rules and regulations. Tribal Resolution T-13-71 states:

Whereas, the Tribe recognizes the public interest in continued use of this area ... the Tribe will maintain existing recreational facilities for public use; will continue to recognize the dedication of that portion included in the Mt. Adams wilderness uses ...

In contrast, the other areas are accessible only for the exclusive use and benefit to the Tribal members. These include areas with a long history of personal and family use. Tribal use is not so much nestled within the context of *recreation*, but instead these areas currently may undergo other uses by the Tribal members, which may include camping. An attempt has been made to map those unique, special areas to prevent irreparable damage. Other reasons for identifying them include the need to maintain important historic, cultural, or scenic values, and to protect certain vegetation, fish, and wildlife resources, or other natural systems or processes.

During each planning period, management emphasis areas are refined as tools such as GIS and aerial imagery are improved. These areas now better approximate on-the-ground geographical references, and follow the contours, slope, and aspect of the land. The Yakama forestry program used historic species composition and stand densities as references for the desired range of stand conditions. The restoration strategy focuses on an ecosystem management approach, which considers the sustainability of all resources. Emphasis is on achieving management objectives at the scale of sub-basins. Silvicultural prescriptions are based on forest habitat types, which are used to classify land according to potential natural vegetation and productive capability, and are modified according to management emphasis area objectives. Silvicultural prescriptions are designed to regulate stand densities to decrease competition after harvests occur, to avoid stressful overstocking prior to the next planned harvest, and to meet the objective of each management area. To obtain an estimate of growth, mortality, volume, and trends for informing the forest plan, 1,286 Continuous Forest Inventory plots were measured. Average conditions for three of the more common forest types are presented in Figure 40.3. To address forest health issues on the Yakama Reservation, stand densities are being reduced through silvicultural treatments that include precommercial thinning, commercial thinning, and prescribed burning, and management practices emphasize a shift toward early-successional, shade-intolerant tree species.

Five management alternatives were developed following a public scoping and analysis process, and ranged from no change in management direction to a complete ecosystem-based management alternative. The alternative selected was a modified ecosystem management alternative. It addresses three issues of concern: the length of time for re-entry into logging units; the social and economic impacts of continuing a 158 MMBF (372,880 m³) harvest level for 10 years followed by a potential sharp reduction in timber volume, revenue, and jobs; and the need for additional mitigation activities to reduce resource impacts. This management alternative asserts high priority to the management of water resources, fisheries, and wildlife habitat on the Yakama Reservation. The calculated allowable annual cut is based on the available commercial areas within the logging units that were delineated for the 2005 Forest Management Plan. Archaeological, cultural, fish, wildlife, soil, and water resources surveys, stand examinations, timber cruises, and market assessments for

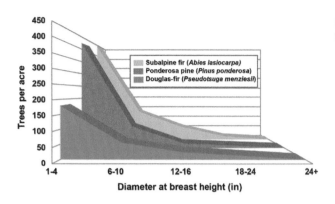

FIGURE 40.3 Diameter distributions representing average conditions for three common forest types found on the Yakama Reservation.

individual timber sales are conducted during the timber sale planning process. The diminution coefficient (q-factor) was important in developing harvest recommendations and ranged from 1.2 to 1.4, depending on forest type. Entry cycles are 10–20 years, again depending on forest types, residual basal area levels range from 40 to 140 ft^2 per ac (9.2–32.2 m^2 per ha), and residual canopy cover levels range from 20% to 80%. An adaptive approach to riparian buffer distances is employed, and a minimum of 50% shade potential (obtained through crown closure or topographic conditions) will be maintained adjacent to stream channels.

OUTCOMES OF THE PLAN

The intent of the 2005 forest management plan was to recover as much value as possible from insect-damaged forests at the beginning of the plan's time horizon, followed by a return to a sustained-yield forest management program (Yakama Nation and the Bureau of Indian Affairs, 2005). In addition to broad goals that represent moving the forest resource toward a resilient state with regard to potential natural disturbances, the outcomes of the preferred alternative suggest that a road management plan, a smoke management plan, and a range management plan would also be developed. Inventory and mapping activities would address other resources of concern (e.g., interior forest and old-growth conditions), and fuel treatments would occur on about 20,000 ac (8,094 ha) annually. The Continuous Forest Inventory indicated the total gross conifer timber volume on trust land within the Yakama Reservation in 2005 was 9 billion board feet (21.23 million m^3). The proposed harvest level under the 2005 plan for the first few years was 158 MMBF (372,880 m^3), but the management direction indicates the harvest levels should range between 120 MMBF (283,200 m^3) and 160 MMBF (377,600 m^3) annually. The annual timber harvest level in 2012 was 67 MMBF (158,120 m^3) and will likely be reduced to 50 MMBF (117,925 m^3) by 2020. The actual amount of timber to be harvested depends on forest growth, disturbances, staffing, lumber markets, and tribal desires.

For long-term management of the western spruce budworm problem, silvicultural treatments that alter stand conditions are proposed. These include prescriptions aimed at invigorating stands through thinnings, altering tree species composition to pines and western larch (*Larix occidentalis*), reducing vertical diversity in fir-dominated stands, and replacing pockets of damaged trees. The main emphasis is to aggressively salvage the imminent mortality of trees, to take proactive measures to reduce the threat of further insect outbreaks, and to promote resistance to future outbreaks. Following commercial harvesting, precommercial thinning is performed by the Yakama Nation Tribal Forestry Program to control species composition and regulate stand density to ensure future productivity. In some areas, there is adequate natural regeneration of the desired species; however, if there is not an adequate seed source then the areas are planted to ensure that the desired species are established at appropriate densities.

Clear-cuts, or small forest openings, where prescribed, should generally not exceed 5 ac (2 ha), although in some cases where ground warming and the stimulation of shrub growth is deemed important, may approach 20 ac (8 ha). In addition, 10% of each sub-basin should contain old-growth forest types that are at least 80 ac (32 ha) in size. In response to the need to manage goshawk habitat, guidelines were developed for the distribution of potential nest areas and successional stage requirements for post-fledging and foraging areas. For example, in foraging areas, mid- to late-successional forest conditions should be maintained across 60% of the affected landscape, with about 20% in a young forest condition, about 10% seedling/sapling forest conditions, and about 10% in grass, forb, or shrub conditions. Guidelines were also developed for the amount of snag and coarse woody debris resources desired in key habitat areas. For big game, hiding cover and foraging areas are important, and guidelines were developed to describe the conditions desired to accommodate these species of wildlife. As an example, hiding (security) buffers 300 ft (91 m) wide need to be maintained around openings that are 8 ac (3.2 ha) or larger in key big game areas. In addition, forest managers need to provide about 25% of the landscape as optimal cover conditions in wildlife wintering areas. The guidelines that relate to old-growth forest types were developed primarily for those wildlife species that utilize tight, closed, and multilayered canopies and were designed to provide some of this type of habitat broadly across the reservation.

DISCUSSION AND CONCLUSIONS

The forests of the Yakama Reservation have changed and will continue to change over time. Without a doubt, the Yakama Reservation looks very different today compared to how it looked 20 years ago; however, the successful silvicultural treatments are moving the species composition and stand densities toward a range of more sustainable conditions. The lessons from the budworm and beetle epidemics must not be forgotten. In many areas, inaction led to defoliation and beetle attacks followed by further deterioration and loss of resources. The pathway to sustainable forestry requires proactive and adaptive management. The long-term management goals are to restore forest health by promoting the development of more open forest stands at low and middle elevations and early-seral species at appropriate densities at upper elevations. Achieving these management goals will ensure that the Yakama Reservation will be maintained as a dependable source of water, food, medicine, spiritual values, employment, and revenue for the Yakama Nation.

Sustainability Issues

Sustainable forest management accommodates environmental protection, economics and social concerns in forest management planning and decision-making processes. While the sustained yield of goods and services is important, in their environmental assessment, the Bureau of Indian Affairs (2005) notes the following:

> The goal of sustainable forest management is to meet the Yakama Nation's present needs and values without compromising the management options of future generations.

With this in mind, the Yakama Nation has a primary goal of maintaining its forest in a healthy condition in order to sustain multiple resources that include spiritual values and cultural resources such as medicines, food plants, water, fish, and wildlife. Broadly speaking, the 2005 forest management plan utilizes an ecosystem approach that considered the sustainability of all resources. Resource sustainability for this land implies that a dependable source of roots, huckleberries, clean water, fish, vigorous trees, animal forage, and clean air will be provided, and that the forest would be resilient to periodic disturbances and changes in the climate (Bureau of Indian Affairs, Yakama Agency, 2005). Cultural, spiritual, economic, and environmental goals are all important. Equity with regard to future generations of people is also important, and proactive management of the forest resources is necessary to achieve these goals.

Plan Implementation Challenges

The silvicultural prescriptions in the forest plan emphasize managing for appropriate tree species and stand densities with regard to the carrying capacity of the land. The long-term goal is to restore the forest to a sustainable condition. The characteristics of that elusive condition will likely change over time as forest pests, climate, and tribal perspectives continue to change. The Yakama Nation Fuels Management Program conducts controlled burns to reduce wildland fire hazards. Prescribed fire can be very beneficial by recycling nutrients, regulating species composition, and adjusting stand density. These thinning, planting, and burning projects provide jobs for tribal members. Unfortunately, funding for prescribed burning projects has been greatly reduced in recent years. Further, the detection of insect outbreaks is critical to the management of the forest, therefore, continuous efforts are needed to understand evolving pest management situations in stand conditions that the various insects prefer.

The actual timber volumes that are harvested from year to year are affected by markets or decisions made during the timber sale planning process that may reduce the land areas and timber volumes available for harvest. For example, timber on steep slopes that are difficult to access may not be harvested if it is not economically feasible to do so at the time of the timber sale. Additional areas may be deferred or excluded from harvest to protect wildlife habitat and cultural resources. The Yakama Tribal Council makes the final decision whether to approve a proposed timber sale. This type of management does create challenges, but rather than adjusting management prescriptions to make a timber sale more economically feasible, the Yakama Nation will defer those areas for future consideration. Under the current revision of the forest management plan, the staff is attempting to identify those deferred areas and make adjustments to the annual allowable cut.

REFERENCES

Bureau of Indian Affairs, Yakama Agency, 2005. Environmental Assessment for the Forest Management Plan. Bureau of Indian Affairs, Yakama Agency, Branch of Forestry, Toppenish, WA.

Uebelacker, M.L., 1984. Timeball: A story of the Yakima People and the Land. The Yakima Nation, Toppenish, WA. 218 p.

Yakama Nation and the Bureau of Indian Affairs, 2005. Forest Management Plan, Yakama Reservation. U.S. Department of the Interior, Bureau of Indian Affairs, Yakama Agency Branch of Forestry, and the Yakama Nation, Toppenish, WA. 259 p.

Chapter 41

French-Severn Forest, Ontario, Canada

Barry Davidson

Westwind Forest Stewardship Inc., Parry Sound, Ontario, Canada

MANAGEMENT SETTING AND BACKGROUND

The French-Severn Forest is located in central Ontario on the southern part of the Canadian Shield, which is known for shallow soils, glacial till, and ancient seabeds. There are distinct geoclimatic differences between this area and the productive farmland that can be found only a few miles to the south. The forest is typified by rugged terrain, hundreds of lakes, and thousands of ponds. The western side of the forest borders Georgian Bay, which is the largest freshwater lake bay in the world and is part of Lake Huron, one of the five Great Lakes. The famous Muskoka Lakes are also found in this forest. These features, along with the fact that the forest is only a two-hour drive from the city of Toronto and within a five-hour drive of several million people in southern Ontario and parts of the United States, make tourism an economic mainstay of the area. The cottaging industry is a main component of the economic health of the region, including the towns of Parry Sound, Huntsville, Bracebridge, and Gravenhurst. The forest is bounded to the east by Canada's oldest and best known provincial park, Algonquin. Several thousand park visitors travel through the French-Severn Forest each year to camp and experience Algonquin Park. It should also be noted that there are eight First Nations communities (aboriginal people who are neither Métis (mixed First Nations and European heritage) nor Inuit (Artic peoples)) within or immediately adjacent to this forest and additional Aboriginal communities that are nearby.

The French-Severn Forest (Figure 41.1) covers 1,281,700 hectares (ha) (3,167,081 acres (ac)); however, almost half is privately owned. When the extensive areas of waterbodies and other non-forested and nonproductive forest areas are removed from the inventory, the Forest Management Plan (FMP) directly applies to about 325,000 ha (803,075 ac) of public production forest. The forest is dominated by two main forest types: the shade-tolerant hardwoods and eastern white pine (*Pinus strobus*) forests.

The first main forest type, shade-tolerant hardwoods, is the most common (Table 41.1). It is dominated by sugar maple (*Acer saccharum*) associated with yellow birch (*Betula allenghaniensis*), American beech (*Fagus grandifolia*), red maple (*Acer rubrum*), red oak (*Quercus rubra*), and black cherry (*Prunus serotina*). Other hardwood species occur incidentally. Eastern hemlock (*Tsuga canadensis*) occurs as scattered individuals, clumps, or large patches within the shade-tolerant hardwood forests. Shade-tolerant hardwoods are managed under the selection or shelterwood silvicultural systems. The second main forest type, eastern white pine, is commonly associated with red pine (*Pinus resinosa*), balsam fir (*Abies balsamea*), white birch (*Betula papyrifera*), and trembling aspen (*Populus tremuloides*). Eastern white pine forests are managed under the shelterwood system (Figure 41.2). Other tree species are found within the shade-tolerant hardwood and pine stands, as well as in mixedwood or near pure stands (e.g., jack pine (*Pinus banksiana*)).

The diversity of tree species and forest types has resulted in a significant level of wildlife species richness. The big game species are: moose (*Alces alces*), white-tailed deer (*Odocoileus virginianus*), and black bear (*Ursus americanus*), which are all abundant. Many neotropical birds nest in the forest, while other species, such as red-shouldered hawk (*Buteo lineatus*) and the great blue heron (*Ardea herodias*), have special significance in the forest. This forest is also home to a number of species recognized under Ontario's Endangered Species Act (2007), the most notable being reptiles that include the eastern massassauga rattlesnake (*Sistrurus catenatus*) and Blanding's turtle (*Emydoidea blandingii*).

Commercial logging has been ongoing in the French-Severn Forest since the mid-to-late 1800s. Early logging supplied white pine to European and emerging American markets. Until the mid-1970s, harvesting focused on the bigger, better, and more valuable trees. In recent years, management efforts have begun to reverse that trend and improve the forest. Hardwood

Forest Plans of North America. http://dx.doi.org/10.1016/B978-0-12-799936-4.00041-2

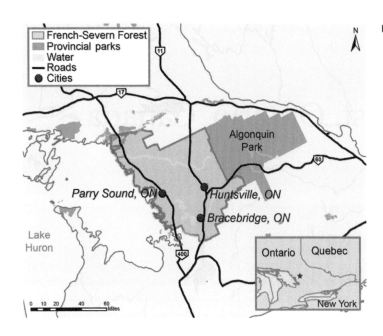

FIGURE 41.1 Location map of the French-Severn Forest.

TABLE 41.1 Forest Area by Forest Unit in the French-Severn Forest in Ontario

Forest Unit	Area (ha)	Area (ac)
Mixed conifer lowlands	6,475	16,000
Mixed conifer uplands	12,372	30,571
General mixedwoods	14,658	36,220
Jack pine	9,887	24,431
Poplar	45,154	111,575
White pine/red pine	78,476	193,914
Shade-tolerant hardwoods	189,879	469,191
Total	356,901	881,902

FIGURE 41.2 Shelterwood harvest in a pine stand of the French-Severn Forest.

management uses tree marking and partial cutting systems like selection and shelterwood that improve the quality, composition, and health of the forest. White pine stands are being renewed through partial harvesting (using the uniform shelterwood silvicultural system), supplemented with site preparation, planting, and tending.

The historic sawlog economy focused on quality log production rather than maximum fiber production. Sawmills within and adjacent to the forest continue to be partially supplied by wood from this forest and receive wood from private forests as well as other management units. The management strategies that improve the health and quality of forests produce a significant amount of low-quality wood that is not suitable for sawlogs or veneer. Limited markets exist for low-quality material; pulpwood is trucked to pulp mills outside the forest, and a significant portion of the harvest is used as fuelwood. There are currently no biomass facilities, such as wood pellet mills, in the area; however, this situation is expected to change in the future.

PLANNING ENVIRONMENT AND METHODOLOGY

The primary parties involved in managing the French-Severn Forest include the Ontario Ministry of Natural Resources and Forestry (MNRF) and the Sustainable Forest License (SFL) holder, Westwind Forest Stewardship Inc. (Ontario Ministry of Northern Development, Mines and Forestry, 2011). In Ontario, there are two main types of SFL holders: (a) those that hold licenses for harvesting and processing, and (b) those that are "co-op" type SFLs, in which the SFL holder plans and oversees forest management activities on behalf of a group of logging and processing companies. Westwind falls into this latter category. Westwind is a unique forest management company, as it is registered as a not-for-profit company and is guided by a community-based Board of Directors with three positions for industry representatives, four positions for representatives from the community-at-large, and one position to represent Aboriginal communities. Westwind was the first forestry company in Canada to attain Forest Stewardship Council certification for a large public forest.

An interdisciplinary planning team is formed by the MNRF District Manager. The plan author is the Westwind forester in charge of planning. The district forester also plays a key role in developing the plan. Additional plan members include biologists, land-use planners, park staff, fire management staff, GIS staff, Aboriginal members, industry members, and a representative from the Local Citizens Committee.

Guiding Laws, Regulations, and Policies

Before any forestry activities can take place in a management unit, there must be an approved forest management plan. The forest management planning system for Ontario's Crown forests is based on a forest policy and legal framework that incorporates sustainability, public involvement, Aboriginal involvement, and adaptive management as key elements. The Crown Forest Sustainability Act of 1994 and the Environmental Assessment Act provide the legislative framework for forest management on Crown land in Ontario (Ontario Ministry of Natural Resources, 2014).

The purpose of the Crown Forest Sustainability Act is to provide for the sustainability of Crown forests. The Act defines forest sustainability as long-term Crown forest health and requires forest management planning to provide for determinations of sustainability based on two fundamental principles:

1. Large, healthy, diverse, and productive Crown forests and their associated ecological processes and biological diversity should be conserved.
2. The long-term health and vigor of Crown forests should be provided for by using forest practices that, within the limits of silvicultural requirements, emulate natural disturbances and landscape patterns while minimizing adverse effects on plant life, animal life, water, soil, air, and social and economic values, including recreational values and heritage values.

To achieve this end, each FMP contains a broad management strategy or Long-Term Management Direction, which balances objectives related to forest diversity, socioeconomics, forest cover, and silviculture.

The provisions of the environmental assessment approval under the Environmental Assessment Act, referred to as the "Declaration Order regarding the Ministry of Natural Resource's Class Environmental Assessment Approval for Forest Management on Crown Lands in Ontario," give MNRF approval to undertake the activity of forest management. These activities comprise the interrelated activities of access, harvest, renewal, and maintenance, subject to the conditions outlined in the declaration order, including requirements for planning. The forest management planning requirements of the Crown Forest Sustainability Act and the provisions of the environmental assessment approval under the Environmental Assessment Act are incorporated into the Forest Management Planning Manual, which was revised in 2009 and which provides the direction for preparing a FMP (Ontario Ministry of Natural Resources, 2009).

Process for the Protection and Maintenance of Social and Environmental Values

Integrated resource management has been a hallmark of management in Ontario for a few decades. Although the term itself is seldom used, planning and carrying out forest management operations is always done with the fundamentals of integrated resource management in mind. There is a societal recognition that the forest provides for many values of which timber supply is but one. In a forest that is so close to large population centers and that provides a backdrop for intensive cottaging, tourism, and outdoor recreational activities, non-timber values and their maintenance is of crucial significance. Forest managers also recognize the role that the forest plays in providing habitat to both terrestrial and aquatic wildlife and overall biological diversity. Very early in the planning process for the 10-year strategic plan, the planning team and Local Citizens Committee participate in a joint meeting to identify forest benefits. These benefits include non-timber forest values.

The FMP relies on an inventory of non-timber forest values to assist in plan development and implementation. MNRF is responsible for maintaining values inventories. At the beginning of a planning period, MNRF undertakes various surveys to make significant improvements to the inventories. For example, helicopter surveys of heronry and osprey nests have been undertaken, often in conjunction with moose aquatic feeding area surveys.

The FMP must provide for the protection and maintenance of both social and environmental values. The impact of forestry operations on social values can generally be described through visual and noise aesthetics. Generally, noise esthetics is limited to the actual forestry operations while visual aesthetics must be considered long term. Also, social values may be affected, either negatively or positively, through the change in access to the forest that is a result of building, upgrading, and maintaining forestry roads. The FMP mainly addresses visual aesthetics through no-cut or partial-cut requirements around such values as cottaging lakes, hiking trails, and portages. Noise concerns are mainly dealt with through best management practices that limit activities during the busiest recreational season (July and August). Partnerships between Westwind and tourism providers identify specific agreements to limit negative impact on those tourism values. Existing forest roads must be left in as good or better condition than they were found. The exception to this is in areas where higher order land-use planning has decreed that areas should be kept more remote. Such higher order plans help guide MNRF's decision on whether the forest industry should leave the road and associated water crossings in an upgraded condition or whether the road system should be actively abandoned by removing water crossings or by installing barriers to passenger vehicle traffic.

Wildlife habitat is addressed in four main ways in forest management planning.

1. For a variety of habitat types, the long-term management direction ensures that the abundance of each habitat type is consistent with the range of natural variation, i.e., at levels consistent with the natural changes in the forest in the absence of harvesting. This verification is achieved through nonspatial modeling.
2. For key wildlife species, spatial modeling ensures that the size and distribution of planned harvest cuts are consistent with the plan's objectives for those species. One challenge of this modeling is that it includes only the land base under jurisdiction of the plan, i.e., Crown land. Private land is not considered even though it may provide excellent habitat.
3. For specific wildlife values, specific prescriptions are developed through *Area of Concern* planning. Wildlife values include heron colonies, raptor nests, turtle and snake habitat, species-at-risk habitat, and water features such as lakes, ponds, streams, and rivers. The prescriptions consider timing of operations (e.g., avoiding nesting periods of a raptor), modifying operations (e.g., limiting the amount of removal of forest cover), access-related restrictions, and no-harvest reserves. These prescriptions are based on the best available science from provincial guidelines, expert opinion, the planning team, and the Local Citizens Committee. However, provincial guidelines are weighted most heavily, especially in consideration of values related to species at risk.
4. Conditions on regular operations limit disturbance to soils and damage to residual trees, or provide direction on how wildlife species and habitat are to be enhanced or protected. For instance, a minimum number of cavity trees, mast trees, supercanopy trees, conifer trees, and veteran trees must be identified and left after harvest. This varies by forest type and silvicultural system. Roughly 40% of the vertebrate wildlife species rely on trees with a hole in them for denning, nesting, refuge, or food storage, and this direction to leave cavity trees directly contributes to the protection of habitat. Deer winter habitat is protected by maintaining a certain amount of conifer cover while encouraging operations in nearby hardwoods that will increase future browse for wintering animals.

In addition to these wildlife and social values, cultural heritage values must be protected from possible negative impacts of forestry operations. The main concern is that activities such as mechanical site preparation or road building may damage artifacts or other types of cultural heritage values of historical and spiritual significance. *Area of Concern* prescriptions consistent with the Forest Management Guide for Cultural Heritage Values (Ontario Ministry of Natural Resources, 2007) are used to minimize this risk.

Public Consultation Process

Public consultation is a vital element of the development of forest management plans in Ontario. This consultation is undertaken in a number of ways. The local MNRF district manager ensures that this consultation follows prescribed methodologies and concerns, questions, and comments are dealt with appropriately. The Local Citizens Committee is comprised of a group of individuals who represent a variety of interest groups and committees. This committee advises the District Manager on concerns and recommendations with respect to the development of the plan and deals with public concerns and resolves disputes. The chair of the Local Citizens Committee participates directly on the planning team. Members of the Local Citizens Committee for the French-Severn plan include the following interest groups: cottagers, municipalities, tourists, trappers, snowmobilers, anglers, hunters, laborers, general public, naturalists, and the forest industry. This committee stays active during the implementation of the FMP and inspects the annual work plans. It also helps the District Manager categorize any proposed amendments to the FMP, which then determines consultation requirements for the amendment proposal.

The public is informed of opportunities for involvement in the development of the FMP through direct mailings to known interested groups or individuals as well as through the local news media. The French-Severn planning team has also used radio spots to advertise open-house discussions to encourage participation. In addition, the Ontario Ministry of the Environment and Climate Change operates an Internet site that posts all significant projects on Crown land. Ontario's *Environmental Registry* must include postings for significant milestones for plan development, including the various formal opportunities for public consultation.

The FMP development schedule centers on opportunities for public consultation. There are five formal opportunities for public consultation that occur during the first 5-year term of the 10-year plan, also known as *Phase I* (Davidson et al., 2009).

1. An invitation to participate helps identify interested parties at the beginning of plan preparation.
2. An open-house discussion provides an opportunity to review strategic planning products, as well as general operational items, such as preferred harvest allocations and primary road planning maps.
3. A second open-house discussion provides operational details on planned harvest, renewal and tending areas, road construction and maintenance, and values protection strategies, including a series of prescriptions used to maintain or enhance special habitat features. These details relate to the first 5 years of operations under the plan.
4. An opportunity to inspect and comment on the Draft FMP. The government also reviews the draft plan and identifies required changes to be made to the draft plan.
5. An opportunity to inspect the final plan.

During the preparation of the operational plan for the second 5-year term (Davidson et al., 2013), the stages of consultation include an open house that highlights planned operations, including any changes from the first 5-year term. This is followed by an inspection period for both the draft plan and the approved plan.

Eight First Nations communities and a number of Métis groups (recognized Aboriginal peoples who are descended from First Nations and Europeans) have interests in activities on the French-Severn Forest. During the early stages of plan development, each community was visited and discussions occurred on how the community wishes to participate in plan development. Some communities chose to take advantage of the opportunity to be a part of the planning team, while others do not play such an active role.

The planning process outlines detailed specific steps to take when concerns arise from the public. Most input relates to questions about locations of planned activities and time frames when work will occur. This is reflective of the abundant amount of private land as well as the many cottagers and other forest users in the forest. Some concerns relate to perceived environmental risks. These concerns are normally addressed by explaining the landscape-level approaches used to ensure that a wide range of habitats will be made available over the next century. The concerns of most people are addressed after they learn that the rules within the plan ensure the protection and enhancement of forest values such as raptor nests, deer wintering areas, species-at-risk habitat and populations, water quality and fishery values, to name a few. As well, the plan stipulates the retention of specific stand-level forest attributes such as cavity trees (trees with holes in them that provide habitat for many species), mast trees (trees that provide substantial food such as red oak acorns, beechnuts, and black cherries), and supercanopy trees (trees that provide important diversity to the forest canopy).

In this forest, very little public opposition to forest management on Crown land occurs. In truth, the level of participation in the public consultation process can be disappointing and attendance at open-house discussions is low. There have been some specific and limited disputes from members of a cottaging group not wishing any forest management to occur on Crown land within a significant distance from the lake that their cottages surround. There have been disputes relative to how recreational trails are protected from possible negative impacts from forest management. These have been dealt with by meeting with the concerned parties, explaining the planned operations, and arriving at solutions that address main concerns. This is known as the *issue resolution process*.

The issue resolution process normally starts with the concerned party meeting with the plan author. If a mutually agreeable solution cannot be reached, the MNRF District Manager must meet with the concerned party. If that does not result in a satisfactory resolution, the MNRF Regional Director becomes involved. If the concerned party does not come forward until late in the planning process, the process may be shortened to bypass the plan author or even the District Manager. This is to avoid delays in plan approval when there is not enough time to go through all stages. During the production of the 2009–2019 10-year plan, issue resolution did occur between a cottaging community and the Regional Director. All parties agreed that winter harvest operations were possible and preferable, and that cut boundaries around the lake would be slightly altered. During the preparation of the second 5-year operational plan (2014–2019), no issues were raised.

In this region of Ontario, one might expect many more public issues to be raised. With the proximity to highly populated areas and a forest that supports one of the largest cottaging environments in the province, one might expect even more disputes. The lack of confrontations is likely due to excellent communications through the various aspects of the public consultation process presented above, which is augmented by individuals from Westwind and the local MNRF office being willing to become actively involved in discussions with many groups and companies throughout plan implementation. Staff members often meet with dogsled tourism providers, snowmobile clubs, trapping associations, cottaging groups, municipalities, and First Nations communities, as well as support or participate in nongovernment organizations, such as watershed councils, and environmental groups, such as the Georgian Bay Biosphere Network. This is routinely done outside the formal public consultation process.

In addition to the involvement of staff in the community, the forest industry also plays a key role in being a good corporate citizen. The forest industry regularly avoids operations on Crown land next to developments on weekends, especially during the summer near cottaging lakes. Roads that are shared by forest users are at least maintained and are normally upgraded by the forest industry.

The fact that greater than 90% of forest management is achieved through the partial cutting means that very little forest is managed through clear-cutting. Regardless of the ecological and silvicultural appropriateness of clear-cutting in many forest types, this practice remains a highly contentious issue, and the forest changes brought about by clear-cutting result in significant aesthetic changes. The forest management plan in this forest region does not avoid clear-cutting due to social pressures, but rather due to the fact that the forest types in this forest are better suited to non-clear-cut silvicultural systems. The main drawback to partial cutting relative to other forest users, such as cottagers, is that the return cycle can be as short as only 10 years and is typically 20–25 years instead of the 80-year rotation typical with clear-cutting. This means that although the disturbance to social values of visual and noise related esthetics is lower, it is more frequent. On the French-Severn Forest, partial cut systems seem to be more compatible to the social values of other forest users.

Strategic Modeling Process

A key consideration in developing the long-term management direction is knowing how forest conditions will change and how objectives (social, environmental, and economic) will be achieved through time. On the French-Severn Forest, the Strategic Forest Management Model (SFMM) was used to provide this insight (Davis, 1993). SFMM is a linear programming model that brings together the planning inventory, growth and yield information, natural disturbance assumptions, succession rules, wildlife habitat matrices, silvicultural costs and revenues, timber yields, cutting frequency, and other management assumptions. The following is a partial list of objectives, goals, and constraints that were modeled:

- Ensure that there is no net loss of white pine, red pine, or oak stands
- Maintain planned silvicultural costs within planned silvicultural expenses
- Ensure a relatively even wood supply
- Maintain an abundance of forest unit area within a range that considers presettlement forest condition and provincial policy
- Maintain harvest areas by forest unit over time at a level that supports current utilization demands
- Meet forest volume requirements by species and product type
- Provide for preferred habitat area within a range of natural variation for selected wildlife species, including species at risk
- Provide for an age-class structure in the forest by forest type that falls within a range of natural variation, with particular emphasis on increasing the amount of older pine and hemlock forests

Often, not all of the objectives can be achieved at the desired levels or within desired time frames. If this occurs, the planning team needs to adjust the strategy until an acceptable solution and acceptable long-term management direction, is obtained.

A key component of the long-term management direction is the available harvest area (AHA), which provides an upper limit on the area that may be harvested in each forest unit. Unlike many other jurisdictions, Ontario regulates harvest by area, not volume. The volume from the available harvest area, called the *available harvest volume*, is comparable to the allowable cut used in other jurisdictions. The AHA has an associated profile (i.e., an age class and stage of management, for clear-cut and shelterwood forest types). The stage of management is used primarily in shelterwood forest units and refers to the seeding cut, first removal, and final removal. For selection forest types, the profile is expressed in terms of quality based on basal area. Managers strive to follow the profile when selecting stands for harvest.

Most of the shade-tolerant hardwood forests are managed under either the selection or shelterwood silvicultural systems. For hardwood forest types managed under selection systems, SFMM simulates rather than optimizes the calculation of harvest areas, using a variation of available area divided by the cutting cycle. Optimization of selection management is possible, but was not pursued given the limitations of SFMM and the current inventory. For all other forest types, the harvest area was optimized, in spite of SFMM limitations, which included the inability to represent different management intensities for shelterwood forest types. Figure 41.3 compares the area of the land base to the AHA by silvicultural system. The AHA for selection and shelterwood systems is proportionately higher than for clear-cut systems because of the shorter interval between harvests; there are more stand entries over a given time period than under the clear-cut system.

In addition to harvest areas and volumes, the analysis supporting the long-term management direction predicts the levels of achievement for non-timber objectives. The forest management plan includes tables and graphs for all nonspatial objectives, including the amount of area over time by forest type and development stage and the provision of preferred habitat for the selected wildlife species.

The achievement of management objectives is based on the full utilization of the AHA, and a sensitivity analysis is conducted to assess the impact of undercutting on objective achievement. The results are relatively minor with obvious reductions in volumes but increases in old forest and in preferred habitat for selected species that prefer older forest conditions. Correspondingly, there are some reductions in the achievement of objectives related to young forest conditions and the amount of preferred habitat for species that prefer younger forest conditions.

The forest management plan also identifies a number of objectives that cannot be modeled with SFMM. Spatial models are used to verify that the habitat for pileated woodpecker (*Dryocopus pileatus*), red-shouldered hawk, moose, and white-tailed deer are maintained within a prescribed range. Spatial modeling is also used to for landscape-level objectives, such as the distribution of old growth as well as disturbance patterns that emulate natural events.

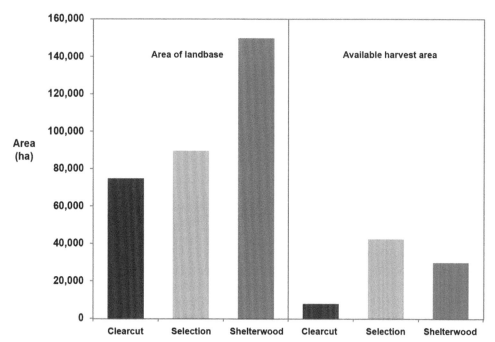

FIGURE 41.3   Land base and available harvest area by silvicultural system for the French-Severn Forest.

OUTCOMES OF THE PLAN

For the French-Severn Forest, the most significant objectives were the provision of old forest for eastern white pine stands and no loss of area of eastern white pine and red pine ecosystems. The selected long-term management direction provides for an increase of eastern white pine-dominated stands in older age classes and results in a modest increase in pine area. Similarly, the amount of eastern hemlock and oak-dominated forests was projected to remain constant or increase slightly, while significant increases in the amount of old eastern hemlock stands are expected to be realized.

In comparing the projected forest with objectives based on a modeled natural disturbance scenario, the preferred habitat for all selected wildlife species was within an acceptable range, as was virtually all forest type and development stage combinations. One acceptable exception was young jack pine stands. A significant amount of jack pine area on Crown land in this forest is not available for management due to location on poor site classes or its protected status. Under a natural disturbance modeling scenario, all jack pine areas are susceptible to frequent fire disturbance. Young jack pine cannot be created at a scale on managed Crown land, as can be observed in a modeled natural disturbance scenario. Therefore, the inability of the plan to achieve this one objective was not considered significant.

The projected harvest levels within the plan were consistent with current industrial demand for timber by species group and by product type. The strategic planning process did reveal that the level of eastern white pine volume would vary and, in the future, would fall below current industrial demand levels. This was the only real fiber supply issue identified in the plan, and is one that should be mitigated by expected declines in white pine demand. During the first five years of the 2009–2019 FMP, the actual utilization of the pine available area was approximately 35%. Although some improvements in markets are expected, it is anticipated that the under-utilization of the harvest area in this plan will help compensate for anticipated minor shortfalls in future planning terms.

One challenge to the implementation of the FMP is the consistent under harvest of the AHA, especially for some forest units. Under achievement of the available harvest area has been the norm in this forest for a number of reasons:

1. Fluctuating markets for some products and species. For example, the eastern white pine market has been depressed in recent years in Ontario.
2. The distance to markets is often economically limiting, especially for pulpwood. There are no nearby pulp mills, so pulpwood utilization is below desired levels.
3. Poor quality stands are often not marketable during poor pulpwood or firewood markets. Planned harvest areas include both high- and low-quality stands to promote forest improvement and prevent high grading. Even when pulpwood markets are strong, the economics of these partial-cut systems rely on quality sawlog production.
4. Weather and seasonal restrictions limit when harvesting can occur. The FMP sets out rules and standards to meet objectives that protect site and stand quality, particularly in partial-cut systems. Harvesting does not occur from winter break-up until mid-June. Wet autumns are quite common, and some years the conditions are so bad that little harvesting can occur during parts of this season. In addition, winters that start late or end early greatly reduce the amount of area that can be harvested in those months.
5. Access is a significant issue in a number of ways. There is very little aggregate (rock) for the western portion of the forest. This limits road-building activities, especially all-season roads. Portions of the forest are fairly remote and rely on winter roads for access. Short and mild winters limit this access. Other land uses (e.g., parks and protected areas) and the large complex of patent lands also limit access. Loggers are often denied access to a private road or are unable to cross a private property to access Crown forests on the other side.

DISCUSSION AND CONCLUSIONS

The French-Severn Forest differs significantly from other management units in Ontario. A community-based, not-for-profit forest management company, Westwind Forest Stewardship Inc., leads forest planning, silvicultural operations, and harvest compliance inspections on behalf of several logging companies. As a pioneer in managing a Forest Stewardship Council-certified public forest in Canada, Westwind works with local communities, businesses, individuals, and government in not only developing but also implementing the plan in a manner that is based on mutual respect of the forest. This approach, along with a highly regulated planning environment, provides for public and Aboriginal involvement in preparing a plan that puts many non-timber objectives at the forefront. The result is a high public use forest where industrial forestry activities take place with minimal public dispute. Because this forest lies within the temperate hardwood forest and relies on tree marking and partial cutting systems, impacts on esthetics are minimized while management objectives to improve the growth and quality of the forest are realized.

Learnings and Insights

Plan development can be impeded by interested parties that are anxious about the various impacts forestry activities may have on their enjoyment of the area. This is especially true in an area so close to large urban areas and where use by cottagers and other recreational groups is extensive. The FMP public consultation process is fair and provides ample opportunity for information gathering, debates, and the resolution of issues. However, it is clear that maintaining good relations with as many groups as possible before, during, and after plan preparation, independent of this formal process, minimizes controversy.

Sustainability Issues

There have been no significant sustainability issues identified during either plan development or plan implementation. Forest unit changes that move stands from one management system to another are a reflection of the need for very detailed and field-supported inventory. However, writing strategic plans every 10 years provides opportunity for continual improvement.

Plan Development Challenges

The most significant challenge in preparing the latest FMP was incorporating restrictive value protection standards for some species at risk. While the plan author and planning team had no authority to deviate from the provincial direction, members of the planning team and the Local Citizens Committee frequently expressed concern regarding this direction, particularly related to Blanding's turtles. Although it is recognized that these turtles are very commonly found throughout a significant portion of the western part of the management unit, actual inventories are incomplete causing uncertainties of a number of planned operations.

Plan Implementation Challenges

An accurate inventory is a key component to any plan. Remote sensing can be an effective method of creating and updating inventories in some forest types; however, in multi-aged forests and forests in which tree quality varies widely, other techniques are required. Although Westwind carries out many field inspections, inspecting each stand that is allocated for harvest operations in a FMP is not possible. When Westwind undertakes a field inspection to develop a forest operations prescription, a number of planned harvest blocks prove not to be operationally feasible or silviculturally ready for treatment. This requires identifying replacement areas. In a forest that normally has 20 or more companies carrying out logging operations, finding replacement areas not already assigned can be difficult.

Ontario has a fairly comprehensive and detailed direction for protecting values. In recent years, that direction for some species considered to be at risk has proved to be very restrictive, and carrying out operations in large zones of the forest can be difficult if not logistically infeasible to harvesting companies and forest managers. Improving science and finding opportunities to work within the provincial parameters will be a continual process during plan implementation.

ADDITIONAL READING AND RESOURCES

This chapter represents a synthesis of the 2009 forest management plan for the French-Severn Forest in Ontario. To view the plan itself, please visit this Internet site, which was available on July 28, 2014:

http://www.efmp.lrc.gov.on.ca/eFMP/viewFmuPlan.do?fmu=360&fid=59006&type=CURRENT&pid=59006&sid=4201&pn=FP&ppyf=2009&ppyt=2019&ptyf=2009&ptyt=2014&phase=P1

If in the future the link to the site appears broken, search the Internet using the title of the plan and the keywords provided.

REFERENCES

Davidson, B., Davidson, M., McDonnell, J., Martin, G., Pamajewon, W., Miles, D., Henry, M., Brenner, K., McNutt, J., Noganosh, W., Johnson, J., Deugo, D., Heidman, L., Fallows, V., Rouse, J., 2009. Forest management plan for the French/Severn Forest (360). Ontario Ministry of Natural Resources, Parry Sound District, Parry Sound, ON. http://www.efmp.lrc.gov.on.ca/eFMP/viewFmuPlan.do?fmu=360&fid=59006&type=CURRENT&pid=59006&sid=4201&pn=FP&ppyf=2009&ppyt=2019&ptyf=2009&ptyt=2014&phase=P1 (Accessed July 28, 2014).

Davidson, B., Henry, M., Heatlie, M., Duquette, G., Aubichon, S., Collins, A., Tran, P., McDonnell, J., Kroes, G., Contlin, S., Zyganiuk, A., Johnson, J., Rouse, J., Michener, G., Williams, E., 2013. Ten-year forest management plan, April 1, 2009 to March 31, 2019 for the French/Severn Forest (360), planned operations for the 2nd 5-year term from April 1, 2014 to March 31, 2019. Ontario Ministry of Natural Resources, Parry Sound District, Parry Sound, ON. http://www.efmp.lrc.gov.on.ca/eFMP/viewFmuPlan.do?fmu=360&fid=59006&type=CURRENT&pid=59006&sid=14709&pn=FP&ppyf=2009&ppyt=2019&ptyf=2014&ptyt=2019&phase=P2 (Accessed July 28, 2014).

Davis, R., 1993. Analyzing Ontario's timber supply with the Strategic Forest Management Model. In: Davis, R. (Ed.), Analytical Approaches to Resource Management. Queen's Printer for Ontario, Toronto. pp. 52–71.

Ontario Ministry of Natural Resources, 2007. Forest Management Guide for Cultural Heritage Values. Queen's Printer for Ontario, Toronto. 75 p.

Ontario Ministry of Natural Resources, 2009. Forest Management Planning Manual for Ontario's Crown Forests. Queen's Printer for Ontario, Toronto. 447 p.

Ontario Ministry of Natural Resources, 2014. Ontario's Policy Framework for Stewardship of Crown Forest Lands. Queen's Printer for Ontario, Toronto. http://www.mnr.gov.on.ca/en/Business/Forests/2ColumnSubPage/STEL02_163862.html (Accessed May 23, 2014).

Ontario Ministry of Northern Development, Mines and Forestry, 2011. Sustainable forest licence No. 542411. Ontario Ministry of Northern Development, Mines and Forestry, Sudbury, ON. http://www.mnr.gov.on.ca/stdprodconsume/groups/lr/@mnr/@forests/documents/document/mnr_e000477.pdf (Accessed May 23, 2013).

Chapter 42

Martel Forest, Ontario, Canada

Robert Keron,[1] Dan Rouillard[1] and Sarah Sullivan[2]

[1]*Ontario Ministry of Natural Resources, Forests Branch, Sault Ste. Marie, Ontario, Canada,* [2]*Tembec, Timmins, Ontario, Canada*

ABBREVIATIONS

AHA   Available Harvest Area
AR   Annual Report
CFSA Crown Forest Sustainability Act

MANAGEMENT SETTING AND BACKGROUND

There are more than 71 million hectares (ha) (175.44 million acres (ac)) of forest in Ontario, 90% of which is owned by the province (Ontario Ministry of Natural Resources, 2014). This area is referred to as Crown forest. Forest management activities are conducted in Ontario on the *Area of Undertaking*, an area of approximately 43.8 million ha (108.2 million ac), of which 27.1 million ha (67.0 million ac) is Crown forest (Ontario Ministry of Natural Resources, 2014). Within the Area of Undertaking, there are three regions of Ministry of Natural Resources and Forestry (MNRF) governance. Each of these regions is further divided into districts. For management purposes, the Crown forest in Ontario is divided into management units. The Martel Forest is a management unit located within the Northeast region, Chapleau District, which is situated in the heart of the Canadian shield (Figure 42.1). The current boundaries of the Martel Forest were established in 2006 and cover a total of 1,191,274 ha (2,943,638 ac). The Crown-managed portion of the forest totals 999,417 ha (2,469,559 ac), of which 918,069 ha (2,268,549 ac) is forested. The remainder of the forest is comprised of Provincial Parks, conservations reserves, and a small area of patent lands, first nation lands, and federal lands. A land patent represented the transfer of land ownership to an original land settler and conferred ownership and a certain amount of rights. The largest community within the forest is the town of Chapleau, Ontario with a population of 2,100.

Since before recorded history, the Martel Forest has been inhabited by the Ojibwe and Cree First Nations who used the forest to support their subsistence lifestyles. The first European settlement in the area was established in 1777 by the Hudson's Bay Company. The fur trade dominated First Nation-European relations for most of the eighteenth and nineteenth centuries. There are a number of binding documents and treaties that define modern relationships between First Nation communities and the Crown. Relevant to the Martel Forest, these documents include the Royal Proclamation of 1763, Treaty 9 of 1905, and the Constitution Act of 1982.

The town of Chapleau was established in 1885 when the Canadian Pacific Railway was built through the area to provide access to the Hudson's Bay Company Trading Post. A fire in 1948 encouraged the government to develop a road so that logging contractors could remove the timber before it rotted. Consequently, Highway 129 was completed. In future years, Highways 101 and 17 were constructed to link Chapleau with Timmins, 200 kilometers (km) (124 miles (mi)) to the east, and Wawa, 140 km (87 mi) to the west.

The landscape of the Martel Forest was heavily influenced by the Laurentide glaciation of the Wisconsinan age (approximately 100,000 years ago). The receding ice sheet halted at Chapleau about 11,000 years ago. This gave rise to a large band of moraines known as the Sultan Scarp, bounded to the south by an extensive outwash plain that covers the south-central portion of the forest. Forest management in this area is relatively easy, with flat terrain and plentiful material for road building. North and south of this broad sand outwash plain, the geology is dominated by eskers, which contain sand, gravel, and till soils. There is an abundance of material for road construction; however, terrain conditions vary significantly resulting in a wide range of growing conditions, which translates into variable silviculture options and

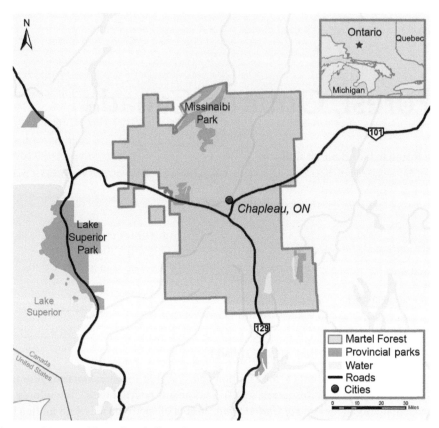

FIGURE 42.1 Location map of the Martel Forest Ontario Crown Lease.

access strategies. The remainder of the Martel Forest is dominated by rolling uplands, broken by occasional ridges of Precambrian granite bedrock (Evans and Cameron, 1984).

The Martel Forest is located within the boreal forest region. The southeast portion of the land base is characterized with some components of the Great Lakes-St. Lawrence forest (Rowe, 1972). The area represents the northern edge of the range for eastern white pine (*Pinus strobus*), red pine (*Pinus resinosa*), sugar maple (*Acer saccharum*), and yellow birch (*Betula alleghaniensis*). The species commonly found in the boreal forest are common in the area, including jack pine (*Pinus banksiana*), black spruce (*Picea mariana*), white spruce (*Picea glauca*), balsam fir (*Abies balsamea*), trembling aspen (*Populus tremuloides*), and white birch (*Betula papyrifera*). There are also populations of less common species, including eastern white cedar (*Thuja occidentalis*), tamarack (*Larix laricina*), red maple (*Acer rubrum*), black ash (*Fraxinus nigra*), largetooth aspen (*Populus grandidentata*), and balsam poplar (*Populus balsamifera*). The Martel forest is predominantly a mixedwoods type community (Figure 42.2), with pure stands of jack pine found on outwash plains and black spruce found on moist organic sites. The forest is stratified for forest management purposes into forest units. The Forest Management Planning Manual (Ontario Ministry of Natural Resources, 2009) defines a forest unit as an aggregate of forest stands that will normally have similar tree species composition, will develop in a similar manner, and will be managed under the same silvicultural system. The Martel Forest Management Plan (FMP) classifies the productive forest into 15 forest units, 14 of which are actively managed for timber production (Figure 42.3) (Sullivan, 2011). Forest units are the fundamental strata used in strategic modeling to forecast yield and model forest dynamics such as natural forest succession, post renewal succession, response to natural disturbance, and habitat.

Between April 1, 2006 and March 31, 2009, more than 1.89 million cubic meters (m³) (66.8 million cubic feet (ft³)) of timber was shipped from this forest to mills in Chapleau, Wawa, Timmins, Englehart, Longlac, Hearst, Sault Ste. Marie, and Espanola (Sullivan, 2011). This timber was used to produce a variety of products, including dimensional lumber, oriented strand board, hardwood plywood and veneer, particleboard, and pulp and paper. The employment levels supported by the production and transportation of these goods are critical to the health of these communities and also to the Aboriginal communities located in the region.

FIGURE 42.2 Rolling terrain and mixedwoods forests common to the Martel Forest Ontario Crown Lease. Photo courtesy of James Cook.

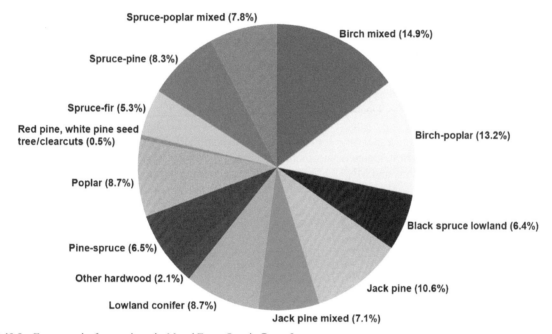

FIGURE 42.3 Forest area by forest unit on the Martel Forest Ontario Crown Lease.

The Martel Forest supports a diversity of wildlife species that depend on a wide range of forest conditions. Many of the species common to the forest are highly valued for providing recreational opportunities such as hunting, wildlife viewing, and commercial opportunities such as tourist outfitting and furbearer trapping. The harvest of game species including black bear (*Ursus americanus*), moose (*Alces alces*), and small game (snowshoe hare (*Lepus americanus*), ruffed grouse (*Bonasa umbellus*), and spruce grouse (*Dendragapus canadensis*)) and some furbearing animals (wolf (*Canus lupus*), lynx (*Lynx canadensis*), coyote (*Canis latrans*), and fox (*Vulpes vulpes*)) is regulated by the MNRF. In 2008, combined hunting expenditures for moose and bear on the Martel Forest made an estimated contribution of $1.995 million Canadian dollars to the local economy (Sullivan, 2011).

The species at risk in Ontario list is determined by the Committee on the Status of Species at Risk in Ontario (an independent committee of experts who report to the Minister of Natural Resources). In 2011, the following 12 species at risk whose mapped ranges overlap the Martel Forest were listed as follows:

- Chimney swift (*Chaetura pelagica*)
- Short-eared owl (*Asio flammeus*)
- Black tern (*Chlidonias niger*)
- Common nighthawk (*Chordeiles minor*)

- Olive-sided flycatcher (*Contopus cooperi*)
- Bald eagle (*Haliaeetus leucocephalus*)
- Canada warbler (*Cardellina canadensis*)
- Monarch butterfly (*Danaus plexippus*)
- Mountain lion or eastern cougar (*Puma concolor couguar*)
- Eastern wolf (*Canis lycaon*)
- Wood turtle (*Glyptemys insculpta*)
- Peregrine falcon (*Falco peregrinus*)

The forest also supports a wide range of recreational and tourism activities with 10 provincial parks and 48 tourist outfitters within or bordering the Martel Forest. The forest also includes nearly half of the Chapleau Crown Game Preserve, which at more than 7,000 square km (1.73 million ac) is the largest animal preserve in the world. The preserve is a source of tourism, drawing nature enthusiasts and fishermen to the area. All forms of hunting and trapping have been forbidden in the preserve since the 1920s. More recently, First Nations have exerted their treaty rights and have been able to hunt within the preserve. All other activities are permitted within the preserve, including timber harvesting. Numerous other resource-based tourist opportunities exist, including kayaking, canoeing, cottaging, and Crown land camping.

PLANNING ENVIRONMENT AND METHODOLOGY

Guiding Laws, Regulations, and Policies

The Martel Forest is currently managed by Tembec under a sustainable forest license (SFL). Tembec, with head offices in Montréal, Québec, has mills in Ontario, Québec, the United States, and France. Tembec's corporate mission is to "be an industry leader in value creation by being the best steward of resources — human, capital and forest" (Tembec, 2014). As an SFL holder in Ontario, Tembec is responsible for carrying out the activities of forest management planning, harvest, access road construction, forest renewal, maintenance, monitoring, and reporting. These are all subject to the Crown Forest Sustainability Act (CFSA) regulations and MNRF approvals. SFLs are granted by the MNRF and are valid for 20 years. The MNRF reviews the license every 5 years to ensure that the terms and conditions of the license have been met (Ontario Ministry of Natural Resources, 2014).

Before any forestry activities can take place in a management unit, there must be an approved forest management plan. The forest management planning system for Ontario's Crown forests is based on a forest policy and legal framework that incorporates sustainability, public involvement, Aboriginal involvement, and adaptive management as key elements.

The CFSA and the Environmental Assessment Act provide the legislative framework for forest management on Crown land in Ontario.

The purpose of the CFSA is to guarantee the sustainability of Crown forests. The Act defines forest sustainability as long-term Crown forest health, and requires forest management planning to provide for determinations of sustainability based on two fundamental principles (S.O. 1994, Chapter 25):

1. Large, healthy, diverse and productive Crown forests, their associated ecological processes, and biological diversity should be conserved.
2. The long-term health and vigor of Crown forests should be provided for by using forest practices that, within the limits of silvicultural requirements, emulate natural disturbances and landscape patterns while minimizing adverse effects on plant life, animal life, water, soil, air, and social and economic values, including recreational values and heritage values.

To achieve this end, each FMP contains a broad management strategy or Long-Term Management Direction (LTMD), which balances objectives related to forest diversity, socioeconomics, forest cover, and silvicultural practices.

The provisions of the environmental assessment approval under the Environmental Assessment Act, referred to as the "Declaration Order regarding the MNRF's Class Environmental Assessment Approval for Forest Management on Crown Lands in Ontario," give the MNRF approval to undertake the activity of forest management. These activities comprise the inter-related activities of access, harvest, renewal, and maintenance, subject to the conditions outlined in the declaration order, including requirements for planning. The forest management planning requirements of the CFSA and the provisions of the environmental assessment approval under the Environmental Assessment Act are incorporated into the Forest Management Planning Manual, which provides the direction for preparing a FMP.

The Forest Management Planning Manual is supported by a series of provincial manuals and guides that outline silvicultural practices and methods to conserve biodiversity and enhance or protect wildlife habitat, esthetics, watersheds, and other values. These guides are regularly reviewed and updated to ensure they stay consistent with current science and policy. Forest management in Ontario is conducted in an adaptive management cycle. The implementation of the Martel Forest plan is monitored to ensure compliance with the approved plan and that the implementation of operations is achieving the desired results, including the impacts on forest ecosystems by operations. Monitoring is accomplished through Tembec's regular supervision and monitoring of its forest operations, including conducting more formal forest compliance inspections and silviculture effectiveness monitoring. The MNRF also performs its own monitoring through various programs, including the Forest Compliance Monitoring and Independent Forest Audit programs. The MNRF also reviews Tembec's Annual Reports, which provide a summary of the previous year's forest operations.

The 10-year forest management plan is implemented through successive Annual Work Schedules and reported to the Ministry using an Annual Report (AR). Following the year-seven AR, the implementation of the plan is assessed. It is determined whether the plan provided for the sustainability of the Crown forest, and recommendations are made for future plans. The next FMP is expected to consider the recommendations of the year-seven AR, changes to the forest condition, and any updates to science and policy (Ontario Ministry of Natural Resources, 2009).

While Provincial direction encourages forest companies in Ontario to seek forest certification by independent third-party organizations, it is up to the individual SFL holder to select the certification standard and to obtain certification (Ontario Ministry of Natural Resources, 2014). In January of 2006, Tembec obtained Forest Stewardship Council (FSC) certification according to the *Canadian Nation Boreal Standard* for the Martel Forest. Maintaining this certification requires a series of annual audits on the Martel Forest to assess conformance to the 10 FSC principles.

A FMP is prepared for a 10-year period for each forest management unit in the province. A plan is prepared in an open and consultative fashion by a Registered Professional Forester with the assistance of a multidisciplinary planning team and a local citizens committee, as well as input from Aboriginal communities, stakeholders, and interested members of the public. The preparation of the plan occurs in two phases. The strategic long-term planning and the planning of the first 5 years of operations occur in Phase I, which includes five stages of public and Aboriginal consultation. Phase II involves the planning of the second 5 years of operations and includes three stages of public and Aboriginal consultation (Ontario Ministry of Natural Resources, 2009).

In the preparation of a FMP, the determination of sustainability involves:

- The development of management objectives that address provincially mandated objective categories and indicators and other desired forest benefits identified by the public, planning team, and Local Citizens Committee
- An assessment of the achievement of management objectives for a period of 160 years
- The development of a proposed long-term management direction that balances the achievement of the management objectives
- A conclusion that the forest management plan provides for the sustainability of the Crown forest on the management unit

In the Martel Forest management plan, a total of 22 management objectives and 29 indicators were identified. These can be found, along with corresponding target levels established through strategic modeling, in Table FMP-9 of the Martel FMP (Sullivan, 2011). These 22 management objectives can be grouped into four main categories: forest diversity, silviculture, forest cover, and social and economic objectives. These objectives are to:

- Move toward a natural disturbance pattern
- Maintain (within natural levels) forest area by forest unit through time
- Maintain (within natural levels) a range of seral stages by forest unit, including mature and old forest areas
- Maintain (within natural levels) red pine and white pine forest unit areas
- Maintain (within natural levels) habitat levels for indicator species
- Provide marten (*Martes americana*) core areas
- Consider and provide moose emphasis areas
- Maintain forest-dependent species at risk existing within the forest
- Minimize impacts on environmental quality
- Minimize impacts on water quality
- Ensure that high-conservation-value forest attributes are identified and considered for preservation
- Keep the forest ecosystems productive and healthy
- Employ cost-effective renewal and tending treatments that will provide a perpetual and continuous flow of quality wood fiber from the Martel Forest

- Reduce herbicide usage through proven science, integrated pest management principles, and Tembec's annual herbicide limit under FSC certification
- Ensure the protection of natural resource features, non-timber uses, and other values dependent on forest cover on the land base
- Conduct forestry practices that allow all resource users to gain benefits from the forest
- Minimize the impact of forest operations on known cultural heritage values
- Supply industrial and consumer wood needs with a continuous supply of roundwood
- Ensure the available forest is protected from sustained deforestation and conversion to other uses
- Provide local First Nations and the public with opportunities for participation in the development of the plan
- Ensure compliance with legislative and regulatory requirements
- Plan road access strategies that balance competing interests (i.e., timber extraction vs. remote tourism)

Once the objectives for the future forest have been specified, target levels for each of the competing objectives are established through a *scoping analysis*, which is an iterative modeling process that involves a series of investigations to provide insight as to what the forest is capable of producing in order to develop realistic and feasible desirable levels for objective indicators. Scoping analyses consider implications on wood supply, forest conditions, habitat, and other non-timber resources for the short term, medium term, and long term (up to 160 years).

Strategic Modeling Approach

The Strategic Forest Management Model (SFMM) (Davis, 1993) has been the standard forest estate model used in Ontario since 1997. SFMM is a linear programming model (Model III formulation) developed by the Ontario Ministry of Natural Resources and Forestry and adapted to suit provincial forest management planning requirements as they have evolved. SFMM is a nonspatial model, that is, the spatial reference of model inputs such as the forest inventory are lost as the model is compiled, modeling processes are not governed by explicit spatial constraints or objectives, and outputs have no spatial reference. During planning, SFMM enables foresters to analyze the relationships between forest condition, silvicultural practices, wood supply, and potential wildlife habitat. This analysis enables planning teams to understand how a forest develops through time and to explore alternative forest management strategies and trade-offs.

Conventionally, planning teams have relied on SFMM exclusively to support the decision support needs of strategic planning. A typical strategic modeling approach would be to develop a base model in SFMM and conduct a scoping analysis that informs the development of specific desired levels for management objectives. Once the planning team agrees on the desired benefits, the associated target levels are used to develop the LTMD, which is a management strategy that achieves a balance of desired forest-level benefits and becomes the benchmark used to assess the sustainability of future management actions.

The primary output of the strategic modeling exercise is the Available Harvest Area (AHA) for the first 10 years of the FMP, which describes the area available for harvest by forest unit and age class. The AHA provides a harvest profile that must be used at the tactical planning level to guide the spatial allocation of harvest areas. This spatial layout is not limited to harvest profile, but must also consider spatial landscape policies and constraints, while being aware of the costs of the proposed pattern of operations, including the existing and proposed infrastructure (e.g., roads, mills).

In practice, spatial allocation issues often limit the ability to fully allocate the harvest profiles defined by the AHA. Infeasibilities are often attributed to natural disturbance pattern emulation objectives, spatial habitat provisions, lack of reasonable road access, and other site-specific spatial constraints. This conventional approach to strategic modeling lacks the ability to evaluate the effect that many of these long-term spatial objectives have on the level of AHA. This presents the possibility of future scenarios where spatial goals cannot be met and long-term sustainability is compromised. Recognizing this shortcoming of the conventional modeling approach used in Ontario, the Martel Forest planning team used an innovative hierarchical modeling approach, which included incorporating spatial and economic constraints into the strategic modeling exercise. The method used by the Martel Forest planning team for developing the LTMD involved the use of a spatially explicit strategic model, Patchworks™ (Spatial Planning Systems, 2009). Patchworks is a spatially explicit geographic information systems (GIS)-based sustainable forest management planning model and uses a goal-programming formulation to address multiple-objective planning problems.

The hierarchical implementation of SFMM and Patchworks employed by the planning team takes advantage of the fact that SFMM is very well suited to scoping and coarse trade-off analyses, and has functionality specifically designed to address Ontario's planning requirements (Rouillard and Moore, 2008). Another important benefit of the hierarchical arrangement of the two models is that SFMM provides Patchworks with a narrowed solution space, which improves the efficiency of the heuristic search process employed.

As a spatially explicit model, Patchworks provides a detailed depiction of the location of stand-level features and proposed management actions. Advantages to the extra effort of tracking stand-level detail are several, but the principal one is that it opens the possibility to embed spatial and operational policies within a long-term strategic management context, thus avoiding some of the previously discussed shortcomings of the current hierarchical approach. Thus, to the extent that spatial policy issues can be embedded within the strategic planning framework, the model can be used to consider strategies that balance strategic, tactical, and operational issues. The ability to address location issues within the modeling approach facilitates the creation of a LTMD that has reasonable operating characteristics and is logistically feasible.

Specific spatial objectives considered for the LTMD for the Martel Forest included silvicultural intensity zones, 10-year operational deferrals, marten core areas, preferred and not-preferred operating areas, moose emphasis areas, and the minimization of the number of small harvest blocks. Silvicultural intensity zones are strategic objectives that identify geographic areas where intensive and elite silviculture treatments would be economically feasible. These were produced using a GIS buffering exercise that used a threshold haul cost to delineate silvicultural intensity zones. Ten-year operational deferrals are strategic zones used to identify areas in the short term where no infrastructure existed or was planned. Included in these deferrals were preferred FSC candidate protected areas. Marten core areas are used as a coarse-filter approach for managing biodiversity. Marten core areas are large contiguous tracts of land that contain an older forest that is preferred by the marten. The details of the precise habitat requirements can be found in Forest Management Guidelines for the Provision of Marten Habitat (Ontario Ministry of Natural Resources, 1996). The guidelines require 10% to 20% of the forest, which has the capability to produce marten habitat, to be maintained in a suitable condition. The spatial analysis allowed for the evaluation of marten habitat much later in the planning process. The planning team discovered that it could increase marten targets from 10% to 11.7% without adversely influencing other objectives. This translated into an additional 14,464 ha (35,741 ac) of marten core areas being allocated.

Stand accessibility was used to define preferred operating areas. In the Martel Forest management plan, cut-blocks were only eligible for harvest if they were accessible by an existing, or future road. The modeling database included a fully connected network of road segments with identified costs for construction, maintenance, and hauling. Cut-block size limits were used as an economic objective, which minimized the proportion of small clear-cuts while still meeting natural disturbance emulation targets. The previous FMP allocated a similar AHA to the 2011 plan (slightly less than 47,000 ha, or 116,137 ac), but allocated this area over nearly twice the amount of cuts (382 planned cuts in the 2006 FMP and 216 planned cuts in the 2011 FMP). The database identified the closest road to each stand and the accumulated costs involved from the site of harvest to the initial existing road segment and then to the mill. Targets were established in Patchworks that limited expenditures on road construction and maintenance and haul costs. The use of road construction and maintenance budget limits forced the Patchworks model to better consolidate harvest blocks. The cut block size limits, in combination with the road construction and maintenance budget limits, forced the Patchworks model to better consolidate harvest blocks. The use of hauling cost targets prevented the model from only harvesting those blocks closest to the mills over the short term. In combination, these targets helped ensure a more balanced and operationally feasible harvest schedule.

Other spatial controls included moving toward a natural landscape disturbance pattern by setting disturbance frequency and area targets by size class. The natural disturbance template selected by the planning team was developed for the ecological site region, as described in the *Forest Management Guide for Natural Disturbance Pattern Emulation* (Ontario Ministry of Natural Resources, 2002). The forest was analyzed to determine the landscape disturbance pattern that existed at the start of planning (2011), at the end of planning if no management occurred (2021), and at the end of planning under the proposed operations (2021).

The first step in developing the operational plan was for operations personnel to review the harvest schedule and determine its feasibility. At this stage, the operational field staff was engaged to confirm the cut-blocks scheduled for harvest or to identify suitable replacement blocks, as required. This is an iterative process that continues until the scheduled harvest area approximates the AHA set by the LTMD.

The next step was to impose this revised harvest schedule within the Patchworks model as a constraint, and then re-solve the model to determine its impact on the achievement of the targets outlined in the LTMD. This was where the spatial constraints are recognized. The changes in levels of target achievement are either deemed acceptable or unacceptable, thereby requiring modifications to the operational plan. These steps were repeated until an acceptable harvest schedule was identified, which resulted in the achievement of a target acceptable to the planning team. The final model scenario resulting from this refinement process confirmed that the refinements made in operational planning did not significantly alter the objective achievement agreed to in the LTMD.

OUTCOMES OF THE PLAN

The hierarchical modeling approach used by the planning team on the Martel Forest helped ensure that a strategic LTMD was developed that achieved an acceptable balance of economic, social, and environmental objectives taking into account spatial and nonspatial strategic, tactical, and operational considerations.

The enhanced modeling procedure used by the planning team identified productive forestland adjacent to established all-weather roads and in close proximity to mills. The purpose was to limit the more expensive silvicultural treatments to these lands allowing the manager to grow more wood on less land. The 10-year operational plan for the Martel Forest allocated 99.3% of the available harvest area as determined by the LTMD. The planned harvest area for every forest unit was within 2% of its available harvest area with the exception of the red pine/white pine forest unit, which exceeded the AHA by 40%. This 40% overharvest represents 11 ha (27 ac) on the landscape. The slight overharvest of red pine/white pine (with respect to total area of the entire Martel Forest) is deemed acceptable because assessing sustainability goes beyond meeting AHA precisely, and involves the balancing of many objectives on the landscape. The total available and planned harvest areas for each forest unit are presented in Table 42.1.

The total amount of marten habitat is forecasted to decline from 280,069 ha (692,051 ac) in 2011 to 203,010 ha (501,638 ac) by 2111. However, suitable marten core areas will be maintained for the 20-, 40-, and 60-year terms at levels of 11.7%, 10.3%, and 10.3%, which fall within the acceptable target range for suitable marten habitat (10 to 20% of the land area), as recommended by the Forest Management Guidelines for the Provision of Marten Habitat (Ontario Ministry of Natural Resources, 1996). This decline is also consistent with the natural decline in marten habitat forecasted in the natural benchmark run.

Natural landscape disturbance pattern emulation is measured by disturbance frequency and area by disturbance size class. The proposed operations show a movement toward the natural landscape disturbance pattern in the frequency of larger cuts (501–1,000 ha, or 1,238–2,471 ac), but also show movement away from the smaller disturbance size

TABLE 42.1 Available and Planned Harvest Areas for Forest Units within the Martel Forest Ontario Crown Lease

Forest Unit	Available Harvest Area (ha)			Planned Harvest Area (ha)		
	Period 1	Period 2	10-year[a]	Period 1	Period 2	10-year[a]
Birch mixed	5,978	6,730	12,708	6,313	6,349	12,662
Birch-poplar	4,704	5,128	9,832	4,870	4,926	9,796
Black spruce lowland	4,151	3,140	7,291	3,593	3,635	7,228
Jack pine	4,514	5,386	9,900	4,896	4,875	9,771
Jack pine mixed	5,901	6,133	12,034	5,985	5,985	11,970
Lowland conifer	4,344	4,120	8,464	4,200	4,202	8,402
Other hardwoods	579	366	945	465	463	928
Pine-spruce	3,467	3,930	7,397	3,674	3,609	7,283
Poplar	5,446	5,333	10,779	5,364	5,429	10,793
Red pine, white pine seed tree/ clearcut	3	24	27	18	20	38
Spruce-fir	1,723	1,398	3,121	1,567	1,516	3,083
Spruce-pine	3,598	2,919	6,517	3,251	3,202	6,453
Spruce-poplar mixed	2,701	2,161	4,862	2,459	2,357	4,816
Total	47,109	46,768	93,877	46,655	46,568	93,223

[a]The sum of periods 1 and 2.

class (11–100 ha, or 27–247 ac). There are plans to reverse this negative movement by manually portioning some of the disturbances in the 101–500 ha (250–1,236 ac) size class into smaller disturbances that would fit in the 11–500 ha (27–1,236 ac) size class. Overall, the evaluation of the frequency of landscape disturbances for all size classes shows there is a considerable movement toward the natural landscape disturbance pattern as compared to pre-2011 activities.

DISCUSSION AND CONCLUSIONS

Forest management planning within the boreal forest of Ontario presents many unique challenges. Ontario's rigorous planning process requires managers to consider multiple competing objectives when developing forest management plans.

Learnings and Insights

The modeling approaches taken by Tembec required significant deviations from the conventional FMP process. Traditionally, there was a separation between the strategic and tactical planning levels, and the FMP process used this gap for part of its public consultation process. However, the use of Patchworks essentially spans the gap between strategic and tactical planning, which complicated the public consultation process that had co-evolved with SFMM. One complication was the AHA, and associated projections on forest area, habitat, and volume through time, were not available until much later in the planning process. More time was spent exploring the effects of spatial objectives and adjusting other targets to find a balanced solution. This proved to require considerably more effort than traditional SFMM analyses.

Sustainability Issues

Since 2006, there has been a significant decline in markets for poplar- and birch-oriented strand board products; however, strong demand still exists for high-quality hardwood veneer and sawlogs. Historically, hardwood veneer and sawlog production was a by-product of the oriented strand board market. As a result of the decline in oriented strand board markets, available volumes of hardwood veneer and sawlog have dropped below traditional production and demand levels. A second complication of the loss of the hardwood-oriented strand board markets is the inability to access spruce, pine, and fir timber within mixedwood stands. The loss of the hardwood market has led to a significant decline in harvest area and the SPF volumes, which would normally be available to local sawmills. The consequence of this on sustainability is clear. First, the socioeconomic sustainability is compromised over large amounts of unutilized AHA. Second, the future forest conditions, levels of future forest by forest unit, and wildlife habitat predicted in the LTMD are dependent on the forest management activities proposed in the plan. Any deviation from the proposed activities means the predictions of future forest conditions based on those actions are no longer valid and long-term sustainability may be altered.

Plan Development Challenges

The spatial nature of many of the objectives in the Martel FMP has challenged the traditional boundaries of strategic modeling in Ontario. These challenges were addressed with a unique two-model approach to strategic modeling. By incorporating the nonspatial SFMM with the spatial Patchworks, significant improvements in strategic planning were achieved. These include a more economical harvest schedule and increased allocated area for focal species habitat when compared to traditional approaches. The implementation of this approach was not without its challenges, including the need to adapt the current planning timeline to accommodate this additional analysis effort. It also highlighted the need to reevaluate the traditional approach to strategic modeling in Ontario and to gain a better appreciation of the differences and possible benefits or limitations of using a spatial model like Patchworks. In spite of the challenges that were encountered while incorporating spatial objectives to better reflect operational realities in the Martel Forest FMP, these efforts should lead to an FMP that better supports long-term sustainability as compared to traditional modeling efforts.

There were fundamental design differences between SFMM and Patchworks that led to differences in their forecasting. The primary difference being, SFMM aggregates data by forest unit and age class, while Patchworks generates detailed polygon level predictions. In SFMM, uncertainties are modeled using simple deterministic rules; for example, response to a harvest area could be that 80% of the area stays in the same forest unit while 20% transfers to another. Patchworks tracks the age, forest unit, and area of each polygon, which requires a modification to the rule sets normally used within SFMM to track the area associated with different stand dynamics, such as natural succession or post-harvest renewal. Since each polygon can only belong to one age class and one forest unit, multiple response rules are not allowed. To allow for consistency between the two models, the initial forest units are stratified based on response rules developed by the planning team, directing the forest unit area to an appropriate pathway. This would divide our example forest unit above into two distinct strata, one of which is 80% of the area of the parent forest unit and transitions to the same forest unit. The other would be 20% of the area of the parent forest unit and would transition to a new forest unit. The spatial allocation of these strata is randomly applied to stand polygons by the Patchworks model based on the area associated with each.

A second technical issue was the fundamental difference in the model formulations between SFMM and Patchworks. SFMM uses a linear programming formulation to optimize a specific objective function subject to a series of constraints. The criticism of this method is that only one management objective (the objective function) is treated as a goal, while all other management objectives are treated as constraints and their target inputs are binding. Patchworks uses a goal-programming formulation that is not limited to a single goal with multiple constraints. Goal programming treats multiple competing objectives as goals and optimizes the deviation from these goals. This allows for a powerful trade-off analysis by the planning team for balancing competing objectives. It was observed that although the modeling parameters were slightly different, the future forest conditions predicted in each model were very similar for the first harvest rotation (i.e., 60–80 years) but had a higher degree of variance later in the modeling horizon (i.e., 100–150 years), which is generally less of a concern in strategic planning efforts.

ADDITIONAL READING AND RESOURCES

This chapter represents a synthesis of the 2011 forest management plan for the Martel Forest. To view the planning documents, please visit this Internet site, which was available on May 20, 2014:

http://www.efmp.lrc.gov.on.ca/eFMP/viewFmuPlan.do?fmu=509&fid=100093&type=CURRENT&pid=100093&sid=8354&pn=FP&ppyf=2011&ppyt=2021&ptyf=2011&ptyt=2016&phase=P1

If in the future the link to the site appears broken, search the Internet using the title of the plan and the keywords provided.

REFERENCES

Davis, R., 1993. Analyzing Ontario's timber supply with the Strategic Forest Management Model. In: Davis, R. (Ed.), Analytical Approaches to Resource Management. Queen's Printer for Ontario, Toronto. pp. 52–71.

Evans, L.J., Cameron, B.H., 1984. Reconnaissance soil survey of the Chapleau-Foleyet area northern Ontario. Queen's Printer for Ontario, Toronto. 49 p.

Ontario Ministry of Natural Resources, 1996. Forest Management Guidelines for the Provision of Marten Habitat. Queen's Printer for Ontario, Toronto. 30 p.

Ontario Ministry of Natural Resources, 2002. Forest Management Guide for Natural Disturbance Pattern Emulation. Version 3.1. Queen's Printer for Ontario, Toronto. 29 p.

Ontario Ministry of Natural Resources, 2009. Forest Management Planning Manual for Ontario's Crown Forests. Queen's Printer for Ontario, Toronto. 447 p.

Ontario Ministry of Natural Resources, 2014. Forest Management Planning in Ontario. Queen's Printer for Ontario, Toronto. http://www.mnr.gov.on.ca/en/Business/Forests/2ColumnSubPage/STEL02_163511.html (Accessed May 20, 2014).

Rouillard, D., Moore, T., 2008. Patching together the future of forest modeling: Implementing a spatial model in the 2009 Romeo Malette Forest Management Plan. The Forestry Chronicle 84, 718–730.

Rowe, J.S., 1972. Forest Regions of Canada. Canadian Forest Service, Department of the Environment, Toronto. 172 p.

Spatial Planning Systems, 2009. What is Patchworks? Spatial Planning Systems, Deep River, ON. http://www.spatial.ca/products/index.html (Accessed April 29, 2014).

Sullivan, S., 2011. Forest Management Plan for the Martel Forest. http://www.efmp.lrc.gov.on.ca/eFMP/viewFmuPlan.do?fmu=509&fid=100093&type=CURRENT&pid=100093&sid=8354&pn=FP&ppyf=2011&ppyt=2021&ptyf=2011&ptyt=2016&phase=P1 (Accessed May 20, 2014).

Tembec, 2014. Vision and Values. Tembec, Montréal, Québec. http://tembec.com/en/company/vision-and-values (Accessed May 8, 2014).

Chapter 43

Prince Albert Forest Management Agreement (FMA), Saskatchewan, Canada

Cam Brown,[1] Dave Knight[2] and Pat Mackasey[3]

[1]Forsite Consultants Ltd., Salmon Arm, British Columbia, Canada, [2]Sakâw Askiy Management Inc., Prince Albert, Saskatchewan, Canada, [3]Government of Saskatchewan, Ministry of Environment, Forest Service, Prince Albert, Saskatchewan, Canada

ABBREVIATIONS

FMA   Forest Management Agreement
FMP   Forest Management Plan
FMPD   Forest Management Planning Document
VOITs Value, Objectives, Indicators, Targets

MANAGEMENT SETTING AND BACKGROUND

The Prince Albert Forest Management Agreement (FMA) covers an area of 3.35 million hectares (ha) (about 8.28 million acres (ac)) in north-central Saskatchewan (Figure 43.1). The area is provincial Crown land with long-term forest management rights assigned to Sakâw Askiy Management Inc. (Sakâw) based on the preparation and approval of a 20-year management plan. Sakâw is a unique partnership of six forest companies with Saskatchewan operations and two First Nations. A FMA covering this area was first assigned to Weyerhaeuser Canada in 1986 when the company purchased the Prince Albert pulp mill, Big River saw mill, and other smaller operations. Weyerhaeuser operated the FMA to supply fiber to its mills until 2006 when it ceased operations and sold off its physical assets. In 2010, Sakâw Askiy Management Inc. acquired the FMA from Weyerhaeuser on behalf of its eight member companies, several of which operate mills in the area. The eight companies are:

- *A.C. Forestry*, a First Nation company working in forestry, logging, hauling, or administrative positions
- *Carrier Forest Products*, which mills softwood lumber at its two mills in Saskatchewan (Prince Albert, Big River)
- *Edgewood Forest Products*, which mills softwood lumber from its mill in Carrot River
- *L & M Wood Products*, a family-owned and operated sawmill (softwood) business
- *Meadow Lake Mechanical Pulp*, which is part of the Paper Excellence Group and produces hardwood pulp using a bleached chemi-thermomechanical (CTMP) pulp process
- *Meadow Lake OSB Limited Partnership*, which is wholly owned by Tolko Industries and uses hardwood species to produce OSB panels
- *Montreal Lake Business Ventures*, a First Nation harvesting and forestry business
- *NorSask Forest Products*, which is owned by the Meadow Lake Tribal Council and operates one of the largest First Nation-owned sawmills in Canada

The FMA is located just north of the city of Prince Albert and contains the smaller communities of Big River, Weyakwin, Candle Lake, Waskesiu Lake, and Montreal Lake. The FMA wraps around the Prince Albert National Park and contains the Candle Lake, Narrow Hills, and Clarence-Steepbank provincial parks. It contains approximately 1.8 million ha (4.45 million ac) of productive forest (53% of FMA), and 1.4 million ha (3.46 million ac) (42% of FMA) of this area is considered available and suitable for forest management (Table 43.1). Typical forest growth rates are 1.0–2.5 m³ per ha (14.3–35.7 ft³ per ac) per year, depending on the tree species and site productivity.

Forest Plans of North America. http://dx.doi.org/10.1016/B978-0-12-799936-4.00043-6

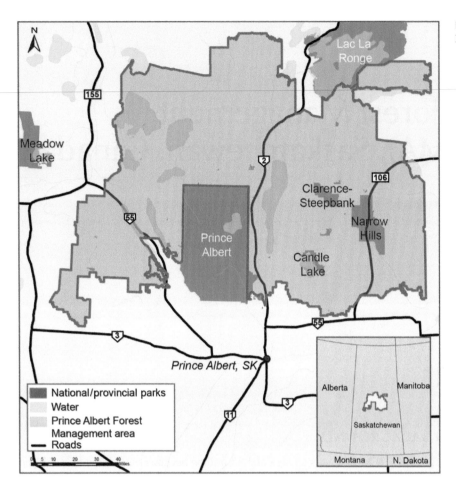

FIGURE 43.1   Location and extent of the Prince Albert FMA area.

Forests consist of pure and mixed stands of aspen (*Populus tremuloides*), white spruce (*Picea glauca*), black spruce (*Picea mariana*), and jack pine (*Pinus banksiana*) on generally gentle terrain. Black spruce, jack pine and aspen stands are the most common (Figure 43.2). Stands are intermixed with significant areas of water features (lakes, muskeg, wetlands). Natural disturbance plays a dominant role in the ecology of the FMA. Fire was a historic disturbance agent that affected large amounts of forest each year and acted as a catalyst for stand renewal and helped to maintain forest health and vigor by controlling insect and disease outbreaks within the forest. Wildfire is now significantly reduced due to suppression practices, and forest harvesting is the dominant method of disturbance and renewal, but fire still remains as an active disturbance agent. Tree damage (windthrow, breakage) as a result of severe wind events also has played a significant role in the history of stand disturbance.

The climate of the FMA is characterized by long cold winters and warm summers. Historical weather station data for several population centers in the area indicate average summer temperatures of 16.6–17.5° Celsius (C) (61.9–63.5° Fahrenheit (F)), average winter temperatures of −18.1 to −20.4 °C (−0.6 to −4.7 ° F), and mean annual precipitation of 400–550 millimeters (mm) (15.7–21.6 inches (in)). Soils and soil landforms within the Prince Albert FMA were largely determined by deposition type after the last continental glaciation. Morainal (undulating) deposition is the most common, with fluvioglacial (glacial outwash), eolian (windblown), and lacustrine (lake bottom) depositions occurring on smaller areas. The upland areas that dominate the FMA area consist of deep, loamy to clayey-textured glacial till, lacustrine deposits, and inclusions of coarse fluvioglacial deposits. Rougher moraine deposits with a large number of small lakes, ponds, and sloughs occupy shallow depressions. Permafrost is very rare and only found in peat lands. Well-drained gray Luvisolic soils are dominant in the region. The FMA falls into the Churchill River Basin and contains four major watersheds (Beaver River, Churchill River, Saskatchewan River, and North Saskatchewan River). These watersheds each ultimately flow into Hudson Bay through the province of Manitoba.

The forest and lakes of the FMA area are a favorite destination for day visitors, overnight visitors, campers, and cottagers from nearby communities and the city of Saskatoon. Beyond the national and provincial parks in and around the FMA, 28 dedicated recreational areas have also been designated. Much of the recreational activity is in the road-accessible areas

TABLE 43.1 Landbase Area Summary of the Prince Albert Forest Management Agreement (FMA) Area in the Province of Saskatchewan

Land Base Element	Total Area (ha)	Total Area (ac)	Effective Area (ha)[a]	Effective Area (ac)[a]	Percent of Total Area	Percent of PFLB[b]
Total crown area	3,349,533	8,276,696	3,349,533	8,276,696		
Less						
Non-FMA lands[c]	49,569	122,485	49,569	122,485	1.5	
Dispositions (buffered and non-buffered)	16,117	39,825	14,279	35,283	0.4	
Treaty land entitlements	3,588	8,866	3,579	8,844	0.1	
Non-forest/non-productive forest	1,527,834	3,775,278	1,492,202	3,687,231	44.5	
Roads, railways, utilities corridors	1,626	4,018	1,207	2,982	0.0	
Productive forest land base (PFLB)			1,788,697	4,419,870	53.4	
Less						
Reserved forest[d]	131,223	324,252	84,790	209,516	2.5	4.7
Subjective leave areas around developments	1,587	3,921	816	2,016	0.0	0.0
Steep slopes	7,247	17,907	3,923	9,694	0.1	0.2
Non-Commercial—Low Density	49,734	122,893	46,692	115,376	1.4	2.6
Non-Commercial—Problem Types	13,552	33,487	5,022	12,409	0.1	0.3
Non-Commercial—"Larchy"	146,307	361,525	121,582	300,429	3.6	6.8
Non-Commercial—Low Site Productivity	96,035	237,302	49,933	123,384	1.5	2.8
Isolated Areas (Uneconomic)	6,124	15,132	6,124	15,132	0.2	0.3
Riparian (lakes, rivers, streams)	53,643	132,552	12,359	30,539	0.4	0.7
Stand Level Retention[e]	—	—	58,298	144,054	1.7	3.3
Timber harvesting land base			1,399,157	3,457,317	41.8	78.2

[a]Effective netdown area represents the area that was actually removed as a result of a given factor. Removals are applied in the order shown above, thus areas removed lower on the list do not contain areas that overlap with factors that occur higher on the list. For example, lake buffers netdown does not include non-forested area.

[b]PFLB (Productive Forest Land Base) refers to the forested crown land area within the FMA. A subset of this area is suitable for timber harvesting.

[c]Indian reserves, Northern Community lands, Patent lands, miscellaneous leases.

[d]Representative area networks, subjective leave areas, recreation areas, Provincial parks.

[e]Insular: 9% gross, 4% net impact.

around lakes adjacent to campsites or cottage subdivisions that have been or are being developed. Popular summer activities include walking, wildlife viewing, fishing, camping, swimming, and boating. Cross-country skiing, ice fishing, and snow-mobiling are common winter activities. The level of activity is much greater during the summer months than at any other time, but the southern part of the FMA area has an extensive network of snowmobile trails that are well used in the winter.

A diverse species mix of animals that include white-tailed deer (*Odocoileus virginianus*), moose (*Alces alces*), elk (*Cervus canadensis*), grouse, ducks, fisher (*Martes pennanti*), and American marten (*Martes americana*), etc.) are hunted and trapped in the FMA for sustenance, recreational, and commercial purposes. Extensive fishing activities also occur, and some lakes support the production of wild rice crops. Forest management in the FMA is focused on sustainable timber extraction while maintaining natural landscape patterns (Figure 43.3), protecting water features, and maintaining suitable habitat for wildlife species. In particular, woodland Caribou (*Rangifer tarandus caribou*) are found in the FMA and are of special concern because of their recent 2012 listing as *threatened* under the Canadian federal government's Species At Risk Act.

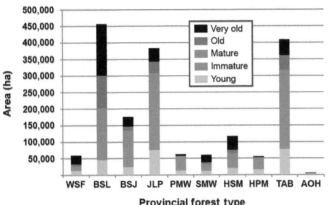

FIGURE 43.2 Productive forest area by provincial forest type and seral stage.

WSF (white spruce or balsam fir dominated softwood stands), BSL (black spruce or tamarack/larch dominated stands), BSJ (black spruce and jack pine dominated mixed softwood stands), JLP (jack or lodgepole pine dominated softwood stands), PMW (pine dominated mixedwood stands), SMW (spruce dominated mixedwood stands), HSM ((hardwood with spruce (bS, wS, bF, and tL) mixedwood)), HPM (hardwood and pine mixedwood), TAB (trembling aspen or white birch dominated hardwood stands), AOH (any other hardwood dominated)

FIGURE 43.3 Large harvested area with internal retention and riparian protection.

PLANNING ENVIRONMENT AND METHODOLOGY

The Province of Saskatchewan's Forest Resource Management Act requires licensees to comply with the Forest Management Planning Document (FMPD). This Act provides direction for the development of Forest Management Plans (FMP) in the province (Saskatchewan Ministry of Environment, 2007). The Act contains detailed descriptions of all steps required to complete the planning process and outlines the documents that must be prepared and approved by the Province. In general, FMPs provide strategic level forest management direction using the principles of ecosystem-based management. They define sustainable rates of harvest under management practices that consider the full range of forest values and include considerable opportunity for input from public, other forest users, and Aboriginal groups.

FMPs are prepared for 20-year periods and renewed every 10 years in the province. Saskatchewan has adopted a results-based regulatory model that focuses on desired environmental and resource management outcomes rather than on prescriptive regulation and processes. Thus, the FMP is designed to provide a vision of the desired future forest and its attributes through a set of values and objectives. Clearly defined indicators and associated targets provide a framework for ongoing evaluation of progress toward desired outcomes.

In general, the FMP process followed the steps outlined below, each requiring staged-approval across a 2-year development period by the province:

1. Establish planning team and *Terms of Reference*
2. Develop FMP work plan and public consultation plan
3. Form a public advisory group

4. Summarize FMA background/context/historical management information
   a. Seek public review and comment
5. Assess sustainability
   a. Identify values, objectives, indicators, and targets
   b. Develop *Silviculture Ground Rules* (acceptable practices and outcomes)
   c. Define desired management practices and alternatives
   d. Assess timber supply through forest estate modeling and draft tactical plans
   e. Seek public review and comments
6. Prepare draft Forest Management Plan
   a. Identify a preferred management strategy
   b. Develop a spatially explicit 20-year tactical plan
   c. Seek public review and comments
7. Finalize the Forest Management Plan for submission and approval
8. Implement the plan (the tactical plan provides guidance)
9. Monitor and report

The use of a public advisory group and three separate public review and comment sessions (serving to ensure the public's interest in the FMA) was understood and addressed in the planning process. Further to this work, the province consulted on the process, at the three same stages of public review, with First Nations and Métis Locals to determine if there were any concerns with the draft FMP deliverables with regard to their respective treaty or Aboriginal rights and traditional uses.

The forest estate modeling process involved a significant effort to update the 2003–2005 forest inventory and to reflect harvesting and natural disturbances that have occurred since the last plan was completed. Then, various datasets were used to identify the subset of the FMA considered eligible and suitable for sustainable forest management. This included identification and exclusion of areas that were ineligible (private land, leases/dispositions, recreation reserves, etc.) or unsuitable (riparian areas, low-site productivity, non-merchantable species, environmentally sensitive areas such as steep slopes, etc). As no growth and yield models are currently approved for use in Saskatchewan, a series of about 5,400 temporary sample plots established in natural stands and linked to inventory polygons were used to develop imperial yield curves for 10 stand types split into strata with 20 yield curves developed. Only plot data from the final net land base were used in the development of yield curves. Unfortunately, no data was available to support enhanced yield curves for actively managed stands so post-harvest stands simply regrow as natural stands, yielding conservative estimates of regrowth.

The development of Values, Objectives, Indicators, and Targets (VOITs) was a key section of the FMP development process because it focused planning team members on defining strategic outcomes and how they would be assessed on an ongoing basis. The document *Defining Sustainable Forest Management in Canada: Criteria and Indicators 2003* (Canadian Council of Forest Ministers, 2003) identified a set of criteria developed at the national level. There are six key criteria, and they form the backbone of the FMP VOITs:

1. Biological diversity.
2. Ecosystem condition and productivity.
3. Soil and water resources.
4. The FMP's role in global ecological cycles.
5. Economic and social benefits.
6. Society's responsibility for sustainable development.

For example, under the *biological diversity* criteria, an objective of "Conservation of biological diversity in the FMA's forests" led to the following example indicators (and associated targets):

- Area by stand types regenerated relative to the area harvested by stand type in any operating year
- Area of old and very old forest by species group
- Size distribution of old and very old forest patches
- Size distribution of harvest events
- Area of residual trees left after harvest expressed as a percentage of harvest area

Other elements of the planning process were aligned with VOITs as needed. For example, the indicator/target requiring regenerated stand type areas to be consistent with pre-harvest stand type areas at a landscape level, provided a clear objective for evaluating the suitability of silviculture practices in the plan.

VOITs were integrated into forest estate modeling assumptions using various indicators and targets accommodated within the Patchworks™ model. Outcomes were explored using scenario analyses. Scenarios were assessed for acceptability using each of the economic, environmental, and social indicators. A spatial 20-year tactical plan was also produced using Patchworks™ and evaluated for acceptability by the Province—particularly, with respect to caribou habitat outcomes. Consideration of caribou habitat was a critical element of the plan, but definitive rules and indicators could not be integrated into the modeling process due to a lack of high-quality habitat mapping and spatial complexity with defining "disturbed" habitat. Thus, the tactical plan was the key link to ensure caribou habitat outcomes were acceptable for the duration of the 20-year plan. Longer-term habitat outcomes were assessed broadly using model outputs for seral stage and patch size metrics.

A key assumption underpinning the plan is that forest harvesting that attempts to mimic natural disturbance patterns will help to ensure that the forest structures are in place to sustain existing plant and wildlife species. This assumption is not made blindly, and it is acknowledged that additional measures may need to be taken to ensure the desired outcomes are achieved, such as access management to limit the amount of active road at any given time and consideration of changing climatic conditions that could make past conditions less relevant.

It should be noted that no stand values or treatment costs were incorporated directly into the forest estate modeling (i.e., economic considerations did not impact harvest scheduling directly). The optimization heuristics employed in the model were used only to maximize harvest volume while achieving appropriate spatial harvest patterns and maintaining non-timber indicator targets. A key element in achieving appropriate spatial harvest patterns was to integrate a basic road network into the model and then seek to minimize the amount of active road use in any given period. This helped to group harvesting into common geographic areas (roadsheds) in each time period. The full permanent road network is open to public traffic at all times, so this modeling technique was only meant to cluster harvesting for tactical planning purposes.

While not yet fully complete, the final plan will be crafted based on the preferred scenario selected from the range of alternatives modeled in Patchworks™. Additional plan information such as wildfire management, forest health management, and monitoring activities will be documented to complete the plan text. Model outputs will be refined to create a spatial 20-year plan that will be used to guide operating plans and forest management activities (general size and location of harvest areas and planned road corridors).

OUTCOMES OF THE PLAN

The outcome of the plan establishes a sustainable rate of harvest for softwood sawlogs (1.03 million m^3 (36.4 million ft^3) per year) and hardwood (1.075 million m^3 (38.0 million ft^3) per year), plus an opportunity to increase short-term harvest levels to capture old stands before they are lost to succession. Softwood pulpwood was forecast as a byproduct of softwood harvest. The rate of harvest was possible after achieving the range of non-timber objectives set out in the indicators. This involved:

- Maintaining important caribou habitat, with a focus on replacing it over time
- Retaining 15% of the area in stands that are in old or very old seral conditions
- Maintaining up to 50% of the old and very old seral forest in interior forest conditions
- Harvesting from the full profile of timber types across the FMA
- Harvesting a range of opening sizes from 10 ha (25 ac) to more than 4,000 ha (9,884 ac) with the retention of internal structure and a mosaic of leave areas (residual uncut patches) similar to what historical large fires would have created
- Regenerating stands successfully within 2 years and confirming free-to-grow status within 14 years
- Maintaining the current balance of hardwood, softwood, and mixed stands on the land base into the future
- Minimizing impacts on caribou habitat by focusing harvesting in already impacted areas and minimizing the amount of open road with industrial traffic on the FMA at any given time

As discussed previously, a 20-year tactical plan is also a key output of the planning process, and the licensees will be expected to follow its direction when developing annual operating plans.

In each year of the 20-year FMP, an Annual Operating Plan will be prepared and input will be sought from the public, stakeholders, and Aboriginal groups. Implementation of the FMP is conducted through annual meetings between the Government of Saskatchewan's Forest Service and the FMA holder, who submit an annual report which measures progress on FMP commitments and results of annual forest operations. Regular monitoring and reporting on the indicators developed in the FMP is part of this process and is used to assess consistency with FMP commitments and to measure progress on targets.

The plan anticipates harvesting and regenerating approximately 16,000 ha (39,536 ac) per year with average per ha yields of 125 m^3 per ha (1,787 ft^3 per ac). Harvest ages are typically between 80 and 100 years old and piece sizes are small (0.15–0.20 m^3 (5.3–7.1 ft^3) per tree). Hardwood stands (e.g., aspen) can be relied upon to regenerate naturally through suckering, while softwoods (e.g., spruce, jack pine) are consistently regenerated through planting.

DISCUSSION AND CONCLUSIONS

At the end of a 2-year planning period, the FMP for the Prince Albert FMA area provides strategic guidance for forest management over at least the next 10 years. It largely focuses on maintaining the natural forest patterns historically created by wildfire in Canada's boreal forest and maintains key habitats for specific wildlife species such as woodland caribou. Commercial timber extraction remains a key management goal and has now largely replaced fire as the dominate stand replacing disturbance on the landscape.

Learnings and Insights

Forest planning is an involved, technical process requiring knowledge in numerous disciplines (biology, ecology, economics, social sciences, etc.). Additionally, it requires obtaining public input on broad forest management strategies and the ability to incorporate this information into the final plan in order to create acceptable strategies for land management. Members of the public typically want to comment on special areas to be protected close to where they live or visit, and some advance broader ideas with good intentions but are not always fully informed. The use of a planning team that includes forest licensees, government employees, the public, Aboriginal representatives, and technical specialists was a fruitful approach to developing a forest management plan because it ensured that all interests were represented throughout the process and allowed for different parties to contribute their areas of strength.

Roads and access management issues were some of the biggest challenges of forest management in the area. The public generally wants roads to remain open and accessible for use, while forest companies want them open only as long as they are needed to harvest and regenerate an area. Furthermore, wildlife biologists typically want to keep the amount of open road to a minimum. Such competing interests almost always result in someone dissatisfied with the local plans.

Maintaining the 2-year timeline for Forest Management Plan development was very difficult. There was often temptation to wait for new imminent information or government policy because it would improve the plan; however, it was rarely as imminent as it was anticipated. In order to maintain timelines, clearly defined cutoff dates for the introduction of new information should be established at the outset of the planning. If this is not the case, it is almost tacit acceptance that timelines can be extended.

Sustainability Issues

The plan specifically identified timber production, wildlife habitat, water quality, visual quality, and biological diversity as key sustainability issues. Modeling implemented targets to ensure the sustainability of each of these factors, and a tactical plan was produced to guide the implementation of harvest activity. Monitoring of indicators is an important part of the plan that will provide feedback on whether operational practices are consistent with the sustainability assumptions built into the plan.

Evolving information on the requirements of woodland caribou and the forest management practices necessary to provide suitable habitat have the potential to change during the term of the plan as significant effort is currently being expended in this area. Depending on the outcome of this work, the management approach assumed under this plan may need to be altered. The regular revisiting of the plan (each decade) is meant to provide the opportunity to adjust management actions as needed before there is a chance of getting too far off course, even though stand rotations are 70–100 years in length.

Plan Development Challenges

The following were some of the challenges encountered during the development of the plan:

1. *Maintaining timelines*: As mentioned earlier, the temptation to wait for better information to arrive made it very difficult to remain on the intended planning timeline.
2. *Evolution of policy during the plan development period*: Important information was lacking in some areas and it was more logical to wait for it to come forth from the government than to develop a separate strategy with different subject experts. This forced the parallel development of FMP strategies and provincial standards and policies.

3. *Lack of information on caribou habitat or management*: The province was actively working on developing a caribou management strategy during the development of the FMP, in response to the 2012 federal government's Recovery Strategy, and it started with extensive data collection and mapping of habitat. Unfortunately, this information was not ready for use in the FMP, so the best available measures were implemented in the plan.
4. *Conflicting views on suitable indicators of sustainability*: In several instances, specific indicators and targets desired by the province to measure sustainability were at odds with the flexibility desired by industry to meet business needs. For example, agreeing on how to ensure a spatial distribution of harvest across the FMA proved challenging.

Plan Implementation Challenges

This FMA has eight different business entities working together to implement the plan, while also acting to meet their specific business needs. Challenges exist because each company produces different products, and each company wants access to different components of mixed species stands (hardwood, softwood, pulp). This requires significant coordination and negotiation between players to get desired volumes at acceptable delivered wood costs because the mills are not located in the same areas and, therefore, have different hauling costs. In this case, the hardwood mills are all on the western side of the FMA, and one of the softwood mills is located on the eastern side. Thus, softwood stands located on the eastern side are most attractive to this mill, but the hardwood incidental volume in these stands is expensive for the hardwood mill, limiting purchases. Additional complications result when different levels of milling capacity are included in the plan. For example, each of the oriented strand board, pulp, and dimension lumber mills requires different raw volume inputs depending on markets, inventories, and other factors. Finding the right mix of stands to supply these needs can be complex. In the recent past, the pulp mill was not operating and the sawlog milling capacity was limited due to economics, so the hardwood operators were forced to focus logging in dominantly hardwood stands. When all mills are operating, the ability to harvest the full profile of the FMA (as per FMP requirements) is greatly increased.

In addition to the economic issues, the desire to implement a wide range of harvest patch sizes will mean educating the public and other stakeholders that larger openings with internal retention are a part of natural landscape patterns, and not just a timber grab. Coordination with other forest values will be the key. For example, guides and outfitters will not want to see large openings that change key areas of their assigned hunting areas. Ongoing communication and education of other stakeholders will be a challenge during plan implementation.

ADDITIONAL READING AND RESOURCES

This chapter is based on the 2015–2035 Forest Management Plan for the Prince Albert FMA in Saskatchewan Canada. To view the plan itself, please visit the following web site, which was available on March 22, 2014:

http://sakaw.ca/management_plan.html

If in the future the link to the site appears broken, search the Internet using the title of the plan and the keywords provided.

REFERENCES

Canadian Council of Forest Ministers, 2003. Defining Sustainable Forest Management in Canada—Criteria and Indicators. Canadian Council of Forest Ministers, Ottawa, ON. http://www.ccfm.org/ci/rprt2005/English/Criteria_and_Indicators_Framework_2005.htm (Accessed April 3, 2014).

Saskatchewan Ministry of Environment, 2007. Forest Management Planning Document—Forest Planning Manual. Government of Saskatchewan, Ministry of Environment, Regina, SK. 251 p, www.environment.gov.sk.ca/Default.aspx?DN=40f6639e-2687-4ec1-8b07-e6e6b2279d54 (Accessed April 3, 2014).

Chapter 44

Rayonier, Inc., Southern United States of America

John Paul McTague[1] and Michael J. Oppenheimer[2]

[1]Rayonier, Inc., Yulee, Florida, USA, [2]Rayonier, Inc., Fernandina Beach, Florida, USA

ABBREVIATIONS

GUI   Graphical User Interface
iSharp   Integrated Sustainable Harvest and Resource Planner
LP   Linear Programming
MPS   Mathematical Programming System
NCASI   National Council for Air and Stream Improvement
REIT   Real Estate Investment Trust
TIMO   Timberland Investment Management Organization

MANAGEMENT SETTING AND BACKGROUND

Rayonier, Inc. currently manages approximately 1.9 million acres (ac) (about 769,000 hectares (ha)) of timberland in the southern United States. Across these forests, 42% of the lands are occupied by fast-growing loblolly pine (*Pinus taeda*) plantations and another 23% contain slash pine (*Pinus elliottii*) plantations. The remaining 35% are made up of natural pine and hardwood stands, typically too wet to plant and intensively manage. Rayonier began acquiring these southern timberlands in 1938 with the construction of the Fernandina Beach, Florida pulp mill. Over the years it expanded, first northward into Georgia and then, more recently, westward into the Gulf states region.

The corporate structure of Rayonier is categorized as a publicly held Real Estate Investment Trust (REIT) that is listed on the New York Stock Exchange. As a REIT, income from the sale of timber is generally not taxed at the corporate level, permitting Rayonier to distribute higher returns to investors in the form of tax-efficient dividends. The timberlands are owned and managed to maximize net present value (NPV). The current management of the Rayonier forest asset in the southern United States is largely unconstrained with respect to long-term supply agreements; however, REIT requirements and stockholder preferences regarding dividends imply an expectation of a steady annual and quarterly flow of income that is derived from selling stumpage and delivered wood.

Long-range forest planning at Rayonier is conducted with the use of constrained optimization tools. Foresters and financial managers are often puzzled about the requirement of constrained optimization, given that the optimal stand-level forest management decision criterion derived by Martin Faustmann in 1849 is widely accepted and used on a regular basis for sustainable production. Gunn (2007) notes that each stand type and site-quality class will have its own optimal regime and yields. Strategies that require a regular flow of wood products have led to harvest scheduling models that use linear programming (LP) or simulation models. Ignoring for a moment the complexities of stands of varying densities and site qualities, it is apparent from the example in Figure 44.1 (top) that the age-class distribution of this forest does not provide a steady flow of yields and income, using the optimal stand-level formulas of Samuelson (1976) and Chang (1998).

Paredes and Brodie (1989) argue that the Faustmann-Samuelson stand-level solution can be used for forest-level analysis; however, it needs to be solved with the appropriate revenue and opportunity costs for each planning period. We are in full agreement with Paredes and Brodie (1989) and believe that the dual decomposition methodology of Hoganson and Rose

Forest Plans of North America. http://dx.doi.org/10.1016/B978-0-12-799936-4.00044-8

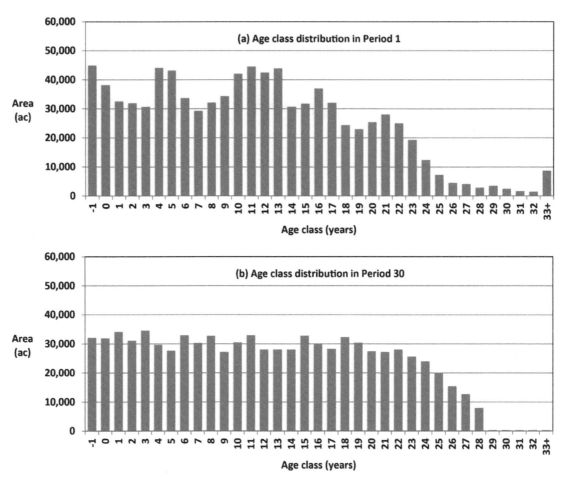

FIGURE 44.1 The age-class distribution of the Rayonier southern timberlands in acres. Panel (a) depicts conditions in planning period 1, prior to the solution of a harvest schedule while panel (b) indicates the age-class distribution following period 30.

(1984) is well suited for that task. Rayonier has elected, however, to solve the harvest scheduling problem with a Model I integer formulation that can be addressed with either linear programming or heuristics because of its computational speed, ease in understanding, repeatability, and ability to condition the outcome to comply with forest certification requirements.

The rapid rise of Timberland Investment Management Organizations (TIMOs) and REITs as a major component of the private corporate landowner class in North America, is mirrored by the decline of the vertically integrated forest products company. In many cases, the production of forest products from TIMO- or REIT-managed forests are no longer obligated to furnishing the input requirements of a specific industrial complex. Most TIMOs and REITs deliberately seek to manage a diverse set of forest tracts over a wide geographic range to reduce risk. Thus, forest plans have become large since they are no longer "mill-centric." The Rayonier harvest scheduling plan for the southern United States encompasses 28,000 forest stands and maximizes NPV for timberlands spanning from Georgia to Oklahoma.

PLANNING ENVIRONMENT AND METHODOLOGY

Rayonier adheres to the standard objectives for forestland management that are promulgated by the Sustainable Forestry Initiative (SFI). Performance Measure 1.1 of the SFI Standard (Section 2) requires that forest management plans include long-term harvest levels that are sustainable and consistent with appropriate growth and yield models (Sustainable Forestry Initiative, 2014). Certification requirements of the SFI program also require that harvest plans address the size, shape, and placements of clear-cut harvest areas for management of visual quality (Performance Measure 5.1).

Harvest Scheduling System

The current Rayonier harvest scheduling system, known as Integrated Sustainable Harvest and Resource Planner (iSharp), represents a clear departure from the strata-based Model II formulation that had been previously used. The decision variable is an integer 0-1 variable indicating the assignment of regime j to stand i. The matrix generation is conducted with proprietary growth and yield models that have prediction and projection capabilities at both the stand and diameter-class levels (McTague, 2012). Rayonier has confidence in procedures that generate the technological coefficients of multiple product yields, since stand-table projection methods are employed to grow the diameter classes of observed inventory. Constraints can be generated at both the enterprise and the operating unit level.

The iSharp system does not make use of accounting rows (a description of which can be found in Bettinger et al., 2009). Acknowledging that accounting rows greatly facilitate the writing of constraints and computing sums in the report writer, it is also recognized that they add unnecessarily to the number of rows in the matrix and the computational time for solving the problem. The iSharp report writer contains multiple tables that include summarizations of treatment areas, cash flows, timber flows, and estimates of residual inventories, and produces graphs of these and other metrics. While an integer formulation is used as the problem architecture, LP is typically used to solve the problem. The integer formulation, however, allows for the potential to solve the problem with heuristics.

The harvest planning procedure used by Rayonier includes the following steps:

1. Advise operating units to update harvest and silvicultural activities within the appropriate databases.
2. Clean and fix the data.
3. Obtain exogenous inputs from the long-range plan. These include sub-regional timber prices, land acquisition and disposition goals, and timber lease requirements.
4. Generate a LP matrix. This involves computing terminal values for regimes, objective function coefficients, and technological coefficients.
5. Solve the LP problem using MOSEK.
6. Verify harvest yields that are suggested.
7. Refine the schedule by including operating unit flow constraints, lease and land sales targets, and maximum opening size constraints.
8. Assess whether targets are met.
9. Uncover infeasibilities in the LP solution and revise right-hand-side values of constraints.
10. Review the harvest plan with operations personnel and executives.
11. Revise if necessary.

The iSharp architecture allows for the net revenue of regimes to vary by sub-region or the location of the operating unit. Figure 44.2 displays the current sub-regions that are modeled in the Rayonier harvest plan. A sub-regional estimate of changing stumpage prices are computed for a 10-year period, based upon predictions from the SRTS model (Abt et al.,

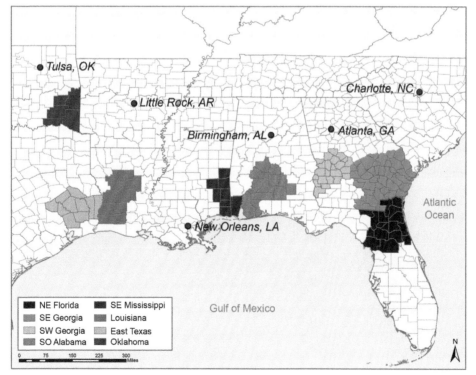

FIGURE 44.2   The eight operating units of the Rayonier southern timberlands.

2000). After 10 years, the revenues and costs in real terms are assumed to remain constant. The objective function coefficients of each regime contain two components: (1) the NPV of the plan in real values for the entire horizon and (2) the terminal value of the optimal stand-level regime in perpetuity. This feature obviates the need for ending inventory constraints and leads to stable plans that are practically invariant to the length of the planning horizon.

Rayonier uses two primary sets of constraints to achieve both the corporate strategic objectives and the SFI certification requirements for sustainability: (1) harvest areas and (2) timber flow, in terms of scheduled harvest tonnage. These are applied throughout the southern and sub-regional levels. The constraints vary by harvest type and, in the case of timber flow, by merchantable product. The matrix rows associated with the flow constraints are sparse, because only harvest or cut volume cells contain nonzero technological coefficients. Standing or inventory volume is computed for only the basic decision variables of the optimal solution, and it is conducted during the post-solution report writing.

Decision variables or regimes are generated for each stand, with treatments and age ranges that are described in Table 44.1. These can be specified by sub-region, by stand type, and by species. Although the rotation length varies among regimes, they typically spans two rotations (i.e., the current rotation and one future rotation). All possible regime combinations are then assigned to each stand using the setup parameters defined in Table 44.1. Restricting the range of rotation ages, minimum years between intermediate and final harvests, and the minimum required harvested yield per unit area aids in reducing the number of regimes to a more manageable number. Irrespective of the start date of the forest plan, it is not uncommon to have stand-specific cutting contracts in place that pre-determine the majority of volume to be extracted in the early phase of the plan. "Pre-blocking," or predetermined future treatments of a stand, curtail even further the potential number of regimes. Although pre-blocked regimes allow for cutting throughout the planning horizon, they must conform to planned treatments during the early phase of the plan. Even with all these regime filters, the Rayonier Southwide model contains more than 5.5 million regimes (matrix columns) covering 28,000 stands.

Prompt reforestation is the policy on all lands managed by Rayonier and it occurs on all leased property of Rayonier, so the regimes for long-term leases are identical to fee land with the exception that no clear-cutting can occur in the last year of the lease. To accommodate the lease expiration, the iSharp system truncates the regime at that final date and assigns a terminal value of zero. Thus, the objective function coefficient of a lease regime contains only the NPV associated with volume harvested during the planning horizon. The rent on the lease is not included in each regime's NPV calculation since it is a fixed cost that will not vary by different harvest scenarios. For fee land terminal values, iSharp determines the optimal single rotation regime for each stand and discounts its Faustmann perpetuity value to the present (calculated with Equation 10.4 on page 280 of Clutter et al. (1983)).

Timber product yields for each harvest within each regime are calculated by projecting each stand's most recent inventory measurement to the harvest age and then simulating the timber removal. In the case of clear-cuts, the simulation is easy because cut volume is identical to standing volume. In the case of thinnings, the simulation represents a row thin with selection from below with a residual basal area or tree density target that varies by sub-region. All of the growth and yield projections and merchandizing wood into products are made using Rayonier's suite of proprietary growth and yield models (McTague, 2012). These models are built into the iSharp system and include treatment response components for thinning and fertilization.

The Rayonier Southwide harvest schedule plan is generated once every two years. For the Southwide model, the planning horizon is comprised of 30, 1-year periods with the first planning period representing the current year. Ongoing harvests and harvests under contract, or in preparation, typically occupy the first two periods, so it is generally the third

TABLE 44.1 Management Regime Guidelines for Harvest Scheduling in Rayonier's Southern Timberlands

Treatment	Stand Type	Age Range (years)	Minimum Number of Years Before Next Harvest	Minimum Required Yield[a] (tons per ac)
Clearcut	Plantation	16–35		
	Natural	30–50		
First thinning	Plantation	10–20	5	20
Second thinning	Plantation	15–25	5	20
Fertilization	Plantation	10–20	5	

[a]Merchantable volume.

planning period and beyond that has strategic planning significance. The Rayonier model is stumpage based so there are few cost assumptions, which vary by sub-region. Rayonier utilizes costs per unit area for site-preparation, planting, and fertilization. All other costs are considered fixed and, therefore, are not included. For land-base valuation and other strategic planning, these fixed costs will be added in the report writing phase of the harvest scheduling. Additional yield assumptions include slight increases in site quality at the inception of the second rotation to represent cultural and genetic gains. These increases are stand-based and are governed by the current stand site index, level of silvicultural inputs, and generation of genetic improvement.

In addition to the growth and yield functions within iSharp, the main components of the system include the following:

1. *Job set-up.* A graphical user interface (GUI) that was developed by Rayonier for controlling the planning horizon, objective function, and other model parameters.
2. *Regime generator.* A GUI that was developed by Rayonier for constructing management regimes and for adding filters to limit availability.
3. *Assumption editor.* A GUI that was developed by Rayonier for adding or editing assumptions.
4. *Constraint editor.* A GUI that was developed by Rayonier for adding or editing constraints.
5. *Front-end/matrix generator.* A process that was developed by Rayonier for associating stands to regimes, for calculating yields and NPVs, and for formatting data for input into the solver.
6. *LP solver.* An available LP solver (MOSEK) that accepts Mathematical Programming System (MPS) formatted input.
7. *Habplan solver.* The National Council for Air and Stream Improvement (NCASI) harvest scheduling software.
8. *Report writer.* A GUI that was developed by Rayonier for generating output reports and graphics.

The job setup, regime generation, assumption, and constraint builders are all simple GUI applications for transferring input items into the iSharp database. The job setup also controls stand-data extraction from the Rayonier integrated GIS/database system into the iSharp database. This is typically performed once, at the beginning of the harvest scheduling exercise. The front-end regime generator component performs the yield and financial computations needed to produce the technological and objective function coefficients of the harvest scheduling problem. The matrix generator then converts the problem matrix into a standard MPS format. The front-end, which only needs to be executed if a regime or assumption is changed, takes approximately 45 minutes (min) of run-time on a Windows 64-bit computer with two 8-core Intel Xeon processors and 128 GB of memory. The matrix generator, which needs to run with each constraint change, requires approximately one hour to complete. The LP solver, a 64-bit version of MOSEK, takes about 45 min to solve the optimal solution for the Rayonier Southwide schedule. The report writer, which does all the post-solution, standing inventory growth and yield calculations and summarization (by period), requires approximately 30 min for execution. Hence, from start to finish, the harvest scheduling task requires three hours of CPU (Central Processing Unit) time to complete all functions. Should a revised run require only a change to a constraint, less than 2.5 hours (hr) are needed to determine the new optimal solution. The ability to quickly solve a problem of this size (i.e., the MPS file exceeds 50 Gb in size) can be partially attributed to the Model I formulation with a sparse matrix that is devoid of accounting row constraints. The solution speed is also improved with other features such as multithread programming in a 64-bit multiprocessor environment.

For spatially constrained problems, which impose SFI adjacency rules, the iSharp system can be integrated with the NCASI Forest Harvest and Habitat Scheduling system, known as Habplan Version 3 (Van Deusen, 1999). The Habplan program uses heuristics to solve spatial objects such as average clear-cut size, or impose spatial constraints, which recognize that adjacent stands cannot be mutually harvested within a 3-year interval. The iSharp system produces all the needed input files for Habplan including multi-objective valuation, volume flow, and spatial components. It then automatically starts Habplan, relinquishes control, and displays the Habplan user interface. Constraints, both flow and spatial, are then entered directly into Habplan, and executable runs are initiated automatically.

Even though Rayonier's land base contains numerous large contiguous blocks exceeding 20,000 ac (8,094 ha), our experience demonstrates that these blocks are diverse enough in age to obviate the need for spatial constraints. For SFI purposes, Rayonier defines "opening size" as all adjacent stands less than 3 years old. Although a formal maximum opening size constraint is not part of the Rayonier model formulation, a posterior spatial inspection of the optimal solution is conducted to ensure that no opening size exceeds 500 ac (202 ha). The few scheduled clear-cuts of the optimal solution that possess openings greater than 500 ac (202 ha) are later divided into smaller parts during the tactical planning phase using 300 ft (91.4 m) or wider buffers. Since Habplan lacks the ability to add such buffers, Rayonier routinely relies exclusively on the LP solutions for solving the Southwide harvest schedule. As for spatial objectives, Rayonier's Southwide average opening size has always been far below the standard SFI average size of 120 ac (49 ha). As a result, additional measures to decrease the opening size have not been needed. Figure 44.3 displays a common landscape mosaic of plantations and natural stands, with a variety of age classes that are prevalent on managed Rayonier timberlands.

FIGURE 44.3 Aerial view of a recently regenerated plantation and surrounding forest stands.

The lower inflation and interest rates experienced over the last decade in the United States has impacted many harvest schedules. As forest plans currently employ a low discount rate, the terminal value of a forest tract at the end of a 30-year planning horizon can be substantial. Rayonier does not employ the use of ending inventory constraints to ensure that the standing inventory is not liquidated at the end of the planning horizon. We feel that ending inventory constraints are linked to the concept of a target-forest, which is prone to inaccuracy, since it is impossible to know a priori sustainable and optimal cutting levels. Rather, Rayonier accounts for the terminal value by employing the perpetual value methodology proposed by Clutter et al. (1983).

Not all of the managed timberlands have a projected cash flow that is modeled in perpetuity. Quite often, timberlands are managed from lands designated as timber deeds or leased land. REITs, such as Rayonier, possess a real-estate division that occasionally places higher-valued land on the market for sale and development, which counterbalances land sales with land and forest acquisitions. The Rayonier harvest schedule model contains a feature that allows for the analysis of terminating leases and the purchase and selling of timberland.

Before the iSharp system was developed, Rayonier employed the Remsoft Woodstock model for its U.S. South harvest scheduling needs. There are two main differences between the iSharp system and the Woodstock model that Rayonier previously used. First, iSharp accommodates a Model I integer formulation that seems appropriate for intensively managed forests, while Woodstock utilizes a Model II strata-based formulation with a subsequent disaggregation of the solution to stands. Remsoft does provide tools to make inputs (regimes) into and outputs (stand lists) from the MPS-based solver seem as though the user is dealing with a Model I formulation. However, routine tasks, such as constraining a stand to a specific regime, require intricate coding. "Exclude" statements and operability masks need to be assigned to each activity and across the entire planning horizon. The second key difference involves terminal values. iSharp uses unconstrained optimal regimes to calculate perpetuity values for the objective function. We have been unable to find a workable procedure in Woodstock for adding terminal values. A proxy, which prevents harvest liquidation of the forest estate at the end of the planning horizon, is the use of ending inventory constraints. Setting the level for these constraints is subjective and requires numerous runs to determine what level of standing inventory by age class is sustainable while simultaneously approaching economic optimality.

Another key feature of the iSharp systems is the method used to evaluate the land at the end of the planning period. The impact of terminal values on the wood flow differences between an unconstrained and constrained run and between a 30-year horizon and a 50-year horizon are quite extraordinary. These differences were witnessed during the beta testing phase of the iSharp system. The testing was conducted using an 180,000 ac (72,845 ha) pine plantation forest in southeast Georgia consisting of both loblolly and slash pine with a fairly typical, although far from uniform, age-class distribution. As a benchmark, Rayonier's Woodstock model for the U.S. South was applied to the same forest. Assumptions, regimes, and yield statistics were kept as similar as possible between the two systems. In the Woodstock unconstrained scenario, the harvest flow starts out fairly uniform; however, the flow becomes quite erratic for second rotation stand harvesting. We noticed a large, scheduled increase in harvests during the last year of the planning horizon, which was a direct effect of the absence of terminal values. Without terminal values, there is no positive net value associated with standing inventory at the end of the planning horizon; and therefore, the optimal decision is to liquidate the estate. To maximize the liquidation value, given the minimum available clear-cut age of 18, Woodstock chose a similar large harvest 19 years earlier, just in time to regenerate and then harvest 18 years later.

OUTCOMES OF THE PLAN

Traditionally, the report writer functions of harvest schedulers have utilized the LP basic solution file and created wood flows, stand lists, and ending inventory reports. These tables were typically exported to .wk1, .xls, or .dbf format so that summaries and distributions could be made. The iSharp system accomplishes these traditional tasks by first populating a number of solution and inventory tables in its SQL server database. Routine stand-level and estate-level reports and exports to Excel are available for items such as the age-class distribution at the end of the planning horizon. More sophisticated outputs can be acquired through simple SQL programming, and might include

- Projected total pine harvest volume from different management scenarios
- Amount of projected pine harvest volume from clear-cutting prescriptions that involve thinnings
- Amount of projected pine harvest volume from clear-cutting prescriptions that do not involve thinnings
- Amount of projected pine harvest volume from thinnings alone
- Amount of projected pine harvest by product grade

To better understand the impact of wood volume flow constraints by period, especially for the unconstrained scenario, age-class distributions are exported from the iSharp system. An example from a U.S. South project is displayed in Figure 44.1. Using the age-class distribution at the inception of the planning horizon, we would expect modest harvest levels during the first few planning periods since few forests exceed the age of 23. Volume flow could increase after the first 3 to 4 years, but the lack of forests in the age 18- and 19-year-old stands would again cause another drop in volume flow later on. The even, or virtually uniform, age-class distribution depicted in Figure 44.1b following period 30, clearly shows the impact that flow constraints have in moving the Southwide Rayonier timberlands to a fully regulated and sustainable forest estate. Beyond year 30, and as the forest estate moves toward a more even-aged distribution, the flow constraints become less binding.

For Rayonier, Inc., total harvest volume by harvest type and period are key flow variables. When scenarios are analyzed, graphics are used to monitor flow changes during the transition from an unconstrained scenario to the final constrained scenario. Of course, NPV is also a variable of paramount importance and we monitor its decrease as a percentage of the unconstrained scenario. The iSharp system produces an unconstrained counterpart to each constrained scenario, so the NPV change due to constraints is always separable from other changes, such as regime or product specifications, that often occur during a harvest scheduling project.

The most important outputs from the harvest scheduler are the cutting plans distributed to the local operational foresters for tactical planning and final implementation. In addition to wood flow summaries, Rayonier uses two key output products to facilitate operational budgeting and planning: (1) lists of harvest treatments by stand and planning year and (2) a GIS shapefile that visually displays the first 5 years of harvests. Stands in the basic solution are listed by plan year and harvest type. Stand attributes, such as species, harvest age, harvest areas, and volume by product are included in both the list of harvest treatments and the GIS shapefile.

DISCUSSION AND CONCLUSIONS

Rayonier, Inc. has recently developed the iSharp system for harvest scheduling processes that relate to its timberland holdings in the U.S. South. The harvest scheduling model has a Model I integer programming structure that accommodates advanced growth and yield processes.

Learnings and Insights

In contrast with the past, the large harvest scheduling system presented here does not entail compromises in model precision. The iSharp system of Rayonier recognizes sub-regional differences in costs and revenue and it uses stand-table projection methods to predict yield at the stand-level for one-year planning periods. The valuation computation in the objective function makes a clear distinction between leased and fee timberland, and the basic solution is fully compliant with near-future cutting contracts. In the past, large harvest scheduling models were notoriously slow. The iSharp system utilizes a sparse matrix, devoid of accounting rows, and it contains procedures that reduce the number of decision variables prior to the generation of the matrix. Thus, the emphasis with the iSharp system is focused on sensitivity analysis and successful deployment at the operational level.

Sustainability Issues

Achieving a sustained flow of harvest products from a regulated forest still endures as the heart of forest management. As Davis (1966) indicated, the forest manager usually works with timber properties that are unbalanced with respect to area by age class, and the opportunity costs of moving to a balanced situation must be considered. The question in hand is how fast the forest estate should progress to a fully regulated state. Better GIS tools and area estimates, efficient forest sampling designs, and modern forest planning models reduce the risk associated with a deviation from the sustainable trajectory to full regulation. At Rayonier, an annual inspection of recent and future activities is used to identify any possible departure from sustainability and the path to a regulated forest estate.

Plan Development Challenges

Scheduling harvest activities for the entire Southern timberlands of Rayonier, using one model, presents challenges. For example, treatment regimes need to be flexible enough to accommodate each operational unit's unique management and market issues, but still need to conform to Rayonier's overall management objective. The interaction of flow constraints among the operational units can also make them difficult to reconcile. While the alternative of modeling each unit separately would be clearer to define and manage, it has the unacceptable drawback of producing a suboptimal solution for the southern U.S. land that Rayonier manages.

Plan Implementation Challenges

As we slowly emerge from the great recession of 2008, changing wood markets have led to some significant divergences from the harvest schedule that is produced every 2 years. The startup of numerous pellet mills and other biofuel facilities has also contributed to variance from the harvest schedule. More recently, log exports have developed that utilize products not included in the current harvest schedule. These types of market changes are difficult to predict, but they can be somewhat ameliorated by making more frequent runs of the harvest schedule. For the southern United States, Rayonier's current frequency of generating a harvest plan is once every 2 years, but it can easily be moved to 1-year frequency if the need arises.

REFERENCES

Abt, R.C., Cubbage, F.W., Pacheco, G., 2000. Southern forest resource assessment using the Subregional Timber Supply (SRTS) model. Forest Products Journal 50 (4), 25–33.

Bettinger, P., Boston, K., Siry, J.P., Grebner, D.L., 2009. Forest Management and Planning. Academic Press, New York. 331 p.

Chang, S.J., 1998. A generalized Faustmann model for the determination of optimal harvest age. Canadian Journal of Forest Research 28, 652–659.

Clutter, J.L., Fortson, J.C., Pienaar, L.V., Brister, G.H., Bailey, R.L., 1983. Timber Management: A Quantitative Approach. John Wiley & Sons, New York. 333 p.

Davis, K.P., 1966. Forest Management: Regulation and Valuation, second ed. McGraw-Hill, New York. 519 p.

Gunn, E.A., 2007. Models for strategic forest management. In: Weintraub, A., Romero, C., Bjørndal, T., Epstein, R. (Eds.), Handbook of Operations Research in Natural Resources. Springer, New York. pp. 317–341.

Hoganson, H.M., Rose, D.W., 1984. A simulation approach for optimal timber management scheduling. Forest Science 30, 220–238.

McTague, J.P., 2012. Stand-level growth and yield models for second rotation loblolly and slash pine plantations in southeastern United States. Rayonier, Inc., Fernandina Beach, FL. Rayonier Forest Resources Research Report. 12-01.

Paredes, G.L., Brodie, J.D., 1989. Land value and the linkage between stand and forest level analysis. Land Economics 65, 158–166.

Samuelson, P.A., 1976. Economics of forestry in an evolving society. Economic Inquiry 14, 462–492.

Sustainable Forestry Initiative, 2014. SFI Standards, An Overview of the Requirements for the SFI 2010-2014 Program. Sustainable Forestry Initiative Inc., Washington, DC. http://www.sfiprogram.org/sfi-standard/sfi-standards/ (Accessed May 1, 2014).

Van Deusen, P.C., 1999. Multiple solution harvest scheduling. Silva Fennica 33, 207–216.

Chapter 45

San Juan National Forest, Colorado, United States of America

Mehmet Demirci,[1] Pete Bettinger[2] and Krista Merry[2]

[1]*Forest Management and Planning Department, General Directorate of Forestry, Ministry of Forest and Water Affairs, Ankara, Turkey,* [2]*Warnell School of Forestry and Natural Resources, University of Georgia, Athens, Georgia, USA*

ABBREVIATIONS

SJNF San Juan National Forest
TRFO Tres Rios Field Office

MANAGEMENT SETTING OR BACKGROUND

Located in the southwestern corner of Colorado, the San Juan National Forest (SJNF) and the Tres Rios Field Office (TRFO) of the Bureau of Land Management (BLM) include portions of the Colorado Plateau and the San Juan Mountains (Figure 45.1). A joint plan was developed for these two areas in 2013, and it encompasses approximately 1,867,800 acres (ac) (about 755,888 hectares (ha)) of the SJNF and approximately 504,400 ac (204,128 ha) of the TRFO. There is a separate record of decision document for both organizations (U.S. Department of Agriculture, Forest Service, 2013). This chapter focuses on the SJNF aspect of the joint plan.

The SJNF was established in 1905 by President Theodore Roosevelt (Bond, 2014). In 1911, a portion of it was transferred to the Durango National Forest, yet the Durango National Forest was incorporated back into the SJNF in 1920. In 1947, a portion of the Montezuma National Forest was also added to the SJNF (Davis, 1983). The SJNF has three district offices, and the area spans pieces of 11 counties. Most of the land area is concentrated in five main counties, with U.S. Forest Services (USFS) and BLM programs accounting for about 6% of the employment and 5% of the income in the five-county area (U.S. Department of Agriculture, Forest Service and U.S. Department of Interior, Bureau of Land Management, 2013a). The San Juan Skyway is a National Forest Scenic Byway, an All-American Road, and a component of the Colorado Scenic and Historic Byway System, and it traverses the National Forest.

The SJNF contains alpine lakes and meadows (Figure 45.2), canyons, waterfalls, and plateaus (often called *tablelands* or *mesas*). Elevations range from 4,900 feet (ft) (1,493 meters (m)) to over 14,000 ft (4,267 m)) (U.S. Department of Agriculture, Forest Service and U.S. Department of Interior, Bureau of Land Management, 2013a). Spruce-fir forests occur at elevations ranging from about 9,000 to 11,500 ft (2,743–3,505 m) in the subalpine climate zone. The most prevalent species are Engelmann spruce (*Picea engelmannii*) and subalpine fir (*Abies lasiocarpa*). Aspen (*Populus* spp.) forests are also abundant in these areas and occur at elevations ranging from about 7,500 to 11,200 ft (2,286–3,414 m). Douglas-fir (*Pseudotsuga menziesii*) and white fir (*Abies concolor*) occur on warmer drier sites at elevations ranging from about 8,500 to 10,000 ft (2,591–3,048 m). Ponderosa pine (*Pinus ponderosa*), Douglas-fir and white fir tree species are often found in the mixed coniferous forests on the mountains, hills, and tablelands (plateaus) of the national forest, where elevations range from about 7,500 to 9,000 ft (2,286–2,743 m). Mountain shrublands that occur in the lower montane and montane climate zones can host Gambel oak (*Quercus gambelii*), mountain mahogany (*Cercocarpus montanus*), serviceberry (*Amelanchier* spp.), squaw apple (*Peraphyllum ramosissimum*) and other trees and shrubs (U.S. Department of Agriculture, Forest Service and U.S. Department of Interior, Bureau of Land Management, 2013a).

Forest Plans of North America. http://dx.doi.org/10.1016/B978-0-12-799936-4.00045-X

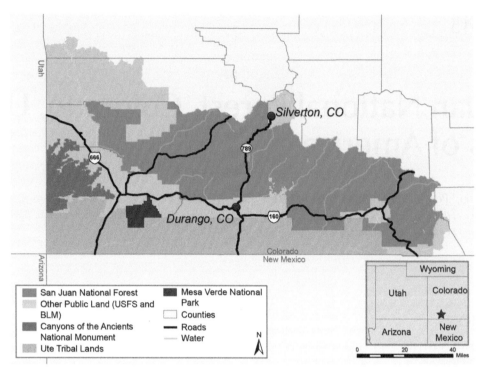

FIGURE 45.1 A map of the land area encompassing the San Juan National Forest and Tres Rios Field Office of the BLM in southwest Colorado.

FIGURE 45.2 A view of a portion of the San Juan National Forest. Courtesy U.S. Forest Service Staff, through Wikimedia Commons.

Some of the most influential factors in shaping the condition and structure of the forests are fire, insect outbreaks, diseases, droughts, and management actions. During the first decade of the twenty-first century, insect outbreaks and diseases had a larger impact, by area, than fire (Rocky Mountain Region & Forest Health Protection, 2010). It has been reported that between 2001 and 2004 in southwest Colorado, up to 90% of the piñon pine (*Pinus edulis*) trees in many areas were killed by a piñon ips beetle (*Ips confusus*) epidemic (Lewis et al., 2005).

In the early twentieth century, wood produced from the SJNF was a significant resource for mining and settlement purposes. From 1960 to 1980, average annual harvest volume was more than 50 million board feet (MMBF) (118,000 m³) per year, but timber harvest levels have declined as a result of depressed market conditions. Only about 49% of the SJNF is capable of growing commercial timber products, and since 1955 only about 19% of the land has supported timber harvesting activities. The most common harvesting methods are shelterwood systems (spruce-fir and ponderosa pine forests), coppice systems (aspen forests), timber stand improvement activities, and sanitation and salvage harvests. Improvement cuts through thinnings are used for fuel reduction and forest restoration purposes (U.S. Department of Agriculture, Forest Service and U.S. Department of Interior, Bureau of Land Management, 2013a). In the 1970s and 1980s, several local mills closed, yet the forest products industry continues to be important to local communities (U.S. Department of Agriculture, Forest Service and U.S. Department of Interior, Bureau of Land Management, 2013b).

PLANNING ENVIRONMENT AND METHODOLOGY

The first comprehensive forest plan for the SJNF was developed in 1976. It was redeveloped in 1983 (U.S. Department of Agriculture, Forest Service, 1983) and significantly amended in 1992 (U.S. Department of Agriculture, Forest Service, 1992). The current, joint forest plan was approved in 2013. The joint plan was based on efforts that were driven by resource data and public values. The data-driven effort provided the decision makers technical analyses of resource conditions and projections of trends for social, economic, and ecological concerns. The planning process involved more than 40 experts from a variety of disciplines that included archaeology, ecology, engineering, fisheries, forestry, hydrology, landscape architecture, mining, wildlife, and recreation, among others.

A public participation process was used to help understand the values people place on the landscape and the typical uses of the public lands. In developing the joint management plan, the USFS and the BLM worked with other federal agencies, state agencies, local governments, tribes, and local communities. Input for the management plan was received from individuals and stakeholder groups through public meetings and outreach efforts. Almost 54,000 letters were received, and only about 1,500 of these were deemed original responses or letters containing original text. Of the substantive comments that were grouped into unique public concern statements, 78 related to minerals and energy, 68 to water, 51 to recreation, 45 to terrestrial wildlife, 44 to air quality, 38 to access and travel management, 34 to livestock and rangeland management, 28 to timber management, 11 related to economics, 4 to climate change, 2 to insects and disease, and 1 to wilderness and wilderness study areas. These concern statements ranged in scope from broad generalities to very specific points regarding the management of the national forest (U.S. Department of Agriculture, Forest Service and U.S. Department of Interior, Bureau of Land Management, 2013c).

The USFS and the BLM planning processes can be described as *adaptive*, where adjustments may be made based on the results of monitoring systems and the social, economic, and ecological outcomes that are measured. Within both agencies, forest planning generally occurs at three hierarchies: (1) the national level, (2) the forest level, and (3) the site-specific level. The USFS national plan is strategic in nature and establishes the goals, objectives, and strategies for the management of lands within the National Forest System. Forest-level management plans are more specific to each national forest, and also contain goals and objectives that are related to desired future conditions. Site-specific plans relate directly to the management actions that are used to meet the goals and objectives of the forest plan (U.S. Department of Agriculture, Forest Service and U.S. Department of Interior, Bureau of Land Management, 2013b).

In general, the planning process included these steps:

1. Identify the issues.
2. Develop the planning criteria.
3. Collect the necessary data and information.
4. Analyze the current management situation.
5. Formulate alternatives for future management.
6. Assess the alternatives.
7. Select a preferred alternative.
8. Finalize the land and resource management plan.
9. Implement the management plan.
10. Monitor the management plan.

Some of the important actions accomplished during the planning process included:

- Defining the desired outcomes of the forest, including the multiple-use goals and the objectives
- Determining the appropriate management actions
- Determining the measures or criteria that would be used to guide day-to-day activities
- Defining the direction for each management area, including the allowable uses, restrictions, and prohibitions
- Designating the research natural areas, areas of critical concern, and other special places
- Recommending some roadless areas for wilderness designation
- Identifying certain segments of rivers for inclusion in the National Wild and Scenic Rivers System
- Defining the area of land suitable for timber harvest activities
- Determining the allowable sale quantity (quantity of timber that may be sold from the area of land suitable for timber harvest activities)

During the planning process, management direction was prepared and presented in terms of desired conditions, objectives, suitability, allowable uses, and finally, the standards and guidelines, or the controls that are used to implement the forest plan strategy. The following list of resources and resource uses were addressed:

- Air quality
- Alternative energy resources (geothermal, wind, solar, biomass)
- Aquatic ecosystems and fisheries
- Heritage and cultural resources
- Insects and diseases
- Invasive species
- Minerals and energy resources
- Paleontological resources
- Rangeland resources
- Recreation resources
- Riparian areas and wetland ecosystems
- Scenery and visual resources
- Special use lands
- Terrestrial ecosystems
- Terrestrial wildlife
- Timber and other forest products
- Water resources

As an example of desired conditions, two of the desired conditions (among others) for the Delores Ranger District geographical area included (from U.S. Department of Agriculture, Forest Service and U.S. Department of Interior, Bureau of Land Management, 2013b)

3.2.1 Public lands continue to function as "working lands." Collaborative forest health and rangeland management practices reduce wildfire hazards, contribute to the viability of private ranch lands, and sustain ecosystem services (including watershed health and wildlife habitat). The local economy benefits from, and contributes to, sustainable resource management, as well as the preservation of open space.

3.2.2 The Dolores River system remains a primary water source in order to meet domestic and agricultural needs while, at the same time, contributing a wide array of recreational, ecological, and aesthetic services. Collaborative efforts support watershed health, instream water quality, scenic assets, healthy native and sport fish populations, rafting and flat water boating opportunities, and flow and spill management below McPhee Dam in support of ecological, recreational, reservoir management, and water rights imperatives.

Four desired conditions (among others) for the Columbine Ranger District geographical area included

3.3.1 The full spectrum of outdoor recreational opportunities, ranging from wilderness settings to in-town access, is provided. This is the result of a collaborative process for the allocation and sharing of uses and stewardship responsibilities designed to protect the quality of the human experience and health of the natural environment.

3.3.2 Extensive heritage resources remain central to the area's economy, culture, and recreational experience. Heritage resources, as well as the natural settings that make these resources so unique, are protected and sustainable.

3.3.3 Destination and resort development, especially along the river corridors, is planned, developed, and managed in order to minimize its impact on the health of surrounding landscapes, natural resources, and communities. This is the result of sustained cooperation from the land management agencies, interested citizens, state and local agencies, and developers.

3.3.4 Oil and gas development is planned, conducted, and reclaimed to a standard commensurate with the ecological, esthetic, and human values attached to the land where the extraction is occurring.

The full list of desired conditions, along with more extensive descriptions of the management direction for specific geographical areas, can be found in the land and resource management plan.

Guiding Laws, Regulations, and Policies

National Forest plans are a requirement of the Forest and Rangeland Renewable Resources Planning Act of 1974, as amended by National Forest Management Act (16 U.S.C. 1600–1614). The SJNF management plan was also prepared in accordance with the 1982 USFS planning regulations and, because this was a joint planning effort, in accordance with the BLM's planning regulations. An assessment of the plan's potential environmental impacts is also required by the National Environmental Policy Act (42 U.S.C. § 4321 et seq.). Other important federal laws that influence the decisions of the SJNF include the Clean Air Act, Clean Water Act, Endangered Species Act, National Historic Preservation Act, Migratory Bird Treaty Act, Multiple-Use and Sustained-Yield Act, and the Mineral Leasing Act, to name a few.

A sustainable ecosystem strategy and a species management strategy were both used to establish the ecological framework for the management plan. These strategies were used to help develop the desired future conditions, objectives, standards, and guidelines for the forest area. Protected areas were viewed as key components, and about 48% of the land within the planning area was allocated in this respect, although a wide range of public uses and management activities also occur on the national forest. The ecosystem management concept, which uses a historical range of variability as a basis, was used to help develop guidance for the maintenance and restoration of certain SJNF ecosystems. The reference period used was the period of time of indigenous settlement (1500s to the late 1800s). Some species that are not adequately recognized or protected through an ecosystem management approach were given special management considerations. During the planning process, indicator species were selected to understand how the management plan alternatives meet their habitat objectives, and ultimately, these would be monitored for plan effectiveness under an adaptive management approach. The indicator species included Abert's squirrel (*Sciurus aberti*), American marten (*Martes americana*), hairy woodpecker (*Picoides villosus*), and Rocky Mountain elk (*Cervus canadensis*). The federally listed endangered terrestrial wildlife species that are found on the SJNF include:

- Canada lynx (*Lynx canadensis*)
- Mexican spotted owl (*Strix occidentalis lucida*)
- Southwestern willow flycatcher (*Empidonax traillii extimus*)
- Uncompahgre fritillary butterfly (*Boloria acrocnema*)

Management activities that can cause impacts to Canada lynx are evaluated based on the Southern Rockies Lynx Assessment (U.S. Department of Agriculture, Forest Service, 2008), and consultation with the U.S. Fish and Wildlife Service is required for project-level management activities that may affect the other three species. Further, management actions that may affect the Gunnison sage-grouse (*Centrocercus minimus*) are guided by the direction provided in the Gunnison Sage-Grouse Rangewide Conservation Plan (Gunnison Sage-Grouse Rangewide Steering Committee, 2005).

Four alternative plans were developed, including one (a "no action" alternative) that represented the continuation of the current management direction under the previous forest plan. A second alternative, deemed the preferred alternative plan, focused on balancing the need for working forests and rangelands and the need for preserving core, undeveloped lands. Management activities, such as timber harvesting and oil and gas development, would be located in areas that already contained roads. A third alternative provided a mix of multiple-use activities, with an emphasis on maintaining the undeveloped character of the national forest. The production of commodities from timber harvesting activities would continue, but might have been secondary to nontimber objectives. A fourth alternative had a similar intent as the third alternative, yet the production of commodities from timber harvesting activities had more importance and on many lands might not have been secondary to other objectives.

Management area allocations (Table 45.1) were created by an interdisciplinary planning team. The public was allowed to express a preference for how these areas should be managed. The planning team then developed a set of guidelines for the intensity of management that might be expected within each management area. To varying degrees, multiple uses can occur within each management area. The primary management area that allows timber production is called the *Active Management* area. A limited amount of timber production is allowed in the *Natural Landscapes with Limited Management*, *High-Use Recreation Emphasis*, *Public and Private Lands Intermix*, and *Highly Developed Areas* management areas (U.S. Department of Agriculture, Forest Service and U.S. Department of Interior, Bureau of Land Management, 2013b).

OUTCOMES OF THE PLAN

The plan alternative that was selected was chosen for its perceived ability to retain and restore the ecological resilience of the national forest and for its ability to provide a broad range of services to society. According to Forest Service (U.S. Department of Agriculture, Forest Service, 2013), the selected alternative

1. Provides a suite of management strategies that are responsive to the issues and concerns of the stakeholders who were involved in the planning process.
2. Establishes objectives for the management of ecosystems, recreational opportunities, and heritage resources.

TABLE 45.1 Management Area Allocations for Four Alternatives Considered in the Development of the San Juan National Forest Management Plan (U.S. Department of Agriculture, Forest Service and U.S. Department of Interior, Bureau of Land Management, 2013a)

Management Area Allocations	Alternative A[a] (ac)	Alternative B[b] (ac)	Alternative C[c] (ac)	Alternative D[d] (ac)
Natural processes dominate	483,869	598,517	1,016,281	497,856
Special areas and unique landscape areas	8,949	91,985	86,295	59,602
Natural landscape, with limited management	755,418	596,119	245,753	710,990
High-use recreation emphasis	148,022	69,864	46,502	79,854
Active management	454,035	451,730	426,507	454,137
Public and private lands intermix	0	49,560	40,679	49,547
Highly developed areas	14,538	7,056	2,814	12,845
Total[e]	1,864,831	1,864,831	1,864,831	1,864,831

[a] A continuation of the management direction under previous USFS and BLM management plans.

[b] A balance of the goals of maintaining working forest and rangelands, retaining core undeveloped lands, and providing and maintaining the full diversity of uses and active recreation opportunities.

[c] A mix of multiple-use activities, with the primary emphasis on maintaining the undeveloped character of the planning area.

[d] A mix of multiple-use activities, with the primary emphasis on producing a higher level of commodity goods and services when compared to the other alternatives.

[e] The planning documents state that the final environmental impact statement covers approximately 1,867,800 ac of land of the San Juan National Forest, even though these management areas total 1,864,861 ac.

3. Provides opportunities for responsible resource use and commodity production from the forest.
4. Protects special areas.
5. Represents the best approach examined for restoring and sustaining healthy forest and rangeland conditions.

The present value of net revenues (discounted public lands revenues minus discounted public lands management costs) is estimated to be around $486 million U.S. dollars (USD); and the broader present value of net benefits (those benefits commonly represented in the marketplace, in addition to nonmarket valuations for recreational use and meat production) is estimated to be around $7.38 billion USD (U.S. Department of Agriculture, Forest Service, 2013). The lands where timber harvests may occur are divided into two classes: lands suitable for timber production and other tentatively suitable lands where timber harvesting activities may eventually occur. According to the plan, the SJNF contains 311,949 ac (126,244 ha) of suitable timberlands and 395,067 ac (159,881 ha) of other tentatively suitable lands. The areas of suitable timberlands and tentatively suitable lands have decreased in size with each subsequent forest plan for the SJNF. Figure 45.3 presents a comparison of suitable timberlands and tentatively suitable lands for timber production for the last three versions of forest plans.

Given the direction provided by the management plan, wood production from the SJNF will likely occur in landscapes with the following characteristics (U.S. Department of Agriculture, Forest Service and U.S. Department of Interior, Bureau of Land Management, 2013b):

- They are located in a wildland urban interface
- They are considered areas with a high risk for insect and disease outbreaks
- They are landscapes with significant dead or dying trees
- They are areas where active management could effectively transition current forest age classes, size classes, tree densities, and tree species to desired conditions
- They are areas treated previously that would help maintain sustainable resource conditions
- They are areas that would improve scenic integrity
- They are areas where wood processing facilities can effectively and economically utilize products

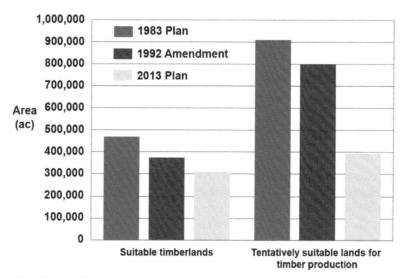

FIGURE 45.3   Comparison of suitable timberlands and tentatively suitable lands for timber production in San Juan National Forest.

To achieve the desired forest conditions, the preferred alternative suggests that approximately 4,500 ac (1,821 ha) per year will receive a mechanical fuel treatment (mowing, mastication, hand thinning), and at least 6,000 ac (2,428 ha) would receive prescribed fire treatments. Thinnings and selection harvests are planned for coniferous forests, and coppice treatments are planned for aspen forests. The long-term sustained yield capacity of the forest *"highest uniform wood yield that may be sustained under specified management intensities consistent with multiple-use objectives after stands have reached desired conditions"* is projected to be about 35.5 MMBF (83,780 m^3) per year for the first decade of the plan, and the allowable sale quantity is projected to be about 19.9 MMBF (46,964 m^3) per year (U.S. Department of Agriculture, Forest Service, 2013).

DISCUSSION AND CONCLUSIONS

The joint USFS-BLM planning effort represented a coordinated land management approach across a great expanse of public land in southwest Colorado. The SJNF management plan serves as a general framework for decision makers, describing the likely timber harvest levels and areas of necessary management actions, while not including site-specific direction on where to implement activities. As with other national forests, the implementation of management actions is subject to budget constraints and other priorities of the forest and the region within which it resides, and project-level planning requires working with the public to examine specific management alternatives for specific areas of land.

Learnings and Insights

During the development of the forest plan, several important issues guided the development of the plan's alternatives:

- The maintenance of working forests and rangelands and the retention of core undeveloped areas
- The management of recreational opportunities couched within a sustainable ecological framework
- The management of special areas and unique landscapes
- The development and management of oil and gas leases

During the planning process, discussions concerning the *working forest* were focused on the national forest's ability to respect the valid and existing rights of people to the forest's resources, on retaining commodity production activities that were important to local communities, and on continuing the historical uses of the land and the forest resources in areas where investments into access and infrastructure were already made. The SJNF has become an important recreational resource; therefore, some of the discussions focused on achieving a balance between the desires of local residents and other visitors who may use these public lands.

The SJNF also contains several areas that may be important with respect to the extraction of oil and gas resources. Local members of the community stressed that management decisions regarding oil and gas leasing opportunities should be coordinated so that the infrastructure required (roads, well pads, and pipelines) would be compatible with the desired conditions for the landscape. A number of comments regarding the plan related to whether new roads should be constructed in areas that are currently undeveloped (U.S. Department of Agriculture, Forest Service and U.S. Department of Interior, Bureau of Land Management, 2013a).

Sustainability Issues

Continuous monitoring and assessment of land and forest resources is required by the National Forest Management Act; therefore, a monitoring program was developed to help the national forest assess implementation success. The monitoring program facilitates adaptive management and meeting the sustainability goals of the national forest (U.S. Department of Agriculture, Forest Service and U.S. Department of Interior, Bureau of Land Management, 2013b). As was suggested earlier, a sustainable ecosystem strategy was developed with the purpose of conserving and managing ecosystems, habitats, and species.

To ensure that the growing stock of the forests will not decline, the Forest Service estimated an annual harvest level that was considered to be sustainable over the long term. Aspects of the management plan also promote the use of appropriate management practices to be conducted after timber harvests or in response to wildfires and natural disasters. For example, the plan suggests that on suitable timberlands, stocking must be adequate within 5 years after a regeneration harvest (U.S. Department of Agriculture, Forest Service and U.S. Department of Interior, Bureau of Land Management, 2013b).

Plan Development Challenges

The plan was developed by two different agencies, the USFS and the BLM, that are located within two different Departments of the United States federal government. While the fundamental issues of land and resource planning are shared by both agencies, certain aspects of some of the common steps in public land planning processes may differ. For example, the SJNF and the TRFO are both guided by different planning regulations and different priorities. Because the plan that was developed applies to both agencies, the format and some of the terminology may vary from other conventional forest management plans within either agency (U.S. Department of Agriculture, Forest Service and U.S. Department of Interior, Bureau of Land Management, 2013b).

Plan Implementation Challenges

As mentioned earlier, the plan provides a general framework for the decision makers, and it does not include any specific projects or activities. Project-level, site-specific, implementation decisions are made after a more detailed analysis and after further public involvement. As of April 2014, the SJNF was working on 40 different projects: 2 projects were proposed, 20 projects were being analyzed, 14 projects had been analyzed, and 4 other projects were on hold. After receiving public comments on these, the USFS identifies and analyzes the issues raised, finalizes the proposed actions, and if necessary, develops alternative approaches to the proposed activities. A draft environmental assessment is published with an additional opportunity for the public to comment on the proposed activities. The USFS then makes a decision whether to implement the proposed action or to implement an alternative approach. The time required for the complete set of steps necessary to implement a specific project may be extensive.

From a wood utilization perspective, a number of issues are monitored at different time scales to ensure that forest management practices are appropriate within the context of the larger southwest Colorado environment and socioeconomic system. As alluded to in this chapter, both even-aged and uneven-aged forest management systems can be employed to meet management objectives. Timber will be harvested from the SJNF only where there is assurance that the forest can be adequately restocked within 5 years after harvest, thus initial stocking needs to be monitored. If even-aged management systems are employed, regardless of forest type, the maximum clear-cut size is 40 ac (16 ha) (U.S. Department of Agriculture, Forest Service and U.S. Department of Interior, Bureau of Land Management, 2013b).

ADDITIONAL READING AND RESOURCES

This chapter represents a synthesis of the 2013 Land and Resource Management Plan for the SJNF in Colorado. To view the plan itself, please visit this Internet site, which was available on June 15, 2014:

http://www.fs.usda.gov/detail/sanjuan/landmanagement/planning/?cid=stelprdb5432707

If in the future the link to the site appears broken, search the Internet using the title of the plan and the keywords provided.

REFERENCES

Bond, A., 2014. History of the San Juan National Forest. U.S. Department of Agriculture, Forest Service, Durango, CO. http://www.fs.usda.gov/Internet/FSE_DOCUMENTS/stelprdb5283889.pdf (Accessed June 13, 2014).

Davis, R.C., 1983. Encyclopedia of American Forest and Conservation History, vol. 2. Macmillan Publishing Company, New York. pp. 743–788.

Gunnison Sage-Grouse Rangewide Steering Committee, 2005. Gunnison Sage-Grouse Rangewide Conservation Plan. Colorado Division of Wildlife, Denver, CO.

Lewis, P., Kauffman, M.R., Huckaby, L.S., Leatherman, D., 2005. Report on the Health of Colorado's Forests 2004. Special Issue: Ponderosa Pine Forests. Colorado Department of Natural Resources, Division of Forestry, Fort Collins, CO. 32 p.

Rocky Mountain Region, Forest Health Protection, 2010. Field Guide to Diseases and Insects of the Rocky Mountain Region. U.S. Department of Agriculture, Forest Service, Rocky Mountain Region, Fort Collins, CO General Technical Report RMRS-GTR-241.

U.S. Department of Agriculture, Forest Service, 1983. San Juan National Forest Land and Resource Management Plan. U.S. Department of Agriculture, Forest Service, Durango, CO.

U.S. Department of Agriculture, Forest Service, 1992. Amended Land and Resource Management Plan for the San Juan National Forest. U.S. Department of Agriculture, Forest Service, Durango, CO.

U.S. Department of Agriculture, Forest Service, 2008. Supplemental Biological Assessment for the Southern Rockies lynx Amendment. U.S. Department of Agriculture, Forest Service, Region 2, Lakewood, CO. 132 p.

U.S. Department of Agriculture, Forest Service, 2013. Record of decision: final Environmental Impact Statement for the San Juan National Forest Land and Resource Management Plan. U.S. Department of Agriculture, Forest Service, Rocky Mountain Region, Golden, CO.

U.S. Department of Agriculture, Forest Service, U.S. Department of Interior, Bureau of Land Management, 2013a. San Juan National Forest Land and Resource Management Plan. Final environmental impact statement. vol. 1. U.S. Department of Agriculture, Forest Service, Rocky Mountain Region/U.S. Department of Interior, Bureau of Land Management, Colorado State Office, Golden, CO/Lakewood, CO.

U.S. Department of Agriculture, Forest Service, U.S. Department of Interior, Bureau of Land Management, 2013b. San Juan National Forest land and resource management plan. Final San Juan National Forest and Proposed Tres Rios Field Office Land and Resource Management Plan. vol. 2. U.S. Department of Agriculture, Forest Service, Rocky Mountain Region/U.S. Department of Interior, Bureau of Land Management, Colorado State Office, Golden, CO/Lakewood, CO.

U.S. Department of Agriculture, Forest Service, U.S. Department of Interior, Bureau of Land Management, 2013c. San Juan National Forest land and resource management plan. Appendices. vol. 3. U.S. Department of Agriculture, Forest Service, Rocky Mountain Region/U.S. Department of Interior, Bureau of Land Management, Colorado State Office, Golden, CO/Lakewood, CO.

Chapter 46

Tongass National Forest, Alaska, United States of America

Pete Bettinger,[1] Krista Merry,[1] Mehmet Demirci[2] and Anna M. Klepacka[3]

[1]*Warnell School of Forestry and Natural Resources, University of Georgia, Athens, Georgia, USA,* [2]*Forest Management and Planning Department, General Directorate of Forestry, Ministry of Forest and Water Affairs, Ankara, Turkey,* [3]*Department of Production Management and Engineering, Warsaw University of Life Sciences, Warsaw, Poland*

MANAGEMENT SETTING AND BACKGROUND

Located in the Alexander Archipelago of southeast Alaska, the Tongass National Forest (Figure 46.1) is the largest national forest in the United States. The total land area of the national forest encompasses nearly 17 million acres (ac) (about 6.8 million hectares (ha)) currently allocated to a number of very large potential land uses. To put this in perspective, the national forest is larger in land area than the state of West Virginia. The national forest, in general, is considered a temperate coastal rainforest (Figure 46.2) with steep coastal mountains and thousands of densely forested islands (Deal, 2007; Everest, 2005). Annual precipitation ranges from 60 to 200 inches (in) (1,520–5,080 millimeters (mm)) per year, and the major tree species include Sitka spruce (*Picea sitchensis*), western hemlock (*Tsuga heterophylla*), Alaska yellow-cedar (*Chamaecyparis nootkatensis*), and red alder (*Alnus rubra*). Almost 21,500 miles (mi) (nearly 34,600 kilometers (km)) of Class I and II streams (the higher-ordered streams) are located on the national forest. Higher-ordered streams are those that are downstream from headwater streams and likely contain resident fish. This region of North America is arguably among the first to be colonized by humans via the Bering Land Bridge (Dixon, 2001), yet remains sparsely populated, and much of it has been only lightly disturbed by humans (Everest, 2005).

Through the treaty with Russia (*Alaska Purchase*) in 1867, control of the area transferred to the United States. During the tenure of President Theodore Roosevelt, both the Alexander Archipelago Forest Research and the Tongass National Forest were proclaimed, and in 1908 they were combined into the Tongass National Forest. Historically, southeast Alaska attracted economic activity due to the proximity of forests and the coast, and this perceived accessibility led to the establishment of about 20 sawmills and 2 large pulp mills by the mid-twentieth century. Around the time of their development, two pulp mills had entered into long-term (50-year) timber supply contracts with the Forest Service. Many of the sawmills have since closed, as have the two pulp mills, due to insufficient supply of wood caused in part by litigation associated with the timber supply contracts (Clark, 2013). A major challenge to the development of a regional economy based on forest resources is the remoteness and accessibility of this area. Even while timber may be available under the planned supply system, transportation costs and the distance to major markets restrict the competitiveness of wood production facilities. Studies have indicated that Alaskan forest product exports might also be sensitive to international market conditions (Boyce and Szaro, 2005). Although in recent decades economic growth in eastern Asia has increased the demand for timber and wood products, economic growth in China has recently decelerated. Sales opportunities within domestic markets are also limited, given the housing market collapse a few years ago. Regulations on road construction may further increase transportation costs, which directly affect the region's competitive position. While they can have both desirable and undesirable outcomes, timber harvests from the Tongass National Forest continue to be important for local communities (Burchfield et al., 2003). However, there are concerns over the use timber sales, regardless of possible financial losses, for economic development purposes (Gorte, 2004).

As a state, Alaska is heavily dependent on economic activity associated with the use of natural resources in a broader sense than the timber industry. Other forms of ecosystem services are important, including (but not limited to) commercial fishing, recreation, and cultural services. The Tongass National Forest is a provider of tourism and recreation opportunities, and a custodian of many of the unique natural amenities and ecosystem values that both attract tourists and enhance the

Forest Plans of North America. http://dx.doi.org/10.1016/B978-0-12-799936-4.00046-1

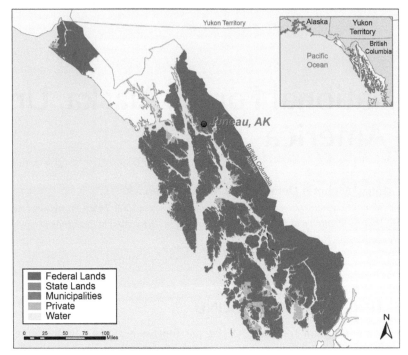

FIGURE 46.1 A map of the land area encompassing the Tongass National Forest.

FIGURE 46.2 An example of the temperate coastal forest resources found within the Tongass National Forest. Photo courtesy of the Tongass National Forest.

quality of life for existing and potential residents. Based on regional, national, and international economic and demographic trends, these roles are likely to be of increasing importance to the economic vitality of the region (Crone, 2005). However, demand for ecosystem services can be difficult to sustain in the context of the slow growth of the general economy.

Over the last few decades, southeast Alaska has experienced significant economic and social change (Crone, 2005). The area remains sparsely populated, with about 100,000 permanent residents, and the regional economy is in some respects disconnected from other parts of the United States. While tourism levels have increased in recent years (Kruger, 2005), timber harvest, fishing and seafood processing, subsistence harvests of wild foods, and recreation (such as hunting) all remain important to the economy. Further, as an economic activity, tourism is highly spatially concentrated, and the one million or more annual visitors arrive mainly on cruise liners. Because of the association with the cruise industry, a typical visitors' average length of stay, and their destination (well-established docking sites), are both limited. Consequently, the local, mixed economy depends heavily on the market and nonmarket value of its natural resources. Assessing or predicting the economic effects of alternative solutions to the region remains a challenge.

PLANNING ENVIRONMENT AND METHODOLOGY

The National Forest Management Act of 1976 requires each U.S. National Forest to prepare a comprehensive land and resource management plan. The first plan for the Tongass National Forest was developed in 1979. These management plans should be revised every 15 years; therefore, the second plan's development process began in 1987 and was revised in 1997. In 2005, the U.S. Court of Appeals for the Ninth Circuit issued a decision (*Natural Resources Defense Council*, et al., *v. United States Forest Service*, et al., *421 F.3d 797 (9th Cir. 2005)*) that noted deficiencies in the process used to develop the revised 1997 plan. These deficiencies were related to analyses of timber demand, effects of actions on roadless areas, and cumulative effects of actions on nonfederal land. A 2008 amendment to the plan was prepared in response to the court's decision. This became the 2008 forest plan (U.S. Department of Agriculture, Forest Service, 2008a) that is described in this chapter. The 2008 plan uses an adaptive management approach, which involves a continuous series of monitoring and evaluation (active learning) processes, and subsequent adjustments of the desired management goals and objectives.

In developing the forest plan, the Forest Service involved other federal and state agencies, communities of southeast Alaska, and tribal governments. It also received input from numerous other individuals and interest groups. In addition, a partnership was formed with Alaska, and the governor of the state cosigned the preamble to the record of decision for the plan (U.S. Department of Agriculture, Forest Service, 2008b).

Guiding Laws, Regulations, and Policies

The management direction for the 2008 plan primarily arises from four sources: (1) multiple-use goals that are established during the planning process, (2) anticipated forest management objectives for the first decade of the plan, (3) management goals, objectives, and desired future conditions for each land allocation, and (4) standards and guidelines for the forest as a whole (U.S. Department of Agriculture, Forest Service, 2008a). Guidance for timber supply, riparian protection, and other local issues was reinforced by the Tongass Timber Reform Act of 1990 (Public Law 101-626) and its amendments. This law states that to "the extent consistent with providing for the multiple use and sustained yield of all renewable forest resources," the Forest Service needs to provide a supply of wood that addresses annual market demand.

In addition to the National Forest Management Act, a number of federal and state laws, regulations, and rules guided the development of the Tongass National Forest plan. The laws included the following: Alaska National Interest Lands Conservation Act of 1980, Alaska Native Claims Settlement Act of 1971, Clean Air Act, Clean Water Act, Coastal Zone Management Act, Endangered Species Act, Forest and Rangeland Renewable Resources Planning Act of 1974, Multiple-Use Sustained-Yield Act of 1960, National Environmental Policy Act, National Historic Preservation Act, Tongass Timber Reform Act of 1990, and Wild and Scenic Rivers Act. Rules that guided the development of the forest plan included the Management Planning Rule of 1982, as it was prepared prior to the creation of the Management Planning Rule of 2011, and the National Roadless Area Conservation Rule of 2001.

The National Roadless Area Conservation Rule was adopted in 2001 to prevent logging and road construction in the unroaded portions of national forest system lands. The Tongass National Forest initially received an exemption from the rule, but in 2011 it was found to be in violation of the Administrative Procedures Act (5 U.S.C. §§ 701–706) (Clark, 2013). The rule was the subject of several legal challenges, and in March 2014 the order was reversed, as it was concluded that the U.S. Department of Agriculture's reasons for the exemption were neither arbitrary nor capricious (U.S. Circuit Court of Appeals for the Ninth Circuit No. 11-35517).

As mentioned, the 2008 forest plan was developed using the 1982 planning regulations (U.S. Department of Agriculture, Forest Service, 2008b). The planning process contained the following phases:

- Identification of public issues, management concerns, resource uses, and development opportunities
- Preparation of planning criteria
- Collection of inventory data and information
- Analysis of the ability of the forest to supply goods and services in response to societal demands
- Formulation of alternatives
- Estimation of the physical, biological, economic, and social effects of implementing each alternative
- Evaluation of alternatives
- Recommendation of the preferred alternative
- Approval of the proposed plan in a record of decision and final environmental impact statement
- Development of a monitoring process

National Forest plans developed using the 1982 regulations required that plans provide, in an environmentally sound manner, multiple-use and sustained yields of goods and services to maximize long-term net public benefits. Net public benefits involve inputs and outcomes that may or may not be quantitatively valued, yet are evaluated in the environmental impact statement using an economic efficiency analysis approach (U.S. Department of Agriculture, Forest Service, 2008b). Thus, its goal was to maximize a social welfare function. This challenged forest planners, as the economic analysis involved using monetary values assigned to market and nonmarket goods and services (where possible), and other qualitative criteria when the assignment of monetary values was not reasonable.

Twenty-four goals and more than 40 objectives were designed for the 2008 forest plan, addressing concerns such as air quality, biodiversity, fisheries, heritage resources, local and regional economies, minerals, plants, recreation, tourism, research, esthetics, water, subsistence, timber, transportation, wetlands, wilderness, wildlife, and others. During the planning process, the Forest Service needed to identify indicator wildlife species, based on the assumption that responses of these species to management actions could reflect probable responses of other species that have similar habitat requirements. A number of fish and wildlife indicator species were selected for this analysis, including the following:

- Red squirrel (*Tamiasciurus hudsonicus*)
- Black bear (*Ursus americanus*)
- Brown bear (*Ursus arctos*)
- Marten (*Martes americana*)
- River otter (*Lontra canadensis*)
- Sitka black-tailed deer (*Odocoileus hemionus sitkensis*)
- Mountain goat (*Oreamnos americanus*)
- Gray wolf (*Canis lupus*)
- Vancouver Canada goose (*Branta canadensis fulva*)
- Bald eagle (*Haliaeetus leucocephalus*)
- Red-breasted sapsucker (*Sphyrapicus ruber*)
- Hairy woodpecker (*Picoides villosus*)
- Brown creeper (*Certhia americana*)
- Dolly Varden char (*Salvelinus malma*)
- Cutthroat trout (*Oncorhynchus clarki*)
- Coho salmon (*Oncorhynchus kisutch*)
- Pink salmon (*Oncorhynchus gorbuscha*)

Land-use designations (Table 46.1) were developed to address the management objectives of the plan. Each of these has its own goals, objectives, standards, and guidelines designed to ensure achievement of the forest plan objectives. Land-use designations are considered either areas of avoidance for transportation and utility systems, or windows of opportunity for the development of these systems (U.S. Department of Agriculture, Forest Service, 1997). Some of the designations do not allow developmental activities, while others do. The primary land-use designations that allow timber management activities are the *Timber Production*, *Modified Landscape*, and *Scenic Viewshed* designations, which encompass approximately 3.4 million acres (1.376 million ha), or about 20% of the national forest. *Scenic River* and *Recreational River* designations also allow timber management activities to be implemented if an adjacent land-use designation also allows these types of activities and if viewshed management guidelines are followed (U.S. Department of Agriculture, Forest Service, 2008b).

During the planning process, 49 forest plan alternatives were considered, some of which were based on alternatives that were considered in previous planning processes. In the development of the environmental impact statement, only seven of these were analyzed in detail. After final approval, a monitoring and evaluation program was implemented to determine how well the plan's objectives were being met and how closely management standards and guidelines were being applied. Due to the vast size of the forest and the complexity of the planning process, a number of analytical methods and models were used to assist in the development of the plan. Spectrum, a linear programming matrix generator developed by the Forest Service (U.S. Department of Agriculture, Forest Service, 2008c), was used to assist in the planning process and measure the economic efficiency of timber management activities. Management objectives, constraints, and other assumptions were developed for the management alternatives, and a suite of management prescriptions was then optimized. Although the only objective used in the final analysis was to *maximize present net value*, other objectives that were considered during the planning process included maximizing timber volume, minimizing harvest from the unroaded areas, and minimizing harvests from old-growth forests. The types of management prescriptions considered were:

1. Minimum levels of custodial maintenance.
2. Clear-cut with no other intermediate treatments before the next clear-cut activity.
3. Clear-cut with a precommercial thinning at age 20 of the regenerated stand.
4. Clear-cut with commercial thinning at ages 70, 80, or 90 of the regenerated stand.
5. Clear-cut with precommercial thinning at age 20 and commercial thinnings at age 60, 70, or 80 of the regenerated stand.
6. Small group selection (0.5–5 ac, 0.2–2 ha) and uneven-aged harvesting activities (50-year entry cycle, 25% removal of volume).

TABLE 46.1 Land allocations of the Tongass National Forest (U.S. Department of Agriculture, Forest Service, 2008b).

Class	Area (ac)	Area (ha)
Wilderness	2,637,292	1,067,297
Wilderness National Monument	3,111,792	1,259,325
Nonwilderness National Monument	166,942	67,561
Land Use Designation II[a]	721,002	291,786
Remote recreation	2,033,665	823,013
Semiremote recreation	3,023,152	1,223,453
Old-growth habitat	1,221,173	494,202
Enacted municipal watershed	45,226	18,303
Research natural area	26,093	10,560
Special interest area	221,176	89,509
Wild river	62,799	25,414
Scenic river	27,133	10,981
Recreational river	27,387	11,083
Experimental forest	31,405	12,709
Scenic viewshed	307,402	124,404
Modified landscape	728,679	294,892
Timber production	2,381,486	963,774
Total	16,773,804	6,788,265

[a] *Congressionally designated areas in Public Law 101-626 (1990), such as the Takutat Forelands and Berners Bay areas, effectively administered as roadless areas.*

7. Old-growth two-aged management (partial cuts), where on the first entry, 75% of the volume is removed and second-growth activities occur within the regenerated stand area beginning at age 80. The residual trees (accounting for the remaining 25% of the volume at the time of first entry) may not be removed (as they typically would under a shelterwood system) in order to provide some level of structural diversity and biological legacy for the resulting stand.

In response to the estimated sizes of the Spectrum problem formulations, three models were developed for each forest plan alternative, each representing one of the old administrative areas of the national forest (Chatham, Ketchikan, Stikine). To further reduce problem formulation size, only those areas classified as suitable for timber production were considered. Timber operability, forest productivity, access (roaded or unroaded), age (or volume) class, management intensity (referred to as a *regulation class*), and value comparison unit (used to recognize spatially variable costs) were used to determine the unique analysis areas (potentially 1,226,016) for the problem formulations. Some of the costs (all U.S. dollars (USD)) assumed in the modeling process included:

- Sale preparation and administration, $32 per thousand board feet ($13.56 per cubic meter ($m^3$))
- Road construction: $185,000 per mi ($114,978 per km)
- Road maintenance: $50,000 per mi ($31,075 per km)
- Planting and regeneration survey costs (on average, where necessary): $24 per ac ($59 per ha)
- Precommercial thinning costs: $550 per ac ($1,359 per ha)

Timber harvest costs assumed in the model varied by stand volume class, logging operability, geographic zone, productivity group, stand age, and management prescription. Timber values were determined using timber appraisal methods appropriate for southeast Alaska. Merchantable volumes were based on U.S. Forest Service Forest Inventory and Analysis plots, permanent study plots, and growth projections from the Forest Vegetation Simulator (Dixon, 2013). Operational constraints were added to the models to ensure that the results obtained were contained within certain forest guidelines and objectives.

These included constraints associated with harvest levels to certain forest strata, logging operability, watershed access, the use of precommercial thinning practices, and other issues of importance. More specifically, a nondeclining yield timber policy was assumed (per the National Forest Management Act of 1976), a sustained yield level (harvest no larger than long-term sustained yield levels) was assumed, and the rotation age was assumed no shorter than 95% of the culmination of mean annual increment (MAI) of each forest types, or 60–170 years based on stand productivity. After the generation of Spectrum results for each forest plan alternative, an estimation of the cumulative effects on fish, wildlife, plants, and other resources was pursued, as well as analyses of the effects on scenic integrity and visual quality (U.S. Department of Agriculture, Forest Service, 2008d).

OUTCOMES OF THE PLAN

The forest plan alternative that was selected as the 2008 plan suggests that about 3.5 million ac (1.42 million ha) of the national forest would fall into land-use designations that allowed the development of natural resources for commodity production, about 13.3 million ac (5.38 million ha) would fall into nondevelopment land-use designations. The forest plan alternative that was selected, based on the analysis processes employed, seemed to best balance the environmental, social, and economic issues that face the management of the Tongass National Forest. The main outcomes of the management plan include those related to commodity production, values experienced by recreationists and other users of the forest resources, ecosystem services (e.g., clean water), and existence values. As with many other forest plans, the project outcomes of the plan are dependent on the annual budget of the forest and other considerations (U.S. Department of Agriculture, Forest Service, 2008b). Interestingly, the amount of timber that the national forest can offer for sale over the life of the 2008 plan (the allowable sales quantity) is unchanged from level described in the 1997 plan, or approximately 267 million board feet (MMBF) (630,120 m^3) per year over a 10-year period (2008–2017). Further, the total suitable land base for timber harvests is similar to the previous plan, or about 773,000 acres (about 312,829 ha) (U.S. Department of Agriculture, Forest Service, 2008b). The estimated timber supply provided by the plan, assuming that the entire allowable sales quantity (267 MMBF) is harvested, consists mainly of spruce and hemlock forest products (Figure 46.3).

The discounted (4% rate) net timber revenue associated with the 2008 forest plan was estimated to be about $44 million USD. Similarly, direct recreation-related discounted revenue was estimated to be around $54 million USD. Because the Tongass National Forest is a popular tourist destination, the estimated discounted recreation and tourism consumer surplus was estimated to be more than $7.6 billion USD. This represents a value placed on recreation goods that are not actually traded in the marketplace and is, thus, not directly comparable to actual revenues received (U.S. Department of Agriculture, Forest Service, 2008b). This also reflects an estimated willingness to pay for recreational experiences above and beyond what people actually pay. Projected annual average direct employment associated with the forest plan over the first decade was 1,343 jobs in the wood products industry and 4,319 jobs in the recreation and tourism industries. The total income associated with the wood products industry was estimated to be $73 million USD and $71.1 million USD for the recreation and tourism industries during the first decade of the forest plan, estimated using multipliers from a 1998 IMPLAN (IMPLAN Group LLC, 2013) model. The amount of roads necessary to implement the plan over a 100-year period is about 8,685 mi (13,974 km). The 2008 plan suggests that about 37.4 mi (60.2 km) of new roads need to be developed every year, on average.

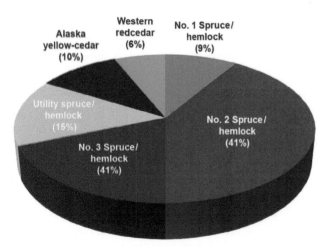

FIGURE 46.3 Estimated annual timber supply from the Tongass National Forest during the second decade of the plan (U.S. Department of Agriculture, Forest Service, 2008a). No. 1 and No. 2 grade logs are suitable for the manufacture of high-grade lumber. No. 1 grade logs also include peeler logs. No. 3 grade logs are suitable for shop-grade and better lumber. Utility-grade logs are suitable for the production of usable pulp or chips due to the presence of rot and other problems.

The reserve network of land provided through the nondevelopment land-use designations was expanded from the 1997 plan to provide a framework to ensure the development and maintenance of suitable habitat for wildlife species. While the other lands facilitate a variety of other functions and activities, the reserve network is meant to protect the integrity of the old-growth forest ecosystem and serve as core areas for old-growth ecological communities. The other lands facilitate important ecological functions as well, such as ecologically valuable structural components (down logs, snags, and large trees) that are important habitat components for certain wildlife species (e.g., marten and goshawk (*Accipiter gentilis*)), and these areas will constitute a mosaic of successional stages from early-seral forests, to second-growth and old-growth forests. Connectivity between reserves is said to be provided by riparian buffers and beach fringe (1,000 ft, 305 m) buffers (U.S. Department of Agriculture, Forest Service, 2008b).

DISCUSSION AND CONCLUSIONS

The management of the Tongass National Forest is of great importance for local communities in southeast Alaska because they heavily depend on the ecosystem services that the forest supplies, which include employment opportunities in the timber and fishing industries, recreation, tourism, mining, and mineral development. Most of the Tongass National Forest is still wild and undeveloped, therefore, the management of natural resources is of great interest regionally, nationally, and internationally.

Learnings and Insights

During the public comment period of the planning process, more than 84,000 pieces of input were received, many of which were based on opinions or preferences regarding the management of the Tongass National Forest. A very small portion (2.5%) of these was classified as unique comments. Thirty-two percent of the comments arose from California, New York, and Florida, and more than 97% were delivered via e-mail. The comments related to modifications of management alternatives, analysis of alternatives not given serious consideration in the draft environmental impact statement, improvements or supplements to analyses processes, and factual corrections. As an example, one set of comments suggested that cumulative effects of management activities in the surrounding landscape, on non-Forest Service lands also be taken into account in the development of the plan. Another comment suggested that the plan should be specific for each island group (or individual island) within the National Forest. As a result of this input, one of the plan alternatives was revised, and a number of improvements were made to the final plan documents (U.S. Department of Agriculture, Forest Service, 2008b).

Sustainability Issues

The Tongass National Forest plan makes numerous references to the sustainability of forest resources. For example, the plan suggests that all wild renewable resources should be managed sustainably (*subsistence* standards and guidelines) and that the timber resource should promote economic sustainability (*timber management* objective). To determine whether new courses of action may be required to respond to changing conditions, implementation monitoring, effectiveness monitoring, and validation monitoring efforts were outlined in the management plan (Marcot et al., 2012). Depending on the issue, various forms of these were to be conducted every 1–5 years for the purpose of determining whether the management guidelines are appropriate for the forest. In past planning efforts, public debate concerning forest management goals for the forest revealed differences in preferences and for the various outcomes that the forest could facilitate, and science-based descriptions of the social and economic dynamics were suggested (Crone, 2005; Everest, 2005).

Management of the Tongass National Forest is a complex process, and it is further complicated by the fact that in southeast Alaska, logging activity and employment in the forest sector have declined over the last few decades. Biomass and bioenergy resources have recently become important issues for energy generation and greenhouse gas emission mitigation efforts. However, the wood-waste, or residue, generated by logging activities currently has little or no market value and thus remains unclaimed. Potentially, this logging by-product could be developed for biomass energy generation, and the transformation of wood-waste into energy could, in turn, create jobs, reduce adverse ecological impacts from unclaimed wood residue (e.g., fires), and introduce an alternative source of energy. Establishing a wood biomass sector also might highlight the sustainability of the wood energy sector and also act as a potential economic, environmental, and social benefit in Alaska (Sikka et al., 2013). In fact, in 2013, a wood pellet plant began operating in Ketchikan, using residues generated from the manufacture of other custom products being produced locally. Wood pellets have since been delivered to the public library and the federal building to be used as a fuel for the heating systems of these facilities (Tongass Forest Enterprises, 2014).

Plan Development Challenges

As with most planning processes involving federal and state forest lands, public input was sought during the development of the 2008 forest plan. A formal comment period was announced in 2007, and in conjunction with this, a number of meetings were held throughout southeast Alaska to allow people to review maps, ask questions, and discuss the draft plan and environmental impact statement. One meeting was also held in Anchorage, and one meeting was held electronically over the Internet. As suggested earlier, from these outreach efforts numerous comments and concerns received are reviewed by the Forest Service. If the comments were substantive, the Forest Service responded in one or more of five ways:

1. By modifying the alternatives considered.
2. By developing and analyzing alternatives that were not given serious consideration in the draft environmental impact statement.
3. By supplementing, improving, or modifying the analyses in the draft environmental impact statement.
4. By making factual corrections.
5. By explaining why the substantive comments did not need further response.

Given the expansive recreational opportunities on the Tongass National Forest and the vast areas of wilderness and national monument designations, assessing both timber and nontimber (environmental, social) values has been difficult. Markets for ecosystem services are relatively nonexistent, even though regulating and cultural services are important. Therefore, the assessment of trade-offs among alternatives was complicated by the mixture of market and nonmarket values that were used to describe the outcomes.

Plan Implementation Challenges

With regard to implementing timber harvest activities, a management strategy (the *Timber Sale Adaptive Management Strategy*) was developed to limit harvests to land areas that are necessary to support demonstrated levels of timber demand. This strategy attempts to ensure that a logical progression of the locations of timber harvests will be followed as the level of timber harvests rise. In other words, if the level of actual timber harvest remains low (below 100 million board feet ($236,000 \, m^3$)), harvests will be restricted to forests within close proximity to the existing road network. As demand (actual timber harvests) increases, other areas of the suitable land base farther from the existing road system will then be accessed.

The amount of harvested timber depends on the demand, which responds to price changes and wood quality. If quality of timber is affected by the location of timber harvest (for example, old growth vs. young growth), then even in years of relatively strong demand for timber and wood, the region may not be able to compete in the global market. Robertson and Brooks (2001) suggested that Alaska was only a marginally competitive source of timber. Therefore, as time passes, the location of harvest may or may not be a market-relevant issue. However, if the timber harvest levels fluctuate, the demand for labor and timber harvesting equipment will change and likely have a destabilizing effect on the local economy. Two studies (Robertson, 1999, 2003) reported that the effects of reduction in timber harvest and its export were generally buffered by growth of government employment, fishing, tourism, and private business. The amount of timber allowed to be harvested every year establishes an upper bound on the potential supply, but actual supplied amount of timber can be lower in response to market demands. Crone (2005) noted that Alaska, as a timber supplier, is sensitive to price swings in international markets. However, it seems that if less timber is harvested than is allowed under the plan, other conservation goals can be exceeded.

Other Interesting Issues Related to the Plan

In 2013, the Secretary of Agriculture issued Memorandum 1044-009 that called for a transition in the timber harvesting program away from old-growth forests in roadless areas to second-growth forests in roaded areas over the next 10-15 years. This action was in response to the 2001 Roadless Rule. In the meantime, the Forest Service will continue to offer a supply of old-growth timber to industries in Alaska while it increases the supply of young-growth. However, Clark (2013) suggests that this action makes it impossible for the Forest Service to supply the amount of wood necessary to attain and support an integrated timber industry in the region, and thus could violate the Tongass Timber Reform Act of 1990. The majority of second-growth forests on the Tongass National Forest are apparently a few decades away from the culmination of MAI (and thus harvest) at this point in time. Thus, there is some concern about potential delays in the development of an economically viable young-growth forest management program. In his memorandum, the Secretary of Agriculture therefore asked the Forest Service to continue to work with the U.S. Congress to "exempt a limited amount of young growth on the Tongass from current requirements that generally restrict harvesting young growth timber until it has reached maximum growth rates" (the culmination of MAI), and to strongly consider whether to amend the forest plan so that additional opportunities to promote the transition to young-growth management could be facilitated.

REFERENCES

Boyce Jr., D.A., Szaro, R.C., 2005. An overview of science contributions to the management of the Tongass National Forest, Alaska. Landscape and Urban Planning 72, 251–263.

Burchfield, J.A., Miller, J.M., Allen, S., Schroeder, R.F., Miller, T., 2003. Social Implications of Alternatives to Clearcutting on the Tongass National Forest. U.S. Department of Agriculture, Forest Service, Pacific Northwest Research Station, Portland, OR, General Technical Report PNW-GTR-575. 28 p.

Clark, J., 2013. Review of Federal Tongass Forest Management Policy 1980–2013. Alaska Department of Natural Resources, Anchorage, AK.

Crone, L.K., 2005. Southeast Alaska economics: a resource-abundant region competing in a global marketplace. Landscape and Urban Planning 72, 215–233.

Deal, R.L., 2007. Management strategies to increase stand structural diversity and enhance biodiversity ion coastal rainforests of Alaska. Biological Conservation 137, 520–532.

Dixon, E.J., 2001. Human colonization of the Americas: timing, technology and process. Quaternary Science Reviews 20, 277–299.

Dixon, G.E., 2013. Essential FVS: A User's Guide to the Forest Vegetation Simulator. U.S. Department of Agriculture, Forest Service, Forest Management Service Center, Fort Collins, CO. 226 p.

Everest, F.H., 2005. Setting the stage for the development of a science-based Tongass land management plan. Landscape and Urban Planning 72, 13–24.

Gorte, R.W., 2004. Below-Cost Timber Sales: An Overview. The Library of Congress, Congressional Research Service, Washington, DC, CRS Report for Congress, Order Code RL32485. 11 p.

IMPLAN Group, LLC, 2013. IMPLAN. IMPLAN Group, LLC, Huntersville, NC. http://implan.com/ (Accessed April 20, 2014).

Kruger, L.E., 2005. Community and landscape change in southeast Alaska. Landscape and Urban Planning 72, 235–249.

Leon, R.J., 2013. Memorandum opinion, Civil Case No. 11-1122 (RJL), State of Alaska et al. v. U.S. Department of Agriculture et al.. United States District Court for the District of Columbia, Washington, DC. 8 p.

Marcot, B.G., Thompson, M.P., Runge, M.C., Thompson, F.R., McNulty, S., Cleaves, D., Tomosy, M., Fisher, L.A., Bliss, A., 2012. Recent advances in applying decision science to managing national forests. Forest Ecology and Management 285, 123–132.

Robertson, G.C., 1999. When the Mill Shuts Down: A Test of the Economic Base Hypothesis in the Small Forest Communities of Southeast Alaska. University of Washington, Seattle, WA, Ph.D. dissertation. 123 p.

Robertson, G.C., 2003. A Test of Economic Base Hypothesis in the Small Forest Communities of Southeast Alaska. U.S. Department of Agriculture, Forest Service, Pacific Northwest Research Station, Portland, OR, General Technical Report PNW-GTR-592. 101 p.

Robertson, G.C., Brooks, D.J., 2001. Assessment of the Competitive Position of the Forest Products Sector in Southeast Alaska, 1985–1994. U.S. Department of Agriculture, Forest Service, Pacific Northwest Research Station, Portland, OR, General Technical Report PNW-GTR-504. 29 p.

Sikka, M., Thorton, T.F., Worl, R., 2013. Sustainable biomass energy and indigenous cultural models of well-being in an Alaska forest ecosystem. Ecology and Society 18 (3), 38.

Tongass Forest Enterprises, 2014. Wood Pellets. Tongass Forest Enterprises, Ketchikan, AK. http://gdyn.akforestenterprises.com/biofuel (Assessed April 20, 2014).

U.S. Department of Agriculture, Forest Service, 1997. Tongass National Forest, Land and Resource Management Plan. U.S. Department of Agriculture, Forest Service, Alaska Region, Juneau, AK. R10-MB-603b.

U.S. Department of Agriculture, Forest Service, 2008a. Tongass National Forest, Land and Resource Management Plan. U.S. Department of Agriculture, Forest Service, Alaska Region, Juneau, AK. R10-MB-603b.

U.S. Department of Agriculture, Forest Service, 2008b. Tongass Land and Resource Management Plan, Final Environmental Impact Statement, Plan Amendment, Record of Decision. U.S. Department of Agriculture, Forest Service, Alaska Region, Juneau, AK. R10-MB-603a.

U.S. Department of Agriculture, Forest Service, 2008c. Spectrum User Guide. U.S. Department of Agriculture, Forest Service, Ecosystem Management Coordination, Planning & Analysis Group, Washington, DC.

U.S. Department of Agriculture, Forest Service, 2008d. Tongass Land and Resource Management Plan, Final Environmental Impact Statement, Plan Amendment. Appendices. vol. 2. U.S. Department of Agriculture, Forest Service, Alaska Region, Juneau, AK. R10-MB-603d.

Chapter 47

Western Oregon Districts, Bureau of Land Management, United States of America

Edward W. Shepard[1,a] and Duane Dippon[2]

[1]*Bureau of Land Management, Department of the Interior, Newberg, Oregon, USA*, [2]*Bureau of Land Management, Department of the Interior, Washington, District of Columbia, USA*

ABBREVIATIONS

CFR	Code of Federal Regulations
EPA	Environmental Protection Agency
FEMAT	Forest Ecosystem Management Assessment Team
FLPMA	Federal Land Policy and Management Act
LSMA	Late-Successional Management Areas
NWFP	Northwest Forest Plan
NSO	Northern Spotted Owl
RMA	Riparian Management Areas
RMPs	Resource Management Plans
TMA	Timber Management Areas
WOPR	Western Oregon Plan Revision

MANAGEMENT SETTING AND BACKGROUND

The United States Bureau of Land Management (BLM) lands included in the Western Oregon Plan Revisions (WOPR) consist of the revested Oregon and California Railroad (O&C) Grant Lands located in western Oregon (Figure 47.1). The approximately 2.4 million acres (ac) (971,280 hectares (ha)) of valuable forestland consist primarily of Douglas-fir (*Pseudotsuga menziesii*) and mixed-conifer species. The U.S. Congress originally granted this land in alternating square-mile sections to the O&C Railroad Company in 1866 in order to construct a railroad from Portland south to the California border. In granting the land, Congress stipulated that the lands be sold to actual settlers in parcels no larger than 160 ac (64.8 ha) and at a price no greater than $2.50 per ac ($6.18 per ha) (Muhn and Stuart, 1988). The railroad violated the terms of the grant and, after lengthy litigation ending with a Supreme Court decision in 1915 (238 U.S.C. 411), Congress revested title of the grant land to the United States with the intent that it be sold as rapidly as possible to help settle the area and provide a tax base for local governments. Poor market conditions and rough terrain limited the sale of these lands. In 1937, Congress passed the Oregon and California Revested Lands Sustained Yield Act (O&C Land Grant Act) of 1937 (43 U.S.C. 1181a) retaining the land in Federal ownership with the mandate that the land administered under the Act

> …*shall be managed for permanent forest production, and the timber thereon shall be sold, cut and removed in conformity with the principal (sic) of sustained yield for the purpose of providing a permanent source of timber supply, protecting watersheds, regulating stream flow and contributing to the economic stability of local communities and industries, and providing recreational facilities.*

The Act further required that receipts be divided evenly between the Federal Government and 18 western Oregon counties (Richardson, 1980). Currently, the land is managed by five BLM Districts and one Field Office.

---

[a] Retired Oregon/Washington State Director

Forest Plans of North America. http://dx.doi.org/10.1016/B978-0-12-799936-4.00047-3

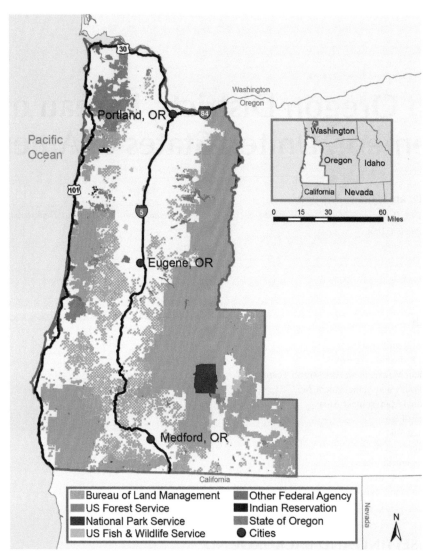

FIGURE 47.1   A map of the land area of the Western Oregon Plan Revision.

This land was managed under a series of management framework plans through the late 1980s. These plans allocated the land into areas available for timber harvest and areas restricted from harvest due to management and environmental reasons (Figure 47.2). Additionally, these lands produced a sustainable harvest level of 1,176 million board feet (MMBF) (2.832 million cubic meters (m³)) per annum.

In the late 1980s, concern began to surface regarding survival of the northern spotted owl (*Strix occidentalis caurina*) (NSO) and other environmental issues, such as retention of old-growth forests. Litigation over timber harvest escalated, prompting the BLM to initiate work on new Resource Management Plans (RMPs). In 1988, the Oregon BLM started development of RMPs to cover the O&C districts under one consistent planning effort (U.S. Department of the Interior, Bureau of Land Management, 1992). Rather than develop plans district by district consecutively over a period of several years, as had been done with the previous plans, the BLM elected to complete all of the plans concurrently. This became the BLM's first ecosystem-based landscape management planning effort. This plan was also the first to use a geographic information system (GIS) to build a "data-driven" planning process through integrated harvest scheduling and geospatial-based environmental cumulative effects analysis. This plan designated NSO habitat blocks and riparian reserves for owls, water quality, and anadromous fish. The plan resulted in altering the allocation of lands from approximately 80% of the land base dedicated to sustained yield timber production and 20% dedicated to conservation (U.S. Department of the Interior, Bureau of Land Management, 1984) to approximately 51% of the landscape allocated for a conservation emphasis and 49% for timber production (U.S. Department of the Interior, Bureau of Land Management, 1992). Some timber harvest could

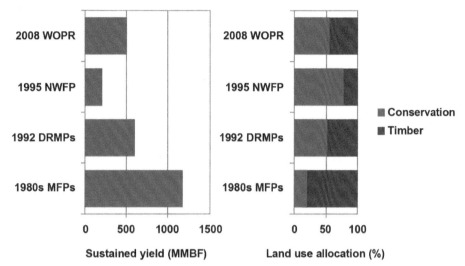

FIGURE 47.2 Comparison of land-use allocations and resulting sustainable harvests for the 2008 BLM management plans.

still be scheduled on approximately half of the lands allocated for conservation as long as the conservation objectives were met, that is, acceleration of stand conditions to that of older forest conditions to provide NSO habitat conditions. The plan produced a sustained-yield timber harvest of 595 MMBF (1.4 million m^3).

The NSO was officially listed as threatened by the U.S. Fish and Wildlife Service (USFWS) in June 1990. Litigation surrounding forest management and the listing of the NSO resulted in court injunctions against the existing BLM and U.S. Forest Service land-use plans for lack of protection for the NSO under the Endangered Species Act (ESA). Through appeals based on the Federal Land Policy and Management Act (FLPMA), National Forest Management Act, and National Environmental Policy Act (NEPA), federal timber harvests came to a near standstill by 1991 and 1992. The draft RMP was released for public comment in the summer of 1992; however, completion of the plan was overtaken by the Presidential election in November 1992. Presidential candidate Bill Clinton, while campaigning in the Pacific Northwest promised that, if elected, he would convene a Forest Summit in the Northwest to seek solutions. Soon after the Clinton Administration took office, work on the BLMs final RMP was suspended. Clinton kept his promise by holding a Forest Summit in Portland in April 1993. The president, vice president, and nearly the entire cabinet spent an entire day meeting with stakeholders from all sides of the issues, local and state governments, and agencies. The result of the summit was the initiation of the Northwest Forest Plan (NWFP).

The NWFP became the Nation's first multiagency landscape plan covering more than 24 million ac (9.71 million ha) of federal forests from the Canadian border south to near Sacramento, California and from the eastside of the Cascade Range west to the Pacific Ocean; the entire range of the NSO in the United States. The plan included 4 National Parks, all or parts of 19 National Forests, and 7 BLM Districts (U.S. Department of Agriculture, Forest Service, and U.S. Department of the Interior, Bureau of Land Management, 1994).

The Forest Ecosystem Management Assessment Team (FEMAT), consisting of approximately 100 scientists and agency personnel, was assigned the task to develop alternatives and propose a plan to be adopted by the affected National Forests and BLM Districts. The GIS support team faced the daunting task of aggregating geospatial data and GIS technology where no standards existed between agencies and even between units within the same agency, leading to numerous analytical difficulties. The GIS team supporting FEMAT had only 45 days to compile the geospatial data and another 45 days to analyze nine alternatives considered by the FEMAT team of scientists. Data and technology inconsistencies led to the GIS support team simplifying the data to the lowest common descriptors for the area of analysis. This aggregation resulted in the summarization of forest conditions and characteristics on a 40 ac (16.2 ha) grid. This process resulted in representing the planning area with a sample of 600,000 points (center points of a 40 ac grid pattern) with simplified attributes. At this scale, roads, streams, and other finer scale features could not be mapped nor directly considered in the comparative analysis of alternatives.

When completed, the NWFP became the first interagency ecosystem-based plan to provide a network of reserves to provide habitat for a late-seral obligate species. It was intended to meet multiple objectives, including recovery of the NSO, marbled murrelet (*Brachyramphus marmoratus*), listed anadromous fish, and hundreds of other species that were thought to require older forest conditions. This was accomplished by establishing reserves on nearly 77% of the land base; providing larger late-seral reserve and greater expanses of riparian reserves. The resulting Forest and Land Management Plans were intended to move away from species-by-species management to managing habitats for numerous species. The NWFP was also intended to meet economic needs by providing a predictable, legally certain, and sustainable harvest of timber from the 23% of the land remaining in the timber base. The BLM's share of the annual harvest level under the NWFP was 203 MMBF (0.48 million m^3).

The NWFP was litigated from the start at both the plan and individual project level of implementation. Although the plan was found to be legally sufficient, this did not stop litigation on individual implementation projects, which continues even now, 20 years after the NWFP was finalized.

Nearly two decades of NWFP monitoring point to areas of success, yet some of the primary objectives are not being met. The NSO population continues to decline, based primarily on competition with the invading barred owl (*Strix varia*) while habitat has increased (Rapp, 2008). Even with nearly 77% of BLM lands dedicated to habitat and old-growth protection under the NWFP, continuous appeals and litigation against most of the plan's implementation projects resulted in very few of the timber objectives being met as intended. As a result, the economic objectives were not met. This led to the counties dependent on the timber receipts being strapped for revenues, resulting in the reduction of county services. The forest products industry dependent on the BLM and Forest Service for timber has not been able to acquire the volume or type of timber promised by the NWFP. This resulted in more litigation by both the counties and industry to force the BLM to meet its economic mandate under the O&C Land Grant Act.

The BLM completed a required decadal evaluation of the RMPs in 2004 and concluded that the objectives of the plans were not being met. BLM regulations (43 Code of Federal Regulations (CFR) 1610.5-5,6) require the agency to amend or revise plans when plan implementation deviates significantly from plan objectives. Also at the time, litigation filed against the NWFP by the timber industry and O&C Counties was moving forward in the District Court for the District of Columbia (American Forest Resource Council, et.al. v. Clarke (Civil No. 94-1031-TPJ (D.D.C.)). The BLM and the industry settled the litigation in late 2003 by agreeing to revise the RMPs. The BLM then began the process of developing a new plan, the WOPR.

PLANNING ENVIRONMENT AND METHODOLOGY

The BLM was directed to initiate a new planning process under provisions included in the Settlement Agreement mentioned above. Key components in that agreement were that the BLM would complete the revision of the new set of RMPs no later than December 31, 2008 and that at least one of the analyzed alternatives would include no reserve areas except as necessary to protect species listed under the ESA. The BLM was directed to use the interpretation of the 1937 O&C Land Grant Act as interpreted by the Ninth Circuit Court of Appeals in the 1990 Headwater v. BLM case (914 F.2d 1174): forestland suitable for timber production would be managed for permanent forest production over other uses. The BLM proceeded under the standard RMP planning process as proscribed by the FLPMA of 1976. As with the 1992 Draft Plans and the NWFP, the Bureau approached this as a landscape planning processes across all BLM lands of western Oregon with one Environmental Impact Statement (EIS) tiered to multiple District and Field Office RMPs. The following steps were used to develop the WOPR:

- Establish project and interdisciplinary team, scientific team, GIS team, and legal team
- Invite cooperating agencies into the process, and provide funding support as needed
- Conduct public scoping to determine the issues and concerns to be addressed
- Conduct a "data-driven" planning process where geospatial analytics are used to compare and contrast alternative development and cumulative environmental analyses
- Incorporate existing information and inventory data including NWFP monitoring data, BLM standard landscape geospatial data across all programs and disciplines, research information, and past implementation experience
- Analyze and summarize the existing management situation and publish for comment
- Develop planning criteria and formulate the purpose and need and publish for comment
- Formulate alternative management strategies and publish for comment
- Prepare draft District/Field Office RMPs and one EIS describing the purpose and need, the affected environment, the alternatives, an analysis of the estimated effects of the alternatives, and a comparison of the effects of the alternatives, resulting in the selection of a preferred alternative
- Prepare proposed District/Field Office RMPs and one final EIS based on the public, cooperators, agency staff, and scientific team feedback on the draft RMPs
- Prepare Records of Decision and approve RMPs
- Implement, monitor, and evaluate building off of the geospatial data consolidated to create the plans

Public involvement was solicited and encouraged throughout the process. Nearly 200 meetings with the public were held. Public comments were also solicited and accepted at several stages throughout the process. Several methods were used to disseminate information to the public and receive comment, including web-based geospatial specific commenting tools, distribution of documents and related geospatial data, news releases, public workshops and meetings, and editorial board visits. There were 17 counties, 10 Oregon state agencies, 4 federal agencies, and one Indian tribe on the cooperating group. The cooperators were involved throughout the entire process. A 60-day governor's consistency review was conducted between the proposed RMP and the Record of Decision.

Guiding Laws, Regulations, and Policies

Management of BLM lands in western Oregon is governed by two major "organic" acts; the O&C Land Grant Act of 1937 (43 U.S.C. §1181a, et seq.) for the 2.1 million ac (850,000 ha) of O&C lands and the FLPMA of 1976 (43 U.S.C. 1701 et seq.) for the 400,000 ac (161,878 ha) of public domain lands, and for the planning process proscribed in FLPMA. In addition, the BLM must comply with many other major acts regulating the management of federal lands including, but not limited to, the NEPA of 1969 (42 U.S.C. 4321 et seq.), the Clean Air Act of 1990 (42 U.S.C. 7418), the Clean Water Act of 1987 (33 U.S.C. 1251), and the ESA of 1973 (16 U.S.C. 1531 et seq.). The BLM planning process is regulated by 43 CFR Part 1600, subpart 1610-Resource Management Planning.

The development of the WOPR had to comply with several federal court decisions that have been made over the last several decades. Most notable in the case of this plan are the August 28, 2003 Settlement Agreement in *AFRC v. Clarke* (Civil No. 94-1031-TPJ [D.D.C.] in the District of Columbia District Court, which required the BLM to revise its RMPs for western Oregon and to consider at least one alternative that did not create reserves, except as necessary to avoid jeopardy to species listed as threatened or endangered under the ESA. The agreement also stipulated that the BLM revision shall be consistent with the interpretation of the O&C Act in *Headwaters, Inc. v. BLM*, 914 F.2d 1174 (9th Cir. 1990) that was one of several court cases, but the most explicit, to state that the O&C Act "…envisions timber production as a dominant use" and not as a multiple-use act.

Other policies that guided the development of the WOPR were the USFWS recovery plans and critical habitat designation for marbled murrelet, NSO, salmonids, and other ESA listed species that were in varying stages of development at the time WOPR was being developed (U.S. Department of the Interior, Bureau of Land Management, 2006).

Planning, Decision, and Analysis Area

The planning area for the WOPR included the five western Oregon BLM districts (Salem, Eugene, Roseburg, Coos Bay, and Medford) and the Klamath Falls Field Office of the Lakeview District. The land was divided into approximately 80,000 stands, each with detailed forest characteristics to support habitat management and harvest scheduling modeling. Between the NWFP and WOPR, the entire hydrologic network on BLM lands had been mapped, including both perennial and intermittent components. Salmonid and fish presence was linked to the hydrology network. Roads, nonforest, special habitat areas, NSO habitat, and areas of environmental concern were mapped. Topology was mapped at a 32.8 feet (ft) (10 meter (m)) scale, the smallest area of analytical interest. The spatial footprint of each of the 80,000 stands was subdivided and allocated to one of the land-use conditions, based on locations of roads, steep slopes, rock, meadows, wetlands and other features that result in areas unsuitable for timber management; or rules affecting the definition and location of riparian management, NSO habitat, or other management rule sets defining each of the alternatives.

The WOPR Interdisciplinary Team used these data to describe current conditions of 18 specific resources described with more than 100 maps, figures, and tables. These products were published in the Analysis of Management Situation (U.S. Department of the Interior, Bureau of Land Management, and Bureau of Land Management, 2005) with the expectation that they would support NEPA-based public outreach and collaboration in the scoping phase of the planning process.

Each of the land-use allocations resulted in forest management strategies for different goals and objectives. The 32.8 ft (10 m) analytical scale allowed detailed modeling of forest stand condition development and its influence on the growth and development of NSO habitat and watershed condition modeling, including salmonid habitat and water quality conditions. The WOPR interdisciplinary team members evaluated alternative impacts on their area of specialty; timber production, wildlife habitat development, water, recreation, fire, and so forth. All had detailed projections over the next 10, 20, 50, and 100 years on which to base their professional opinions. Because of the checkerboard pattern of land ownership (alternating one mile (1.61 km) square sections of land (Figure 47.3)) in western Oregon, the analysis considered projected management regimes on non-BLM lands based on the type of owners (industrial timber land, state, nonindustrial forestland, National Forest, etc.). The planning decisions made were for BLM administered land only and did not govern non-BLM

FIGURE 47.3 A landscape view of the O&C checkerboard landscape with privately owned and intermixed lands in western Oregon.

administered lands. Although the planning was completed simultaneously for all of the planning area to allow for regional level analysis, decisions were made separately for each district to allow for local conditions. These analytical products in the form of maps, figures, and tables were included in the Draft RMPs documents that were sent out for review and comment. The geospatial data used to support the draft plans were also listed on the Internet for review (metadata) and download (geodatabases) to the interested public.

The planning team coined a term for its approach to developing and analyzing the plan. The team called it a "data-driven" planning process since it had detailed resources data feeding science-based models on which each change in an alternative and the differences in alternatives could be measured and analyzed. This approach to planning was much more detailed and analytical than what supported the FEMAT team of scientists. The WOPR interdisciplinary team of specialists identified its management questions, proposed analytical models, data to be used, products to be generated, and references to related science and published the set of planning assumptions under the Proposed Planning Criteria and State Director Guidance (U.S. Department of the Interior, Bureau of Land Management, 2006).

Vision, Goals, and Objectives

The vision for the WOPR was (U.S. Department of the Interior, Bureau of Land Management, 2006)

The BLM will manage the natural resources under its jurisdiction in western Oregon to contribute to the social well-being of the human population and to help enhance and maintain the ecological health of the environment.

Three major goals, each with its own desired objectives or outcomes, were established as follows:

Goal 1: Maintain healthy forest ecosystems with habitat that will support populations of native species and protection of riparian areas and water.

- Manage the lands within the landscape to contribute to conservation needs of special status species (including ESA Listed Species) and ecosystems on which they depend
- Provide clean waters that support viable fish and wildlife populations, domestic water use, safe drinking water, functional riparian areas, and recreation use
- Maintain the capacity of soils to function for sustained timber yield
- Prevent introduction of invasive species and provide for their control to minimize impacts to economic, ecological, and human health
- Restore fire-resilient stands and protect communities at risk from uncharacteristic wildfire
- Identify, designate, and protect areas of critical environmental concern
- Protect public lands and their resources from mineral entry uses

Goal 2: Provide a sustainable supply of timber and other forest products that will maintain the stability of local and regional economies, and contribute valuable resources to the national economy on a predictable and long-term basis.

- Maintain permanent forest production in conformity with the principles of sustained yield
- Annually declare and sell timber in an amount equal to the sustained-yield capacity of the forested lands
- Provide for multiple uses on the public domain lands
- Acquire adequate legal access to public lands for forest management activities and the removal of federal timber

Goal 3: Provide amenities that enhance communities as places to live and work.

- Protect cultural resources for the benefit of present and future generations
- Provide a broad spectrum of recreational opportunities
- Make public lands available for special uses and needed rights-of-way
- Protect public health and welfare by mitigating the impacts of air pollution emissions from wildland and prescribed fire on air quality and visibility

OUTCOMES OF THE PLAN

The planning process analyzed four action alternatives and the no-action alternative (continuation of the NWFP on BLM lands) (U.S. Department of the Interior, Bureau of Land Management, 2007, 2008). From the analysis of these alternatives, a proposed RMP was constructed and chosen as the final decision by Records of Decision for each of the districts. The landscape was divided into one of four land-use allocations: Late-Successional Management Areas (LSMA), Riparian Management Areas (RMA), Timber Management Areas (TMA), and Withdrawn lands. Each had customized forest management prescriptions.

LSMAs were allocated as large, connected blocks of habitat suitable for NSOs and marbled murrelets. These areas were based on interactive collaborative mapping exercises with specialists from the USFWS and on the final NSO recovery plans and critical habitat that were in place at the time (later withdrawn by the USFWS). These areas, which included 22% of the land base, were to be managed primarily to grow older complex forests with no commercial harvest except as a by-product of treatments, such as density management, to accelerate development of complex older forest conditions.

RMAs were allocated to maintain and promote the development of mature or structurally complex forests and provide for the riparian and aquatic conditions that supply streams with shade, sediment filtering, leaf litter, and large wood and root masses that stabilize stream banks. The RMA widths were one site-potential tree height (average of 180 ft (54.9 m)) on each side of the stream channel on perennial and intermittent fish-bearing streams with an exclusion of any silvicultural activities within 60 ft (18.3 m) of the bank. Intermittent nonfish bearing streams were allocated a width of one-half of one site-potential tree on each side of the channel and a 35 ft (10.7 m) silvicultural exclusion. As with the LSMAs, silvicultural treatments were allowed (outside of the exclusion zones) to advance the objectives of the RMA; no sustainable harvest of timber was planned for the RMA. The RMAs comprised 9% of the area.

TMAs were forests allocated to achieve a high level of continuous timber production on a sustained-yield basis. The TMAs were further subdivided in three categories; Uneven-Aged TMAs, TMAs, and Deferred TMAs. The Uneven-Aged TMAs consisted of approximately 6% of the area and were located in the drier forests of southwest Oregon to be managed using single-tree and group-selection harvest methods to promote fire-resilient forests. The TMAs consisted of 26% of the land base and would be managed using even-aged management on an 80- to 100-year rotation. Together, these two TMA allocations would produce 502 MMBF (1.18 million m^3) of wood on a sustained-yield basis. The Deferred TMAs were mapped areas within the other two TMA allocations where substantially all of the existing levels of older and more structurally complex multilayered conifer forests would have timber harvest deferred for 15 years to provide additional habitat for the NSO while the USFWS studied the interaction between the NSO and barred owl. The Deferred TMA consisted of approximately 7% of the land base.

Withdrawn lands included lands within the National Landscape Conservation System and were congressionally withdrawn (e.g., wilderness or wild and scenic rivers), administratively withdrawn (e.g., national monuments), lands unsuitable for timber management (e.g., steep slopes, soil fertility, etc.), or designated for other uses (e.g., recreation areas). Withdrawn lands comprise approximately 29% of the area. The aggregated allocations and the resulting sustainable harvest are depicted in Figure 47.2.

DISCUSSION AND CONCLUSIONS

The new plans were focused on minimizing permanent reserves, to optimize required conservation balanced by use, except for Congressional Reserves (e.g., wilderness, Wild and Scenic Rivers, and Areas of Critical Environmental Concern). In addition to these permanent reserves, land-use allocations were focused on defining an optimized mix of geospatially designated lands dedicated to late-successional management, aquatic and watershed function, and sustainable timber production.

Learnings and Insights

An unexpected result from this more detailed, model-based approach found that the continuation of the no-action alternative of the NWFP would result in growing NSO habitat and old growth conditions faster in the 77% conservation network than could be removed from the remaining 23% of the landscape given harvest sustainability and forest age-class model constraints, contrary to opinions expressed in the NWFP. This finding provides one reason that the decadal Monitoring Study had concluded that the loss of NSO and old-growth habitat was less than expected by the NWFP, resulting in more functional habitat than anticipated (Rapp, 2008).

The WOPR planning team found that better data and improved geospatial technology and models led to a more efficient allocation of the land base than the NWFP. Riparian Management Zones were defined by in-depth watershed analyses that modeled landslide-prone areas and potential natural contribution of large wood to streams. Riparian Management Zones were also defined by modeled habitat suitability index for multiple species of anadromous fish. LSMA locations and footprints were defined by existing areas of functioning habitat, modeled in-growth of habitat, and modeling rules from USFWS on the science of defining size and distribution of a network of islands of habitat for NSO. This more detailed data allowed a BLM and USFWS team of NSO experts and GIS specialists to collaborate on optimal size and placement of a NSO reserve system. In the final alterations of the proposed alternative, the BLM adopted additional management allocations for NSO, water resources, and salmonid ESA-listed species.

Modeling outcomes suggested that ecological functionality of LSMAs would increase over existing conditions in time. The plans resulted in approximately 55% of the landscape being designated for conservation while 45% was allocated to timber production for a sustained harvest of 502 MMBF (1.18 million m^3). Partial stand thinning and harvesting would occur on much of the conservation lands to meet nontimber objectives, but these volumes were not included in the calculated sustainable harvest level (an estimated additional 86 MMBF (0.2 million m^3) could be harvested for 2 decades from density management and forest health treatments, and then the volume would decline in subsequent decades). The WOPR increased the level of timber harvest versus the NWFP with more efficient data, science-driven defined conservation areas, and better economic results (tied to O&C Land Grant Act) contributing to economic stability of local communities and industries.

Sustainability Issues

Managing landscapes for ecological objectives can result in sustained timber harvests, however, managing the same landscapes for multiple, individual species under ESA makes defining sustainability nearly impossible to quantify as protecting individual species sites trump landscape goals and conditions. External events, including the retraction or revision of related ESA Critical Habitat or Recovery Plans, can have major adverse effect on the planning process. Changes in the political environment in the form of a change of administration at either the state or national level can adversely affect a controversial planning process. Land-use plans with an intended life span of 15–20 years can be easily upset when political philosophies are tied to short-term election cycles. Changing politics can make multiyear planning processes difficult if not impossible. Using land planning and the NEPA process to resolve an issue with deep-seated, polarized positions is not automatic and can fail without consistent support of political and agency leadership.

Plan Development Challenges

Data development for large-scale planning efforts such as this can be extensive. However, there is an important planning benefit of standard, maintained, detailed resources data that can be aggregated to multiple scales and landscape. The resource issues and ESA-related habitats exist at multiple scales, and the ability to work with detailed data is extremely valuable. Additionally, the cost and time it took to prepare the data for the WOPR planning process was less than 10% that of previous planning efforts and it was more complete with greater detail. Therefore, data development was not necessarily a challenge to the BLM plan development, because the WOPR planning effort benefited from implementation actions under the NWFP. The NWFP mandated extensive watershed-based geospatial data gathering and analyses resulting in a major increase and densification of mapped streams and roads. Implementation monitoring requirements resulted in the BLM mapping and tracking all restoration and management activities in a consistent enterprise GIS. Thus, the WOPR planning process started with 95% of the required data ready, reducing time delays and significantly reducing the cost to support a data-driven planning effort. The detailed geospatial data allowed the use of current science and science-derived models to predict change and evaluate alternatives. This allowed the interdisciplinary team to better understand the relevant ecological conditions and to describe cumulative effects.

The BLM funded positions in other agencies to have dedicated staff to support collaboration. In retrospect, the collaborative multiagency engagement did not work as anticipated. The BLM maintained control over the planning effort, which did not resonate with some of the cooperating agencies. A better means of managing engagement over multiple years and with diversified interests could benefit from improved dispute-resolution processes.

The USFWS and the National Marine Fisheries Service are focused on defining threats across the entire range of a listed species. Thus, the landscape approach to planning that the BLM implemented was used to map land-use allocations and general management prescriptions. However, these other agencies also wanted to identify any and all implementation actions that could limit or improve habitat, both in the short run and the long term. The proper scale for plan consultation (the broader, more generally described landscape scale, or the narrower, more specifically analyzed project level) was a key question. Land management agencies do not possess the capability and are not intended to map out all of their potential actions over the life of a land-use plan (15–20 years) with certainty. This inability for land management and regulatory agencies to agree on scale and scope for consultations under section 7(a) (2) of the ESA will continue to lead to imperfect planning efforts.

Plan Implementation Challenges

After working through the entire NEPA process beginning in the spring of 2005, the plans were released for final public review in the summer of 2008. However, a decision by the USFWS to withdraw the 2008 NSO Recovery Plan and Critical Habitat designation resulted in undercutting the entire multiyear planning investment. In addition, after the fall presidential election, the Governor of Oregon, the USFWS, the National Marine Fisheries Service, and the EPA all expressed concerns with the proposed plans for various reasons, despite being involved from the start as cooperators. The BLM determined that the plans did not implement any action on the ground that would affect ESA-listed species and informed the USFWS and the National Marine Fisheries Service of the no-effect determination and that consultation under § 7(a) (2) of the ESA would occur at the project level. The regulatory agencies originally agreed to this, but later withdrew their agreement.

By the end of 2008, litigation was filed immediately after signing the Records of Decision by environmental interest groups. After the election, the WOPR was never implemented with the plans being withdrawn by the Secretary of the Interior citing legal error in July 2009. It was never determined by a court that an error existed. The government agreed with the claim of the new plaintiffs and did not contest their claim. An alternative strategy could have directed the BLM to correct any possible error(s) by amending the WOPR versus withdrawing the entire plan, thereby reducing costs by using existing analysis and speeding up the revision process.

If the plans had been implemented, new NSO Recovery Plan and Critical Habitat designations would likely have required an immediate revision or amendment of the WOPR plans. Project-level ESA consultation would have redefined WOPR plan management requirements, potentially leading to further plan amendments as the plans were implemented. Withdrawal of WOPR put BLM back under the NWFP, along with continued litigation. Based on the analyses of the No Action Alternative (i.e., the NWFP), the BLM had already determined that the objectives of the NWFP-related RMPs were not being met, leaving the BLM no alternative but to start new RMPs. A new western Oregon planning effort was initiated in 2012 and is currently in process.

Other Interesting Aspects of the Plan

The WOPR Interdisciplinary team published the Analysis of Management Situation (2005) and the State Director Guidance (2006) and distributed the documents for public comment; but very little useable, substantive comment was received from the public, environmental interest NGOs, state of Oregon, tribes, counties, and timber industry on these documents. The team was surprised and disappointed that neither the Analysis of Management Situation nor the State Director Guidance generated significant suggestions from the broader stakeholder community regarding better data, scientific information, or models to be applied. It appears that greater effort will be necessary to use these initial planning documents to better engage interested parties in collaboration earlier in the planning process. While using the best geospatial data, science, and modeling processes available should be necessary components of planning; however, their use does not guarantee success.

Public involvement and collaboration have limits. Repeated FLPMA and NEPA planning efforts on the same issues and within the same landscape appeared to result in reduced public willingness to invest in a process that was judged not likely to result in a tangible compromise. Even with the large number of meetings and outreach efforts, there was no real public groundswell in support for or against the plan. The environmental community's involvement was just enough to have legal standing in the process and to allow it use of BLM's protest process. Throughout the FLPMA and NEPA processes, the BLM was unsuccessful in resolving the differences and moving the primary interested parties, O&C counties, timber industry, and environmental interest groups to compromise. Political support, or lack thereof, mirrored this continued impasse. Without buy-in to the public involvement-based planning process by groups on all sides of the debate, just following the FLPMA and NEPA process does not ensure a successful outcome.

The BLM is required to meet its regulatory (ESA, Clean Water Act, NEPA, etc.) as well as organic act (O&C Land Grant Act, FLPMA) mandates. Part of the FLPMA/NEPA process is the active and in-depth collaboration with other federal agencies, tribes, and state government. In hindsight, the BLM's approach to collaboration did not result in an interactive dialog that resulted in significant compromise and resolution by all of the parties involved. This may be due to federal regulatory agencies not viewing their legally derived mandates as necessarily requiring them to assist the BLM meet its legal mandates. This could be especially true if there is an expectation that, in the interest of protecting a listed species, it is better served by no new plan, but by the maintenance of the status quo. The current structure of federal laws puts agencies trying to implement their various laws and direction at odds with one another. A clearer understanding of how the various agency mandates interact may help with interagency collaboration under the NEPA processes.

The withdrawal of the 2008 NSO Recovery Plan (U.S. Fish and Wildlife Service, 2008) undercut the entire multiyear planning investment. If land management plans are to be tiered to ESA recovery plans, both need to be managed as dependent, interrelated efforts. The failure of one affects the other. The success of both is required.

REFERENCES

Muhn, J., Stuart, H.R., 1988. Opportunity and Challenge, the Story of BLM. U.S. Government Printing Office, Washington, D.C.

Rapp, V., 2008. Northwest Forest Plan—The First 10 Years (1994–2003): First Decade Results of the Northwest Forest Plan. U.S. Department of Agriculture, Forest Service, Pacific Northwest Research Station, Portland, OR, General Technical Report PNW-GTR-720.

Richardson, E., 1980. BLM's Billion-Dollar Checkerboard Managing the O&C Lands. Forest Historical Society, Santa Cruz, CA.

U.S. Department of Agriculture, Forest Service, U.S. Department of the Interior, Bureau of Land Management, 1994. Record of Decision for Amendments to Forest Service and Bureau of Land Management Planning Documents Within the Range of the Northern Spotted Owl. U.S. Department of Agriculture, Forest Service/U.S. Department of the Interior, Bureau of Land Management, Portland, OR.

U.S. Department of the Interior, Bureau of Land Management, 1984. Timber management for the 1980s. U.S. Department of the Interior, Bureau of Land Management, Oregon State Office, Portland, OR.

U.S. Department of the Interior, Bureau of Land Management, 1992. Executive Summary—Western Oregon Draft Resource Management Plans/Environmental Impact Statements. U.S. Department of the Interior, Bureau of Land Management, Oregon State Office, Portland, OR.

U.S. Department of the Interior, Bureau of Land Management, 2006. Western Oregon Plan Revisions—Proposed Planning Criteria and State Director Guidance. U.S. Department of the Interior, Bureau of Land Management, Oregon State Office, Portland, OR.

U.S. Department of the Interior, Bureau of Land Management, 2007. Draft Environmental Impact Statement for the Revisions of the Resource Management Plans of the Western Oregon Bureau of Land Management Districts. U.S. Department of the Interior, Bureau of Land Management, Oregon State Office, Portland, OR.

U.S. Department of the Interior, Bureau of Land Management, 2008. Final Environmental Impact Statement for the Revision of the Resource Management Plans of the Western Oregon Bureau of Land Management Districts. U.S. Department of the Interior, Bureau of Land Management, Oregon State Office, Portland, OR.

U.S. Department of the Interior, Bureau of Land Management, Bureau of Land Management, 2005. Western Oregon Plan Revisions—Analysis of Management Situation. U.S. Department of the Interior, Bureau of Land Management, Oregon State Office, Portland, OR.

U.S. Fish and Wildlife Service, 2008. Final Recovery Plan for the Northern Spotted Owl (*Strix occidentalis caurina*). U.S. Fish and Wildlife Service, Portland, Oregon.

Chapter 48

Whitefeather Forest, Ontario, Canada

Aaron Palmer[1] and Robert Keron[2]

[1]Haileybury, Ontario, Canada, [2]Ontario Ministry of Natural Resources, Forests Branch, Sault Ste. Marie, Ontario, Canada

ABBREVIATIONS

DCHS   Dynamic Caribou Habitat Schedule
WFCRMA   Whitefeather Forest Community Resource Management Authority

MANAGEMENT SETTING AND BACKGROUND

The Whitefeather Forest (Figure 48.1) is a located in northwestern Ontario within the boreal forest region. The management unit covers an area of 1,175,726 hectares (ha) (2.9 million acres (ac)) of land, consisting of 767,038 ha (1.895 million ac) of managed Crown land, 407,829 ha (1.007 million ac) of Provincial Parks (unmanaged Crown land), 840 ha (2,076 ac) of federal lands (First Nations Reserve), and 19 ha (47 ac) of private lands. Of the total area of Crown lands (managed and parks), 914,042 ha (2.26 million ac) is productive forest area. Water and nonforested land account for 151,440 ha (374,208 ac) and nonproductive forest totals 109,385 ha (270,290 ac) (Palmer, 2012).

Prior to 2004, the area received minimal fire suppression. This, coupled with minimal human and industrial disturbance, has resulted in a large intact forest with a landscape pattern characterized by large disturbances. The forest is typical of the northern boreal forest (Figure 48.2) and is dominated by coniferous trees such as jack pine (*Pinus banksiana*), black spruce (*Picea mariana*), white spruce (*Picea glauca*), balsam fir (*Abies balsamea*), eastern white cedar (*Thuja occidentalis*), and tamarack (*Larix laricina*). The major hardwood species found on the landscape include trembling aspen (*Populus tremuloides*) and white birch (*Betula papyrifera*). This forest is stratified for forest management purposes into forest units. The Forest Management Planning Manual (Ontario Ministry of Natural Resources, 2009) defines a forest unit as an aggregate of forest stands that will normally have similar species composition, will develop in a similar manner, and will be managed under the same silvicultural system. Forest units are the fundamental strata used in strategic modeling to forecast yield and model forest dynamics such as natural forest succession, post-renewal succession, response to natural disturbance, and habitat. Of the 22 northwest regional standard forest units, 17 are present on the Whitefeather Forest. These standard forest units were amalgamated into nine plan forest units for the purpose of strategic modeling and operational planning. The productive forest area of each is presented in Figure 48.3.

The Whitefeather Forest is home to a variety of animal life. Subsistence living and commercial fur trapping provide for the livelihood of many people in the forest. Game species such as black bear (*Ursus americanus*), moose (*Alces alces*), and small game (snowshoe hare (*Lepus americanus*), ruffed grouse (*Bonasa umbellus*), and spruce grouse (*Dendragapus canadensis*)) are common to the forest. Furbearing mammals are an important component of the forest. There are 22 trap lines within the forest, which provide limited income for 304 people. The primary targets of fur trappers include North American beaver (*Castor canadensis*), weasel (*Mustela* spp.), North American river otter (*Lontra canadensis*), muskrat (*Ondatra zibethicus*), and fisher (*Martes pennanti*).

There are many species at risk that occur, or are thought to occur within the Whitefeather Forest. Species listed as endangered, threatened, or of special concern include:

- Woodland caribou (*Rangifer tarandus caribou*)
- Wolverine (*Gulo gulo*)
- Lake sturgeon (*Acipenser fulvescens*)

Forest Plans of North America. http://dx.doi.org/10.1016/B978-0-12-799936-4.00048-5

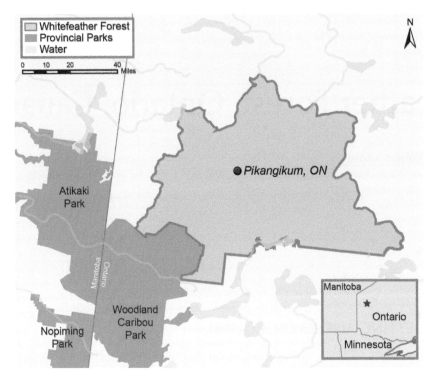

FIGURE 48.1 Map of the Whitefeather Forest.

FIGURE 48.2 Coniferous forests within the Whitefeather Forest.

- Whip-poor-will (*Caprimulgus vociferous*)
- Bald eagle (*Haliaeetus leucocephalus*)
- Golden eagle (*Aquila chrysaetos*)
- American white pelican (*Pelecanus erythrorhynchos*)
- Common nighthawk (*Chordeiles minor*)
- Snapping turtle (*Chelydra serpentina*)
- Short-eared owl (*Asio flammeus*)
- Mountain lion or cougar (*Puma concolor*)
- Canada warbler (*Wilsonia Canadensis*)
- Olive-sided flycatcher (*Contopus cooperi*)
- Barn swallow (*Hirundo rustica*)

The Whitefeather Forest is the traditional territory of Pikangikum First Nation (Whitefeather Forest Management Group, 2014). Pikangikum is an Ojibway First Nation and member of Treaty 5 and the Independent First Nations organization. The community of Pikangikum is located on Pikangikum Lake on the Berens River system, which is part of the Hudson's Bay

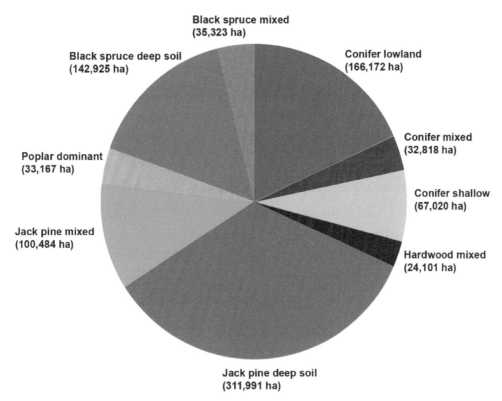

FIGURE 48.3 Area of Crown productive forest by forest unit, Whitefeather Forest.

drainage system. This is a remote access community with a population of just over 2,400 (Whitefeather Forest Management Group, 2014). The government of Pikangikum is composed of a Chief, Deputy Chief, and nine First Nation Councillors. Consensus-based decision making is very important in the community. While the chiefs and the Council are democratically elected, they are supported in decision making by the broader community, with great weight attributed to the wisdom of the Esteemed Elders of Pikangikum. There are very few roads in the forest, with the majority of the area accessible only by air, boat, or snowmobile. Consequently, economic opportunities for the residents of Pikangikum are very few. The residents are committed to retaining the Ojibway way of life and boast nearly a 100% retention rate of Ojibway fluency. The majority of the population still lives a subsistence lifestyle, off the land, and away from the community for a significant portion of the year. Most households are supported by subsistence activities such as hunting, fishing, trapping, and gathering. Consequently, traditional employment numbers are very low (Whitefeather Forest Management Group, 2014).

Pikangikum people have maintained the biological diversity of the Whitefeather Forest and, in many cases, nurtured greater abundance and diversity on the landscape. This has been achieved through customary indigenous resource stewardship practices that have several important implications on contemporary development of the Whitefeather Forest (Pikangikum First Nation, 2006). These include:

1. Pikangikum customary stewardship responsibilities are derived from a sacred trust between the Creator and the people of Pikangikum. The delegation of these responsibilities would deny the sacred trust and be a rejection of the gifts of the Creator.
2. The livelihood use of the ancestral lands is key to their ecological preservation. By using the land for their sustenance, the people of Pikangikum are preserving their lands.
3. The people of Pikangikum have always managed their ancestral lands as a whole, never dividing their lands into zones to be set aside for development or preservation.
4. Pikangikum customary stewardship approaches have always been adapted to contemporary opportunities within an indigenous livelihood context. The tools they use to engage in their livelihoods will change, but they will continue to be guided by their customary stewardship practices.

Feeling they were being left out of the benefits of commercial logging to their south, the people of Pikangikum opened a dialog with the Ontario Ministry of Natural Resources (MNR) over the potential for commercial development of their forest. It was decided that the community must be in control of any development of its lands, and that this development must be driven by Pikangikum traditional stewardship practices and indigenous knowledge. The community discussed the costs and benefits of such an endeavor and came to a consensus that the economic development was necessary for the well-being of its youth and community (Pikangikum First Nation, 2006). This decision was the beginning of the process for community-based land-use planning and environmental assessment coverage for the Whitefeather Forest.

Today, the forest is managed by the Whitefeather Forest Community Resource Management Authority (WFCRMA) under the authority of a Sustainable Forest License and under the guidance of the Esteemed Elders of Pikangikum. The ultimate development and approval of the 2012–2022 Whitefeather Forest Management Plan provides potential for a bright economic future for the people of Pikangikum and a framework for the economic development of other First Nation lands and resources.

PLANNING ENVIRONMENT AND METHODOLOGY

Guiding Laws, Regulations, and Policies

Management of Crown forests in Ontario has traditionally been governed by the laws, regulations, and policies of the MNR. The management of the Whitefeather Forest is unique in that it is governed by a combination of MNR laws, regulations, policies, and the land-use strategy and traditional knowledge of Pikangikum First Nation.

From the perspective of the MNR, forest management activities on the Whitefeather Forest presented a significant legal hurdle. Prior to 2009, forest management activities on Crown land in Ontario were conducted only on the *Area of Undertaking*, an area of approximately 43.8 million ha (108.23 million ac). Within the *Area of Undertaking*, forest management planning is regulated by a forest policy and legal framework that incorporates sustainability, public involvement, Aboriginal involvement, and adaptive management as key elements.

The Crown Forest Sustainability Act and the Environmental Assessment Act provide the legislative framework for forest management on Crown land in Ontario. The purpose of the Crown Forest Sustainability Act is to provide for the sustainability of Crown forests. The Act defines forest sustainability as long-term Crown forest health and requires forest management planning to provide for determinations of sustainability based on two fundamental principles:

1. Large, healthy, diverse, and productive Crown forests and their associated ecological processes and biological diversity should be conserved.
2. The long-term health and vigor of Crown forests should be provided for by using forest practices that, within the limits of silvicultural requirements, emulate natural disturbances and landscape patterns while minimizing adverse effects on plant life, animal life, water, soil, air, and social and economic values, including recreational values and heritage values.

Environmental assessment coverage for forest management activities on the Area of Undertaking is provided by Declaration Order MNR-71, granted in June 2003 under the Environmental Assessment Act. With 55 conditions, the approval covers a range of activities, including the building of access roads, harvest of trees, renewal of the forest, and conducting forest maintenance operations (Ontario Ministry of Natural Resources, 2014). Declaration Order MNR-71 does not cover forest management activities on the Whitefeather Forest, which lies north of the Area of Undertaking. In order for forest management activities to occur in the Whitefeather Forest, environmental assessment coverage would need to be extended beyond the Area of Undertaking.

This extension of coverage was granted in April 2009 by Declaration Order MNR-74, which granted environmental assessment coverage to forest management activities on Crown lands in the Whitefeather Forest in the vicinity of Pikangikum First Nation in northwestern Ontario. Declaration Order MNR-74 contains 52 conditions, including the same forest management planning requirements that are contained in Declaration Order MNR-71 as well as additional specific planning direction for the Whitefeather Forest. This additional planning direction addresses the incorporation of Pikangikum First Nation's customary stewardship practices, the provision of a continuous supply of habitat for woodland caribou, and a strategic approach to the planning of forest access roads in forest management planning for the Whitefeather Forest (Ontario Ministry of Natural Resources, 2014).

The forest management planning requirements of the Crown Forest Sustainability Act, the provisions of the environmental assessment approval under the Environmental Assessment Act, and the additional planning requirements for the Whitefeather Forest dictated by MNR-74 are incorporated into the Forest Management Planning Manual (Ontario Ministry of Natural Resources, 2009), which provides the direction for preparing a FMP. The Forest Management Planning Manual is supported by a series of provincial manuals and guides that outline silvicultural practices and methods to conserve biodiversity and enhance or protect wildlife habitat, aesthetics, watersheds, and other values. These guides are regularly reviewed and updated to ensure they stay consistent with current science and policy.

The forest management planning process for the Whitefeather Forest is heavily reliant on the Traditional Knowledge of Pikangikum people. This is incorporated through the "Keeping of the Land" land-use strategy. Pikangikum people have a relationship to the land and water that is at the heart of their culture and way of life. It is their belief that the ancestral lands on which they have been placed have been provided to them for the purpose of sustaining them as a people. This belief is at the core of their "Keeping the Land" strategy.

"Keeping the Land" began with a vision brought forth by the Esteemed Elders of Pikangikum. Their vision is: "A future in which the People of Pikangikum are able to maintain our ancestral stewardship responsibilities for Keeping the Land for the continued survival and well-being of Pikangikum people" (Pikangikum First Nation, 2006). The strategy carries forward the ancestral stewardship responsibilities for the land occupied by the Pikangikum First Nation and is used to "keep the land" as it has been, in all its diversity and abundance, since time immemorial (Pikangikum First Nation, 2006). In this context, land refers to the earth, water, all that is under the earth, and all that lives on the earth or in the waters. The goals of "Keeping the Land" are to:

- Ensure that Pikangikum First Nation customary stewardship responsibilities for "Keeping the Land" guide the protection and orderly development of lands and resources
- Secure resource-based economic development opportunities for the people of Pikangikum
- Harmonize proposed new land uses with existing and customary land-use practices of the people of Pikangikum

Plan Development Process

The FMP for the Whitefeather Forest was prepared for the 10-year period from 2012 to 2022. The plan was prepared in an open and consultative fashion by a Registered Professional Forester with the assistance of a multidisciplinary planning team, the Whitefeather Forest Initiative Steering Group (represented by Pikangikum Elders), and the Red Lake Local Citizens Committee. The preparation of the plan occurred in two phases. The strategic long-term planning and the planning of the first 5 years of operations occurred in Phase I, which included five stages of public and Aboriginal consultation. Phase II will involve the planning of the second 5 years of operations and will include three stages of public and Aboriginal consultation (Ontario Ministry of Natural Resources, 2009). Planning for this phase is slated to begin in 2015. The strategic long-term plan is referred to as the Long Term Management Direction (LTMD). In developing the LTMD, the planning team balanced multiple competing objectives, or values, for the landscape.

Two sources of values have been used in the development of the Whitefeather Forest Management Plan. The first was an indigenous knowledge values database produced and maintained by the WFCRMA for the community of Pikangikum, and the second consisted of values obtained through the Natural Resource Values Information System. During the development of the LTMD, Pikangikum indigenous knowledge was used in two significant areas. First, the Dynamic Caribou Habitat Schedule (DCHS) was developed under the guidance of Pikangikum Elders through their intimate knowledge of historic fire-cycles and mosaic patterns. Second, primary road corridors were strategically identified by Pikangikum Elders to minimize road construction, protect remoteness, and avoid culturally or ecologically sensitive areas (Palmer, 2012).

In operational planning, Pikangikum Elders directed the use of a "light-footprint" approach to harvest, renewal, and tending. This approach prescribes the use of mechanized cut-to-length harvesting methods. Renewal efforts would use prescribed burns as a site-preparation tool and aerial seeding and natural regeneration in an effort to achieve post-fire stocking densities necessary to produce high-quality timber. Indigenous knowledge provided the planning team with the locations of significant cultural and ecological values within and adjacent to the Whitefeather Forest. Each of these areas was assigned an Area of Concern prescription with the aid of indigenous knowledge experts. Indigenous knowledge also identified caribou habitat and known sites of caribou use.

In the preparation of an FMP, the determination of sustainability involved:

- The development of management objectives that address provincially mandated objective categories and indictors and other desired forest and benefits identified by the public, planning team, and local citizens committee
- An assessment of the achievement of management objectives for a period of 160 years
- The development of a proposed LTMD that balances the achievement of the management objectives
- A conclusion that the forest management plan provides for the sustainability of the Crown forest on the management unit

In the Whitefeather Forest management plan a total of 22 management objectives were identified. These can be found, along with corresponding target levels in Table FMP-9 of the Whitefeather FMP. These 22 management objectives can be grouped into four main categories: Forest diversity, silviculture, forest cover, and social and economic objectives. These objectives are to:

1. Provide wood fiber to create economic and employment opportunities through resource-based tribal enterprises, with the primary beneficiary being the Pikangikum people.
2. Contribute to the economic viability of the fur industry and trapping in or adjacent to the Whitefeather Forest through the protection of associated values.
3. Provide opportunities for the involvement of indigenous knowledge experts in the development of the Forest Management Plan.
4. Create opportunities for adjacent First Nations involvement in FMP development.
5. Have the local citizens committee effectively participate in the development of the FMP.
6. Emulate natural disturbance and landscape patterns characteristic of the management unit.
7. Maintain or move toward a natural range of forest composition and age distribution.
8. Maintain forest function for wildlife habitat in the management unit.
9. Support healthy, self-sustaining fish and wildlife populations.
10. Provide wood fiber to create economic and employment opportunities through resource-based tribal enterprises for the forest economy of Ontario.
11. Conduct forestry in the Planning Area in a manner that will not negatively impact non-timber forest products.
12. Identify and maintain cultural values and sites of special significance through specific protection or management practices.
13. Develop a road-use strategy to increase permanent road density within the general use area while maintaining remoteness as a defining feature of the Whitefeather Forest.
14. Ensure all interested parties are involved in strategic access planning.
15. Contribute to the economic viability of mining and mineral exploration in or adjacent to the Whitefeather Forest through the review of road-use strategies.
16. Maintain Pikangikum people's relationship to the land as a cultural landscape.
17. Promote natural regeneration through the use of prescribed burning and prescribed fire, where appropriate.
18. Employ timber harvesting and silviculture operations techniques that result in a light footprint on the land.
19. Sustain and do no harm to water bodies, headwaters, and aquatic ecosystems.
20. Achieve an excellent compliance record for all activities (harvesting, renewal, access, tending) with all other forest management activities.
21. Ensure the productive land base is maintained over time, enabling the people of Pikangikum to continue exercising their responsibilities as keepers of the land.
22. Provide a continuous, predictable, and sustainable supply of quality timber required by wood processing facilities that receive wood from the Whitefeather Forest.

A key consideration in developing the LTMD is knowing how forest conditions will change and how the management objectives will be achieved through time. On the Whitefeather Forest, the primary model used for strategic analysis was the Strategic Forest Management Model (SFMM) (Davis, 1993). SFMM is a linear programming model (model III formulation) that brings together the planning inventory, growth and yield information, natural disturbance assumptions, succession rules, wildlife habitat matrices, silvicultural costs and revenues, timber yields, cutting frequency, and other management assumptions. SFMM is a nonspatial model that determines the timber production capacities of the forest under different management strategies. It also provides the levels of wildlife habitat abundance in the forest through time.

Strategic modeling in Ontario uses an iterative process. A base model is built in SFMM and a scoping analysis is performed that informs the development of specific desired levels for management objectives. Once the planning team agrees on the desired benefits, the associated target levels are used to develop the LTMD, which is a management strategy that achieves a balance of desired forest-level benefits and becomes the benchmark used to assess the sustainability of future management actions.

The primary output of the strategic modeling exercise is the available harvest area for the first 10 years of the FMP, which describes the area available for harvest by forest unit and age class. The available harvest area provides a harvest profile that must be used at the tactical planning level to guide the spatial allocation of harvest areas. This spatial layout is not limited to harvest profile, but must also consider spatial landscape policies and constraints, while being aware of the costs of their proposed pattern of operations, including the existing and proposed infrastructure (e.g., roads, mills). Often, not all of the objectives can be met at desired levels or within desired time frames. If this occurs, the planning team needs to adjust the strategy until an acceptable solution, and acceptable LTMD is obtained.

In the Whitefeather Forest, there is very little existing infrastructure. Spatial concerns were mainly centered around the location of new forest roads and the provision for a continuous supply of caribou habitat. A Caribou Task Team was assembled to develop the DCHS, which would provide a continuous supply of habitat for woodland caribou. The Caribou Task Team developed the DCHS through a balance of MNR data, indigenous knowledge, and scientific literature. Present and future caribou habitat was mapped through the delineation of large landscape patches based on vegetation, soils, forest age classes, and landforms. Knowledge of known habitat use (such as calving sites), travel routes, and wintering grounds were also included. A DCHS was developed based on this assembled information with guidance from the Esteemed Elders of Pikangikum and MNR planning staff.

The operational planning was performed under the direction of "Keeping the Land," which directs that the landscape needs of woodland caribou are addressed and that silviculture practices restore caribou habitat where it has been harvested. Operational planning also considered the 1999 MNR caribou guide, which provides site-specific recommendations for caribou habitat needs. Caribou habitat was considered in five ways:

1. *Area of Concern Prescriptions*—Operational prescriptions were developed to protect known caribou calving sites and migration routes.
2. *Conditions on Regular Operations*—In the event of the discovery of an unknown calving site, the company will immediately stop operations and contact the MNR for further direction.
3. *Silvicultural Ground Rules*—Rules were developed that promote natural regeneration through prescribed burns, restore caribou habitat where it has been harvested, and emulate the natural disturbance pattern at a landscape scale.
4. *Assessment of Regeneration Success*—The goal is that 100% of the sites that have the potential to produce woodland caribou habitat will be regenerated to the target Forest Unit of the Silvicultural Ground Rule. Indigenous knowledge practitioners will have their expertise incorporated into regeneration assessment through opportunities to participate in regeneration surveys.
5. *Compliance*—Forest operations compliance inspections will monitor compliance of Area of Concern prescriptions and Conditions on Regular Operations. They will also assess the strategy on minimizing operational roads and ensuring road rehabilitation where required.

Primary road corridors were located with the assistance of the Whitefeather Forest Initiative Steering Group as represented by the Esteemed Elders of Pikangikum. The DCHS was created before the road network and was a major driver in the location of the forest roads. Other considerations were the locations of enhanced management areas, dedicated protected areas, water bodies, wetlands, and exposed bedrock. A final consideration was whether a road could be used to replace an existing winter road for access to First Nation communities. The planning team tried to favor the creation of all season primary roads, to promote year-round forestry operations, and to maximize employment and economic opportunities.

The harvesting operations necessary to achieve the LTMD were based on the direction of the Esteemed Elders of Pikangikum for a light footprint approach to forest management. This dictates the use of cut-to-length harvesting methods in order to minimize the impact of machinery on the landscape. Prescribed burning is the preferred method of regenerating harvested stands. There is also an emphasis on natural regeneration and aerial seeding to promote the high-density stocking typical of post-fire natural regeneration.

OUTCOMES OF THE PLAN

The strategic modeling effort for the Whitefeather Forest produced a LTMD that satisfied Pikangikum's "Keeping the Land" strategy as well as Ontario's rigorous planning requirements. The FMP established an available harvest area that provides economic and employment opportunities for Pikangikum people while harmonizing indigenous knowledge and western science. The planned harvest area is regulated by forest unit and limited by the available harvest area determined in the LTMD. For the Whitefeather FMP, 96% of the available harvest area was allocated for harvest over the course of the 10-year FMP. Within individual forest units, this allocation of the available harvest area varied from a low of 71% in the Conifer Mixed Forest Unit to a high of 98% in the Jack Pine Deep Soil Forest Unit (Palmer, 2012). The available and planned harvest levels for the 10-year Whitefeather Forest FMP can be seen in Table 48.1.

The DCHS allocated large tracts of land for harvest on a rotating schedule. Each tract is available for a 20-year period, with the exception of term 1, which is available for a 10-year period. The schedule provides for a continuous supply of caribou habitat in both amount and arrangement through time. The DCHS provides for a 120-year rotation, with areas harvested in the first period (2012–2022) being eligible for harvest again in 2132–2142.

TABLE 48.1 Available and Planned Harvest Area, Whitefeather Forest, Ontario

Forest Unit	Available (ha)	Available (ac)	Planned (ha)	Planned (ac)
Conifer lowland	8,046	19,882	7,794	19,259
Conifer mixed	438	1,082	315	778
Conifer shallow	553	1,366	516	1,275
Hardwood mixed	419	1,035	372	919
Jack pine deep soil	8,150	20,139	7,954	19,654
Jack pine mixed	5,339	13,193	5,137	12,694
Poplar dominant	446	1,102	423	1,045
Black spruce deep soil	21,542	53,230	20,594	50,888
Black spruce mixed	1,310	3,237	1,265	3,126
Total	46,243	114,266	44,370	109,638

The levels of caribou habitat were assessed with the Ontario Landscape Tool for three types of caribou habitat. This tool measures and compares the available habitat through time against a natural range of variation. The amount of caribou refuge habitat was found to be higher than the natural range of variation and was shown to be increasing in later terms. Caribou winter (used) habitat was shown to be at the minimum levels at the start of planning. These levels were expected to increase in later periods of time to a point where the natural range of variation was exceeded in terms 7 through 16. Caribou winter (preferred) habitat was at the minimum target level at the start of planning. By term 7, the habitat level reaches the upper bound of natural variation where it remains for the remainder of the planning horizon.

The management plan calls for the construction of 220 kilometers (137 miles) of primary roads during the first 10-year planning period. These roads have been located according to Pikangikum indigenous knowledge in a manner that avoids ecologically and culturally significant areas and promotes the remoteness of the Whitefeather forest. Two primary roads (Pikangikum and Branchwood) are expected to be used as replacements for winter road access to Pikangikum and other First Nation communities.

DISCUSSION AND CONCLUSIONS

The Whitefeather Forest FMP was a unique planning initiative designed to benefit the economic development of the First Nation community of Pikangikum. Its development and implementation provided many unique challenges compared to other areas of North America.

Learnings and Insights

The Whitefeather Forest FMP is a groundbreaking piece of collaborative planning between Pikangikum First Nation and the MNR. The collaborative process is encouraging because it shows that indigenous knowledge and western science can be merged in a successful and mutually respectful way.

One example of this collaboration took place during planning for harvest, access, renewal, and tending operations. Maps were developed that depicted all of the values in the indigenous knowledge database in the vicinity of proposed harvest areas and road corridors. Each proposed harvest allocation was then reviewed by the Pikangikum indigenous knowledge experts with knowledge of the particular area. The indigenous knowledge experts worked with the plan author and the community liaison to develop Area of Concern prescriptions to protect the natural resource features, land uses, or values identified by the community. Some of the values that were assigned Area of Concern prescriptions included natural mineral licks, bear fishing sites, and caribou migration routes.

The forest management plan provides a framework for other First Nation communities in Ontario to develop natural resources according to their own customary stewardship practices, allowing them to be in control of their own resources and economic futures.

Sustainability Issues

The most significant sustainability issue for the Whitefeather Forest is the current health of the forest products sector in northwestern Ontario. Mill closures and temporary shutdowns have eliminated the intended markets for fiber from the Whitefeather Forest. Consequently, no trees have been harvested under the FMP to date, although road upgrades have been conducted to improve access on the management unit.

Plan Development Challenges

Throughout the development of the DCHS, the caribou task team had difficulty achieving balance between MNR data, scientific literature, and indigenous knowledge. As the task team members worked toward achieving balance, they were faced with challenges that could not be resolved at the planning team level and had to be resolved through direction from the FMP Steering Committee.

Another challenge emerged later in the planning process involving planned operations and their impact on resource-based tourism operators. Of the 21 tourism operators within or adjacent to the Whitefeather Forest, three identified concerns with the proposed harvest allocations and/or roads. The concerns were related to the creation of new access to remote lakes, visibility of harvest areas from tourism lakes, and the timing of operations. To address these concerns, three Resource Stewardship Agreements were signed between Whitefeather and the resource-based tourism outfitters, which outlined prescriptions to protect the tourism values without compromising the objectives in the forest management plan.

Plan Implementation Challenges

The implementation of the Whitefeather FMP has been hindered by many factors. The decline in demand for forest products has led to the closure of several of the mills that were intended markets for this forest. With the recent improvement of the forest industry in Ontario, there is hope that these markets may open up to wood fiber provided by the Whitefeather Forest.

A more difficult problem is the remote nature of the management unit and the long transport distances from harvest area to mill. Transportation costs make up the largest cost in the forest products supply chain, and the long distances from the Whitefeather Forest to mills may make harvesting activities difficult to economize.

Nonetheless, Pikangikum Elders continue promoting their message that the strong northern boreal wood characteristic of the Whitefeather Forest be utilized in such a way that provides the highest value and best end-use opportunities in the wider marketplace. As such, the WFCRMA continues to investigate opportunities for supplying the highest quality logs to an integrated value-added forest products manufacturing facility to be established by Pikangikum First Nation in the Whitefeather Forest at or adjacent to Pikangikum.

ADDITIONAL READING AND RESOURCES

This chapter represents a synthesis of the 2012 Forest Management Plan for the Whitefeather Forest and the Pikangikum First Nation *Keeping the Land* strategy. To view the Forest Management Plan, please visit this Internet site, which was available on May 20, 2014:

http://www.efmp.lrc.gov.on.ca/eFMP/viewFmuPlan.do?fmu=994&fid=100106&type=CURRENT&pid=100106&sid=11582&pn=FP&ppyf=2012&ppyt=2022&ptyf=2012&ptyt=2017&phase=P1

To view the *Keeping the Land* strategy, please visit this Internet site, which was available on May 20, 2014:

http://www.whitefeatherforest.ca/wp-content/uploads/2008/06/land-use-strategy.pdf

If in the future the links to these sites appear broken, search the Internet using the titles or keywords provided.

REFERENCES

Davis, R., 1993. Analyzing Ontario's timber supply with the Strategic Forest Management Model. In: Davis, R. (Ed.), Analytical Approaches to Resource Management. Queen's Printer for Ontario, Toronto. pp. 52–71.

Northwest Region Caribou Task Team, 1999. A Management Framework for Woodland Caribou Conservation in Northwestern Ontario. Ontario Ministry of Natural Resources, Toronto.

Ontario Ministry of Natural Resources, 2009. Forest Management Planning Manual for Ontario's Crown Forests. Queen's Printer for Ontario, Toronto. 447 p.

Ontario Ministry of Natural Resources, 2014. Forest Management Planning in Ontario. Queen's Printer for Ontario, Toronto. http://www.mnr.gov.on.ca/en/Business/Forests/2ColumnSubPage/STEL02_163511.html (Accessed May 20, 2014).

Palmer, A., 2012. 2012–2022 Forest Management Plan for the Whitefeather Forest—Final. Ontario Ministry of Natural Resources, Northwest Region, Red Lake District, Red Lake, ON. http://www.efmp.lrc.gov.on.ca/eFMP/viewFmuPlan.do?fmu=994&fid=100106&type=CURRENT&pid=100106&sid=11582&pn=FP&ppyf=2012&ppyt=2022&ptyf=2012&ptyt=2017&phase=P1 (Accessed June 19, 2014).

Pikangikum First Nation, 2006. Keeping the Land: A Land Use Strategy. Pikangikum First Nation, Pikangikum, ON. http://www.whitefeatherforest.ca/wp-content/uploads/2008/06/land-use-strategy.pdf (Accessed May 20, 2014).

Whitefeather Forest Management Group, 2014. Welcome to the Whitefeather Forest Initiative. Pikangikum First Nation, Pikangikum, ON. http://whitefeatherforest.ca/ (Accessed May 20, 2014).

Chapter 49

Synopsis of Forest Management Plans of North America

Kevin Boston,[1] Krista Merry,[2] Donald L. Grebner,[3] Chris Cieszewski,[2] Pete Bettinger[2] and Jacek P. Siry[2]

[1]Department of Forest Engineering, Resources and Management, Oregon State University, Corvallis, Oregon, USA, [2]Warnell School of Forestry and Natural Resources, University of Georgia, Athens, Georgia, USA, [3]Department of Forestry, Mississippi State, Mississippi, USA

SYNOPSIS

In this book, we have collected 48 case studies of forest plans from throughout North America. We believe this is the most comprehensive set of forest planning case studies gathered using a common template. We are very grateful for the effort that many people (authors, editors, and support staff) put into preparing these chapters describing their approach to forest planning, especially since they were constrained by our timeline and our common template for all of the examples. The landowners were selected through outreach to professional organizations, calls to management organizations and to forestry consultants, and contacts within our personal networks. While perhaps not a representative sample of all potential management plans, we believe that these describe the range of management plans being created in the North American region. In other words, we employed a case study approach rather than a particular statistical design to sample these examples from the larger population of all forest management organizations in the region. Therefore, it is important for one to avoid making strong and significant inferences about the larger population of all forest plans in North America. As a result, this synthesis is limited to the case studies included in the text.

The properties described in this book have a tremendous range of size, from 58 acres (ac) to 17 million ac (23 hectares (ha) to 6.8 million ha). They represent a variety of ownership types such as family forests, municipal forests, city and county forests, state forests, public and private university research forests, urban forests, federally managed forests, tribal forests, and large publicly traded corporate forests. Vertically integrated companies were once very numerous in the United States, but now they are a rare organizational structure, mainly due to changes in the U.S. tax codes over the last two decades.

OWNERSHIP ENTITIES

There are a variety of forest ownership entities in North America. The Canadian forests described are primarily owned by the Crown, or federal and provincial governments. The goals and objectives guiding the management of these lands are heavily influenced by both public input regarding the desired future condition of these resources and the needs of local communities for the resources to provide employment and income. Recreation and wildlife habitat values feature prominently in these plans. Our attempt to locate authors to write about a private forest management plan, particularly in the eastern part of the country, was unsuccessful. We hope that this does not leave the reader with the impression that the entire country is composed of the Crown forests. Some of the chapters in this book represent municipal forests or collaborations with timber companies managed through tree farm licenses, and one represents an association with a First Nation (Chapter 48).

The majority of the Mexican forests are variations of community or private forests that are managed through their own forms of governance. Las Bayas Forest (Chapter 18) represents the management plan for a university-managed forest. These chapters provide a glimpse of the varied landowner groups that have developed management plans for forests of Mexico and the objectives of local communities that influence the development of management plans. These chapters also expose some of the challenges facing land ownership in Mexico, such as safety and security. Further, these chapters illustrate the current concern for silvicultural practices that best facilitate the sustainability of dry-site pine forests.

The United States arguably has the most diverse set of ownership entities represented in this book. The federally managed forests represent a variety of agencies under three different cabinet offices: (1) the Department of Agriculture with lands managed by the Forest Service, (2) the Department of Interior with lands managed by the Bureau of Land Management, National Park Service, Fish and Wildlife Service, and Bureau of Indian Affairs, and (3) the Department of Defense in coordination with the Bureau of Land Management to manage forest lands on military installations. There have been some changes in the management structure of the federal lands, and one example is a locally created conservation district, the Trinity County Resource Conservation District, in the Weaverville Community Forest (Chapter 28). This entity uses a 10-year stewardship contract to enhance the local community involvement in the management of federally managed forests lands. This book also provides only one example of tribal forestry in the United States with the Yakama Nation (Chapter 40). As a sovereign nation, the Yakama people have flexibility to manage their lands based on their tribal governance with consultation and technical advice from the Bureau of Indian Affairs. We believe that there are many more examples of tribal management.

There is a relatively new ownership category that has recently begun to have an influence on forest planning in the United States. These are the lands managed by nongovernmental organizations (NGOs). Examples within this book include the Conservation Fund Garcia River Forest in northern California (Chapter 19) and The Nature Conservancy ownership of the South Willapa Bay in southwestern Washington (Chapter 26). The City of San Francisco example (Chapter 32) also describes the challenges facing urban forestry programs, the problems encountered as municipal budget cuts occurred, and the development of a local nonprofit organization to assist with some of the forestry objectives. Although these examples are situated in the western United States, NGOs own and manage land throughout the country and may be involved in forestry programs where they do not have an ownership interest. These organizations have brought a new version of multiple-use management; they have both a strong conservation emphasis with a goal to remain working forests. Since these are nonprofit organizations, they have some relief from the various taxes imposed on individuals. Additionally, they can raise funds through external donations from foundations and individuals to support restoration management. It will be interesting to follow the trajectory of these types of forestland management plans in the future.

The majority of the privately owned properties in the United States described in this book are either a Real Estate Investment Trusts (REITs), or Limited Liability Corporations (LLCs). Both of these entities provide tax-efficient business opportunities to the landowners as well as some additional protection from liability claims beyond the assets of the REIT or LLC. There are some held as traditional family trusts that facilitate the ease of transfer between generations of owners. Typically, these trusts are the smaller areas that are privately held LLCs as their business entity. Examples of this type of ownership include the Ross Forests in Tennessee (Chapter 14) and the Eddyville Tree Farm in Oregon (Chapter 2). Other private land ownership groups may be incorporated as trusts, such as the Sessoms Timber Trust (Chapter 25), which helps maintain property in the family ownership through multiple generations. Larger ownerships that once were part of vertically integrated forest products companies may employ the REIT structure, which is a specialized trust providing significant tax advantages for the revenue generated from these forests. Examples of these include Weyerhaeuser (Chapter 39) and Rayonier (Chapter 44). Lastly, while we have reached out to several Timberland Management Organizations (TIMOs), which often manage forestlands for institutional investors such as pension funds, we were not successful in our attempt to have them develop a chapter for this book.

THE RANGE IN MANAGEMENT OBJECTIVES

As expected, the dominant uses vary greatly among forests located within North America and its smaller regions. Government forests (local, state, and national) are typically managed according to mandates created from their respective countries' legislatures authorizing the management of public lands. These enabling acts often define a set of goals and objectives for these forests. Private forests, on the other hand, reflect the varied preferences of the owners. It is also possible, as it appears to be the case in Mexico, that the management of both public and private forests is guided by the same enabling acts.

There seems to be similar enabling legislation for the national forests in all three countries. All have their own requirements for public forests to be managed sustainably. An example of these enabling acts is the Federal Land Policy Management Act for the U.S. Bureau of Land Management lands in western Oregon (Chapter 47). Similar government enabling legislation allowed for the creation of the Blue Ridge Parkway (Chapter 16) from Virginia to North Carolina through the Industrial Recovery Act and congressional authorization of the National Park Service to manage the Parkway. Additionally, the National Forest Management Act provides a directive for those lands managed by the U.S. Forest Service. Canadian forests held by the Crown are managed according to the requirements of the Forest Sustainability Act. The management plans of two Mexican forests (Chapters 9 and 13) cite the national forest management act that guides the development of much of the forests in Mexico. The Yakama Reservation (Chapter 40) is our only example for tribal forest planning in the United States. Its management goals are guided by the Yakama Nation Tribal Council, which determines the management direction for this forest. Its management direction involves multiple-use management that includes goals for big-game habitat, forest health, old-growth forests, revenue and employment, water quality, and threatened and endangered species. Several plans include a significant component of traditional culture in the management of forestlands, including the Yakama Reservation (Chapter 40), Whitefeather Forest (Chapter 48), Ejido Borbollones (Chapter 9), Indigenous Community of Nuevo San Juan Parangaricutiro (Chapter 20), and San Pedro El Alto Community Forest (Chapter 23).

One surprise was the variety in direction for the management of local government-owned forests. Typically, these properties have broad goals such as the management for "greatest permanent value" found in the state forests in northwestern Oregon (Chapter 36) or management for the "public good" as in Bayfield County, Wisconsin (Chapter 30). Some municipalities have ecological purposes as the centerpiece of their management direction, such as the City of Arcata (Chapter 8), whose citizens want to accelerate the development of late-seral conditions in their forests while not allowing their community forest to be a drain on the city's financial reserves. Forests of other municipalities (cities, counties, towns), such as Shirley Town Forest (Chapter 7), are guided by traditional forest management principles with the goal to produce revenue for their owners while maintaining multiple-use opportunities on these forests. Rib Lake School Forest (Chapter 6) seeks to be a working sustainable forest while serving as an environmental classroom for the community. Other state lands are managed by the state forestry agencies with the goal to fund public schools or hospitals, such as the South Puget Planning Unit in Washington (Chapter 38).

Forests that are associated with institutions of higher education (both public and private), such as those managed by the State University of New York College of Environmental Science and Forestry (SUNY-ESF) (Chapter 11), Harvard University (Chapter 10), Yale University (Chapter 29), or Durango State University (Chapter 18) are guided by research and educational goals. Forest management activities typically are used to facilitate or complement these goals, yet, can often be employed to address forest health or financial issues. The Mission Municipal Forest (Chapter 22) in Canada also has been a significant contributor to the local community.

The majority of the privately held forests described are in the United States. These lands are primarily regulated by individual state forest practices rules or best management practices that can differ significantly among the 50 states. Certain federal regulations, such as the Clean Water Act and the Endangered Species Act, have components that influence forest management actions on private forestlands, but there is no national private forest management act. The range of uses on the private lands is varied. The large corporate forests are managed with an emphasis on the production of income and the processes used by large corporate owners to maximize their revenues are best described in management plans developed by Weyerhaeuser (Chapter 39) and Rayonier (Chapter 44); however, both firms have adopted the Sustainable Forestry Initiative (SFI) to document and promote their sustainable practices.

There was significant variability in the management direction for the smaller lands, often referred to as nonindustrial lands or lands not associated with a processing facility. Some of the properties are managed primarily for recreation purposes. The Willow Break property in Mississippi (Chapter 15) is a good example of this type of ownership. The land was enrolled in the Wetland Reserve Program, and land management activities can only be undertaken if they are beneficial to wildlife. Other properties have a more multiple-use approach. The Eddyville property in Oregon (Chapter 2) aims to contribute to the generation of income for the Newton family, but there are added benefits of producing habitat for wildlife. Many nonindustrial owners exhibit a real pride in the ownership of their property (e.g., Chapters 5 and 12).

PLANNING METHODS USED TO PREPARE THE PLANS

Within the chapters of this book, the analytical methods used to prepare forest management plans varied greatly. They range from ad hoc stand-level management assessments and traditional methods, such as area control, to the most modern assessments that required the development of customized computerized applications. In the United States, the larger, corporate ownerships were the ones using high-level mathematical programming techniques such as linear programming and other scheduling algorithms (heuristics). In Canada, the Crown properties all use a combination of sophisticated computer applications, including linear programming and heuristic methods, to develop their forest plans. All of the Crown land chapters suggest the use of commercial forest planning software such as Woodstock, the Strategic Forest Management Model, or Patchworks™ systems to aid the development of their management plans. The uniformity of forest planning laws in Canada has encouraged the development of commercial applications that meet the Canadian government requirements for harvest scheduling. Due to their size, these organizations have the capacity and resources, computational power and personnel, to apply the most modern methods to formulate their harvest schedules. These systems link geographic or spatial data with forest growth and yield information and often with forecasted market data in the development of plans. These organizations also view forest planning as a necessary part of the business management cycle and consider the development of these sophisticated planning tools as a necessary element for enhancing or maintaining their competitive positions.

In Mexico, uneven-aged or continuous-cover management dominates the forest management. However, the even-aged forests of the San Pedro El Alto Community Forest (Chapter 23) use an area-volume check when computing the harvest levels. These systems are a variant of volume control as growth information is continually collected from plots used to support the determination of harvest levels using computer systems that assist in organizing the information necessary for forest planning.

For many of the smaller properties, the most common method used to determine the harvest schedule was a variant of area control. Because the properties do not have the scale to provide a harvest in every year, the land managers often attempt to develop a balanced age-class distribution that will provide an opportunity to harvest (and earn revenue) in every time period (e.g., decade). This allows landowners to take full advantage of good markets as they arise. The McPhail Tree Farm (Chapter 12) describes the problem of arranging a property to develop a mixture of age classes that will provide a sustainable cash flow for the landowners.

Some of the chapters within this book describe other methods used to organize harvests. The Arcata Community Forest (Chapter 8) is an example of using desired future conditions to determine when and how to harvest stands of trees. The target volumes for this forest have been determined through the need to achieve conditions that restore late-seral, multiple layers of forest structure. Another example is the Blue Ridge National Park (Chapter 16), where aesthetics, combined with the desire to preserve the natural and cultural heritage of the Blue Ridge Park Way, guide the development of actions and outcomes. The planning method involved factors that were primarily qualitative elements that were used to compare the alternatives against one another.

The variety of methods used to create these forest plans provide data points demonstrating the need to continue to teach a full range of forest management techniques to young forest managers. Students should begin with a holistic approach, one that begins with the goals for the property, proceeds through the creation of alternatives, and then compares or evaluates these alternatives against one another. Courses on forest planning should not be reduced solely to lessons on how to operate specific computer software, and students should be exposed to the broad use of classical approaches such as area-control, volume-control, or area-volume check.

SUSTAINABILITY ISSUES THAT HAVE ARISEN FROM THE PLANS

There were many chapters that described the role of certification in the forest management planning process. The three most common systems described in the book are the American Tree Farm System (ATFS), Forest Stewardship Council (FSC), and Sustainable Forestry Initiative (SFI). The owners of several nonindustrial forests have the honor of being named ATFS Tree Farmer of the Year. There was a common thread among the certification system that the recognition provided by the certification system added to the legitimacy of the owners management of the property. Some, such as the Michael Tree Farm in East Texas (Chapter 3), used the ATFS to assist in obtaining funding from organizations such as the U.S. Department of Agriculture Natural Resource Conservation Service through their Environmental Quality Incentives Program.

The SFI certification program has been adopted by larger REITs in the United States. Rayonier has included additional requirements on the size, shape, and spatial arrangement of harvest units and has incorporated these requirements into its customized algorithm. Weyerhaeuser has used SFI to recognize its environmental performance and set the standards for managing biodiversity on its lands in North Carolina.

Those chapters prepared by the two NGOs, the Conservation Fund (Chapter 19) and The Nature Conservancy (Chapter 26), both use FSC certification as a method to promote their sustainable practices, despite the fact that the properties are in California and Washington, respectively, and both states have significant forest practices rules and enforcement of those rules. Additionally, organizations that practice forestry in politically sensitive areas, such as the Arcata Community Forest (Chapter 8) in the redwoods region of northern California, have selected to participate in FSC certification to create an environment that allows for adaptive management of the plan with third-party oversight to validate the process with a skeptical public. There are many others that have not adopted certification due to the cost but hope to in the future.

Many chapters in this book provided limited explicit information on *sustainability issues*, perhaps because of the impression that this section was meant for discussing issues that were negatively affecting the sustainability of resources. We were, in fact, looking for *any* thoughts the authors might have had with regard to sustainability. Therefore, limited information in this section likely implies that the authors thought that the plan they described was actively guiding the property to a sustainable state or maintaining the property as such.

The relative paucity of information on *sustainability issues* may simply result from the fact that most chapters in this book indicate in their introductory sections that sustainability is of primary importance for their owners and managers in developing forest plans and managing forest lands. Therefore, in their minds, sustainability is of paramount importance and by no means is neglected. Furthermore, many forest owners and managers apply very stringent rules for ensuring forest sustainability including, for example, the requirement that the volume of harvest cannot exceed the of volume growth and that harvested parcels must be successfully regenerated before adjacent forests can be harvested. These requirements may go beyond and above by what is required by forest laws in the respective countries and even by forest certification programs.

LESSON LEARNED FROM THE PLANS

Although often described within the body of each chapter rather than the section on *Lessons* within the Discussion, the lessons learned were varied among the case studies. Many authors suggest that the loss of markets has harmed their ability to achieve the management goals for their property. Sometimes this was described as a short-term loss of markets, yet in others cases (Tongass National Forest in Chapter 46, for example), it involved a longer socioeconomic evolution of the region within which the forest is situated. For example, the Nature Conservancy ownership in the South Willapa Bay, Washington (Chapter 26) describes how the Great Recession in 2007–2009 resulted in a decline in the markets for western hemlock, dramatically reducing the ability to profit from the wood extracted by scheduled harvests. Other landowners have found a systemic loss of markets for products. The Boy Scouts of America's Camp No-Be-Bo-Sco in New Jersey (Chapter 1) laments that there are no local markets within the region for forest products beyond firewood for the low-quality wood that it wants to remove to enhance the use of the property. The Pike Lumber Company in Indiana (Chapter 4) discusses how the lack of a pulpwood market is preventing the harvest of small-diameter trees, a key element in its plan to produce more high-quality logs. The Sessoms Timber Trust (Chapter 25) is one of the few chapters that describes the impact of biomass markets for creating new demand for pulpwood logs and producing wood pellets for European markets.

With the continual changes in the forest industry structure in North America, including consolidation, reduced and less competitive markets may result, and the ability of forest landowners to achieve their long-term goals may be diminished. It may be easy where volume flow is regulated by the desires of the landowner to ignore the importance of changes in markets on the achievement of other objectives of a forest plan. However, now as in the past, tree harvesting is the main tool that foresters and landowners have at their disposal to influence the structure and composition of forests. Harvesting operations are expensive, and in most forests included in this book, budget constraints are common. Without the ability to sell harvested wood and generate income that can be used to offset the cost of these operations, they may be delayed or even abandoned. Even if there were no budgetary constraints and one could harvest trees without the need to sell them, in all likelihood, most of these trees will have to be utilized to mitigate fire and disease risks. These are some of the reasons why well-functioning wood markets are so important for conducting management operations and achieving sustainability.

Forest plans are guides for action, yet the achievement of goals is often influenced by external factors outside the control of forest landowners. For example, the recruitment and retention of the logging workforce is necessary to allow landowners to generate revenue from their forests, and one or two of the chapters in this book alluded to concerns about the reduction in the logging capacity of various regions of North American and the associated increasing age of the logging workforce. Other chapters describe how new and developing markets for forest resources, such as those involving ecosystem services, may be able to enhance the ability of landowners to meet their objectives. For example, the Willow Break property (Chapter 15) is managed primarily for hunting and other recreation purposes, and management actions are allowed if they improve wildlife habitat. The Conservation Fund Garcia River Forest in California (Chapter 19) has been qualified as a carbon offset project through the California Climate Action Reserve and may be able to obtain additional income by selling carbon offsets. These offer some hope to a wider range of services. While the market for carbon credits remains immature at this time, it may offer additional revenue to support forest management plans in the future. And while markets for non-timber forest products offer some promise, at this time, they are unable to replace markets for wood products.

REGULATORY COMPLIANCE

One of the benefits for having management plans is that they often present to others a transparent set of activities for, and the anticipated outcomes from, a forest. For many of the publicly managed forests, both in Canada and in the United States, a plan is a document that can be used by an entity, the owner, or the license holder, to support the regulatory process. However, this benefit does not always occur. The western Oregon Bureau of Land Management described the frustration with northwestern U.S. planning processes and the associated federal laws, as the multiple requirements among these needed to be better coordinated. Private land management organizations, such as the Green Diamond Resource Company (Chapter 35), have incorporated into their forest planning process functions to support the development of habitat conservation plans and other types of plans necessary for the management of land (e.g., a storm water plan). Benefits derived from the regulatory certainty are one of the reasons for this approach.

As regulatory requirements and the number of stakeholders increase, so does the complexity of forest plans and their development. As demonstrated in this book, complex forest plans may take years to develop and can be quite expensive, requiring large professional staff and expensive plan development tools. In some cases, the complexity could be such that despite best efforts, actionable forest plans fail to be developed.

The belief that benefits from integrated planning are worthwhile is not universally held. For example, in the development of at least one forest plan it was suggested that the requirements for the development of a habitat conservation plan did not seem to contribute to the improvement of the environmental, social, and economic outcomes of the forest (e.g., Northwest Oregon State Forests, Chapter 36). Other chapters describe how landowners are considering, but not yet convinced that using the information developed during the forest planning process can help obtain forest certification. In the future, as conservation issues continue to impact private or public forest practices, it is possible that the process of developing a forest plan may play a larger role in other processes aimed at easing regulatory burdens or addressing sustainability issues. However, both the regulated (the landowner) and regulator (or certification auditor) may need to develop an improved understanding of the potential of a forest plan to facilitate these efforts.

CONCLUSION

The case studies provided within the chapters of this book provide support for the ultimate purposes of forest plans: they are guides for action; they provide documentation and information that describe the goals and objectives for a property; and they may suggest the management actions that might be used to pursue the goals and objectives. One comment repeated periodically throughout chapters in the book suggests that the purpose of forest plans is to create a visible document that fosters improved communication among the interested parties. Often, particularly on smaller properties, a forest management plan provides a specific description of the type of activities and the predicted output from these activities on the landscape. Many of the chapters contained within this book allude to the fact that there is a continual need for foresters and land managers who are able to integrate information from a variety of resources into a single directive that can be implemented by a land management group. Therefore, in today's society, the need for skillful development and implementation of a forest plan remains important for the successful and sustainable management of forests in North America.

Index

Note: Page numbers followed by *f* indicate figures, *t* indicate tables.

Printed and bound by CPI Group (UK) Ltd, Croydon, CR0 4YY

08/05/2025

01865029-0003